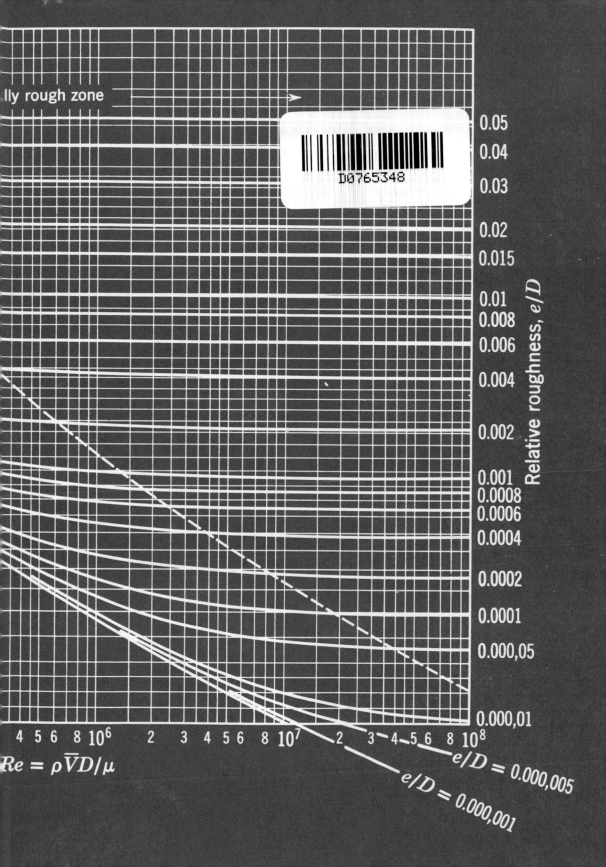

lly rough zone

0.05
0.04
0.03

0.02
0.015

0.01
0.008
0.006

0.004

0.002

0.001
0.0008
0.0006
0.0004

0.0002

0.0001

0.000,05

0.000,01

Relative roughness, e/D

4 5 6 8 10^6 2   3 4 5 6 8 10^7 2   3 4 5 6 8 10^8

$Re = \rho \overline{V} D / \mu$

$e/D = 0.000,005$

$e/D = 0.000,001$

introduction to fluid mechanics

introduction to fluid mechanics

ROBERT W. FOX
ALAN T. McDONALD

School of Mechanical Engineering
Purdue University

John Wiley & Sons, Inc.
New York
London
Sydney
Toronto

Library of Congress Cataloging in Publication Data:

Fox, Robert W.   1934—
    Introduction to fluid mechanics.

    1. Fluid mechanics.   I. McDonald, Alan T., joint author.   II. Title.

TA357.F69   532   73-5842

ISBN 0-471-27035-0

Printed in the United States of America

10 9 8 7 6

preface

This textbook was written for an introductory fluid mechanics course at the junior level. The decision to begin writing our own book reflected our dissatisfaction with existing textbooks. The decision to complete it resulted from our experience with earlier versions that were used in our own classes at Purdue University during the past three years.

Our objective has been to make the material interesting, easy to read, and easy to understand. All of it has been tested in the classroom by two generations of students to ensure that we have reached these goals. Thus we are convinced that students who read the book will understand it.

Because the material is easy to understand, the instructor can depart from conventional "lecture" teaching techniques. He can use classroom time to bring in current outside material, to solve example problems, and to explain the complexities of the assigned homework problems. Thus he can use each class period in the manner that seems most appropriate. Since there is no quantum of material that must be covered during a certain hour, the instructor can spend time in reviewing or amplifying any area in which he senses that students are having difficulty.

This flexibility permits the instructor to have much more interaction with the class than is possible through rigid adherence to a fixed schedule of formal lectures. It also permits the student to relate to the instructor and the material, since the student knows that class time is available to answer questions that are currently "bugging" him. These positive features are explained in Section 1–5, "Note to Students."

The material has been carefully selected. It includes all of the topics that we feel can be covered thoroughly in a 15-week (one-semester) course with three 50-minute lectures per week. At Purdue, we cover the entire book, with the exception of the starred sections. A laboratory (meeting two hours each week) is used for demonstrations, films, and simple experiments. A sample course outline is included in the *Solutions Manual* that accompanies this book. Desirable prerequisites are introductory courses in rigid body dynamics and heat power or thermodynamics, and mathematics through integral calculus. The thermodynamics course may be taken concurrently.

We believe that people—students and teachers alike—learn by doing; that is, by applying the principles they have learned to a variety of problems. Throughout the book, the emphasis is on starting from fundamentals to develop solutions. This approach is also emphasized for student problem solving, and we recommend a standard format for homework and examination solutions (see Section 1–5). This format requires that the problem statements be read and understood before the solution is attempted. It requires that basic equations be written and assumptions listed before numerical calculations are started. Thus, the student is forced to think about both the problem and the equations before he becomes involved in the detailed work of numerical substitution.

The text contains many example problems. We selected them carefully to illustrate matters that, in our experience, have troubled students. Solutions to these examples have been prepared to demonstrate good solution technique and to explain troublesome points. A different physical layout with wider margins has been used to make these solutions as visible and legible as possible.

For further clarification and demonstration of basic principles in fluid mechanics, many fine instructional films and film loops are available. We have referred to these films in the text where their use is appropriate; a complete list of suppliers and titles is included in Appendix C.

When our students have finished this course, we expect them to be able to apply the basic equations to a variety of problems, including new problems that they have not encountered before. We also emphasize physical understanding throughout, in order to make students aware of the variety of phenomena that can occur in fluid flow situations. By minimizing the number of "magic formulas" and emphasizing the fundamental approach, we believe that students will feel more confident in their ability to apply the material and will be able to reason out solutions to rather challenging problems.

Our experience with this material (in the form of notes) over a three-year period at Purdue University has been very encouraging. However, we recognize that no single approach can satisfy all needs. Therefore we welcome criticisms and suggestions from interested readers or users of this book.

We are indebted to several of our colleagues, especially to John A. Brighton of the Pennsylvania State University and James P. Johnston of Stanford University, for providing constructive and cogent reviews of an earlier version of this book. We have incorporated many of their suggestions. However, we are responsible for errors of fact or omission.

ALAN T. McDONALD
ROBERT W. FOX

contents

** Sections so marked may be omitted without loss of continuity in the text material.

introduction to fluid mechanics

introduction

In beginning the study of any subject, a number of questions come to mind immediately. Among those which a student in the first course in fluid mechanics may well wonder about are the following:

What is fluid mechanics all about?

Why do I have to study it?

Why should I want to study it?

How does it relate to subject areas with which I am already familiar?

In this chapter we will try to provide at least a qualitative answer to these and similar questions and thus an introduction to the subject.

1–1 Definition of a Fluid

Since fluid mechanics is a science dealing with the behavior of fluids at rest and in motion, it is logical to begin our study of the subject with a definition of the term fluid.

A fluid is a substance that deforms continuously under the application of a shearing (i.e. tangential) stress no matter how small the shearing stress.

Thus, according to the physical forms in which matter exists, fluids comprise the liquid and gas (or vapor) phase. The distinction between a fluid and the remaining possible state of matter (i.e. the solid state) is clear if one compares a fluid as defined above with the behavior of a solid. A solid is a substance that deforms when a shear stress is applied, but it does not continue to deform.

In Fig. 1.1 the behavior of a solid (Fig. 1.1*a*) and a fluid (Fig. 1.1*b*) under the action of a constant shear force are contrasted. In Fig. 1.1*a* the shear force is applied to the solid through the upper of two plates to which the solid has been bonded. When the shear force is applied to the plate, the block is deformed as shown. From our previous work in mechanics, we know that, provided the elastic limit of the solid material is not exceeded, the deformation is directly proportional to the applied shear stress, τ, where $\tau = F/A$ and A is the area of the surface in contact with the plate.

Now let us repeat the experiment using a fluid between the plates. In order to observe the behavior of the fluid we use a dye marker to outline a fluid element as shown by the solid lines (Fig. 1.1*b*). Upon application of the force, F, to the upper plate, we notice that the fluid element continues to deform as long as the force is applied. The shape of the fluid element, at successive instants of time, $t_2 > t_1 > t_0$, is shown by the dashed lines in Fig. 1.1*b*, which represent the positions of the dye markers. Note also that the fluid in direct contact with the solid boundary has the same velocity as the boundary itself; that is, there is no slip at the boundary. This is an experimental fact based on numerous observations of fluid behavior.[1]

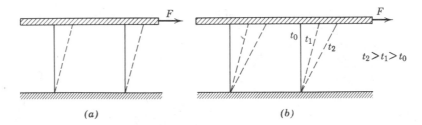

Fig. 1.1 *Behavior of (a) solid and (b) fluid, under the action of a constant shear force.*

Because a fluid deforms continuously under the application of a shear stress, we may alternatively define a fluid as a substance which, when at rest, cannot sustain a shear stress.

[1] The no slip condition is demonstrated in the film loops S-FM003, *Shear Deformation of Viscous Fluids*, and S-FM006, *Boundary Layer Formation*. These loops were produced by Educational Services, Inc., Watertown, Mass., and the National Committee for Fluid Mechanics Films. The films are distributed by Encyclopaedia Britannica Educational Corporation. (A complete list of fluid mechanics film titles and sources is given in Appendix C.)

1–2 Scope of Fluid Mechanics

Having defined a fluid and noted the characteristics that distinguish it from a solid, we might ask the question: "Why study fluid mechanics?" One answer is, "It is a required course in the curriculum!" While this may be a true and logical answer, it is not the only logical answer.

A knowledge and understanding of the basic principles and concepts of fluid mechanics are essential in the analysis and design of any system in which a fluid is the working medium. The design of virtually all means of transportation requires an application of the principles of fluid mechanics. Included are aircraft for both subsonic and supersonic flight, ground effect machines, hovercraft (now in service for channel crossings between France and England), vertical takeoff and landing aircraft requiring minimum runway length, surface ships, submarines, and automobiles. In recent years automobile manufacturers have given more consideration to aerodynamic design. This has been true for some time for the designers of both racing cars and boats. The design of propulsion systems for space flight as well as for toy rockets is based on the principles of fluid mechanics. The collapse of the Tacoma Narrows Bridge some years ago is evidence of the possible consequences of neglecting the basic principles of fluid mechanics.[2] It is commonplace today to perform model studies to determine the aerodynamic forces on and flow fields around buildings and structures. These include studies of skyscrapers, baseball stadiums, smokestacks, and shopping plazas.

The design of all types of fluid machinery including pumps, fans, blowers, compressors, and turbines clearly requires a knowledge of the basic principles of fluid mechanics. Lubrication is an area of considerable importance in fluid mechanics. Heating and ventilating systems for private homes, large office buildings, and underground tunnels, and the design of pipeline systems are further examples of technical problem areas requiring a knowledge of fluid mechanics. The circulatory system of the body is essentially a fluid system. It is not surprising then that the design of artificial hearts, heart-lung machines, breathing aids, and other such devices must rely on the basic principles of fluid mechanics.

Even some of our recreational endeavors are directly related to fluid mechanics. The slicing and hooking of golf balls can be explained by the principles of fluid mechanics (although they can be corrected only by a golf pro!).

The list of applications of the principles of fluid mechanics could be extended considerably. Our main point here is that fluid mechanics is not a subject studied for purely academic interest; rather, it is a subject with widespread importance both in our every-day experiences and in modern technology.

Clearly we cannot hope to consider in detail even a small percentage of these and other specific problems of fluid mechanics. Instead, the purpose of this text

[2] For dramatic evidence of aerodynamic forces in action, see the Ohio State University film, *Collapse of the Tacoma Narrows Bridge.*

scope of fluid mechanics

is to present the basic laws and associated physical concepts that provide the basis or starting point in the analysis of any problem in fluid mechanics.

1–3 Basic Equations

An analysis of any problem in fluid mechanics necessarily begins, either directly or indirectly, with statements of the basic laws governing the fluid motion. These laws, which are independent of the nature of the particular fluid, are:

(a) Conservation of mass.
(b) Newton's second law of motion.
(c) The first law of thermodynamics.
(d) The second law of thermodynamics.

Clearly not all of these laws are always required in the solution of any one problem. In some problems, it is necessary to bring into the analysis additional relations, in the form of constitutive equations describing the behavior of physical properties of fluids under given conditions.

It is obvious that the basic laws with which we shall deal are the same as those used in mechanics and thermodynamics. Our task will be to formulate these laws in forms suitable for solution of fluid flow problems and to apply them to the solution of a wide variety of problems.

It should be emphasized that there are, as we shall see, many apparently simple problems in fluid mechanics that cannot be solved by totally analytical means. In such cases we must resort to experiments and experimental observations.

1–4 Methods of Analysis

As we have indicated, the basic laws that are employed in the analysis of problems in fluid mechanics are the same ones that you have used previously in your earlier studies of thermodynamics and basic mechanics. From these earlier studies you will recall that the first step in solving a problem is to define the system that you are attempting to analyze. In basic mechanics, extensive use was made of the free body diagram. In thermodynamics you referred to the system under analysis as either a closed system or an open system. In this text we shall employ the terms system and control volume. The importance of defining the system or control volume to which the basic equations are to be applied in the analysis of a problem cannot be overemphasized. At this point it is wise to review the basic difference between a system and a control volume.

1–4.1 SYSTEM AND CONTROL VOLUME

A system is defined as a fixed, identifiable quantity of mass; the system boundaries separate the system from the surroundings. The boundaries of the system may be fixed or movable, but no mass crosses them.

In the familiar piston-cylinder assembly from thermodynamics, Fig. 1.2, the gas in the cylinder is considered as the system. If a high temperature source is brought

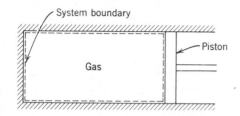

Fig. 1.2 *Piston-cylinder assembly.*

in contact with the left end of the cylinder, the piston will move to the right; the boundary of the system thus moves. From our study in thermodynamics we know that heat and work may cross the boundaries of the system, but the quantity of matter within the system boundaries remains fixed, that is, there is no mass transfer across the system boundaries.

Example 1.1

A radiator of a steam heating system has a volume of 0.7 ft³. When the radiator is filled with dry saturated steam at a pressure of 20 lbf/in², all valves to the radiator are closed. How much heat will have been transferred to the room when the pressure of the steam is 10 lbf/in²?

EXAMPLE PROBLEM 1.1

GIVEN:

Steam radiator with volume, $\forall = 0.7$ ft³, is filled with dry saturated steam at 20 psia and then valves are closed. As a result of heat transfer, pressure in radiator drops.

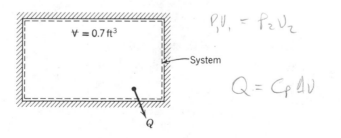

methods of analysis

FIND:

Amount of heat transferred when pressure reaches 10 psia.

SOLUTION:

We are dealing with a system.

At: State ① $\forall = 0.7\,\text{ft}^3$ $p_1 = 20\,\text{psia}$ dry saturated steam

    State ② $\forall = 0.7\,\text{ft}^3$ $p_2 = 10\,\text{psia}$

Basic equation: First law for the system: $Q_{12} - W_{12} = E_2 - E_1$

Assumptions: $W_{12} = 0$ since no shaft work is done and $\forall = $ constant

             $E = U$ since system is stationary

Under these assumptions, the first law statement reduces to

$$Q_{12} = U_2 - U_1 = m(u_2 - u_1)$$

Thus, to determine Q_{12}, we must determine m, u_2, and u_1.

At state ① $p_1 = 20$ psia and from steam tables for dry saturated steam

$$u_1 = 1081.9\frac{\text{Btu}}{\text{lbm}} \qquad v_1 = 20.089\frac{\text{ft}^3}{\text{lbm}}$$

Since

$$v = \frac{\forall}{m} \qquad m_1 = \frac{\forall}{v_1} = 0.7\,\text{ft}^3 \times \frac{\text{lbm}}{20.089\,\text{ft}^3} = 0.0348\,\text{lbm}$$

For steam

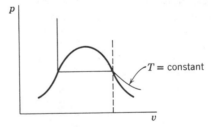

Consequently, when p is reduced at constant v, then state ② must lie in the wet region.

Thus at state ②

$$v_2 = v_1 = v_{2f} + xv_{2fg} \quad \text{and} \quad x = \frac{v_1 - v_{2f}}{v_{2fg}}$$

From the steam tables at $p_2 = 10$ psia,

$$v_{2f} = 0.017\frac{\text{ft}^3}{\text{lbm}}, \qquad v_{2fg} = 38.40\frac{\text{ft}^3}{\text{lbm}}$$

$$x = \frac{v_1 - v_{2f}}{v_{2fg}} = \frac{20.089 - 0.017}{38.40} = \frac{20.072}{38.40} = 0.524$$

EXAMPLE PROBLEM 1.1 (continued)

Then

$$u_2 = u_{2f} + x u_{2fg} = 161.14 + 0.524(911.1) = 640\frac{Btu}{lbm}$$

Finally,

$$Q = m(u_2 - u_1) = 0.0348 \, lbm \times (640 - 1081.9)\frac{Btu}{lbm} = -15.4 \, Btu \qquad\qquad Q_{12}$$

$$\left.\begin{array}{c}\text{The minus sign} \\ \text{indicates heat} \\ \text{transfer out of} \\ \text{system}\end{array}\right\}$$

$$\left\{\begin{array}{l}\text{The purpose of this problem was to review the use of:} \\ \text{(i) the first law of thermodynamics for a system, and} \\ \text{(ii) the steam tables.}\end{array}\right\}$$

In your earlier mechanics courses you made extensive use of the free body (system approach). This was logical because you were dealing with an easily identifiable rigid body. However in fluid mechanics, we are normally concerned with the flow of fluids through devices such as compressors, turbines, pipelines, nozzles, and so on. In these cases it is difficult to focus attention on a fixed identifiable quantity of mass. It is much more convenient, for purposes of analysis, to focus attention on a volume in space through which the fluid flows, that is, to use the control volume approach.

A control volume is an arbitrary volume in space through which fluid flows. The geometric boundary of the control volume is called the control surface. The control surface may be real or imaginary; it may be at rest or in motion. Figure 1.3

Fig. 1.3 *Fluid flow through a pipe.*

shows a possible control surface for use in analyzing the flow through a pipe. Here the inside surface of the pipe, a real physical boundary, comprises part of the control surface. However, the vertical portions of the control surface are

7   **methods of analysis**

imaginary—that is, there is no corresponding physical surface. These imaginary boundaries are selected arbitrarily for accounting purposes. Since the location of the control surface has a direct effect on the accounting procedure in applying the basic laws, it is extremely important that the control surface be clearly defined before beginning any form of analysis.

1–4.2 DIFFERENTIAL VERSUS INTEGRAL APPROACH

The basic laws that we apply in our study of fluid mechanics can be formulated in terms of infinitesimal or finite systems and control volumes. As you might suspect, the equations will look different in each case. In the first case the resulting equations are differential equations; in the latter case we might call them global equations, that is, equations governing the gross behavior of the flow. Both approaches are important in the study of fluid mechanics and both will be developed in the course of our work. The choice of approach in any problem depends on the problem.

Assuming we can solve the equations, the differential approach (i.e. the use of differential equations governing the motion) provides a means of determining the detailed (i.e. point by point) behavior of the flow.

Frequently in the problems under study the information sought does not require a detailed knowledge of the flow. We are often interested in the gross behavior of a device and, hence, can use the integral formulation of the basic laws. The integral formulation, that is, the use of finite systems or control volumes, is usually easier to treat analytically. Since mechanics and thermodynamics deal with the formulation of the basic laws in terms of finite systems, these formulations are the basis for deriving the control volume equations in Chapter 4.

1–4.3 METHODS OF DESCRIPTION

Mechanics and thermodynamics deal almost exclusively with systems. Thus you have made extensive use of the basic equations applied to a fixed, identifiable quantity of mass. Later, in attempting to analyze thermodynamic devices, you often found it necessary to go to a control volume (open system) analysis. Clearly the type of analysis depends on the problem. Where it is easy to keep track of identifiable elements of mass (for example, in particle mechanics) one utilizes a method of description that follows the particle. This is often referred to as the Lagrangian method of description.

Consider for example the application of Newton's second law to a particle of fixed mass, m. Mathematically, we can write Newton's second law for a system of mass, m, in one of the following forms:

$$\Sigma \vec{F} = m\vec{a} \qquad (1.1a)$$

or

$$\Sigma \vec{F} = m\frac{d\vec{V}}{dt} = \frac{d}{dt}(m\vec{V}) \qquad (1.1b)$$

or

$$\Sigma \vec{F} = m\frac{d^2\vec{r}}{dt^2} = \frac{d^2}{dt^2}(m\vec{r}) \qquad (1.1c)$$

where $\Sigma \vec{F}$ is the sum of all external forces acting on the system, $\vec{a}$ is the acceleration of the center of mass of the system, $\vec{V}$ is the velocity of the center of mass of the system, and $\vec{r}$ is the position vector of the center of mass of the system relative to a fixed coordinate system.

Thus, in describing the motion of a particle in a rectangular coordinate system:

$$\vec{F} = \hat{\imath}F_x + \hat{\jmath}F_y + \hat{k}F_z \qquad (1.2a)$$

$$\vec{a} = \hat{\imath}a_x + \hat{\jmath}a_y + \hat{k}a_z \qquad (1.2b)$$

$$\vec{V} = \hat{\imath}u + \hat{\jmath}v + \hat{k}w \qquad (1.2c)$$

$$\vec{r} = \hat{\imath}x + \hat{\jmath}y + \hat{k}z \qquad (1.2d)$$

where F_x, F_y, F_z, are the components of $\vec{F}$ in the x, y, and z directions respectively, a_x, a_y, a_z, are the components of $\vec{a}$ in the x, y, and z directions respectively, u, v, w are the components of $\vec{V}$ in the x, y, and z directions respectively, and x, y, z are the coordinates of the particle.

Since the particle moves under the action of the force, $\vec{F}$, we recognize that its position, $\vec{r}$, velocity, $\vec{V}$, and acceleration, $\vec{a}$, are in general functions of time. That is

$$\vec{r} = \vec{r}(t) \quad \text{and, hence,} \quad x = x(t), \qquad y = y(t), \qquad z = z(t)$$

$$\vec{V} = \vec{V}(t) \quad \text{and, hence,} \quad u = u(t), \qquad v = v(t), \qquad w = w(t)$$

$$\vec{a} = \vec{a}(t) \quad \text{and, hence,} \quad a_x = a_x(t), \qquad a_y = a_y(t), \qquad a_z = a_z(t)$$

From elementary mechanics we know further (as has already been implied in Eqs. 1.1) that

$$\vec{a} = \frac{d\vec{V}}{dt} \quad \text{and} \quad a_x = \frac{du}{dt}, \qquad a_y = \frac{dv}{dt}, \qquad a_z = \frac{dw}{dt}$$

$$\vec{V} = \frac{d\vec{r}}{dt} \quad \text{and} \quad u = \frac{dx}{dt}, \qquad v = \frac{dy}{dt}, \qquad w = \frac{dz}{dt}$$

Example 1.2

A body is thrown vertically upward with an initial velocity of 40 ft/sec. Neglecting air resistance, determine the maximum height to which it will rise, and the time required for it to reach the maximum height.

9   **methods of analysis**

GIVEN:

A body which, at $t = 0$, has $x = 0$ and $\vec{V} = u_0\hat{\imath} = 40\hat{\imath}$ ft/sec.
Neglect air resistance.

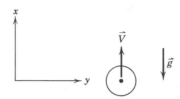

FIND:

(a) The maximum height to which the body will rise.
(b) The time required to reach maximum height.

SOLUTION:

Basic equations:

$$\Sigma\vec{F} = m\vec{a}$$

$$\Sigma F_x = ma_x \qquad a_x = \frac{d^2x}{dt^2} \qquad u = \frac{dx}{dt}$$

Drawing a free body diagram

$$\vec{V} = u\,\hat{\imath} \qquad\qquad \Sigma F_x = ma_x$$

$$-W = -mg = ma_x = m\frac{d^2x}{dt^2}$$

$$\vec{W} = -mg\,\hat{\imath} \qquad \therefore \quad \frac{d^2x}{dt^2} = -g$$

Integrating with respect to time between limits of 0 and t, we have

$$\frac{dx}{dt}\bigg]_t - \frac{dx}{dt}\bigg]_{t=0} = -gt \quad \text{or} \quad \frac{dx}{dt} = u_0 - gt$$

Integrating again with respect to time between limits of 0 and t, we have

$$x - x_0 = u_0t - \tfrac{1}{2}gt^2 \quad \text{or} \quad x = u_0t - \tfrac{1}{2}gt^2$$

The maximum height is reached when $u = dx/dt = 0$. At this condition

$$\frac{dx}{dt} = 0 = u_0 - gt$$

Thus the time required to reach the maximum height is given by

$$t = \frac{u_0}{g} = 40\frac{\text{ft}}{\text{sec}} \times \frac{\text{sec}^2}{32.2 \text{ ft}} = 1.24 \text{ sec} \qquad\qquad\qquad t_{x_{max}}$$

The maximum height is obtained from

$$x = u_0 t - \tfrac{1}{2}gt^2 \quad \text{with} \quad t = \frac{u_0}{g}$$

$$\therefore \quad x_{max} = u_0\frac{u_0}{g} - \frac{1}{2}g\left(\frac{u_0}{g}\right)^2 = \frac{1}{2}\frac{u_0^2}{g} = \frac{1}{2} \times (40)^2 \frac{\text{ft}^2}{\text{sec}^2} \times \frac{\text{sec}^2}{32.2 \text{ ft}}$$

$$x_{max} = 24.9 \text{ ft} \qquad\qquad\qquad\qquad\qquad\qquad\qquad\qquad x_{max}$$

$\left\{\begin{array}{l}\text{This problem is intended as a reminder of the method of description used in particle}\\ \text{mechanics. Note that the velocity, } u, \text{ is a function of time in this method of description.}\end{array}\right\}$

We may consider a fluid to be composed of a very large number of particles whose motion must be described; keeping track of the motion of each fluid particle becomes a horrendous bookkeeping problem. Consequently a particle description becomes unmanageable. Therefore, we find it convenient to use a different type of description. Since we will often deal with control volume analyses, it is convenient to use the field, or Eulerian, description. This method of description focuses attention on the properties of a flow at a given point in space as a function of time. Thus, in using the Eulerian method of description, the properties of a flow field are described as functions of space coordinates and time. We shall see in Chapter 2 that this method of description is a logical outgrowth of the assumption that fluids may be treated as continuous media.

1–5 Note to Students

The goal of this textbook is to provide a clear, concise introduction to the subject of fluid mechanics. It is written for the student. It is our strong feeling that classroom time should not be devoted to a statement of textbook material by the instructor. Instead, the time should be used to amplify the textbook material through discussion of related material and the application of basic principles to the solution of problems. The necessary conditions for accomplishing this goal are: (1) a clear, concise presentation of the fundamentals that you, the student, can read and understand, and (2) your willingness to read the text material before coming to class. We have assumed responsibility for meeting the first condition. You must assume responsibility for satisfying the second condition. There will no doubt be times when we fall short of satisfying these objectives. If so, we would appreciate hearing of these shortcomings either directly or through your instructor.

It goes without saying that an introductory text is not all-inclusive. Your instructor undoubtedly will expand on the material presented and introduce additional new material. We encourage you to refer to the many other available fluid mechanics textbooks listed in the bibliography. Where another text presents a particularly good discussion of a given topic, we will refer to it directly. We assume that you have had prior courses in thermodynamics, statics and dynamics, and integral and differential calculus. No attempt will be made to restate this subject material; however, the pertinent aspects of this previous study will be reviewed briefly when appropriate.

It is our strong belief that one learns best by *doing*. This is true whether the subject under study is fluid mechanics, thermodynamics, or golf. The fundamentals in any of these cases are few, and mastery of them comes through practice. *Thus it is extremely important, in fact essential, that the student work problems.* A large number of problems are included at the end of each chapter to provide the opportunity to gain facility in the application of the fundamentals to the solution of problems. You should avoid the temptation to adopt a "plug and chug" approach to the solution of problems. Most of the problems are of a nature that such an approach will lead to difficulty or simply will not work. In the solution of problems we strongly recommend that you proceed using the following logical steps:

1. State briefly and concisely (in your own words) the information given.
2. State the information that you have to find.
3. Draw a schematic of the system or control volume to be used in the analysis. Be sure to label the boundaries of the system or control volume and label appropriate coordinate directions.
4. Give the appropriate mathematical formulation of the *basic* laws that you consider necessary to effect a solution to the problem.
5. List the simplifying assumptions that you feel are appropriate in the problem.
6. Carry through the analysis to the point where it is appropriate to substitute numerical values.
7. Substitute numerical values (using a consistent set of units) to obtain a numerical answer.
8. Check the answer and the assumptions made in effecting the solution to make sure they are reasonable.

9. Label the answer.

In your initial work this problem format may seem unnecessary. However, such an approach to the solution of problems will lead to fewer errors, save time, and permit a clearer understanding of the limitations of a particular solution. This format is used in all example problems presented in this text.

1–6 Dimensions and Units

Engineering problems are solved to answer specific questions. It goes without saying that the answer must include units. (It makes a difference whether a pipe diameter required is 1 inch or 1 foot!) Consequently it is appropriate at this point to present a brief review of dimensions and units. We say review because the topic is familiar from your earlier work in thermodynamics.[3]

The word dimension is used to refer to any measurable quantity; such quantities include length, time, and velocity. In terms of a particular dimensional system all measurable quantities can be subdivided into two groups—primary quantities and secondary quantities. Primary quantities are those for which we set up arbitrary scales of measure; secondary quantities are those quantities whose dimensions are expressible in terms of the dimensions of the primary quantities.

1–6.1 SYSTEMS OF DIMENSIONS

There are three basic systems of dimensions, that is, there are three basic ways of specifying the primary dimensions. These systems use primary dimensions of:

(a) Mass $[M]$, length $[L]$, time $[t]$, temperature $[T]$.
(b) Force $[F]$, length $[L]$, time $[t]$, temperature $[T]$.
(c) Force $[F]$, mass $[M]$, length $[L]$, time $[t]$, temperature $[T]$.

Any valid equation that relates physical quantities must be dimensionally homogeneous, that is, each term in the equation must have the same dimensions. We recognize that Newton's second law ($\vec{F} \propto m\vec{a}$) relates the four dimensions, F, M, L, and t. Thus in system (a), force $[F]$ is a secondary dimension and the constant of proportionality in Newton's second law is dimensionless. In system (b), mass $[M]$ is a secondary dimension and again the constant of proportionality in Newton's second law is dimensionless. In system (c), both force $[F]$ and mass $[M]$ have been selected as primary dimensions. In this case the constant of

[3] Refer to the text you used in your beginning thermodynamics course. If it is not available, you will find a discussion of dimensions and units presented in each of these texts: W. C. Reynolds, *Thermodynamics* (New York: McGraw-Hill, 1968), pp. 15–19; G. J. Van Wylen and R. E. Sonntag, *Fundamentals of Classical Thermodynamics* (New York: John Wiley, 1965), pp. 22–27; K. Wark, *Thermodynamics*, 2nd ed. (New York: McGraw-Hill, 1971), pp. 7–10; M. W. Zemansky and H. C. Van Ness, *Basic Engineering Thermodynamics* (New York: McGraw-Hill, 1966), pp. 33–36.

proportionality, g_c, in Newton's second law (written $\vec{F} = m\vec{a}/g_c$) is not dimensionless. The dimensions of g_c must in fact be $[ML/Ft^2]$ for the equation to be dimensionally homogeneous. The value of the constant of proportionality depends on the selection of the units of measure for each of the primary quantities.

1–6.2 SYSTEMS OF UNITS

The primary quantities in a particular system of measure are expressed in terms of units. For each basic way of selecting the primary dimensions, there is more than one way of selecting the units of measure. We will present only the most common engineering system of units in use for each of the basic systems of dimensions.

(a) MLtT

SI, which is the abbreviation in many languages for Systéme International d'Unités, is an extension and refinement of the traditional metric system. More than 30 countries have declared it to be the only legally accepted system. The United States will most likely be forced to adopt it within the next decade.

In the SI system of units, the unit of mass is the kilogram, the unit of length is the meter, the unit of time is the second, and the unit of temperature is the degree Kelvin. Force is a secondary dimension, and its unit, the newton, is defined from Newton's second law as

$$1 \text{ newton} \equiv \frac{1 \text{ kg-m}}{\sec^2}$$

Even as SI is broadly adopted, the gram will often be used as a basic unit of mass, probably until a new name is adopted for the kilogram.

In the Absolute Metric system of units, the unit of mass is the gram, the unit of length is the centimeter, the unit of time is the second, and the unit of temperature is the degree Kelvin. Since force is a secondary dimension, the unit of force, the dyne, is defined in terms of Newton's second law as

$$1 \text{ dyne} \equiv \frac{1 \text{ gm-cm}}{\sec^2}$$

(b) FLtT

In the British Gravitational system of units, the unit of force is the pound (lbf), the unit of length is the foot, the unit of time is the second, and the unit of temperature is the degree Rankine. Since mass is a secondary dimension, the unit of mass, the slug, is defined in terms of Newton's second law as

$$1 \text{ slug} \equiv \frac{1 \text{ lbf}}{1 \text{ ft/sec}^2} = \frac{1 \text{ lbf-sec}^2}{\text{ft}}$$

(c) FMLtT

In the English Engineering system of units, the unit of force is the pound force (lbf), the unit of mass is the pound mass (lbm), the unit of length is the foot, the unit of time is the second, and the unit of temperature is the degree Rankine. Since both force and mass are chosen as primary dimensions, Newton's second law is written as

$$\vec{F} = \frac{m\vec{a}}{g_c}$$

A force of one pound (1 lbf) is the force that gives a pound mass (1 lbm) an acceleration equal to the standard acceleration of gravity on earth, 32.17 ft/sec². From Newton's second law we see that

$$1 \text{ lbf} \equiv \frac{1 \text{ lbm} \times \left(32.17 \text{ ft/sec}^2\right)}{g_c}$$

or

$$g_c \equiv \frac{32.17 \text{ ft-lbm}}{\text{lbf-sec}^2}$$

Since a force of 1 lbf accelerates 1 lbm at 32.17 ft/sec², it would accelerate 32.17 lbm at 1 ft/sec². A slug is also accelerated at 1 ft/sec² by a force of 1 lbf. Therefore

$$1 \text{ slug} \equiv 32.17 \text{ lbm}$$

In this text we shall use both the Absolute Metric (MLtT) and the British Gravitational (FLtT) systems of units. In either case, the constant of proportionality in Newton's second law is dimensionless and has a value of unity. Consequently, Newton's second law is written as $\vec{F} = m\vec{a}$.

The acceleration of gravity is 32.17 ft/sec². Thus an object with a mass of 1 lbm at rest in a location where $g = 32.17$ ft/sec² experiences a force in the English Engineering system of 1 lbf. This force is defined as the weight of the object. In general, then

$$W = \frac{mg}{g_c}$$

in this system.

In the British Gravitational system, the weight of an object is given by

$$W = mg$$

Example 1.3

A certain liquid has a density of 1.5 slug/ft³. Determine the specific weight, the specific volume, and the specific gravity of the liquid on the earth and on the

dimensions and units

moon. The gravitational acceleration on the moon is 5.47 ft/sec². What is the density of the liquid on the moon in lbm/ft³?

EXAMPLE PROBLEM 1.3

GIVEN:

Liquid of density, $\rho = 1.5$ slug/ft³.
Gravitational acceleration on the moon is 5.47 ft/sec².

FIND:

(a) Specific weight, γ, specific volume, v, and specific gravity of the liquid (1) on the earth, and (2) on the moon.
(b) Density of the liquid on the moon in units of lbm/ft³.

SOLUTION:

Basic definitions: Specific weight $\quad \gamma = \dfrac{\text{weight}}{\text{volume}} = \rho g$

$\qquad\qquad\qquad\qquad$ Specific volume $\quad v = \dfrac{1}{\rho}$

$\qquad\qquad\qquad\qquad$ Specific gravity $\quad SG = \dfrac{\rho}{\rho_{H_2O}} = \dfrac{\gamma}{\gamma_{H_2O}}$

Properties: $\qquad g_{earth} = 32.2$ ft/sec² $\quad g_{moon} = 5.47$ ft/sec²
$\qquad\qquad\qquad \rho_{H_2O} = 1.94$ slug/ft³.

(a) *Specific weight*

$$\gamma_{earth} = \rho g_{earth} = 1.5 \frac{\text{slug}}{\text{ft}^3} \times 32.2 \frac{\text{ft}}{\text{sec}^2} \times \frac{1\ \text{lbf-sec}^2}{\text{ft-slug}} = 48.3 \frac{\text{lbf}}{\text{ft}^3} \quad\longleftarrow \gamma_{earth}$$

$$\gamma_{moon} = \rho g_{moon} = 1.5 \frac{\text{slug}}{\text{ft}^3} \times 5.47 \frac{\text{ft}}{\text{sec}^2} \times \frac{1\ \text{lbf-sec}^2}{\text{ft-slug}} = 7.20 \frac{\text{lbf}}{\text{ft}^3}$$

$\left\{\begin{array}{l}\text{Recall from } \vec{F} = m\vec{a} \\ 1\ \text{lbf} = 1\ \text{slug} \times 1\ \text{ft/sec}^2\end{array}\right\}\quad \gamma_{moon}$

Specific volume

$$v_{earth} = \frac{1}{\rho_{earth}} = \frac{1}{1.5\ \text{slug/ft}^3} = 0.667 \frac{\text{ft}^3}{\text{slug}} \quad\longleftarrow v_{earth}$$

$$v_{moon} = \frac{1}{\rho_{moon}} = \frac{1}{1.5\ \text{slug/ft}^3} = 0.667 \frac{\text{ft}^3}{\text{slug}} \quad\longleftarrow$$

$\left\{\begin{array}{l}\text{Mass is not dependent} \\ \text{on } g \text{ (weight is)}\end{array}\right\}\quad v_{moon}$

EXAMPLE PROBLEM 1.3 (continued)

Specific gravity

$$SG_{earth} = \frac{\rho}{\rho_{H_2O}} = \frac{1.5 \text{ slug/ft}^3}{1.94 \text{ slug/ft}^3} = 0.773 \longleftarrow \qquad SG_{earth}$$

$$SG_{moon} = \frac{\rho}{\rho_{H_2O}} = \frac{1.5 \text{ slug/ft}^3}{1.94 \text{ slug/ft}^3} = 0.773 \longleftarrow \qquad SG_{moon}$$

(b) Density of liquid in lbm/ft^3

$$\rho = 1.5\frac{\text{slug}}{\text{ft}^3} \times \frac{32.2 \text{ lbm}}{\text{slug}} = 48.3\frac{\text{lbm}}{\text{ft}^3} \longleftarrow \qquad \left\{ \begin{array}{c} \text{Since mass is not dependent on } g, \\ \rho_{earth} = \rho_{moon} \end{array} \right\} \qquad \rho$$

$$\{1 \text{ slug} = 32.2 \text{ lbm}\}$$

The purpose of this problem was to:
  (i) review the definitions of density, specific weight, specific volume, and specific gravity,
  (ii) point out the difference between weight and mass, and
  (iii) illustrate two different units of mass.

problems

1.1 A number of common substances are listed below:

Tar   Sand
"Silly Putty"   Jello
Modeling Clay   Toothpaste
Wax   Shaving cream

Some of these materials exhibit characteristics of both solid and fluid behavior under different conditions. Explain and give examples.

1.2 Both gases and liquids satisfy the definition of a fluid given in Section 1–1. What characteristics distinguish these two states of matter?

1.3 Look up definitions of the term "fluid" in a dictionary and an encyclopedia. Compare these with the definition given in Section 1–1. Are all three equivalent?

1.4 Give a word statement of each of the four basic conservation laws stated in Section 1–3.

1.5 Does a sealed bottle of compressed gas constitute a *system* or a *control volume*? Why? Answer the same questions for a dentist's drill powered by compressed air, an internal combustion engine, and a mercury-in-glass thermometer.

1.6   Air trapped in a bicycle tire pump is compressed suddenly to $\frac{1}{5}$ of its original volume. Both heat transfer and friction may be neglected as a first approximation. Determine the final temperature of the air in the pump, if the initial temperature is 70 F.

1.7   The catapult on an aircraft carrier has a piston diameter of 2 ft and a stroke of 100 ft. Initially the cylinder contains steam at 250 psia, 600 F in a volume of 100 ft^3. Assume heat transfer and friction are negligible as the piston moves through its stroke.

      Determine the temperature and pressure of the steam at the end of its stroke. What fraction of the energy content of the steam was given up during the expansion process?

1.8   A compressed air tank in a service station holds 5 ft^3 of compressed air at 125 psig. Determine the amount of energy required to compress this much air isothermally from atmospheric pressure, assuming a frictionless process. (Note that the release of this much energy would be catastrophic if the tank were ruptured!)

1.9   A projectile is fired with velocity, $\vec{V}_0$, and elevation angle, θ, above the horizon. Air resistance may be neglected.

      Express the range of the projectile in terms of V_0 and θ. Determine the angle that gives maximum range.

1.10   Parametric equations for the motion of a particle are

$$x = A\cos\omega t$$
$$A > B$$
$$y = B\sin\omega t$$

      Determine the velocity and acceleration of the particle as functions of time. Indicate the particle trajectory in the xy plane on a sketch, and locate the point(s) of maximum velocity and acceleration.

1.11   Very small particles moving in fluids experience a drag force proportional to velocity. Consider a particle of net weight, W, dropped in a fluid. The particle experiences a drag force, kV, where V is the particle speed.

      Determine the time required for the particle to accelerate from rest to 95 percent of its terminal speed, V_t, in terms of k, W, and g.

1.12   What is the dimensional representation of the quantities:

(a) Power   (f) Momentum
(b) Pressure   (g) Shear stress
(c) Modulus of elasticity   (h) Specific heat
(d) Angular velocity   (i) Thermal expansion coefficient
(e) Energy

Assume basic dimensions of M, L, t, and T.

1.13　What is the dimensional representation of the quantities:

(a) Power
(b) Pressure
(c) Modulus of elasticity
(d) Angular velocity
(e) Energy

(f) Moment of a force
(g) Momentum
(h) Shear stress
(i) Strain

Assume basic dimensions of F, L, t, and T.

1.14　Air at a pressure of 40 psia and a temperature of 70 F is moving with a velocity of 100 ft/sec. Calculate:

(a) The kinetic energy per unit mass of the air.
(b) The kinetic energy per unit volume of the air.
(c) The total kinetic energy of 2 lbm of the air.
(d) The total kinetic energy of 2 slugs of the air.

1.15　Evaluate a conversion factor between the unit of acceleration in the English Engineering system of units, ft/sec², and that in the SI system, m/sec².

1.16　The stylus of a high quality stereo system is adjusted to balance a mass of 0.75 gm on a small equal arm beam balance. What force (in lbf) does the stylus exert on a record? What is the force expressed in dynes?

1.17　The unit of pressure in the SI system is the Newton per square meter (N/m²). How many pounds force per square inch (psi) correspond to 1 N/m²?

1.18　A container weighs 3.22 lbf when empty. When filled with water at 95 F, the mass of the container and its contents is 1.95 slugs.

Find the weight of water in the container, and its volume in ft³, using data from Appendix A.

1.19　Evaluate the change in specific weight of mercury, in lbf/ft³, as its temperature changes from 70 to 90 F. Use the data in Appendix A.

1.20　Compare the specific weight of air at 14.7 psia, 59 F, in lbf/ft³, with that of water at 4 C.

1.21　Determine the density of mercury and of water, in gm/cm³, at 4 C. What is the specific gravity of mercury at this temperature?

1.22　Four vectors are defined by the equations

$$\vec{r}_1 = 2\hat{\imath} - \hat{\jmath} + \hat{k}$$

$$\vec{r}_2 = \hat{\imath} + 3\hat{\jmath} - 2\hat{k}$$

$$\vec{r}_3 = -2\hat{\imath} + \hat{\jmath} - 3\hat{k}$$

$$\vec{r}_4 = 3\hat{\imath} + 2\hat{\jmath} + 5\hat{k}$$

Determine scalars a, b, and c such that

$$\vec{r}_4 = a\vec{r}_1 + b\vec{r}_2 + c\vec{r}_3$$

19　**problems**

1.23 Two vectors, $\vec{a}$ and $\vec{b}$, are given by the expressions

$$\vec{a} = xy\hat{i} + y^2\hat{j} + 2\hat{k}$$
$$\vec{b} = x^2\hat{i} - xy\hat{j} + z\hat{k}$$

and a scalar, ϕ, is given by

$$\phi = \frac{x^2}{2} - \frac{y^2}{2}$$

Evaluate each of the following:

(a) $\vec{a} \cdot \vec{b}$

(b) $\vec{a} \times \vec{b}$

(c) $\partial\vec{a}/\partial x$

(d) $\nabla\phi$

(e) $\nabla \cdot \vec{a}$

(f) $\nabla \times \vec{b}$

1.24 Two vectors, $\vec{c}$ and $\vec{d}$, are given by the expressions

$$\vec{c} = xyz\hat{i} + 2\hat{j} + y^2\hat{k}$$
$$\vec{d} = x^2\hat{i} + y^2\hat{j} + x\hat{k}$$

and a scalar, ψ, is given by

$$\psi = xy$$

Evaluate each of the following:

(a) $\vec{c} \cdot \vec{d}$

(b) $\vec{c} \times \vec{d}$

(c) $\partial\vec{c}/\partial x$

(d) $\nabla\psi$

(e) $\nabla \times \vec{c}$

(f) $\nabla \cdot \vec{d}$

1.25 Two vectors, $\vec{r}$ and $\vec{s}$, are given by the expressions

$$\vec{r} = x^2y\hat{i} + z^2\hat{k}$$
$$\vec{s} = xz\hat{i} + xyz\hat{j} + x^2y\hat{k}$$

and a scalar, ζ, is given by

$$\zeta = \tfrac{1}{2}(x^2 - y^2)$$

Evaluate each of the following:

(a) $\vec{r} \cdot \vec{s}$

(b) $\vec{r} \times \vec{s}$

(c) $\partial\vec{s}/\partial z$

(d) $\nabla \times \vec{r}$

(e) $\nabla \cdot \vec{s}$

(f) $\nabla^2\zeta$

fundamental concepts

In Chapter 1 we indicated that our study of fluid mechanics will build on earlier studies in mechanics and thermodynamics. To develop a unified approach, we will need to review some familiar topics and to introduce some new concepts and definitions. The purpose of this chapter is to develop these fundamental concepts.

2–1 Fluid as a Continuum

In our definition of a fluid, no mention was made of the molecular structure of fluids. All fluids are composed of molecules in constant motion. However, in most engineering applications we are interested in the gross or average (i.e. macroscopic) effects of many molecules. It is these macroscopic effects that we can perceive and measure. We thus treat a fluid as virtually an infinitely divisible substance (i.e. as a continuum) and do not concern ourselves with the behavior of individual molecules.

The concept of a continuum is the basis of classical fluid mechanics. The continuum assumption is valid in treating the behavior of fluids under normal conditions. However, it breaks down whenever the mean free path of the molecules (approximately 2×10^{-6} in. for air at STP)[1] becomes the same order of magnitude

---

[1] STP (Standard Temperature and Pressure) for air are 59 F and 14.696 psia, respectively.

as the smallest significant characteristic dimension of the problem. In problems such as rarefied gas flow (e.g. as encountered in flights into the upper reaches of the atmosphere) we must abandon the classical laws of fluid mechanics in favor of the microscopic and statistical points of view.

As a consequence of the continuum assumption, each fluid property is assumed to have a definite value at each point in space. Thus fluid properties such as density, temperature, velocity, and so on, are considered to be continuous functions of position and time.

To illustrate the concept of a property at a point, consider the manner in which we determine the density at a point. A region of fluid is shown in Fig. 2.1. We are interested in determining the density at the point C, whose coordinates are x_0, y_0,

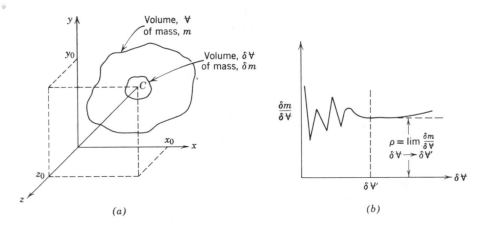

Fig. 2.1 *Definition of density at a point.*

and z_0. The density is defined as mass per unit volume. Thus the mean density within the volume $\forall$ would be given by $\rho = m/\forall$. Clearly this will not in general be equal to the value of the density at point C. To determine the density at point C we must select as small a volume, $\delta\forall$, as possible surrounding the point C and determine the ratio of the mass, δm, within the volume to the volume. The question is how small can we make the volume? Let us answer this question by plotting the ratio $\delta m/\delta\forall$ and allowing the volume to shrink continuously in size. Assuming that the volume $\delta\forall$ is initially relatively large (but still small compared to the volume, $\forall$) a typical plot of $\delta m/\delta\forall$ would appear as in Fig. 2.1b. The average density tends to approach an asymptotic value as the volume is shrunk to enclose only homogeneous fluid in the immediate neighborhood of the point C. When $\delta\forall$ becomes so small that it contains only a small number of molecules, it becomes impossible to fix a definite value for $\delta m/\delta\forall$; the value will vary erratically as molecules cross in and out of the volume. Thus there is a lower limiting value of $\delta \forall$, designated $\delta \forall'$ in Fig. 2.1b, allowable for use in the definition of fluid density at a point. The density at a point is then defined as

$$\rho \equiv \lim_{\delta \forall \to \delta \forall'} \frac{\delta m}{\delta \forall} \tag{2.1}$$

Since the point C was an arbitrary point in the flow field of interest, the density at any point could be determined in a like manner. Thus, if density determinations were made simultaneously at an infinite number of points in the flow, we would obtain an expression for the density distribution as a function of the space coordinates, that is, $\rho = \rho(x, y, z)$, at the given instant in time. Clearly the density at a point may vary with time as a result of work done on or by the fluid and/or heat transfer to the fluid. Thus the complete representation of density (i.e. the field representation) is given by

$$\rho = \rho(x, y, z, t) \tag{2.2}$$

2–2 Velocity Field

In the previous section we saw that the continuum assumption led directly to a field representation of fluid properties. We looked specifically at the density field. Since the density is a scalar quantity, requiring only the specification of a magnitude for a complete description, the field represented by Eq. 2.2 is a scalar field.

In dealing with fluids in motion we shall necessarily be concerned with the description of a velocity field. Referring again to Fig. 2.1a, the fluid velocity at point C is defined as the instantaneous velocity of the center of gravity of the volume $\delta \forall'$ instantaneously surrounding point C. Thus if we define a fluid particle as the small mass of fluid of fixed identity of volume $\delta \forall'$, we are defining the velocity at point C as the instantaneous velocity of the fluid particle which, at a given instant, is passing through the point C. The velocity at any point in the flow field is defined in a like manner. At a given instant of time the velocity field, $\vec{V}$, is a function of the space coordinates x, y, z, that is, $\vec{V} = \vec{V}(x, y, z)$. The velocity at a given point in the flow field may vary from one instant of time to another. Thus the complete representation of velocity (i.e. the velocity field) is given by

$$\vec{V} = \vec{V}(x, y, z, t) \tag{2.3}$$

If fluid properties at a point in a field do not change with time, the flow is termed a steady flow. Stated mathematically, the definition of steady flow is

$$\frac{\partial \eta}{\partial t} = 0$$

where $\eta = \eta(x, y, z, t)$ represents any fluid property. For steady flow,

$$\frac{\partial \rho}{\partial t} = 0, \quad \text{or} \quad \rho = \rho(x, y, z)$$

and

$$\frac{\partial \vec{V}}{\partial t} = 0, \quad \text{or} \quad \vec{V} = \vec{V}(x, y, z)$$

Thus in steady flow, properties may change from one point in the field to another, but they must remain constant with time at any given point.

2–2.1 ONE-, TWO-, AND THREE-DIMENSIONAL FLOWS

We have seen that the general expression for the velocity field, Eq. 2.3, indicates that the velocity field is a function of three space coordinates and time. Such a flow field is termed three-dimensional (it is also unsteady) because the value of the velocity at any point in the flow field depends on the three coordinates required to locate the point in space.

Not all flows are three dimensional. Consider, for example, the flow of water through a long straight pipe of constant cross section. At a point far from the entrance to the pipe the velocity distribution may be given by

$$u = u_{max}\left[1 - \left(\frac{r}{R}\right)^2\right] \tag{2.4}$$

This profile is shown in Fig. 2.2, where cylindrical coordinates r, θ, and x are used to locate any point in the flow field. Since the velocity field is a function of r only, that is, it is independent of the coordinates x and θ, this is a one-dimensional flow.

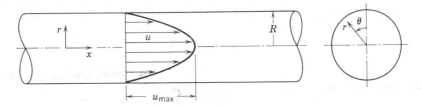

Fig. 2.2 *Example of one-dimensional flow.*

A flow is classified as one-, two-, or three-dimensional depending on the number of space coordinates required to specify the velocity field.[2]

[2] Many authors choose to classify a flow as one-, two-, or three-dimensional on the basis of the number of space coordinates required to specify all fluid properties. In this text, we shall base the number of dimensions on the number of space coordinates required to specify the velocity field only.

An example of a two-dimensional flow is illustrated in Fig. 2.3; the velocity distribution is depicted for a flow between diverging straight walls that are imagined to be infinite in extent (in the z direction). Since the channel is considered to be infinite in the z direction, the velocity field will be identical in all planes perpendicular to the z axis. The velocity field is consequently a function only of the space co-ordinates x and y; the flow field is thus classified as two-dimensional.

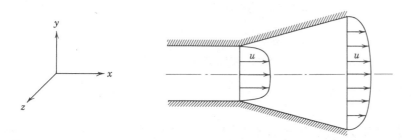

Fig. 2.3 *Example of a two-dimensional flow.*

As you might suspect, the complexity of analysis increases considerably with the number of dimensions of the flow field. The simplest to analyze is one-dimensional flow. For many problems encountered in engineering, a one-dimensional analysis proves adequate to provide approximate solutions of engineering accuracy.

Since all fluids satisfying the continuum assumption must have a zero velocity at a solid surface to satisfy the no-slip condition, most flows are inherently two or three dimensional. Consequently for purposes of analysis it is convenient to introduce the notion of uniform flow at a given cross section. In a flow that is uniform at a given cross section, the velocity is constant across any section normal to the flow. Under this assumption,[3] the two-dimensional flow of Fig. 2.3 becomes the flow shown in Fig. 2.4. In the flow of Fig. 2.4, the velocity is a function of x alone. Thus the velocity field is one-dimensional. (Other properties, e.g. density or pressure, also may be assumed uniform at a section, if desired.)

The term "uniform flow field" (as opposed to uniform flow at a cross section) is used to describe a flow in which the magnitude and direction of the velocity vector are constant, i.e. independent of all space coordinates, throughout the entire flow field.

[3] Convenience alone does not justify this assumption; often results of acceptable accuracy are obtained. Sweeping assumptions such as uniform flow at a cross section should always be reviewed carefully to be sure they provide a reasonable analytical model of the real flow.

velocity field

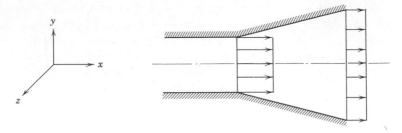

Fig. 2.4 *Example of uniform flow at a section.*

2–2.2 PATHLINES, STREAKLINES, AND STREAMLINES

In the analysis of problems in fluid mechanics, it is frequently advantageous to depict the nature of a flow field graphically. Such a representation is provided by the use of pathlines, streaklines, and streamlines.[4]

A pathline is the path or trajectory traced out by a moving fluid particle. To determine a pathline we might identify a fluid particle at a given instant of time, for example, by the use of dye and, then, take a long exposure photograph of its subsequent motion. The line traced out by the particle is a pathline.

On the other hand, we may choose to focus our attention on a fixed location in space and identify, again by the use of dye, all fluid particles passing through this point. After a short period of time we then have a number of identifiable fluid particles in the flow, all of which had, at some time, passed through one fixed location in space. The line joining these fluid particles is defined as a streakline.

Streamlines are lines drawn in the flow field so that at a given instant of time they are tangent to the direction of flow at every point in the flow field. The shape of the streamlines may vary from instant to instant if the flow velocity is a function of time, that is, if the flow is unsteady. Since the streamlines are tangent to the velocity vector at every point in the flow field, there can be no flow across a streamline.

In steady flow, the velocity at each point in the flow field remains constant with time and, consequently, the streamlines do not vary from one instant to the next. This implies that a particle located on a given streamline will remain on the same streamline. Furthermore, consecutive particles passing through a fixed point in space will be on the same streamline and, subsequently, will remain on this streamline. Thus in a steady flow, pathlines, streaklines, and streamlines are identical lines in the flow field. In the case of unsteady flow, pathlines, streaklines, and streamlines do not coincide.

---

[4] Pathlines, streaklines and streamlines are demonstrated in the film loops: S-FM047, *Pathlines, Streaklines, Streamlines and Timelines in Steady Flow,* and S-FM048, *Pathlines, Streaklines and Stream-lines in Unsteady Flow.* These two loops are taken from the film *Flow Visualization,* S. J. Kline, principal.

Example 2.1

A velocity field is given by $\vec{V} = 2y\hat{i} + \hat{j}$; the units of velocity are ft/sec and y is given in feet.

(a) Is this flow field one-, two-, or three-dimensional? Why?
(b) Determine the velocity components u, v, w, at the point (1, 2, 0).
(c) Determine the slope of the streamline through the point (1, 2, 0).

EXAMPLE PROBLEM 2.1

GIVEN:

Velocity field $\vec{V} = 2y\hat{i} + \hat{j}$ (units of velocity are ft/sec).

FIND:

(a) Dimensions of the flow field.
(b) Velocity components u, v, w at point (1, 2, 0).
(c) Slope of streamline at point (1, 2, 0).

SOLUTION:

(a) A flow is classified as one-, two-, or three-dimensional depending on the number of space coordinates required to specify the velocity field.
Since the given velocity field is a function of one space coordinate, the flow field is one-dimensional.

$\longleftarrow$ _____

(b) The velocity field, $\vec{V} = \hat{i}u + \hat{j}v + \hat{k}w$. Since $\vec{V} = 2y\hat{i} + \hat{j}$, then

$$u = 2y$$
$$v = 1$$
$$w = 0$$

At the point (1, 2, 0)

$$u = 4\ \text{ft/sec} \qquad v = 1\ \text{ft/sec} \qquad w = 0$$

$\longleftarrow$ _____

(c) Streamlines are lines drawn in the flow field such that, at a given instant of time, they are tangent to the direction of flow at every point in the flow field. Consequently, the slope of the streamline at the point (1, 2, 0) is such that the streamline is tangent to the velocity vector at the point.

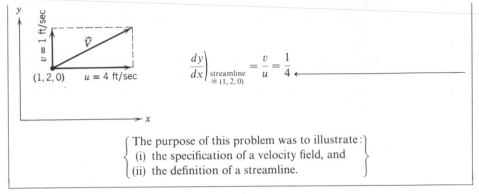

$$\frac{dy}{dx}\bigg)_{\substack{\text{streamline} \\ @ (1,2,0)}} = \frac{v}{u} = \frac{1}{4}$$

$$\left\{ \begin{array}{l} \text{The purpose of this problem was to illustrate:} \\ \text{(i) the specification of a velocity field, and} \\ \text{(ii) the definition of a streamline.} \end{array} \right\}$$

2–3 Stress Field

We know that stresses in a medium result from forces acting on some portion of the medium. The concept of stress provides a convenient means by which we can describe the manner in which forces acting on the boundaries of the medium are transmitted through the medium. Since force and area are both vector quantities, we would anticipate that the stress field will not be a vector field. We shall show that, in general, nine quantities are required to specify the state of stress in a fluid and, hence, that stress is a tensor quantity of second order.

2–3.1 SURFACE AND BODY FORCES

Two types of forces are encountered in the study of continuum fluid mechanics. These are body and surface forces. All external forces, developed without physical contact, and distributed over the volume of the fluid, are termed body forces. Gravitational and electromagnetic forces are examples of body forces arising in a fluid. The only body force which we will consider is the body force due to gravity. The gravitational body force acting on an element of volume, $d\forall$, is given by $\rho \vec{g} \, d\forall$ where ρ is the density (mass per unit volume) and $\vec{g}$ is the local gravitational acceleration. Thus the gravitational body force per unit volume is $\rho \vec{g}$ and the gravitational body force per unit mass is $\vec{g}$.

Example 2.2

A body-force distribution is given as

$$\vec{B} = 10x\hat{\imath} + 15\hat{\jmath}$$

lbf per unit mass (slug) of material acted on. The density (slug/ft³) of the material is given by

$$\rho = x + y^2 + z^3$$

where x, y, and z are given in feet.
What is the resultant body force on the material in the region shown?

EXAMPLE PROBLEM 2.2

GIVEN:

Body-force distribution per unit mass

$$\vec{B} = 10x\hat{\imath} + 15\hat{\jmath} \quad \text{(lbf/slug)}$$

Density of material in volume shown

$$\rho = x + y^2 + z^3 \quad \text{(slug/ft}^3\text{)}$$

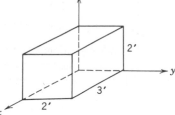

FIND:

Resultant body force on material in volume shown.

SOLUTION:

Body force per unit mass is $\vec{B}$.
Multiplying by mass per unit volume (ρ) gives force per unit volume $= \rho\vec{B}$.
Thus the body force, $d\vec{F}_B$, on an element of volume $d\forall$ is

$$d\vec{F}_B = \rho\vec{B}\, d\forall$$

and

$$\vec{F}_B = \int_\forall \rho\vec{B}\, d\forall = \int_{x=0}^{3} \int_{y=0}^{2} \int_{z=0}^{2} \rho\vec{B}\, dz\, dy\, dx$$

$$= \int_{x=0}^{3} \int_{y=0}^{2} \int_{z=0}^{2} (x + y^2 + z^3)(10x\hat{\imath} + 15\hat{\jmath})\, dz\, dy\, dx$$

$$= \int_{x=0}^{3} \int_{y=0}^{2} \int_{z=0}^{2} \{10(x^2 + xy^2 + xz^3)\hat{\imath} + 15(x + y^2 + z^3)\hat{\jmath}\}\, dz\, dy\, dx$$

Integrating first with respect to z,

$$\vec{F}_B = \int_{x=0}^{3} \int_{y=0}^{2} [10(x^2z + xy^2z + \tfrac{1}{4}xz^4)\hat{\imath} + 15(xz + y^2z + \tfrac{1}{4}z^4)\hat{\jmath}]_{z=0}^{2}\, dy\, dx$$

$$= \int_{x=0}^{3} \int_{y=0}^{2} \{10(2x^2 + 2xy^2 + 4x)\hat{\imath} + 15(2x + 2y^2 + 4)\hat{\jmath}\}\, dy\, dx$$

29   **stress field**

Integrating now with respect to y,

$$\vec{F}_B = \int_{x=0}^{3} [10(2x^2y + \tfrac{2}{3}xy^3 + 4xy)\hat{\imath} + 15(2xy + \tfrac{2}{3}y^3 + 4y)\hat{\jmath}]_{y=0}^{2} \, dx$$

$$= \int_{x=0}^{3} \{10(4x^2 + \tfrac{16}{3}x + 8x)\hat{\imath} + 15(4x + \tfrac{16}{3} + 8)\hat{\jmath}\} \, dx$$

Finally, integrating with respect to x,

$$\vec{F}_B = [10(\tfrac{4}{3}x^3 + \tfrac{8}{3}x^2 + 4x^2)\hat{\imath} + 15(2x^2 + \tfrac{16}{3}x + 8x)\hat{\jmath}]_{x=0}^{3}$$

$$= 10(36 + 24 + 36)\hat{\imath} + 15(18 + 16 + 24)\hat{\jmath}$$

$$\vec{F}_B = 960\hat{\imath} + 870\hat{\jmath} \text{ lbf} \qquad\qquad \vec{F}_B$$

$\left\{\begin{array}{l}\text{Although admittedly artificial, this problem is included to:} \\ \text{ (i) illustrate the calculation of a resultant body force, and} \\ \text{(ii) review the procedure for evaluating triple integrals over a volume.}\end{array}\right\}$

Surface forces include all forces acting on the boundaries of a medium through direct contact. Thus, in considering the stress field in a continuum, we must also be concerned with the surface forces.

2–3.2 STRESS AT A POINT

Consider the element of area, $\delta\vec{A}$, at the point C, acted on by the force, $\delta\vec{F}$, as shown in Fig. 2.5. (The magnitude of $\delta\vec{A}$ is the area of the element, and the direction is normal to the surface.) The stress at a point is defined as

$$\text{Stress} \equiv \lim_{\delta\vec{A}\to 0} \frac{\delta\vec{F}}{\delta\vec{A}} \qquad\qquad (2.5)$$

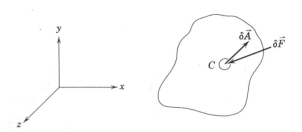

Fig. 2.5 *Definition of stress.*

The definition of stress requires that we evaluate the ratio of the two vectors $\delta \vec{F}$ and $\delta \vec{A}$. In order to perform this operation, let us consider what we know about vectors. The vector $\delta \vec{A}$ can be resolved into three components, that is,

$$\delta \vec{A} = \hat{\imath}\, \delta A_x + \hat{\jmath}\, \delta A_y + \hat{k}\, \delta A_z$$

A simple case is shown in Fig. 2.6 where the element of area $\delta \vec{A}$ is the element ABC.

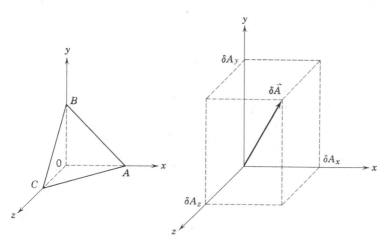

Fig. 2.6 *Element of area.*

The component δA_x is the x component of $\delta \vec{A}$, that is, the projection of $\delta \vec{A}$ on the x axis. Its magnitude is the projection of ABC on the yz plane (area OBC), that is, on the plane perpendicular to the x axis.

Similarly, δA_y is the y component of $\delta \vec{A}$, and its magnitude is the projection of ABC on the xz plane (area OCA), that is, on the plane perpendicular to the y axis.

δA_z is the z component of $\delta \vec{A}$, and its magnitude is the projection of ABC on the xy plane (area OAB), that is, on the plane perpendicular to the z axis.

The vector force, $\delta \vec{F}$, acting at the point C of Fig. 2.5 also can be resolved into three components, that is,

$$\delta \vec{F} = \hat{\imath}\, \delta F_x + \hat{\jmath}\, \delta F_y + \hat{k}\, \delta F_z$$

In defining the stress at a point we may then consider the components δF_x, δF_y, and δF_z of the force $\delta \vec{F}$ at the point C acting on the components δA_x, δA_y, and δA_z of the area $\delta \vec{A}$ at the point C. Thus in defining the stress field, Eq. 2.5 is replaced by a series of nine equations, for we have three stress components (resulting from each of the three force components δF_x, δF_y, and δF_z) acting on each of the three components δA_x, δA_y, and δA_z of the area $\delta \vec{A}$.

In order to keep things straight it is helpful to use a notation that will enable us to delineate both the plane on which the stress is acting and the direction in which the stress is acting. A double subscript notation provides such a convenient description. Thus T_{ij} denotes the stress acting on an i plane in the j direction, where i and j each can stand for x, y, or z. Using this notation we can write the definition of stress (Eq. 2.5) in terms of components as

$$T_{ij} \equiv \lim_{\delta A_i \to 0} \frac{\delta F_j}{\delta A_i} \tag{2.6}$$

Equation 2.6 really represents nine scalar equations, since the subscripts i and j can take on the values x, y, z. For example,

$$T_{xy} \equiv \lim_{\delta A_x \to 0} \frac{\delta F_y}{\delta A_x}$$

defines the stress on an x plane in the y direction. Then in order to specify the state of stress in a fluid we must specify the nine components of the stress field

$$\begin{bmatrix} \sigma_{xx} & \tau_{xy} & \tau_{xz} \\ \tau_{yx} & \sigma_{yy} & \tau_{yz} \\ \tau_{zx} & \tau_{zy} & \sigma_{zz} \end{bmatrix}$$

where σ has been used to denote a normal stress and shear stresses are denoted by the symbol τ. The notation for designating stress is shown in Fig. 2.7.

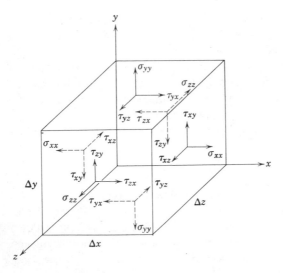

Fig. 2.7 *Notation for stress.*

Referring to the infinitesimal element shown in Fig. 2.7, we see that there are six planes (two x planes, two y planes, and two z planes) on which stresses may act. In order to designate the plane of interest we could use terms like front and back, top and bottom, or left and right. However it is more logical to name the planes in terms of the coordinate axes. The planes are named and denoted as positive or negative according to the direction of the outward drawn normal to the plane. Thus the top plane, for example, is a positive y plane and the back plane is a negative z plane.

It is also necessary to adopt a sign convention for the stress. A stress component is considered positive when the direction of the stress component and the plane on which it acts are both positive or both negative. Thus $\tau_{yx} = 5$ lbf/in.^{2} represents a shear stress on a positive y plane in the positive x direction or a shear stress on a negative y plane in the negative x direction. In Fig. 2.7 all stresses have been drawn as positive stresses. Stress components are negative when the direction of the stress component and the plane on which it acts are of opposite sign.

The specification of the stress field (a tensor field), is considerably more involved (at least it appears that way now) than that of the velocity field (a vector field), or the density field (a scalar field).

2–4 Newtonian Fluid: Viscosity

2–4.1 NEWTONIAN FLUID

We have defined a fluid as a substance that will deform continuously under the action of a shearing stress. In the absence of a shear stress, there will be no deformation. Fluids may be broadly classified according to the relation between the applied shear stress and the rate of deformation of the fluid. Fluids in which the shear stress is directly proportional to the rate of deformation are termed Newtonian fluids. Most common fluids such as water, air, and gasoline are closely Newtonian under normal conditions. The term non-Newtonian fluid classifies all fluids in which the shear stress is not directly proportional to the rate of deformation.

Many common fluids exhibit non-Newtonian behavior. Two familiar examples are toothpaste and Lucite[5] paint. The latter is very "thick" when in the can, but becomes "thin" when sheared by brushing. In this way, a lot of paint is carried by the brush, to minimize the number of times it must be dipped. Toothpaste behaves like a "fluid" when squeezed from the tube. However, it does not run out by itself when the cap is removed. There is a threshold or yield stress below which toothpaste behaves as a solid. Strictly speaking, our definition of a fluid is valid only for materials that have zero yield stress. Non-Newtonian fluids will not be treated in this text.

Consider the behavior of a fluid element between the two infinite plates shown in Fig. 2.8. The upper plate moves at constant velocity, δu, under the influence of a

[5] Trademark, DuPont Corporation.

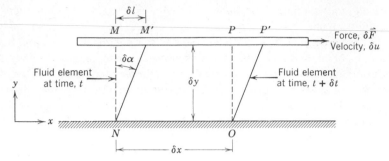

Fig. 2.8 *Deformation of a fluid element.*

constant applied force, $\delta\vec{F}$. The shear stress, τ_{yx}, applied to the fluid element is given by

$$\tau_{yx} = \lim_{\delta A_y \to 0} \frac{\delta F_x}{\delta A_y} = \frac{dF_x}{dA_y}$$

where δA_y is the area of the fluid element in contact with the plate. In the time increment δt the fluid element is deformed from position *MNOP* to position *M'NOP'*. The rate of deformation of the fluid is then given by

$$\text{Deformation rate} = \lim_{\delta t \to 0} \frac{\delta\alpha}{\delta t} = \frac{d\alpha}{dt}$$

The fluid is Newtonian if $\tau_{yx} \propto d\alpha/dt$.

If we wish to calculate shear stresses in a fluid, it is certainly desirable to formulate an expression for $d\alpha/dt$ in terms of more readily measurable quantities. This can be done easily. The distance, δl, between the points *M* and *M'* is given by

$$\delta l = \delta u\,\delta t$$

or alternatively, for small angles,

$$\delta l = \delta y\,\delta\alpha$$

Equating these two expressions for δl, gives

$$\frac{\delta\alpha}{\delta t} = \frac{\delta u}{\delta y}$$

and on taking the limit of both sides of the equality we obtain

$$\frac{d\alpha}{dt} = \frac{du}{dy}$$

Thus, if the fluid of Fig. 2.8 is Newtonian,

$$\tau_{yx} \propto \frac{du}{dy} \tag{2.7}$$

The shear stress acts on a plane tangent to the direction of flow.

2–4.2 VISCOSITY

If one considers the deformation of two different Newtonian fluids, say molasses and water, one recognizes that different fluids will deform at different rates under the action of the same applied shear stress. Molasses exhibits a much larger resistance to deformation than water. Thus we say it is much more viscous. The constant of proportionality in Eq. 2.7 is the absolute (or dynamic) viscosity, μ. Thus in terms of the coordinates of Fig. 2.8, Newton's law of viscosity is given by

$$\tau_{yx} = \mu \frac{du}{dy} \tag{2.8}$$

Note that since the dimensions of τ are $[F/L^2]$ and the dimensions of du/dy are $[1/t]$, then μ has the dimensions of $[Ft/L^2]$. Since the dimensions of force, F, mass, M, length, L, and time, t are related by Newton's second law of motion, the dimensions of μ can also be expressed as $[M/Lt]$. When fluid viscosity is expressed in terms of the British Gravitational system of units, it has units of lbf-sec/ft^2. In the Absolute Metric system of units, the basic unit of μ is called a poise (poise $\equiv$ gm/cm-sec). Calculation of viscous shear stress is illustrated in Example 2.3.

In fluid mechanics the ratio of absolute viscosity, μ, to the density, ρ, often arises. This ratio is given the name kinematic viscosity and is represented by the symbol, v. Since density has the dimensions $[M/L^3]$, the dimensions of v are $[L^2/t]$. In the Absolute Metric system of units, the unit for v is a stoke (stoke $\equiv$ cm^2/sec).

Viscosity arises in real fluids from molecular motions. Molecules from regions of high bulk velocity collide with molecules moving with lower bulk velocity, and vice versa. These collisions transport momentum from one region of fluid to another. Since the random molecular motions are affected by the temperature of the medium, viscosity is also a function of temperature. Viscosity data for a number of common Newtonian fluids are given in Appendix A.

Example 2.3

An infinite plate is moved over a second plate on a layer of oil as shown in the diagram. For small gap width, d, we assume a linear velocity distribution in the oil.

35   **Newtonian fluid: viscosity**

The following information is given

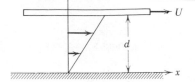

$U = 1$ ft/sec $d = 0.001$ ft

$\mu = 0.7$ centipoise $\rho = 1.4$ slug/ft^3

Calculate:

(a) The fluid viscosity in lbf-sec/ft^2.
(b) The kinematic viscosity of the fluid in ft^2/sec.
(c) The shear stress on the upper plate.
(d) The shear stress on the lower plate.
(e) Indicate the direction of the shear stress for parts (c) and (d).

EXAMPLE PROBLEM 2.3

GIVEN:

Linear velocity profile in the fluid between infinite parallel plates as shown.

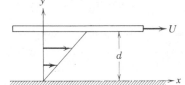

$U = 1$ ft/sec $d = 0.001$ ft $\rho = 1.4$ slug/ft^3

$\mu = 0.7$ centipoise (1 poise = 1 gm/cm-sec)

FIND:

(a) μ in units of lbf-sec/ft^2 (b) v in units of ft^2/sec (c) τ on upper plate
(d) τ on lower plate   (e) direction of stress in parts (c) and (d).

SOLUTION:

Basic equation: $\tau_{yx} = \mu \dfrac{du}{dy}$ Definition: $v = \dfrac{\mu}{\rho}$

(a) $\mu = 0.7$ centipoise $\times \dfrac{1 \text{ poise}}{100 \text{ centipoise}} \times \dfrac{1 \text{ gm}}{\text{cm-sec-poise}} \times \dfrac{2.2 \text{ lbm}}{1000 \text{ gm}} \times \dfrac{\text{slug}}{32.2 \text{ lbm}}$

$\times \dfrac{30.5 \text{ cm}}{\text{ft}}$

$= 0.146 \times 10^{-4} \dfrac{\text{slug}}{\text{ft-sec}} \times \dfrac{1 \text{ lbf-sec}^2}{\text{slug-ft}}$ $\left\{ \begin{array}{l} \text{Recall from } \vec{F} = m\vec{a} \\ 1 \text{ lbf} = 1 \text{ slug} \times 1 \text{ ft/sec}^2 \end{array} \right\}$

$\mu = 0.146 \times 10^{-4} \dfrac{\text{lbf-sec}}{\text{ft}^2}$

(b) $v = \dfrac{\mu}{\rho} = 0.146 \times 10^{-4} \dfrac{\text{lbf-sec}}{\text{ft}^2} \times \dfrac{\text{ft}^3}{1.4 \text{ slug}} \times \dfrac{1 \text{ slug-ft}}{\text{lbf-sec}^2} = 1.04 \times 10^{-5} \text{ ft}^2/\text{sec}$ $\overset{v}{\longleftarrow}$

(c) $\tau_{\text{upper}} = \tau_{yx\,\text{upper}} = \mu \dfrac{du}{dy}\bigg|_{\text{upper}}$

    Since u is a linear function of y, du/dy is a constant

$$\frac{du}{dy} = \frac{\Delta u}{\Delta y} = \frac{U - 0}{d - 0} = \frac{U}{d} \qquad \therefore \quad \frac{du}{dy}\bigg|_{\text{upper}} = \frac{du}{dy}\bigg|_{\text{lower}} = \frac{U}{d}$$

$\tau_{\text{upper}} = \mu \dfrac{du}{dy}\bigg|_{\text{upper}} = \mu \dfrac{U}{d} = 0.146 \times 10^{-4} \dfrac{\text{lbf-sec}}{\text{ft}^2} \times \dfrac{1 \text{ ft}}{\text{sec}} \times \dfrac{1}{10^{-3} \text{ ft}} = 0.0146 \text{ lbf/ft}^2$ $\overset{\tau_{\text{upper}}}{\longleftarrow}$

(d) $\tau_{\text{lower}} = \mu \dfrac{du}{dy}\bigg|_{\text{lower}} = \mu \dfrac{U}{d} = 0.0146 \text{ lbf/ft}^2$ $\overset{\tau_{\text{lower}}}{\longleftarrow}$

(e) Direction of shear stress on upper and lower plates.

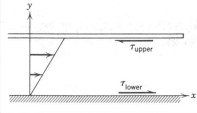

$\left\{ \begin{array}{l} \text{The upper plate is a negative } y \text{ surface, so} \\ \text{positive } \tau_{yx} \text{ acts in the negative } x \text{ direction.} \end{array} \right\}$ (e)

$\left\{ \begin{array}{l} \text{The lower plate is a positive } y \text{ surface, so} \\ \text{positive } \tau_{yx} \text{ acts in the positive } x \text{ direction.} \end{array} \right\}$

2–5 Description and Classification of Fluid Motions

In Chapter 1 we listed a wide variety of typical problems encountered in fluid mechanics and outlined our method of approach to the subject. Before proceeding with our detailed study, we will attempt a broad classification of fluid mechanics on the basis of observable physical characteristics of flow fields. Since there is much overlap in the types of flow fields encountered, there is no universally accepted classification scheme. One possible classification is shown in Fig. 2.9.

2–5.1 VISCOUS AND INVISCID FLOWS

The main subdivision indicated is between inviscid and viscous flows. In an inviscid flow the fluid viscosity, μ, is assumed to be zero. Clearly such flows do not exist; however, there are many problems wherein such an assumption will simplify the analysis and, at the same time, lead to meaningful results. (While simplification of the analysis is always desirable, the results must be reasonably accurate if the solution is to be of value.) Within the area of inviscid flow we may consider problems

37   **description and classification of fluid motions**

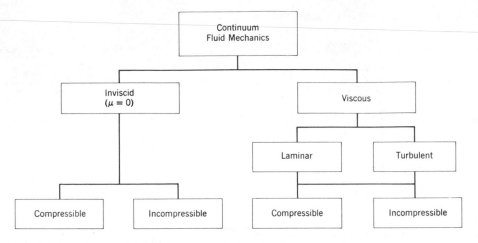

Fig. 2.9 *Possible classification of continuum fluid mechanics.*

in two broad classes. Flows in which variations of density are small and relatively unimportant are termed incompressible. Flows in which density variations play a dominant role, as in high speed gas flows, are called compressible flows. We shall consider both of these subgroups within the area of inviscid flows.

All fluids possess viscosity and, consequently, viscous flows are of paramount importance in the study of continuum fluid mechanics. Although we shall study viscous flows in some detail later, it might be helpful to give a few examples of viscous flow phenomena.

In our discussion following the definition of a fluid (Section 1-1), we noted that in any viscous flow, the fluid in direct contact with a solid boundary has the same velocity as the boundary itself, that is, there is no slip at the boundary. For the one-dimensional viscous flow of Fig. 2.8, the shear stress was given by Eq. 2.8.

$$\tau_{yx} = \mu \frac{du}{dy} \tag{2.8}$$

Since the fluid velocity at a solid surface in a moving fluid is zero, but the bulk fluid is moving, velocity gradients must be present in the flow. These gradients produce shear stresses, which in turn affect the motion.

As a practical case, we might ask what is the nature of the fluid motion around a thin wing or ship hull. Such a flow might be represented crudely by the flow over a flat plate, as shown in Fig. 2.10. The flow approaching the plate is of uniform velocity, U_∞. We are interested in providing a qualitative picture of the velocity distribution normal to the plate at various locations along the length of the plate. Two such stations are denoted by x_1 and x_2. Consider first the location x_1. In

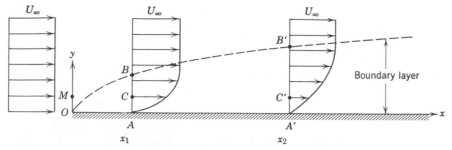

Fig. 2.10 *Incompressible laminar viscous flow over a semi-infinite flat plate.*

order to arrive at a qualitative picture of the velocity distribution, we start by labeling the y coordinates at which the velocity is known.

From the no-slip condition, we know the velocity at point A must be zero; hence we have one point on the velocity profile. Can we locate any other points on the profile? Let us stop for a minute and ask ourselves, "What is the effect of the plate on the flow?" The plate is stationary and, therefore, exerts a retarding force on the flow; it slows the fluid in the neighborhood of the surface. If we are at a y location sufficiently far from the plate, say at point B, the flow at the point will not be influenced by the presence of the plate. Providing the pressure does not vary in the x direction (as is the case for flow over a semi-infinite flat plate) the velocity at point B will be U_∞. It seems reasonable to expect the velocity to increase smoothly and monotonically from the value $u = 0$ at $y = 0$ to $u = U_\infty$ at $y = y_B$. The profile has been so drawn; thus at some point, C, intermediate between points A and B, the velocity has a value that lies between zero and U_∞, that is, for $0 \leq y \leq y_B$, then $0 \leq u \leq U_\infty$. From these characteristics of the velocity profile and our definition of the shear stress, we see that within the region $0 \leq y \leq y_B$, shear stresses are present; for $y > y_B$, the velocity gradient is zero and, hence, there are no shear stresses present.

What about the velocity profile at the station $x = x_2$? Is it exactly the same as the profile at $x = x_1$? A look at Fig. 2.10 suggests that it is not. At least it has not been drawn that way! While it is qualitatively the same, why is it not exactly the same? We might guess that the plate would influence a greater region of the flow field as we move farther down the plate. Looking again at the profile at location x_1, we see that the slower moving fluid adjacent to the plate exerts a retarding force on the faster moving fluid above it. We can see this by considering the shear stress on the plane through the point C. Since we are interested in the stress exerted on the faster moving fluid above the plane, we are looking for the direction of the shear stress on a negative y plane through the point C. From Eq. 2.8 we see that τ_{yx} on the plane through point C has a positive numerical value; consequently the shear stress must be in the negative x direction.

description and classification of fluid motions

In establishing the qualitative picture of the velocity profile at $x = x_2$, we recognize the no slip condition requires that the velocity at the wall be zero; this fixes the velocity at A' as zero. Since, at location x_1, the slower moving fluid exerts a retarding force on the fluid above it, we would expect the distance out to the point where the velocity is U_∞ to be increased at location x_2; that is $y_{B'} > y_B$. Furthermore it is reasonable to expect that $u_{C'} < u_C$.

From our qualitative picture of the flow field, we see that we can divide the flow into two general regions. In the region adjacent to the boundary, shear forces are present; this region is called the boundary layer.[6] Outside the boundary layer the velocity gradient is zero and, hence, the shear stresses are zero. In this region we may use inviscid flow theory in analyzing the flow.

For a given freestream velocity, U_∞, the size of the boundary layer will depend on the properties of the particular fluid. Since the shear stress is directly proportional to the viscosity, we expect the size of the boundary layer to depend on the viscosity of the fluid. In Chapter 8, we will develop expressions for determining the rate of boundary layer growth.

Before leaving our discussion of the viscous flow over a semi-infinite flat plate, we should stop and reflect on two points. We have implied that the shear stress on a y plane was given by the one-dimensional expression for τ, Eq. 2.8. Is the flow one-dimensional? A look at Fig. 2.10 indicates that the velocity, u, is a function of both x and y; thus the flow is two-dimensional and, hence, τ_{yx} as given by Eq. 2.8 can only be considered as approximate.[7]

In our qualitative description of the flow field, we were only concerned about the behavior of the x component of velocity, that is, the velocity, u. What about the y component of velocity, that is, the velocity, v? Is it zero throughout the flow field? To answer this question consider the streamlines of the flow. Rather than consider all possible streamlines, let us consider the particular streamline through the point M. On recalling the definition of a streamline, that is, a line drawn tangent to the velocity vector at every point in the flow, our first inclination might be to depict the streamline through M as a straight line parallel to the x axis. In doing this we would be in error for we would neglect the fact that there can be no flow across a streamline. Because there can be no flow across a streamline, the mass flow between adjacent streamlines (or between a streamline and a solid boundary) must be a constant. Thus in considering the incompressible viscous flow of Fig. 2.10, we recognize that the streamline through the point M cannot be a straight line parallel to the x-axis. The streamline through the point M must move continuously away from the x-axis as we move down the plate. Furthermore, we note that the streamline through M crosses the dashed line we have used to denote the edge of the boundary layer. Consequently we conclude that the edge of the boundary layer

---

[6] The formation of a boundary layer is demonstrated in the film loop, S-FM006, *Boundary Layer Formation.*

[7] As it turns out, this is a very good approximation for laminar flow.

is not a streamline, and there is flow into the boundary layer as we move down the plate. Indeed if the boundary layer is to grow, there must be flow across the edge of the boundary layer.

We have used the incompressible flow over a semi-infinite flat plate to establish a qualitative picture of the viscous flow over a solid boundary. In that example, we had only to consider the effect of shear forces on the flow field; the pressure was constant throughout the flow field. Now let us consider a steady flow field (the incompressible flow over a cylinder) wherein both pressure forces and viscous forces are important. For steady flow, we recognize that pathlines, streaklines, and streamlines are all identical. If we were to employ some means of flow visualization, we would find the flow field to be of the general character shown in Fig. 2.11a.[8]

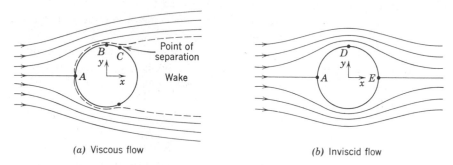

(a) Viscous flow   (b) Inviscid flow

Fig. 2.11 *Qualitative picture of flow over a cylinder.*

We see that the streamlines are symmetric about the x axis. The flow along the central streamline impinges on the cylinder at point A, divides, and flows around the cylinder. Point A on the cylinder is called a stagnation point. As in flow over a flat plate, a boundary layer develops in the neighborhood of the solid surface. The velocity distribution outside the boundary layer can be determined from the spacing of the streamlines. Since there can be no flow across a streamline, we would expect the flow velocity to increase in regions where the spacing between streamlines decreases. Conversely, an increase in streamline spacing implies a decrease in flow velocity.

If we consider for a moment the incompressible flow around a cylinder assuming an inviscid flow, as shown in Fig. 2.11b, then the flow is symmetric about both the x and y axes. The velocity around the cylinder increases to a maximum at point D and then decreases as we move further around the cylinder. For inviscid flow, an increase in velocity is accompanied by a decrease in pressure and conversely.

[8] The details of the flow will depend on the various flow properties. For all but very low velocity flows, the qualitative picture will be as shown.

Thus in the case of an incompressible inviscid flow, the pressure along the surface of the cylinder decreases as we move from point A to point D and, then, increases again in going from point D to point E. Since the flow is symmetric with respect to both the x and y axes, we would also expect the pressure distribution to be symmetric with respect to these axes. This is indeed the case.

Since there are no shear stresses present in an inviscid flow, the pressure forces are the only forces we need consider in determining the net force on the cylinder. The symmetry of the pressure distribution leads to the conclusion that for an inviscid flow, there is no net force on the cylinder in either the x or y directions. The net force in the x direction is termed the drag. Thus for an inviscid flow over a cylinder, we are led to the conclusion that the drag is zero; this conclusion is contrary to experience, for we know that all bodies experience some drag when placed in a real flow. In treating the inviscid flow over a body we have, by the definition of inviscid flow, neglected the presence of the boundary layer. Let us go back and look again at the real flow situation.

In considering the real flow, Fig. 2.11a, we will consider the boundary layer to be thin. Since the boundary layer is taken to be thin, it is reasonable to assume that the behavior of the velocity and pressure fields outside the boundary layer is qualitatively the same as in the inviscid flow case. Then consider a fluid element inside the boundary layer at a position between points A and B. Since the pressure decreases continuously between points A and B, the fluid element experiences a net pressure force in the direction of flow. Over the region between A and B, this net pressure force is sufficient to overcome the resisting shear force and motion of the element in the flow direction is maintained.

Now consider an element of fluid inside the boundary layer at a location on the back side of the cylinder beyond point B. Since the pressure increases in the direction of flow, the fluid element experiences a net pressure force in a direction opposite to its direction of motion. At some point around the cylinder the momentum of the fluid in the boundary layer is insufficient to carry the element further into the region of higher pressure. The fluid layers adjacent to the solid surface will be brought to rest and the flow will separate from the surface;[9] the point at which this occurs is called the point of separation. Boundary layer separation results in the formation of a relatively low pressure region behind a body; the region exhibiting a deficiency of momentum behind a body is called the wake. Thus, for separated flow over a body, there is a net unbalance of pressure forces in the direction of flow; this results in a pressure drag on the body. The greater the size of the wake behind a body, the greater is the pressure drag.

It is logical to ask how one might reduce the size of the wake and thus reduce the pressure drag. Since a large wake results from boundary layer separation, which

[9] The flow over a variety of models, illustrating flow separation, is demonstrated in the following NCFMF film loops: S-FM012, *Flow Separation and Vortex Shedding*; S-FM004, *Separated Flows — Part I*; and S-FM005, *Separated Flows — Part II*.

in turn is related to the presence of an adverse pressure gradient (increase of pressure in the direction of flow), a reduction of the adverse pressure gradient should delay the onset of separation and, hence, reduce the drag.

Streamlining of a body reduces the magnitude of the adverse pressure gradient. In effect we are spreading a given pressure rise over a larger distance and hence reducing the magnitude of the pressure gradient. If, for example, a gradually tapered section is added to the cylinder of Fig. 2.11, the flow field would appear qualitatively as shown in Fig. 2.12. By streamlining the shape of the body, we are able to delay the onset of separation; although we have increased the surface area of the body and, hence, the total shear force acting on the body, the drag is significantly reduced.[10]

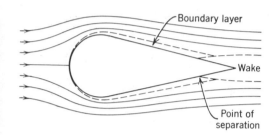

Fig. 2.12 *Flow over a streamlined object.*

Flow separation may also occur in internal flows (flows through ducts) as a result of rapid or abrupt changes in duct geometry.[11]

2–5.2 LAMINAR AND TURBULENT FLOWS

The classification of viscous flows into laminar and turbulent regimes is related to the macroscopic behavior of the flow. A laminar flow is one in which the fluid flows in laminae or layers; there is no macroscopic mixing of adjacent fluid layers. A thin filament of dye injected into a laminar flow appears as a single line; there is no dispersion of dye throughout the flow, except the slow dispersion due to molecular motion. On the other hand, a dye filament injected into a turbulent flow quickly disperses throughout the flow field; the line of dye breaks up into myriad entangled threads of dye. This behavior of turbulent flow is due to small velocity fluctuations superimposed on the mean motion of a turbulent flow; the macroscopic mixing of fluid particles from adjacent layers of fluid results in rapid dispersion of the dye. Whether a flow is laminar or turbulent depends on the properties of

---

[10] The effect of streamlining a body is demonstrated in the film loop, S-FM004, *Separated Flows Part I.*

[11] Examples of separation in internal flows are shown in the film loop, S-FM015, *Incompressible Flow Through Area Contractions and Expansions.*

description and classification of fluid motions

the particular flow. In the case of flow through a pipe, the nature of the flow (laminar or turbulent) is determined by the value of the Reynolds number, $Re = \rho D \bar{V}/\mu$ where ρ is the density of the fluid, D is the pipe diameter, $\bar{V}$ is the average flow velocity, and μ is the viscosity of the fluid.

Although for $Re > 2300$ the flow is often turbulent and for $Re < 2300$ the flow is laminar, there is, in reality, no single value of the Reynolds number at which the flow changes from laminar to turbulent. There is a range of values of Re over which the flow will change from laminar to turbulent. We might expect that flow conditions at the pipe inlet would affect the value of the Reynolds number at which the change (i.e. the transition) from laminar to turbulent flow would occur; this is indeed the case. Thus it is more meaningful to talk of a lower value of Reynolds number, Re_{lower}, below which the flow is always laminar and a higher value, Re_{higher}, above which the flow is always turbulent, regardless of the inlet conditions to the pipe.

The straight filament of smoke rising from a cigarette in still surroundings gives a clear picture of laminar flow. As the smoke continues to rise, it breaks up into a random, haphazard motion; this is an example of turbulent flow.[12]

Boundary layer flows may also be laminar or turbulent; the definitions of laminar and turbulent flows given earlier also apply to boundary layer flows. As we shall see in later chapters, the details of a flow field may be significantly different depending on whether the boundary layer is laminar or turbulent.

2–5.3 COMPRESSIBLE AND INCOMPRESSIBLE FLOWS

Flows in which variations in density are negligible are termed incompressible; when density variations within a flow are not negligible, the flow is called compressible. If one considers the two states of matter, liquid and gas, included within the definition of a fluid, one is tempted to make the general statement that all liquid flows are incompressible flows and all gas flows are compressible flows. For virtually all practical cases the first portion of the statement is correct—that is, all liquid flows are essentially incompressible. However, gas flows may also be considered incompressible provided the flow velocities are small relative to the speed of sound in the fluid; the ratio of the flow speed, V, to the speed of sound, c, in the fluid is defined as the Mach number, M, that is,

$$M = \frac{V}{c}$$

For values of $M < 0.3$, changes in density are only about 2 percent of the mean value. Thus the flow can be treated as incompressible; a value of $M = 0.3$ in air at standard conditions corresponds to a velocity of approximately 330 ft/sec.

[12] Several examples illustrating the nature of laminar and turbulent flows are shown in the film loop S-FM008, *The Occurrence of Turbulence.*

Compressible flows occur frequently in engineering applications. Common examples include compressed air systems used to power shop tools and dental drills, transmission of gases in pipelines at high pressure, and pneumatic or fluidic control and sensing systems. Compressibility effects are very important in the design of modern high speed aircraft and missiles, power plants, fans, and compressors.

Substantial changes in fluid density can occur in compressible flows. These density changes are accompanied by changes in fluid temperature. Acoustic and pressure wave phenomena are responsible for many effects, including "tuning" of intake and exhaust passages in high performance engines, and the "sonic boom" produced by aircraft in supersonic flight.

Shocks, across which fluid properties such as pressure and density change abruptly, can occur in supersonic flows under some conditions. The flow rate in compressible flow can also be choked, or limited, by passage design. These, and other features of compressible flows, will be treated in detail in Chapters 9 and 10.

problems

2.1   The air density field in the vicinity of a powerplant exhaust is given by

$$\rho = \rho_0 + \frac{\Delta\rho}{2}\left[\frac{r_0}{r_0 + (x^2 + y^2)^{1/2}} + e^{-kz}\right]$$

Is the field one-, two-, or three-dimensional? Is it steady or unsteady?

2.2   The density field in the exhaust pipe of a Diesel engine is given by

$$\rho = a[1 + be^{-cx}\cos(\omega t)]$$

Is the field one-, two-, or three-dimensional? Is it steady or unsteady?

2.3   For the velocity fields given below, determine:

(a) Whether the flow is one-, two-, or three-dimensional, and why.
(b) Whether the flow is steady or unsteady, and why.

(The quantities a, b, and c are constants.)

a. $\vec{V} = [ae^{-bx}]\hat{\imath}$
b. $\vec{V} = [ax^2 e^{-bt}]\hat{\imath}$
c. $\vec{V} = ax^2\hat{\imath} + bx\hat{\jmath}$
d. $\vec{V} = ax\hat{\imath} + bx^2 e^{-ct}\hat{\jmath}$
e. $\vec{V} = ax\hat{\imath} + bx^2\hat{\jmath} - cx^2\hat{k}$
f. $\vec{V} = ax\hat{\imath} - by\hat{\jmath}$
g. $\vec{V} = (ax + t)\hat{\imath} + by^2\hat{\jmath}$

h. $\vec{V} = ax\hat{\imath} + by\hat{\jmath} + cx^2\hat{k}$
i. $\vec{V} = ax\hat{\imath} + by^2\hat{\jmath} + cyt\hat{k}$
j. $\vec{V} = ax^2\hat{\imath} + by\hat{\jmath} + cxz\hat{k}$
k. $\vec{V} = ax^2\hat{\imath} + bxz\hat{\jmath}$
l. $\vec{V} = ax\hat{\imath} - by\hat{\jmath} + (t - cz)\hat{k}$
m. $\vec{V} = a(x^2 + y^2)^{1/2}(1/z^3)\hat{k}$

2.4 Give an example of one-, two-, and three-dimensional flow fields. Illustrate with sketches.

2.5 A velocity field is given by

$$\vec{V} = 2x\hat{\imath} + 2y\hat{\jmath} + xyt\hat{k}$$

Determine the number of dimensions of the flow field. Is it steady? Find the slope of the streamline through the point $(x, y, z) = (1, 2, 0)$ at time $t = 0$.

2.6 A velocity field is given by

$$\vec{V} = 2y\hat{\imath} + x\hat{\jmath} + 2\hat{k}$$

The units of velocity are ft/sec.
Determine the number of dimensions of the flow field. Is it steady? Determine the velocity components u, v, w, at the point $(1, 2, 0)$. Determine the slope in the xy plane of the streamline through the point $(1, 2, 0)$.

2.7 The velocity field

$$\vec{V} = ax\hat{\imath} - by\hat{\jmath}$$

can be interpreted to represent flow in a corner.
Find an equation for the flow streamlines. Plot several streamlines in the first quadrant, including the one which passes through the point $(x, y) = (0, 0)$.

2.8 The velocity distribution in a certain region is given by

$$\vec{V} = 2x\hat{\imath} - y\hat{\jmath} + (3t - z)\hat{k}$$

Determine the number of dimensions of the flow field. Is it steady? Find the equation of the streamline through the point $(x, y, z) = (1, 1, 3)$ at $t = 0$ and $t = 1$.

2.9 Parametric equations for the position of a particle in a flow field are given as

$$x_p = c_1 e^{at}$$

and

$$y_p = c_2 e^{-bt}$$

Find the equation of the pathline for a particle located at $(x, y) = (1, 2)$ at $t = 0$. Compare with a streamline through the same point, found in Problem 2.7.

2.10 A tornado can be represented in polar coordinates by the velocity field

$$\vec{V} = -\frac{a}{r}\hat{\imath}_r + \frac{b}{r}\hat{\imath}_\theta$$

where $\hat{\imath}_r$ and $\hat{\imath}_\theta$ are unit vectors in the r and θ directions, respectively.

Recall that an element of distance along the θ direction is $dl = r\,d\theta$. Show that streamlines have equations of the form

$$r = ce^{-a/b\theta}$$

that is, they form logarithmic spirals.

2.11 The velocity field in the region shown in the diagram below is given by

$$\vec{V} = 10y\hat{\jmath} + 5\hat{k}$$

For a unit depth perpendicular to the diagram, an element of area ① may be represented by $dz(-\hat{\jmath})$ and an element of area ② by $dy(-\hat{k})$. (Note that both are drawn *outward* from the control volume; hence the minus sign.)

(a) Find an expression for $\vec{V} \cdot d\vec{A}_①$

(b) Evaluate $\iint_{A_①} \vec{V} \cdot d\vec{A}_①$

(c) Find an expression for $\vec{V} \cdot d\vec{A}_②$

(d) Find an expression for $\vec{V}(\vec{V} \cdot d\vec{A}_②)$

(e) Evaluate $\iint_{A_②} \vec{V}(\vec{V} \cdot d\vec{A}_②)$

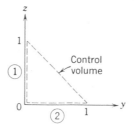

2.12 The area shown shaded is in a flow where the velocity field is given by

$$\vec{V} = -x\hat{\imath} + y\hat{\jmath} + \hat{k}$$

Write a vector expression for an element of the shaded area. Evaluate the following integrals over the shaded area.

(a) $\iint \vec{V} \cdot d\vec{A}$

(b) $\iint \vec{V}(\vec{V} \cdot d\vec{A})$

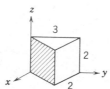

2.13 The shaded area is in a flow where the velocity field is given by

$$\vec{V} = x\hat{\imath} + y\hat{\jmath}$$

Evaluate the following integrals over the shaded area

(a) $\iint \vec{V} \cdot d\vec{A}$

(b) $\iint \vec{V}(\vec{V} \cdot d\vec{A})$

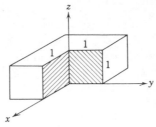

2.14 The velocity distribution for laminar flow in a long circular tube is given by the one-dimensional expression

$$\vec{V} = u\hat{\imath} = u_{max}\left[1 - \left(\frac{r}{R}\right)^2\right]\hat{\imath}$$

For this profile, use the area vector $d\vec{A} = dA\hat{\imath} = 2\pi r\, dr\hat{\imath}$ to evaluate

(a) $\iint \vec{V} \cdot d\vec{A}$

(b) $\iint \vec{V}(\vec{V} \cdot d\vec{A})$

for the tube cross section.

2.15 The density distribution in the fluid column shown is given by

$$\rho = \rho_0(1 + ky)$$

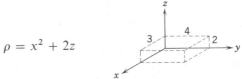

where $\rho_0 = 0.00238$ slug/ft^3 and $k = 0.10$ ft^{-1}.
Compute the body force in lbf acting on the volume.

2.16 A body force distribution is given as $\vec{B} = 16x\hat{\imath} + 10\hat{\jmath}$ per unit mass of the material acted on.
The density of the material is given as

$$\rho = x^2 + 2z$$

Determine the resultant body force on the material in the region shown.

2.17   Use double index notation to label the 6 shear stresses shown.

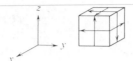

2.18   On the element shown, indicate all possible stresses represented by

$$\tau_{zx} = -10 \text{ lbf/ft}^2$$

$$\sigma_{yy} = 15 \text{ lbf/in.}^2$$

2.19   The following stress levels are known to exist at a point in a continuous medium:

$$\sigma_{xx} = 200 \text{ psi}$$

$$\sigma_{yy} = -100 \text{ psi}$$

$$\tau_{xy} = \tau_{yx} = 30 \text{ psi}$$

Find the magnitudes of the normal and shear stresses acting on plane AA.

2.20   According to Eq. 2.8, there is a linear relation between shear stress (τ_{yx}) and shear rate (du/dy) for Newtonian fluids. If shear stress versus shear rate is plotted, the slope of the resulting straight line equals the fluid viscosity.

Non-Newtonian fluids do not give such a simple result. Sketch a plot of shearing stress versus shear rate for (a) toothpaste and (b) Silly Putty. Contrast their behavior with that of a Newtonian fluid.

2.21   Obtain the conversion factor for converting the viscosity, μ, from units of dynes, seconds, and centimeters (i.e. poises) to units of pounds force, seconds, and feet. Check your answer using data from Appendix A.

2.22   A block weighing 5 lbf and 8 in. square slides down a smooth incline on a thin film of oil. The slope is 30° from the horizontal. The oil is SAE 30 at 70 F, the film is 0.001 in. thick, and the velocity profile may be assumed linear. Calculate the terminal velocity of the block.

2.23   A block weighing 10 lbf and having dimensions 10 in. on each edge is pulled up an inclined surface on which there is a film of SAE 10 oil at 110 F. If the velocity of the block is 5 ft/sec and the oil film is 0.001 in. thick, find the force $\vec{F}$. Assume the velocity distribution in the oil film to be linear. The surface is inclined at an angle of 15° from the horizontal.

2.24   The velocity distribution for laminar flow between parallel plates is given by

$$\frac{u}{u_{max}} = 1 - \left(\frac{2y}{h}\right)^2$$

where h is the distance separating the plates, and the origin is placed midway between the plates. Consider a flow of water at 60 F, with $u_{max} = 1$ ft/sec, and $h = 0.020$ in. Calculate the shear stress on the upper plate, and give its direction.

2.25   A concentric cylinder viscometer is formed by rotating the inner member of a pair of closely fitting cylinders. The gap must be made small so that a linear velocity profile will exist. Consider cylinders of 3 in. diameter and 6 in. height, with a clearance gap of 0.001 in. width, filled with castor oil at 90 F. Determine the torque required to turn the inner cylinder at 250 rpm.

fluid statics

By definition, a fluid must deform continuously when a shearing stress of any magnitude is applied. Absence of relative motion (and thus, angular deformation) implies absence of shearing stress. Fluids either at rest or in uniform motion therefore are able to sustain only normal stresses. Analysis of these "flow" cases is thus appreciably simpler than for fluids undergoing angular deformation (see Section 5–4).

Mere simplicity does not justify our study of a subject. Normal forces transmitted by fluids are of importance in many practical situations. Using the principles of hydrostatics we can compute forces on submerged objects, develop instruments for measuring pressures, and deduce properties of our atmosphere. Another important use is the application to hydraulic brakes in our automobiles.

In a static fluid, or a fluid in uniform motion, a fluid particle retains its identity for all time. Since there is no relative motion within the fluid, a fluid element does not deform. We may apply Newton's second law of motion to evaluate the reaction of the particle to the applied forces.

3–1 *The Basic Hydrostatic Equation*

Our primary objective is to obtain an equation that will enable us to determine the pressure field within the fluid. To do this we choose a differential element of mass, dm, with sides dx, dy, and dz as shown in Fig. 3.1. The element is stationary relative to the rectangular coordinate system shown. The coordinate system may be accelerating with a uniform acceleration, $\vec{a}$, or it may be at rest or moving with constant velocity (in which case $\vec{a} = 0$). When the coordinate system is in motion, we assume that all transient effects (e.g. sloshing) have died out. This guarantees that there will be no relative motion among fluid particles.

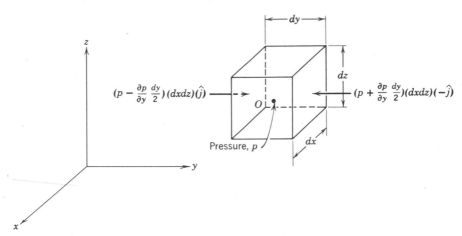

Fig. 3.1 *Differential fluid element and pressure forces in y direction.*

From our previous discussion we recall that two general types of forces may be applied to a fluid. These are body forces and surface forces. The only body force that must be considered in most engineering problems is due to gravity. In some situations body forces due to electric or magnetic fields might be present, but for simplicity they will not be considered in this text. Then for a differential fluid element, the body force, $d\vec{F}_B$, is

$$d\vec{F}_B = \vec{g}\,dm = \vec{g}\rho\,d\forall$$

where $\vec{g}$ is the local gravity vector, ρ the density, and $d\forall$ the volume of the element. In Cartesian coordinates $d\forall = dx\,dy\,dz$, so

$$d\vec{F}_B = \rho\vec{g}\,dx\,dy\,dz$$

In a static fluid no shear stresses can be present. Thus the only surface force is the pressure force. Since the pressure varies with position within the fluid, the net pressure force that results from this variation can be evaluated by summing the forces that act on the faces of the fluid element.

The pressure at the center, O, of the element is p. To determine the pressure at each of the six faces of the element, we must use a Taylor series expansion of the pressure about the point O. [Note that $p = p(x, y, z)$, that is, pressure is a field quantity, too.]

The pressure forces acting on the two y surfaces of the differential element are shown in Fig. 3.1. Each pressure force is a product of three terms. The first is the magnitude of the pressure, obtained by expanding the pressure at the center of the element in a Taylor series about the center. The magnitude is multiplied by the area of the face to give the pressure force, and a unit vector is introduced to indicate direction. Note also in Fig. 3.1 that the pressure force on each face acts *on* the face. In other words, a positive pressure corresponds to a *compressive* stress.

Stresses and forces on the other faces of the element are obtained in the same way. Combining all such forces gives the total surface force acting on the element. Thus

$$d\vec{F}_S = \left(p - \frac{\partial p}{\partial x}\frac{dx}{2}\right)(dy\,dz)(\hat{\imath}) + \left(p + \frac{\partial p}{\partial x}\frac{dx}{2}\right)(dy\,dz)(-\hat{\imath})$$

$$+ \left(p - \frac{\partial p}{\partial y}\frac{dy}{2}\right)(dx\,dz)(\hat{\jmath}) + \left(p + \frac{\partial p}{\partial y}\frac{dy}{2}\right)(dx\,dz)(-\hat{\jmath})$$

$$+ \left(p - \frac{\partial p}{\partial z}\frac{dz}{2}\right)(dx\,dy)(\hat{k}) + \left(p + \frac{\partial p}{\partial z}\frac{dz}{2}\right)(dx\,dy)(-\hat{k})$$

Collecting and cancelling terms,

$$d\vec{F}_S = \left(-\frac{\partial p}{\partial x}\hat{\imath} - \frac{\partial p}{\partial y}\hat{\jmath} - \frac{\partial p}{\partial z}\hat{k}\right) dx\,dy\,dz$$

or,

$$d\vec{F}_S = -\left(\frac{\partial p}{\partial x}\hat{\imath} + \frac{\partial p}{\partial y}\hat{\jmath} + \frac{\partial p}{\partial z}\hat{k}\right) dx\,dy\,dz \tag{3.1a}$$

The term in parentheses is called the gradient of the pressure or simply the pressure gradient, and is designated in rectangular coordinates as

$$\text{grad } p \equiv \nabla p \equiv \left(\hat{\imath}\frac{\partial p}{\partial x} + \hat{\jmath}\frac{\partial p}{\partial y} + \hat{k}\frac{\partial p}{\partial z}\right)$$

$$\nabla p \equiv \left(\hat{\imath}\frac{\partial}{\partial x} + \hat{\jmath}\frac{\partial}{\partial y} + \hat{k}\frac{\partial}{\partial z}\right)p$$

The gradient can be viewed as a vector operator; taking the gradient of a scalar field gives rise to a vector field. Using this designation, Eq. 3.1a can be written as

$$d\vec{F}_S = -\text{grad } p\,(dx\,dy\,dz) = -\nabla p\,dx\,dy\,dz \tag{3.1b}$$

From Eq. 3.1b,

$$\text{grad } p = \nabla p = -\frac{d\vec{F}_S}{dx\,dy\,dz}$$

In words, the gradient of pressure is the negative of the surface force per unit volume due to pressure. We note that the level of pressure is not important in evaluating the net pressure force. Instead, what matters is the rate at which pressure changes occur with distance, the "pressure gradient." We will find this term very useful throughout our study of fluid mechanics, not just in fluid statics.

Since no other kinds of force may be present in a static fluid, we can combine the formulations for surface and body forces that we have developed to obtain the total force acting on a fluid element. Thus

$$d\vec{F} = d\vec{F}_B + d\vec{F}_S = (\rho\vec{g} - \text{grad } p)dx\,dy\,dz$$

or on a unit volume basis

$$\frac{d\vec{F}}{d\forall} = \frac{d\vec{F}}{dx\,dy\,dz} = \rho\vec{g} - \text{grad } p \tag{3.2}$$

In a static fluid, or a fluid in uniform motion, a fluid particle retains its identity as time passes. It does not deform since there is no relative motion within the fluid. For these reasons we may apply Newton's second law of motion to evaluate the reaction of the particle to the applied forces. We may use the result also to evaluate the pressure field that results from a given uniform motion.

For a fluid particle, Newton's second law gives $d\vec{F} = \vec{a}\,dm = \vec{a}\rho\,d\forall$, or

$$\frac{d\vec{F}}{d\forall} = \rho\vec{a}$$

Substituting for $d\vec{F}/d\forall$ from Eq. 3.2 we obtain

$$-\text{grad } p + \rho\vec{g} = \rho\vec{a} \tag{3.3}$$

Let us review briefly our derivation of this equation. The physical significance of each term is indicated below

$$-\text{grad } p \qquad\qquad +\rho\vec{g} \qquad\qquad\qquad = \rho\vec{a}$$

$$\left\{\begin{array}{l}\text{Pressure force}\\ \text{per unit volume}\\ \text{at a point}\end{array}\right\} + \left\{\begin{array}{l}\text{Body force per}\\ \text{unit volume}\\ \text{at a point}\end{array}\right\} = \left\{\begin{array}{ll}\text{Mass per} & \text{Acceleration}\\ \text{unit} & \times \text{ of}\\ \text{volume} & \text{fluid particle}\end{array}\right\}$$

This is a vector equation, which means that it really consists of three component equations that must be satisfied individually. Expanding into components, we find

$$-\frac{\partial p}{\partial x} + \rho g_x = \rho a_x \qquad x \text{ direction}$$

$$-\frac{\partial p}{\partial y} + \rho g_y = \rho a_y \qquad y \text{ direction} \qquad (3.4)$$

$$-\frac{\partial p}{\partial z} + \rho g_z = \rho a_z \qquad z \text{ direction}$$

Example 3.1

As a result of a promotion, you are transferred from your present location. Your wife insists that you transport her fish tank in the back of the station wagon. The tank is 12 in. × 24 in. × 12 in. How much water should you leave in the tank to be reasonably sure that it will not spill over during the trip?

EXAMPLE PROBLEM 3.1

GIVEN:

Fish tank 12 in. × 24 in. × 12 in. partially filled with water to be transported in an automobile.

FIND:

Allowable depth of water for reasonable assurance that it will not spill during the trip.

SOLUTION:

The first step in effecting a solution is to formulate the problem, that is, to translate the general problem into something more specific.

We recognize that there will be motion of the water surface as a result of the car traveling over bumps in the road, going around corners, etc. However, we will assume that the main effect on the water surface is due to linear accelerations (and decelerations) of the car; that is, we will neglect sloshing.

Thus we have reduced the problem to one of determining the effect of a linear acceleration on the free surface. We have not yet decided on the orientation of the tank relative to the direction of motion. Choosing the x coordinate in the direction of motion, should we align the tank with the long side parallel, or perpendicular, to the direction of motion?

If there will be no relative motion in the water, we must assume we are dealing with a constant acceleration, a_x. What is the shape of the free surface under these conditions?

55   **the basic hydrostatic equation**

Let us restate the problem to answer the original questions without making any restrictive assumptions at the outset.

GIVEN:

Tank partially filled with water (to a depth d in.) subject to constant linear acceleration, a_x. Tank height is 12 in., length parallel to direction of motion is b in. Width perpendicular to direction of motion is c in.

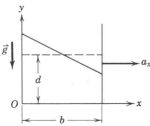

FIND:

(a) Shape of free surface under constant a_x.
(b) Allowable water height, d, to avoid spilling as a function of a_x and tank orientation.
(c) Optimum tank orientation and allowable depth.

SOLUTION:

Basic equation: $-\nabla p + \rho \vec{g} = \rho \vec{a}$

$$-\left(\hat{\imath}\frac{\partial p}{\partial x} + \hat{\jmath}\frac{\partial p}{\partial y} + \hat{k}\frac{\partial p}{\partial z}\right) + \rho(\hat{\imath}g_x + \hat{\jmath}g_y + \hat{k}g_z) = \rho(\hat{\imath}a_x + \hat{\jmath}a_y + \hat{k}a_z)$$

Since p is not a function of z, $\partial p/\partial z = 0$. Also, $g_x = 0$, $g_y = -g$, $g_z = 0$ and $a_y = a_z = 0$.

$$\therefore \quad -\hat{\imath}\frac{\partial p}{\partial x} - \hat{\jmath}\frac{\partial p}{\partial y} - \hat{\jmath}\rho g = \hat{\imath}\rho a_x$$

Writing the component equations:

$$\frac{\partial p}{\partial x} = -\rho a_x$$

$$\frac{\partial p}{\partial y} = -\rho g$$

{ Recall that a partial derivative means that all other independent variables are held constant in the differentiation. }

The problem now is to find an expression for $p = p(x, y)$. This would enable us to find the equation of the free surface. But wait, perhaps we don't have to do that.

Since the pressure, $p = p(x, y)$, the difference in pressure between two points (x, y) and $(x + dx, y + dy)$ is

$$dp = \frac{\partial p}{\partial x}dx + \frac{\partial p}{\partial y}dy$$

EXAMPLE PROBLEM 3.1 (continued).

Since the free surface is a line of constant pressure, then along the free surface, $dp = 0$ and

$$0 = \frac{\partial p}{\partial x} dx + \frac{\partial p}{\partial y} dy = -\rho a_x\, dx - \rho g\, dy$$

$$\therefore \quad \frac{dy}{dx}\bigg|_{\text{free surface}} = -\frac{a_x}{g} \quad \{\text{The free surface is a straight line.}\}$$

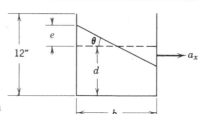

Referring to the diagram at the right.

d = original depth

e = height above original depth

b = tank length parallel to direction of motion

$$e = \frac{b}{2}\tan\theta = \frac{b}{2}\left(-\frac{dy}{dx}\right)_{\text{free surface}} = \frac{b}{2}\frac{a_x}{g} \quad \left\{\text{Only valid for } d \geq \frac{b}{2}\right\}$$

Since we want e to be smallest for a given a_x, the tank should be aligned with b as small as possible. We should align the tank with the long side perpendicular to the direction of motion, that is, choose $b = 12$ in.

With $b = 12$ in.

$$e = 6\frac{a_x}{g} \text{ in.}$$

Maximum allowable value of $e = 12 - d$ in.

$$\therefore \quad 12 - d = 6\frac{a_x}{g} \quad \text{and} \quad d_{\text{max}} = 12 - 6\frac{a_x}{g}$$

Assuming maximum a_x to be $\frac{2}{3} g$ then allowable $d = 8$ in.
To allow a margin of safety, perhaps we should select $d = 6$ in.

$\left.\begin{array}{l} \text{This problem has been included to demonstrate:} \\ \text{(i) not all problems are clearly defined, nor do they have a unique answer, and} \\ \text{(ii) the application of the equation, } -\nabla p + \rho\vec{g} = \rho\vec{a}. \end{array}\right\}$

Recall that a steady acceleration has been assumed in this problem. The car would have to be driven very carefully.

Example 3.2

A cylindrical container, partially filled with a liquid, is rotated at a constant angular velocity, ω, about its axis as shown in the diagram. After a short period of time there is no relative motion, that is, the liquid rotates with the cylinder as if the system were a rigid body. Determine the shape of the free surface.

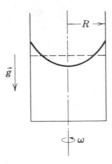

EXAMPLE PROBLEM 3.2

GIVEN:

A cylinder of liquid in solid body rotation with angular velocity, ω, about its axis.

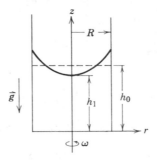

FIND:

The shape of the free surface.

SOLUTION:

It is convenient to use a cylindrical coordinate system, r, θ, z. Since there is circumferential symmetry in this problem, the pressure p will not be a function of θ, that is, $p = p(r, z)$.

Note: The free surface is a surface of constant pressure; the problem is to find the equation of the surface.

Since $p = p(r, z)$, the differential change, dp, in pressure between two points with coordinates (r, θ, z) and $(r + dr, \theta, z + dz)$ is given by

$$dp = \left.\frac{\partial p}{\partial r}\right)_z dr + \left.\frac{\partial p}{\partial z}\right)_r dz$$

Consequently we see that we need to obtain expressions for $\partial p/\partial z)_r$ and $\partial p/\partial r)_z$. This we can do by writing Newton's second law in the z and r directions, respectively, for an infinitesimal fluid element.

From Eq. 3.4 we have, for the z direction

$$-\frac{\partial p}{\partial z}\bigg)_r + \rho g_z = \rho a_z$$

Since $g_z = -g$ and $a_z = 0$, then $\partial p/\partial z)_r = -\rho g$.

To obtain an expression for $\partial p/\partial r)_z$ we apply Newton's second law in the r direction to a suitable differential element.

The pressure at the center of the element is p.

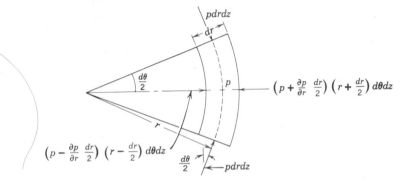

The forces acting in the $r\theta$ plane on the element are shown in the diagram.

Writing Newton's second law in the r direction we have

$$\Sigma\,dF_r = a_r\,dm = a_r\rho\,d\mathbb{V} = -\omega^2 r\rho\,d\mathbb{V} = -\omega^2 r\rho r\,d\theta\,dr\,dz$$

From the figure

$$\Sigma\,dF_r = \left(p - \frac{\partial p}{\partial r}\frac{dr}{2}\right)\left(r - \frac{dr}{2}\right)d\theta\,dz - \left(p + \frac{\partial p}{\partial r}\frac{dr}{2}\right)\left(r + \frac{dr}{2}\right)d\theta\,dz + 2p\,dr\,dz\sin\frac{d\theta}{2}$$

Expanding and canceling like terms, recognizing $\sin d\theta/2 = d\theta/2$ (small angles)

$$\Sigma\,dF_r = d\theta\,dz\left\{p\cancel{r} - p\frac{\cancel{dr}}{2} - r\frac{\partial p}{\partial r}\frac{dr}{2} + \frac{\partial p}{\cancel{\partial r}}\cancel{\left(\frac{dr}{2}\right)^2} - p\cancel{r} - p\frac{\cancel{dr}}{2} - r\frac{\partial p}{\partial r}\frac{dr}{2} - \frac{\partial p}{\cancel{\partial r}}\cancel{\left(\frac{dr}{2}\right)^2} + \cancel{p\,dr}\right\}$$

$$\Sigma\,dF_r = d\theta\,dz\left\{-r\frac{\partial p}{\partial r}\,dr\right\}$$

Then

$$-r\frac{\partial p}{\partial r}\,dr\,d\theta\,dz = -\omega^2 r\rho r\,d\theta\,dr\,dz$$

Dividing through by $d\forall = r \, d\theta \, dr \, dz$

$$-\frac{\partial p}{\partial r} = -\rho\omega^2 r$$

or

$$\frac{\partial p}{\partial r} = \rho\omega^2 r$$

Since

$$dp = \frac{\partial p}{\partial r}\Big|_z dr + \frac{\partial p}{\partial z}\Big|_r dz$$

Then

$$dp = \rho\omega^2 r \, dr - \rho g \, dz$$

To obtain the pressure difference between a reference point (r_1, z_1) where the pressure is p_1 and the arbitrary point (r, z) where the pressure is p we must integrate

$$\int_{p_1}^{p} dp = \int_{r_1}^{r} \rho\omega^2 r \, dr - \int_{z_1}^{z} \rho g \, dz$$

$$p - p_1 = \frac{\rho\omega^2}{2}(r^2 - r_1^2) - \rho g(z - z_1)$$

Taking the reference point on the cylinder axis at the free surface

$$p_1 = p_{atm} \qquad r_1 = 0 \qquad z_1 = h_1$$

then

$$p - p_{atm} = \frac{\rho\omega^2 r^2}{2} - \rho g(z - h_1)$$

Since the free surface is a surface of constant pressure ($p = p_{atm}$), the equation of the free surface is given by

$$0 = \frac{\rho\omega^2 r^2}{2} - \rho g(z - h_1)$$

or

$$z = h_1 + \frac{(\omega r)^2}{2g}$$

The equation of the free surface is a parabola with vertex on the axis at $z = h_1$.
We can solve for the height h_1 under conditions of rotation in terms of the original surface height, h_0, in the absence of rotation.

To do this we use the fact that the volume of fluid must remain constant.

$$\text{Volume of fluid (no rotation)} = \pi R^2 h_0$$

$$\text{Volume of fluid (with rotation)} = \int_0^R \int_0^z 2\pi r \, dz \, dr$$

$$= \int_0^R 2\pi z r \, dr$$

$$= \int_0^R 2\pi \left(h_1 + \frac{\omega^2 r^2}{2g} \right) r \, dr$$

$$= 2\pi \left[h_1 \frac{r^2}{2} + \frac{\omega^2 r^4}{8g} \right]_0^R$$

$$= \pi \left[h_1 R^2 + \frac{\omega^2 R^4}{4g} \right]$$

Then

$$\pi R^2 h_0 = \pi \left[h_1 R^2 + \frac{\omega^2 R^4}{4g} \right]$$

and

$$h_1 = h_0 - \frac{(\omega R)^2}{4g}$$

Finally,

$$z = h_0 - \frac{(\omega R)^2}{4g} + \frac{(\omega r)^2}{2g}$$

$$z = h_0 - \frac{(\omega R)^2}{2g} \left[\frac{1}{2} - \left(\frac{r}{R} \right)^2 \right] \qquad\qquad z(r)$$

$\left\{\begin{array}{l}\text{This problem illustrates the application of Newton's second law to a differential element}\\ \text{and the physical behavior of a liquid with a free surface undergoing solid body rotation.}\end{array}\right\}$

3–2.1 THE BASIC PRESSURE-HEIGHT RELATION

For a static fluid, no acceleration is present. Then Eqs. 3.4 reduce to the component equations:

$$-\frac{\partial p}{\partial x} + \rho g_x = 0$$

$$-\frac{\partial p}{\partial y} + \rho g_y = 0$$

$$-\frac{\partial p}{\partial z} + \rho g_z = 0$$

To simplify still further, it is logical to choose a coordinate system such that the gravity vector is aligned with one of the axes. If the coordinate system is chosen such that the z axis is directed vertically, then $g_x = 0$, $g_y = 0$, and $g_z = -g$. Under these conditions, the component equations become

$$\left.\begin{array}{c} \dfrac{\partial p}{\partial x} = 0 \\[2mm] \dfrac{\partial p}{\partial y} = 0 \\[2mm] \dfrac{\partial p}{\partial z} = -\rho g \end{array}\right\} \qquad (3.5)$$

Equations 3.5 indicate that under the assumptions made, the pressure is independent of coordinates x and y; it depends on z alone. Since p is thus a function of a single variable, a total derivative may be used instead of a partial derivative. With these simplifications, Eqs. 3.5 finally reduce to

$$\frac{dp}{dz} = -\rho g \equiv -\gamma \qquad (3.6)$$

Restrictions: (1) Static fluid
(2) Gravity is the only body force
(3) The z axis is vertical

This equation is the basic pressure-height relation of fluid statics. It is subject to the restrictions noted and, therefore, it must be applied only where these restrictions are reasonable for the physical situation. To determine the pressure distribution in a static fluid, Eq. 3.6 may be integrated and appropriate boundary conditions applied.

Example 3.3

On a certain day, the variation of atmospheric pressure with altitude in the vicinity of 10,000 ft altitude is given by $p = p_0 - az^{1/2}$, where

$$p_0 = \text{pressure at sea level, 14.696 psia}$$

$$a = 10 \text{ lbf/ft}^2/\text{ft}^{1/2}$$

$$z = \text{altitude above sea level, ft}$$

The acceleration of gravity may be assumed constant. Determine:
(a) The specific weight of air at 10,000 ft.
(b) The air temperature at 10,000 ft.

EXAMPLE PROBLEM 3.3

GIVEN:

On a certain day, the pressure in the neighborhood of $z = 10,000$ ft is given by

$$p = p_0 - az^{1/2}$$

$$p_0 = 14.696 \text{ psia} \qquad a = 10 \text{ lbf/ft}^2/\text{ft}^{1/2}$$

FIND:

(a) Specific weight of air at $z = 10,000$ ft.
(b) Air temperature at $z = 10,000$ ft.

SOLUTION:

Basic equation: $\dfrac{dp}{dz} = -\gamma = -\rho g$

Assumptions: Treat air as an ideal gas, $p = \rho RT$
$\qquad\qquad\quad g = \text{constant} = 32.2 \text{ ft/sec}^2$

$$\frac{dp}{dz} = -\gamma \qquad \gamma = -\frac{dp}{dz} = -\frac{d}{dz}(p_0 - az^{1/2}) = -(-a\tfrac{1}{2}z^{-1/2})$$

$$= \frac{a}{2\sqrt{z}} = \frac{1}{2}\frac{10 \text{ lbf}}{\text{ft}^2 \text{ ft}^{1/2}} \frac{1}{\sqrt{10,000 \text{ ft}}}$$

$$\gamma = 0.05 \text{ lbf/ft}^3$$

$$p = \rho RT \qquad\qquad T = \frac{p}{\rho R}$$

$$p = p_0 - az^{1/2} = 14.696\frac{\text{lbf}}{\text{in.}^2} \times \frac{144\ \text{in.}^2}{\text{ft}^2} - 10\frac{\text{lbf}}{\text{ft}^2\ \text{ft}^{1/2}}\sqrt{10,000\ \text{ft}}$$

$$p = 2115\frac{\text{lbf}}{\text{ft}^2} - 1000\frac{\text{lbf}}{\text{ft}^2} = 1115\frac{\text{lbf}}{\text{ft}^2}$$

$$\gamma = \rho g \qquad\qquad \rho = \frac{\gamma}{g} = 0.05\frac{\text{lbf}}{\text{ft}^3} \times \frac{\text{sec}^2}{32.2\ \text{ft}} \times \frac{\text{slug-ft}}{\text{lbf-sec}^2} \quad \left\{ \begin{array}{l} \text{Recall from } \vec{F} = m\vec{a} \\ 1\ \text{lbf} = 1\ \text{slug} \times 1\ \text{ft/sec}^2 \end{array} \right\}$$

$$\rho = \frac{0.05\ \text{slug}}{32.2\ \text{ft}^3}$$

$$T = \frac{p}{\rho R} = 1115\frac{\text{lbf}}{\text{ft}^2} \times \frac{32.2\ \text{ft}^3}{0.05\ \text{slug}} \times \frac{\text{lbm} - \text{R}}{53.3\ \text{ft-lbf}} \times \frac{\text{slug}}{32.2\ \text{lbm}} \quad \left\{ \begin{array}{l} \text{Recall} \\ 1\ \text{slug} = 32.2\ \text{lbm} \end{array} \right\}$$

$$T = 419\ \text{R}$$

$$\left\{ \begin{array}{l} \text{This problem is included to illustrate the use of} \\ \text{(i) the basic pressure-height relation for a static fluid, and} \\ \text{(ii) the equation of state for an ideal gas.} \end{array} \right\}$$

3–2.2 INTEGRATION OF THE BASIC PRESSURE-HEIGHT RELATION

Although ρg is defined as $\rho g \equiv \gamma$, it has been written as ρg in Eq. 3.6 to emphasize that both ρ and g must be considered variables. In order to integrate for the pressure distribution, assumptions must be made about the variations in ρ and g.

For most fluid statics problems encountered in practical engineering situations, the variation in g will be negligible. Only for a situation such as computing very precisely the pressure change over a large elevation difference would the variation in g need to be included. For our purposes we will assume g to be constant with elevation at any given location.

In many practical engineering problems the variation in ρ will be appreciable, and accurate results will require that it be accounted for. Several kinds of variation are easy to treat analytically. The simplest is the idealization of an incompressible fluid.

a. Incompressible fluid

For an incompressible fluid, $\rho = \rho_0 = $ constant. Then for constant gravity,

$$\frac{dp}{dz} = -\rho_0 g = -\gamma_0 = \text{constant}$$

To determine the pressure variation we must integrate the above equation and apply appropriate boundary conditions. If the pressure at the reference level, z_0, is designated as p_0, then the pressure, p, at location z is found by integration

$$\int_{p_0}^{p} dp = -\int_{z_0}^{z} \gamma_0 \, dz$$

$$p - p_0 = -\gamma_0(z - z_0) = \gamma(z_0 - z)$$

For liquids, it is often convenient to take the origin of the coordinate system at the free surface (reference level) and to measure distances as positive downward

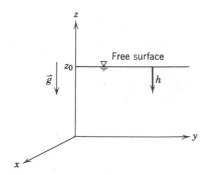

Fig. 3.2 *Coordinates for determination of pressure variation in a static liquid.*

from the free surface, as shown in Fig. 3.2. With h measured positive downward, then,

$$z_0 - z = h$$

and

$$p - p_0 = \gamma h \qquad (3.7)$$

This form of the basic pressure-height relation is often used to solve manometer problems. Students sometimes have trouble analyzing multiple tube manometer situations. The following rules of thumb may help:

1. Any two points at the same elevation in a continuous length of the same liquid will be at the same pressure.
2. Pressure increases as one goes *down* a liquid column (remember the pressure change on diving into a swimming pool!).

Example 3.4

Water flows through pipes A and B. Oil, with specific gravity 0.8, is in the upper portion of the inverted U. Mercury (specific gravity 13.6) is in the bottom of

the static fluid

the manometer bends. Determine the pressure difference, $p_A - p_B$, in units of lbf/in.²

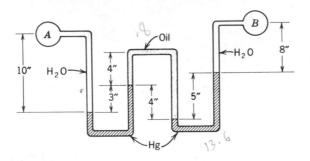

EXAMPLE PROBLEM 3.4

GIVEN:

Multiple tube manometer as shown. Specific gravity of oil is 0.8; specific gravity of mercury is 13.6.

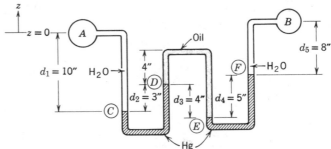

FIND:

The pressure difference, $p_A - p_B$.

SOLUTION:

Basic equations: $\dfrac{dp}{dz} = -\gamma$ $SG = \dfrac{\gamma}{\gamma_{H_2O}}$

$$dp = -\gamma\, dz \quad \text{and} \quad \int_{p_1}^{p_2} dp = -\int_{z_1}^{z_2} \gamma\, dz$$

For γ = constant

$$p_2 = p_1 + \gamma(z_1 - z_2)$$

Beginning at point A and applying the equation between successive points around the manometer:

$$p_C = p_A + \gamma_{H_2O}d_1$$

$$p_D = p_C - \gamma_{Hg}d_2$$

$$p_E = p_D + \gamma_{oil}d_3$$

$$p_F = p_E - \gamma_{Hg}d_4$$

$$p_B = p_F - \gamma_{H_2O}d_5$$

$$p_A - p_B = (p_A - p_C) + (p_C - p_D) + (p_D - p_E) + (p_E - p_F) + (p_F - p_B)$$

$$= -\gamma_{H_2O}d_1 + \gamma_{Hg}d_2 - \gamma_{oil}d_3 + \gamma_{Hg}d_4 + \gamma_{H_2O}d_5$$

Since specific gravity $= \dfrac{\gamma}{\gamma_{H_2O}}$

$$p_A - p_B = -\gamma_{H_2O}d_1 + 13.6\gamma_{H_2O}d_2 - 0.8\gamma_{H_2O}d_3 + 13.6\gamma_{H_2O}d_4 + \gamma_{H_2O}d_5$$

$$= \gamma_{H_2O}(-d_1 + 13.6d_2 - 0.8d_3 + 13.6d_4 + d_5)$$

$$= \gamma_{H_2O}(-10 + 40.8 - 3.2 + 68 + 8) \text{ in.}$$

$$= \gamma_{H_2O} \times 103.6 \text{ in.}$$

$$= 62.4\frac{\text{lbf}}{\text{ft}^3} \times 103.6 \text{ in.} \times \frac{1 \text{ ft}}{12 \text{ in.}} \times \frac{1 \text{ ft}^2}{144 \text{ in.}^2}$$

$$p_A - p_B = 3.74 \text{ lbf/in.}^2 \qquad\qquad\qquad\qquad p_A - p_B$$

b. Barotropic Fluid

A barotropic fluid is one for which ρ is a function of pressure alone. For such a fluid $\rho = \rho(p)$ and

$$\frac{dp}{\rho(p)} = -g\,dz$$

which may be integrated to obtain

$$\int_{p_0}^{p} \frac{dp}{\rho(p)} = -g(z - z_0)$$

When a particular form of the function $\rho = \rho(p)$ is chosen, the integration can be carried out analytically or numerically. (Sea water, although only slightly compressible, is approximately barotropic if at constant temperature.)

c. Density a function of elevation

For this case, $\rho = \rho(z)$. Then

$$dp = -g\rho(z)\,dz$$

and

$$p - p_0 = -g\int_{z_0}^{z} \rho(z)\,dz$$

which also may be integrated once a specific function $\rho(z)$ is chosen.

In all cases the technique is the same—that is, the variables must be separated in Eq. 3.6. Integration then permits calculation of the pressure. Several specific examples are given in the next section.

3–3 The Standard Atmosphere

Several International Congresses for Aeronautics have been held so that aviation experts around the world might better be able to communicate. The outcome of one such Congress was an internationally accepted definition of the Standard Atmosphere. At sea level, conditions of the U.S. Standard Atmosphere are:

$p = 29.92$ in. Hg $= 2116.2$ lbf/ft^2 $\gamma = 0.07651$ lbf/ft^3

$T = 59$ F $= 519$ R $\mu = 3.719 \times 10^{-7}$ lbf-sec/ft^2

$\rho = 0.002378$ slug/ft^3

The temperature profile for the U.S. Standard Atmosphere is shown in Fig. 3.3. If air is assumed to be an ideal gas, knowledge of the temperature distribution suffices to determine the pressure variation, as shown in the following subsections:

a. Temperature Varying Linearly with Elevation: $T = T_0(1 + mz)$

According to Eq. 3.6,

$$dp = -\rho g\,dz$$

For an ideal gas

$$\rho = \frac{p}{RT} = \frac{p}{RT_0(1 + mz)}$$

Substituting for ρ and separating variables

$$\frac{dp}{p} = -\frac{g}{RT_0}\frac{dz}{(1 + mz)}$$

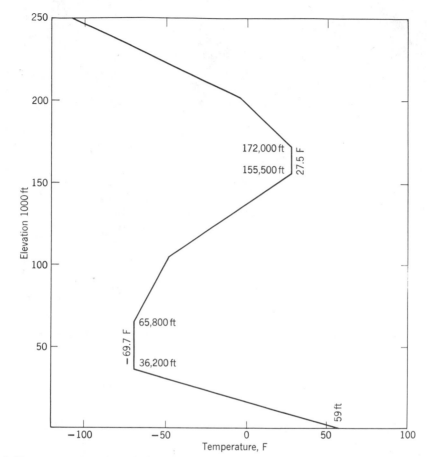

Fig. 3.3 *Temperature variation with altitude in the U.S. standard atmosphere.*

Integrating between the reference level z_0 where the pressure is p_0 and the elevation z where the pressure is p

$$\ln \frac{p}{p_0} = -\frac{g}{mRT_0} \ln \left[\frac{1 + mz}{1 + mz_0} \right]$$

or

$$\frac{p}{p_0} = \left[\frac{1 + mz}{1 + mz_0} \right]^{-g/mRT_0}$$

b. *Temperature Constant with Elevation:* $T = T_0$

According to Eq. 3.6,

$$dp = -\rho g \, dz$$

the standard atmosphere

For an ideal gas,

$$\rho = \frac{p}{RT} = \frac{p}{RT_0}$$

Substituting for ρ and separating variables

$$\frac{dp}{p} = -\frac{g}{RT_0}dz$$

Integrating between the reference level z_0 where the pressure is p_0 and the elevation z where the pressure is p

$$\ln\frac{p}{p_0} = -\frac{g}{RT_0}(z - z_0)$$

or

$$\frac{p}{p_0} = e^{-(g/RT_0)(z-z_0)}$$

The two equations derived above are for very special cases. They should not be memorized, as they were derived only to illustrate a method of solution. A table containing property values for the U.S. Standard Atmosphere may be found in Appendix A.

3–4 Absolute and Gage Pressures

Pressure values must be stated with respect to a reference pressure level. If the reference level is a vacuum, pressures are termed "absolute," as shown in Fig. 3.4.

Most pressure gages actually read a pressure *difference*—the difference between the measured pressure level and the ambient level (usually atmospheric pressure).

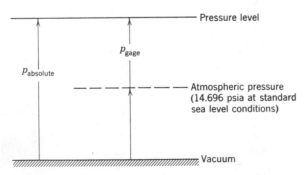

Fig. 3.4 *Absolute and gage pressures, showing reference levels.*

70 **3/fluid statics**

Pressure levels measured with respect to atmospheric pressure are termed "gage pressures."

Absolute pressures must be used in all calculations with the ideal gas or other equations of state. Thus

$$p_{absolute} = p_{gage} + p_{atmosphere} \tag{3.8}$$

Atmospheric pressure may be obtained from a barometer, in which the height of a mercury column is measured. The measured height may be converted to engineering units using Eq. 3.7. For precise work, temperature and altitude corrections must be applied to the measured level. (Data for the specific weight of mercury are given in Appendix A.)

3–5 Hydraulic Systems

Hydraulic systems are characterized by very high pressures. In general, hydraulic fluid may be considered incompressible. (However, in problems where time response to an unsteady input must be determined, both fluid compressibility and elasticity of the boundary structure must be included in the analysis. Such analyses are beyond the scope of this text.)

As a consequence of the high system pressure, hydrostatic variations are often neglected. Thus, negligible error will result if the 1000 psi developed at the master cylinder of a racing car is not corrected for the −0.6 psi developed between the cylinder and wheel due to the 1.2 "G" deceleration!

3–6 Hydrostatic Forces on Submerged Surfaces

Now that we have determined the manner in which the pressure varies in a static fluid, we can examine how these pressures produce forces on surfaces submerged in a liquid.

In order to determine completely the force acting on a submerged surface, we must specify:

1. The magnitude of the force.
2. The direction of the force.
3. The line of action of the resultant force.

We shall consider both plane and curved submerged surfaces.

3–6.1 HYDROSTATIC FORCE ON *PLANE* SUBMERGED SURFACES

A plane submerged surface, on whose upper face we wish to determine the resultant force, is shown in Fig. 3.5. The coordinates have been chosen so that the surface lies in the *xy* plane.

hydrostatic forces on submerged surfaces

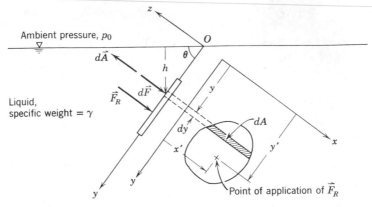

Fig. 3.5 *Plane submerged surface.*

Since there can be no shear stresses in a static fluid, the hydrostatic force on any element of the surface must act normal to the surface. The pressure force acting on an element of the upper surface, $d\vec{A}$, is given by

$$d\vec{F} = -p\,d\vec{A} \tag{3.9}$$

Since the positive direction of the vector $d\vec{A}$ is the outward drawn normal to the area, the negative sign in Eq. 3.9 indicates that the force, $d\vec{F}$, acts *against* the surface, that is, in a direction opposite to that of $d\vec{A}$. The *resultant* force acting on the surface is found by summing the contribution of the infinitesimal forces over the entire area. Thus

$$\vec{F}_R = \int_A -p\,d\vec{A} \tag{3.10}$$

In order to evaluate the integral in Eq. 3.10, the pressure, p, and the element of area, $d\vec{A}$, must both be expressed in terms of the same variables. The basic pressure-height relation for a static fluid can be written as

$$\frac{dp}{dh} = \gamma$$

where h is measured positive downward from the liquid free surface. Then, if the pressure at the free surface ($h = 0$) is p_0, we may integrate the pressure-height relation to obtain an expression for the pressure, p, at any depth, h. Thus

$$\int_{p_0}^{p} dp = \int_{0}^{h} \gamma\,dh$$

and hence

$$p = p_0 + \int_0^h \gamma \, dh$$

This expression for p can then be substituted into Eq. 3.10. The geometry of the surface is expressed in terms of x and y; since the depth h is expressible in terms of y ($h = y \sin \theta$), the integration can be carried out to determine the resultant force.

The point of application of the resultant force must be such that the moment of the resultant force about any point is equal to the moment of the distributed force about the same point. If the position vector, from an arbitrary origin of coordinates to the point of application of the resultant force, is designated as $\vec{r}'$, then

$$\vec{r}' \times \vec{F}_R = \int \vec{r} \times d\vec{F} = -\int_A \vec{r} \times p \, d\vec{A} \qquad (3.11)$$

Referring to Fig. 3.5, $\vec{r}' = \hat{\imath}x' + \hat{\jmath}y'$, $\vec{r} = \hat{\imath}x + \hat{\jmath}y$, and $d\vec{A} = dA\hat{k}$. Assuming the unknown force to be positive, then $\vec{F}_R = F_R\hat{k}$. (If the solution for F_R results in a negative value, as it will in this case, then the force acts in the opposite direction.) Substituting into Eq. 3.11,

$$(\hat{\imath}x' + \hat{\jmath}y') \times F_R\hat{k} = \int (\hat{\imath}x + \hat{\jmath}y) \times d\vec{F} = -\int_A (\hat{\imath}x + \hat{\jmath}y) \times p \, dA\hat{k}$$

Therefore

$$-\hat{\jmath}x'F_R + \hat{\imath}y'F_R = \int_A \hat{\jmath}xp \, dA - \hat{\imath}yp \, dA$$

This is a vector equation, so the components must be equal. Thus

$$y'F_R = -\int_A yp \, dA \qquad (3.12a)$$

and

$$x'F_R = -\int_A xp \, dA \qquad (3.12b)$$

where x' and y' are the coordinates of the point of application of the resultant force. Note that Eqs. 3.10 and 3.11 can be used to determine the magnitude of the resultant force and its point of application on any plane submerged surface. They do not require that the specific weight γ be a constant nor that the free surface of the fluid be at atmospheric pressure. Equations 3.10 and 3.11 are statements of basic principles familiar to you from previous courses in physics and statics:

(a) The resultant force is the sum of the infinitesimal forces (Eq. 3.10).
(b) The moment of the resultant force about any point is equal to the moment of the distributed force about the same point (Eq. 3.11).

Example 3.5

The rectangular gate, AB, is 5 ft wide ($w = 5$ ft) and 10 ft long ($L = 10$ ft). The gate is hinged at B. Neglecting the weight of the gate,

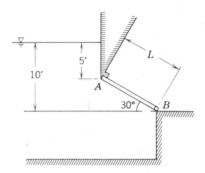

(a) Calculate the force exerted against the stop at A.
(b) Find the location of the single force, $\vec{F}_R$, which is equivalent to the total force of the water on the gate.

EXAMPLE PROBLEM 3.5

GIVEN:

Rectangular gate, AB, hinged at B, width, $w = 5$ ft, length, $L = 10$ ft. Neglect weight of gate.

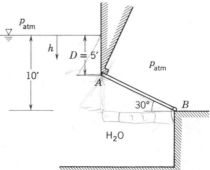

EXAMPLE PROBLEM 3.5 (continued)

FIND:

(a) Force against stop at A.
(b) Point of application of resultant force due to the water.

SOLUTION:

Basic equations: $\vec{F}_R = -\displaystyle\int p\,d\vec{A}$

$$\frac{dp}{dz} = -\gamma \qquad \frac{dp}{dh} = \gamma$$

Moment of a force $\vec{M} = \vec{r} \times \vec{F}$ or $d\vec{M} = \vec{r} \times d\vec{F}$

For equilibrium $\Sigma \vec{M}_B = 0$

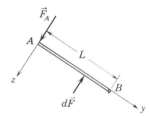

Considering the gate hinged at B as a free body, lying in the xy plane, with coordinates as shown, then:

$\vec{F}_A$, the force of the stop on the gate, acts along the line $y = 0$ as shown, and $d\vec{F}$ is an element of the distributed force due to the water acting as shown.

Note: If in calculating the distributed force due to the water, we do not include the effect of atmospheric pressure at the free surface, then we do not need to include the force due to atmospheric pressure acting on the top of the gate.

Moment about B of $\vec{F}_A = \vec{r} \times \vec{F}_A = -L\hat{j} \times F_A\hat{k}$ $\left\{ \begin{array}{l} \text{Origin of coordinates is at } A \\ \qquad \therefore \quad \vec{r}_{BA} = -L\hat{j} \end{array} \right\}$

Moment about B of total
distributed force $\quad = \displaystyle\int \vec{r} \times d\vec{F} = \int_A -(L - y)\hat{j} \times (-p\,d\vec{A})$

$$= \int_A (L - y)\hat{j} \times p\,d\vec{A} = \int_A (L - y)\hat{j} \times p\,dA\hat{k}$$

$\Sigma\vec{M}_B = 0 \qquad \therefore \quad -L\hat{j} \times F_A\hat{k} + \displaystyle\int_A (L - y)\hat{j} \times p\,dA\hat{k} = 0$

$$-LF_A\hat{i} + \hat{i}\int_A (L - y)p\,dA = 0$$

for sure

and

$$F_A = \frac{1}{L} \int_A (L - y) p \, dA \tag{1}$$

The element of area, dA, of the gate is $dA = w \, dy$, where w is the gate width.

$$\therefore \quad F_A = \frac{1}{L} \int_0^L (L - y) pw \, dy$$

We now need p as a function of y in order to perform the integration.

From the basic pressure-height relation,

$$\frac{dp}{dh} = \gamma \qquad \therefore \quad dp = \gamma \, dh \quad \text{and} \quad \int_{p_a}^p dp = \int_0^h \gamma \, dh$$

Assuming $\gamma = $ constant

$$p = p_a + \gamma h \quad \{\text{This gives } p = p(h). \text{ We need } p = p(y).\}$$

From the diagram,

$$h = D + y \sin 30° \quad \text{where } D = 5 \text{ ft}$$

$$\therefore \quad p = p_a + \gamma(D + y \sin 30°)$$

When we wrote the moment equation ($\Sigma M_B = 0$), we did not include the moment of the force due to atmospheric pressure acting on the top of the gate. Consequently, in figuring the moment due to the water, we should not include the effect of the atmospheric pressure at the free surface. Hence the pressure due to the water alone is

$$p = \gamma(D + y \sin 30°)$$

Thus,

$$F_A = \frac{1}{L} \int_0^L (L - y) pw \, dy = \frac{w}{L} \int_0^L (L - y)\gamma(D + y \sin 30°) \, dy$$

$$= \frac{w\gamma}{L} \int_0^L (DL + Ly \sin 30° - Dy - y^2 \sin 30°) \, dy$$

$$= \frac{w\gamma}{L} \left[DLy + \frac{L}{2} y^2 \sin 30° - \frac{D}{2} y^2 - \frac{y^3}{3} \sin 30° \right]_0^L$$

$$= \frac{w\gamma}{L} \left[DL^2 + \frac{L^3}{2} \sin 30° - \frac{DL^2}{2} - \frac{L^3}{3} \sin 30° \right]$$

$$F_A = \frac{w\gamma}{L} \left[\frac{DL^2}{2} + \frac{1}{6} L^3 \sin 30° \right]$$

With $w = 5$ ft, $D = 5$ ft, $L = 10$ ft, $\gamma = 62.4 \text{ lbf/ft}^3$,

$$F_A = \frac{5 \text{ ft}}{10 \text{ ft}} \times 62.4 \text{ lbf/ft}^3 \left[\frac{5 \text{ ft} \times 100 \text{ ft}^2}{2} + \frac{1}{6} 1000 \text{ ft}^3 \times \frac{1}{2} \right] = 10{,}400 \text{ lbf}$$

F_A is the force of the stop on the gate. The force on the stop is then $-10{,}400$ lbf

$$\vec{F}_A \text{ on stop} = F_A \hat{k} = -10{,}400\hat{k} \text{ lbf} \qquad \{\text{force in } -z \text{ direction}\} \quad \overleftarrow{\qquad} \quad \vec{F}_A$$

To find the location of the single force, $\vec{F}_R$, which is equivalent to the total force of the water, we recognize that the point of application of the resultant force must be such that the moment of the resultant force about point B equals the moment of the distributed force about point B. That is,

$$\vec{r}' \times \vec{F}_R = \int \vec{r} \times d\vec{F}$$

Designating the y coordinate of the location of $\vec{F}_R$ as y', then

$$-(L - y')\hat{j} \times F_R\hat{k} = \int -(L - y)\hat{j} \times (-p\, d\vec{A})$$

or

$$-(L - y')\hat{j} \times F_R\hat{k} = \int_A (L - y)\hat{j} \times p\, dA\hat{k} \qquad \left\{ \begin{array}{c} \text{Unknown } \vec{F}_R \\ \text{is assumed} \\ \text{positive} \end{array} \right\}$$

and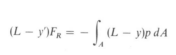

$$(L - y')F_R = -\int_A (L - y)p\, dA$$

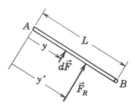

We have already evaluated the integral on the right-hand side in solving part (a). From equation (1),

$$\int_A (L - y)p\, dA = F_A L$$

$$\therefore \quad (L - y')F_R = -F_A L \quad \text{and} \quad y' = L + \frac{F_A L}{F_R}$$

To determine y', we must first evaluate F_R.

$$\vec{F}_R = F_R\hat{k} = -\int_A p\, d\vec{A} = -\int_A p\, dA\hat{k}$$

From our earlier efforts,

$$p = \gamma(D + y \sin 30°)$$

$$dA = w\, dy$$

$$\therefore \quad F_R = -\int_A p \, dA = -\int_0^L \gamma(D + y \sin 30°)w \, dy = -w\gamma\left[Dy + \frac{y^2}{2}\sin 30°\right]_0^L$$

$$F_R = -w\gamma\left[DL + \frac{L^2}{2}\sin 30°\right]$$

With $w = 5$ ft, $D = 5$ ft, $L = 10$ ft, $\gamma = 62.4$ lbf/ft^3

$$F_R = -5 \text{ ft} \times 62.4 \text{ lbf/ft}^3\left[5 \text{ ft} \times 10 \text{ ft} + \frac{100 \text{ ft}^2}{2} \times \frac{1}{2}\right] = -23{,}400 \text{ lbf} \quad \left\{\begin{array}{l}F_R \text{ acts in} \\ \text{the negative} \\ z \text{ direction}\end{array}\right\}$$

Finally,

$$y' = L + \frac{F_A L}{F_R} = 10 \text{ ft} + \frac{10{,}400 \text{ lbf}}{(-23{,}400) \text{ lbf}} \times 10 \text{ ft} = 5.55 \text{ ft} \longleftarrow \qquad\qquad y'$$

Since the gate is of uniform width, w,

$$x' = \frac{w}{2} = 2.5 \text{ ft}$$

$\left\{\begin{array}{l}\text{This problem illustrates the method of applying the basic equations to the calculation of} \\ \text{forces on plane submerged surfaces.}\end{array}\right\}$

3–6.2 HYDROSTATIC FORCE ON *CURVED* SUBMERGED SURFACES

Although slightly more involved, determining the hydrostatic force on a curved surface is no more difficult than determining the force on a plane submerged surface. The hydrostatic force on an infinitesimal element of a curved surface, $d\vec{A}$, acts normal to the surface. However, the differential pressure force on each element of the surface acts in a different direction owing to the surface curvature. Accounting for this change in direction makes the problem a little more involved.

What do we normally do when we wish to sum a series of force vectors acting in different directions? The usual procedure is to sum the components of the vectors relative to a convenient coordinate system.

Consider the curved surface shown in Fig. 3.6. The pressure force acting on the element of area $d\vec{A}$ is given by

$$d\vec{F} = -p \, d\vec{A} \tag{3.9}$$

The resultant force is again given by

$$\vec{F}_R = -\int_A p \, d\vec{A} \tag{3.10}$$

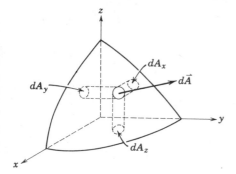

Fig. 3.6 *Curved submerged surface.*

where we can write

$$\vec{F}_R = \hat{\imath}F_{R_x} + \hat{\jmath}F_{R_y} + \hat{k}F_{R_z}$$

where F_{R_x}, F_{R_y}, and F_{R_z} are the components of $\vec{F}_R$ in the x, y, and z directions respectively.

To evaluate the components of the force in a given direction we take the dot product of the force with the unit vector in the given direction. Then,

$$F_{R_x} = \int dF_x = \vec{F}_R \cdot \hat{\imath} = \int d\vec{F} \cdot \hat{\imath} = -\int_A p \, d\vec{A} \cdot \hat{\imath}$$

$$= -\int_A p \, dA \cos \theta_1 = \pm \int_{A_x} p \, dA_x$$

$$F_{R_y} = \int dF_y = \vec{F}_R \cdot \hat{\jmath} = \int d\vec{F} \cdot \hat{\jmath} = -\int_A p \, d\vec{A} \cdot \hat{\jmath}$$

$$= -\int_A p \, dA \cos \theta_2 = \pm \int_{A_y} p \, dA_y \qquad (3.13)$$

$$F_{R_z} = \int dF_z = \vec{F}_R \cdot \hat{k} = \int d\vec{F} \cdot \hat{k} = -\int_A p \, d\vec{A} \cdot \hat{k}$$

$$= -\int_A p \, dA \cos \theta_3 = \pm \int_{A_z} p \, dA_z$$

where

θ_1 is the angle between $d\vec{A}$ and $\hat{\imath}$
θ_2 is the angle between $d\vec{A}$ and $\hat{\jmath}$
θ_3 is the angle between $d\vec{A}$ and $\hat{k}$
$dA_x = dA \cos \theta_1$ is the projection of the area element dA on a plane perpendicular to the x axis

79 **hydrostatic forces on submerged surfaces**

$dA_y = dA \cos \theta_2$ is the projection of the area element dA on a plane perpendicular to the y axis

$dA_z = dA \cos \theta_3$ is the projection of the area element dA on a plane perpendicular to the z axis

The sign (plus or minus) in the last equality of Eqs. 3.13 depends on the magnitude of the angle between the vector $d\vec{A}$ and the unit vectors. To aid in visualizing the relations of Eqs. 3.13, consider the curved surface of constant width (in the x direction) shown in Fig. 3.7, where the element of area, $d\vec{A}$, has been enlarged.

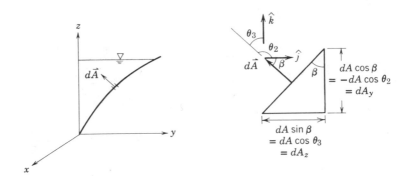

Fig. 3.7 *Element of curved submerged surface.*

From the rules of vector operations, $d\vec{A} \cdot \hat{j} = (dA)(1)(\cos \theta_2)$. From trigonometry $\cos \theta_2 = \cos(\pi - \beta) = -\cos \beta$. Then, $dA \cos \beta = -dA \cos \theta_2 = dA_y$ is the projection of dA on a plane perpendicular to the y axis. Thus,

$$F_{R_y} = \int dF_y = \vec{F}_R \cdot \hat{j} = \int d\vec{F} \cdot \hat{j} = -\int_A p \, d\vec{A} \cdot \hat{j}$$

$$= -\int_A p \, dA \cos \theta_2 = +\int_{A_y} p \, dA_y \qquad (3.14)$$

From the rules of vector operations, $d\vec{A} \cdot \hat{k} = (dA)(1)(\cos \theta_3)$. From trigonometry $\cos \theta_3 = \cos(\pi/2 - \beta) = \sin \beta$. Then, $dA \sin \beta = dA \cos \theta_3 = dA_z$ is the projection of dA on a plane perpendicular to the z axis. Thus,

$$F_{R_z} = \int dF_z = \vec{F}_R \cdot \hat{k} = \int d\vec{F} \cdot \hat{k} = -\int_A p \, d\vec{A} \cdot \hat{k}$$

$$= -\int_A p \, dA \cos \theta_3 = -\int_{A_z} p \, dA_z \qquad (3.15)$$

In general the component of the resultant force in the l direction is given by

$$F_{R_l} = \pm \int_{A_l} p \, dA_l \qquad (3.16)$$

where dA_l is the projection of the area element dA on a plane perpendicular to the l direction, and the sign (plus or minus) depends on the magnitude of the angle θ, between $d\vec{A}$ and the unit vector in the l direction.

We see that finding the resultant force on a curved submerged surface requires the determination of each of the components of the resultant force. This necessitates application of Eq. 3.16 three times at most. As with forces on plane submerged surfaces, the integration in Eq. 3.16 can be carried out only after the pressure, p, and the element of area, dA_l, are expressed in the same variables of integration.

In working with cylindrical surfaces, that is, surfaces with a constant radius of curvature, then $dA = wR \, d\theta$, where

$$R = \text{radius of cylindrical surface}$$
$$w = \text{width of surface}$$

In these cases it is often easier to use θ as the variable of integration. Then

$$F_{R_l} = -\int_A p \, dA \cos\theta = -\int_{\theta_1}^{\theta_2} p \cos\theta \, w R \, d\theta$$

where θ is the angle between $d\vec{A}$ and the unit vector in the l direction.

To specify completely the resultant force acting on a curved surface we must specify the line of action of the force. Since we have solved for the components of the resultant force, we must then specify the line of action of each of the components. The line of action of each component of the resultant force is found by recognizing that the moment of the resultant force about a given point must be equal to the moment of the distributed force. Thus we proceed just as we did with a plane surface.

Recall that for a plane surface the location of the resultant force was found by applying Eq. 3.11

$$\vec{r}' \times \vec{F}_R = \int \vec{r} \times d\vec{F} = -\int_A \vec{r} \times p \, d\vec{A} \qquad (3.11)$$

To find the line of action of each of the components of the resultant force on a curved surface, we would write

$$\vec{r}'_x \times \hat{\imath} F_{R_x} = \int \vec{r} \times dF_x \hat{\imath}$$

$$\vec{r}'_y \times \hat{\jmath} F_{R_y} = \int \vec{r} \times dF_y \hat{\jmath} \qquad (3.17)$$

$$\vec{r}'_z \times \hat{k} F_{R_z} = \int \vec{r} \times dF_z \hat{k}$$

81   **hydrostatic forces on submerged surfaces**

where $\vec{r}'_x$, $\vec{r}'_y$, and $\vec{r}'_z$ are vectors to the lines of action of the components of the resultant force in the x, y, and z directions, respectively. Expressions for dF_x, dF_y, and dF_z are given by Eqs. 3.13.

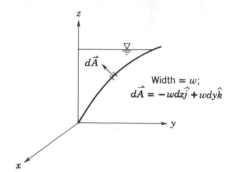

Fig. 3.8 *Curved submerged surface of uniform width.*

Width = w;
$d\vec{A} = -w\,dz\hat{j} + w\,dy\hat{k}$

Referring to Fig. 3.8 (same geometry as Fig. 3.7), we wish to find the lines of action of the components F_{R_y} and F_{R_z}. Since the gate is of uniform width, w, the value of x' will be $w/2$ for both F_{R_y} and F_{R_z}. Thus we need to find z' and y'. To find z' we apply Eq. 3.11, writing,

$$z'\hat{k} \times F_{R_y}\hat{j} = \int z\hat{k} \times dF_y\hat{j}$$

From Eq. 3.14, $dF_y = p\,dA_y$ so that

$$z'\hat{k} \times F_{R_y}\hat{j} = \int z\hat{k} \times p\,dA_y\hat{j}$$

and

$$z'F_{R_y} = \int_{A_y} zp\,dA_y$$

To find y' we apply Eq. 3.11, writing

$$y'\hat{j} \times F_{R_z}\hat{k} = \int y\hat{j} \times dF_z\hat{k}$$

From Eq. 3.15, $dF_z = -p\,dA_z$, so that

$$y'\hat{j} \times F_{R_z}\hat{k} = -\int y\hat{j} \times p\,dA_z\hat{k}$$

and

$$y'F_{R_z} = -\int_{A_z} yp\,dA_z$$

82   **3/fluid statics**

Consequently, there is nothing new required to specify completely the resultant force on a curved submerged surface. In determining the components and their corresponding lines of action, we proceed for each component just as we did for plane submerged surfaces. Because we are dealing with a curved surface, the lines of action of the components of the resultant force will not necessarily coincide. Therefore the complete resultant may be expressed as a force plus a couple.

Example 3.6

The gate shown has a constant width, w, of 5 ft. The equation of the surface is $x = y^2/4$. The depth of liquid, with a specific weight of 60 lbf/ft³, to the right of the gate is 4 ft. Find:

(a) The moment about the point O due to the liquid.
(b) The components F_{R_x} and F_{R_y} of the resultant force due to the liquid and the line of action of each.

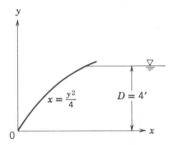

EXAMPLE PROBLEM 3.6

GIVEN:

Gate of constant width, $w = 5$ ft.
Equation of surface in xy plane is $x = y^2/4$.

Liquid with $\gamma = 60$ lbf/ft³ stands at depth, D, of 4 ft to the right of the gate.

FIND:

(a) Moment about 0 due to the liquid.
(b) F_{R_x}, F_{R_y} and line of action of each.

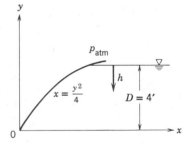

hydrostatic forces on submerged surfaces

SOLUTION:

Basic equations: $\vec{F}_R = -\displaystyle\int p\,d\vec{A}$

$$\frac{dp}{dy} = -\gamma \quad \text{or} \quad \frac{dp}{dh} = \gamma$$

Moment of a force, $\vec{M} = \vec{r} \times \vec{F}$ or $d\vec{M} = \vec{r} \times d\vec{F}$

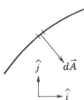

For equilibrium $\Sigma \vec{M}_0 = 0$.

$$\vec{F}_R = \hat{\imath} F_{R_x} + \hat{\jmath} F_{R_y}$$

$$F_{R_x} = \int dF_x = \vec{F}_R \cdot \hat{\imath} = \int d\vec{F} \cdot \hat{\imath} = -\int_A p\,d\vec{A} \cdot \hat{\imath} = -\int_{A_x} p\,dA_x$$

$$F_{R_y} = \int dF_y = \vec{F}_R \cdot \hat{\jmath} = \int d\vec{F} \cdot \hat{\jmath} = -\int_A p\,d\vec{A} \cdot \hat{\jmath} = \int_{A_y} p\,dA_y$$

Moment of F_{R_x} about z axis

$$= y'\hat{\jmath} \times F_{R_x}\hat{\imath} = \int y\hat{\jmath} \times dF_x\hat{\imath} = -\int_{A_x} y\hat{\jmath} \times p\,dA_x\hat{\imath}$$

Moment of F_{R_y} about z axis

$$= x'\hat{\imath} \times F_{R_y}\hat{\jmath} = \int x\hat{\imath} \times dF_y\hat{\jmath} = \int_{A_y} x\hat{\imath} \times p\,dA_y\hat{\jmath}$$

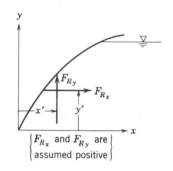

$\left\{ \begin{array}{l} F_{R_x} \text{ and } F_{R_y} \text{ are} \\ \text{assumed positive} \end{array} \right\}$

Moment about 0 due to liquid, $\vec{M}_0 =$ moment of $F_{R_x} +$ moment of F_{R_y}

$$\vec{M}_0 = -\int_{A_x} y\hat{\jmath} \times p\,dA_x\hat{\imath} + \int_{A_y} x\hat{\imath} \times p\,dA_y\hat{\jmath} = \hat{k}\int_{A_x} yp\,dA_x + \hat{k}\int_{A_y} xp\,dA_y$$

$$\vec{M}_0 = M_0\hat{k} = \hat{k}\left[\int_{A_x} yp\,dA_x + \int_{A_y} xp\,dA_y\right] \quad \{M_0 \text{ is assumed positive.}\}$$

$$\therefore \quad M_0 = \int_{A_x} yp\,dA_x + \int_{A_y} xp\,dA_y$$

For constant gate width, w, $dA_x = w\,dy$, $dA_y = w\,dx$ and

$$M_0 = \int_0^4 ypw\,dy + \int_0^4 xpw\,dx$$

We now need a suitable expression for the pressure, p, over the surface of the gate in order to perform the integration.

$$\frac{dp}{dh} = \gamma, \quad dp = \gamma\,dh \quad \text{and} \quad \int_{p_a}^p dp = \int_0^h \gamma\,dh$$

Assuming $\gamma = $ constant,

$$p = p_a + \gamma h$$

Since atmospheric pressure acts on both the top of the gate and the free surface of the liquid, there is no net contribution of the atmospheric pressure force. Thus, in determining the moment due to the liquid we take $p = \gamma h$, and

$$M_0 = \int_0^4 y\gamma hw\,dy + \int_0^4 x\gamma hw\,dx$$

We now need an expression for $h = h(y)$ and $h = h(x)$ along the surface of the gate. Along the surface of the gate, $h = D - y$. Since the equation of the gate surface is $x = y^2/4$, then along the gate $y = 2x^{1/2}$ and thus h can also be written as $h = D - 2x^{1/2}$. Substituting the appropriate equations for h into the expression for M_0, we obtain

$$M_0 = \int_0^4 y\gamma(D - y)w\,dy + \int_0^4 x\gamma(D - 2x^{1/2})w\,dx$$

$$M_0 = \gamma w\left[\frac{D}{2}y^2 - \frac{y^3}{3}\right]_0^4 + \gamma w\left[\frac{D}{2}x^2 - \frac{4}{5}x^{5/2}\right]_0^4 = \gamma w\left[8D - \frac{64}{3} + 8D - \frac{4}{5}32\right]$$

With $\gamma = 60\ \text{lbf/ft}^3$, $w = 5$ ft, $D = 4$ ft,

$$M_0 = 60\ \text{lbf/ft}^3 \times 5\ \text{ft}\left[16 \times 4\ \text{ft}^3 - \frac{64}{3}\ \text{ft}^3 - \frac{4}{5}32\ \text{ft}^3\right]$$

$$M_0 = 5120\ \text{ft-lbf}$$ $\left\{\begin{array}{l}\text{positive value means}\\ \text{counter clockwise}\end{array}\right\}$ M_0

The equations needed to solve for the components F_{R_x}, F_{R_y}, and their lines of action, that is, x' and y', have already been written above.

$$F_{R_x} = -\int_{A_x} p\,dA_x = -\int_{A_x} \gamma h\,dA_x = -\int_0^4 \gamma hw\,dy = -\gamma w\int_0^4 h\,dy = -\gamma w\int_0^4 (D - y)\,dy$$

$$= -\gamma w\left[Dy - \frac{y^2}{2}\right]_0^4 = -\gamma w[16 - 8] = -60\ \text{lbf/ft}^3 \times 5\ \text{ft} \times 8\ \text{ft}^2$$

$$F_{R_x} = -2400 \text{ lbf} \xleftarrow{\hspace{5cm}} \qquad \left\{ \begin{array}{l} F_{R_x} \text{ acts to} \\ \text{the left} \end{array} \right\} F_{R_x}$$

$$y'\hat{j} \times F_{R_x}\hat{i} = \int y\hat{j} \times dF_x\hat{i} = -\int y\hat{j} \times p\, dA_x\hat{i}$$

$$\therefore \quad -y'F_{R_x} = \int_{A_x} yp\, dA_x \quad \text{and} \quad y' = -\frac{1}{F_{R_x}} \int_{A_x} yp\, dA_x$$

$$y' = -\frac{1}{F_{R_x}} \int_0^4 ypw\, dy = -\frac{1}{F_{R_x}} \int_0^4 y\gamma hw\, dy = -\frac{w\gamma}{F_{R_x}} \int_0^4 y(D-y)\, dy = -\frac{w\gamma}{F_{R_x}} \left[\frac{D}{2}y^2 - \frac{y^3}{3} \right]_0^4$$

$$= -\frac{w\gamma}{F_{R_x}} \left[8D - \frac{64}{3} \right]$$

$$y' = \frac{-5 \text{ ft}}{-2400 \text{ lbf}} \times 60 \text{ lbf/ft}^3 \left[32 - \frac{64}{3} \right] \text{ft}^3 = \frac{300}{2400} \times \frac{32}{3} \text{ ft} = 1.33 \text{ ft} \xleftarrow{\hspace{2cm}} \qquad y'$$

$$F_{R_y} = \int_{A_y} p\, dA_y = \int_{A_y} \gamma h\, dA_y = \int_0^4 \gamma hw\, dx = \gamma w \int_0^4 h\, dx = \gamma w \int_0^4 (D - 2x^{1/2})\, dx$$

$$F_{R_y} = \gamma w[Dx - \tfrac{4}{3}x^{3/2}]_0^4 = \gamma w[16 - \tfrac{32}{3}] = 60 \text{ lbf/ft}^3 \times 5 \text{ ft} \times \tfrac{16}{3} \text{ ft}^2 = 1600 \text{ lbf} \xleftarrow{\hspace{1.5cm}} \qquad F_{R_y}$$

$$x'\hat{i} \times F_{R_y}\hat{j} = \int x\hat{i} \times dF_y\hat{j} = \int x\hat{i} \times p\, dA_y\hat{j}$$

$$\therefore \quad x'F_{R_y} = \int_{A_y} xp\, dA_y \quad \text{and} \quad x' = \frac{1}{F_{R_y}} \int_{A_y} xp\, dA_y$$

$$x' = \frac{1}{F_{R_y}} \int_0^4 xpw\, dx = \frac{1}{F_{R_y}} \int_0^4 x\gamma hw\, dx = \frac{w\gamma}{F_{R_y}} \int_0^4 x(D - 2x^{1/2})\, dx = \frac{w\gamma}{F_{R_y}} \left[\frac{D}{2}x^2 - \frac{4}{5}x^{5/2} \right]_0^4$$

$$= \frac{w\gamma}{F_{R_y}} \left[8D - \frac{128}{5} \right]$$

$$x' = \frac{5 \text{ ft}}{1600 \text{ lbf}} \times 60 \text{ lbf/ft}^3 \left[32 - \frac{128}{5} \right] \text{ft}^3 = \frac{300}{1600} \times \frac{32}{5} \text{ ft} = 1.2 \text{ ft} \xleftarrow{\hspace{2cm}} \qquad x'$$

Check: Moment about 0 due to liquid, $\vec{M}_0$ = moment of F_{R_x} + moment of F_{R_y}

$$\vec{M}_0 = y'\hat{j} \times F_{R_x}\hat{i} + x'\hat{i} \times F_{R_y}\hat{j}$$

$$M_0\hat{k} = -y'F_{R_x}\hat{k} + x'F_{R_y}\hat{k}$$

$$M_0 = -y'F_{R_x} + x'F_{R_y} = -(1.33 \text{ ft})(-2400 \text{ lbf}) + 1.2 \text{ ft}(1600 \text{ lbf})$$

$$= 3200 \text{ ft-lbf} + 1920 \text{ ft-lbf} = 5120 \text{ ft-lbf}$$

$$\left. \begin{array}{l} \text{This problem illustrates the method of applying the basic equations to the calculation of} \\ \text{forces on curved submerged surfaces.} \end{array} \right\}$$

Example 3.7

The open tank shown is filled with water to a depth of 10 ft. Determine the magnitude and lines of action of the vertical and horizontal components of the resultant force of the water on the curved part of the tank bottom.

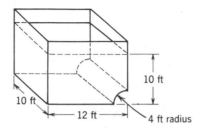

10 ft

10 ft

— 12 ft —

4 ft radius

EXAMPLE PROBLEM 3.7

GIVEN:

Tank of width, $w = 10$ ft, filled with water to a depth of 10 ft.

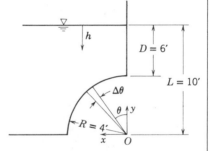

FIND:

Components, F_{R_z}, F_{R_y} (and their lines of action), of resultant force due to water, on the curved portion of the tank.

SOLUTION:

Basic equations:

$$\vec{F}_R = -\int p \, d\vec{A}$$

$$\frac{dp}{dy} = -\gamma \qquad \frac{dp}{dh} = \gamma$$

$$\text{Moment of a force, } \vec{M} = \int \vec{r} \times d\vec{F} \quad \text{or} \quad d\vec{M} = \vec{r} \times d\vec{F}$$

$$F_{R_x} = \int dF_x = \vec{F}_R \cdot \hat{\imath} = \int d\vec{F} \cdot \hat{\imath} = -\int p \, d\vec{A} \cdot \hat{\imath} = -\int p \, dA(1) \cos(90 - \theta)$$

$$F_{R_x} = -\int p \, dA \sin \theta \quad \left\{ \begin{array}{l} \text{Since } \theta \text{ is a logical variable of} \\ \text{integration we will use it.} \end{array} \right\}$$

hydrostatic forces on submerged surfaces

For a circular surface, $dA = wR\,d\theta$, and hence

$$F_{R_x} = -\int_0^{\pi/2} pwR\,d\theta \sin\theta$$

We now need $p = p(\theta)$

$$\frac{dp}{dh} = \gamma \quad \text{or} \quad dp = \gamma\,dh \quad \text{and} \quad \int_{p_{atm}}^p dp = \int_0^h \gamma\,dh$$

For $\gamma = $ constant, $p = p_{atm} + \gamma h$. Since we want the resultant force due to the water (p_{atm} acts on the bottom of the curved surface, as well as on the free surface of the water), we take $p = \gamma h$. Thus

$$F_{R_x} = -\int_0^{\pi/2} \gamma hwR \sin\theta\,d\theta$$

We now need h as a function of θ. At

$$\left.\begin{array}{ll} \theta = 0 & h = D = L - R \\ \theta = \pi/2 & h = L \end{array}\right\} \qquad \therefore \quad h = L - R\cos\theta$$

so that $p = \gamma h = \gamma(L - R\cos\theta)$.

$$F_{R_x} = -\int_0^{\pi/2} \gamma(L - R\cos\theta)wR \sin\theta\,d\theta = -\gamma wR \int_0^{\pi/2} (L - R\cos\theta)\sin\theta\,d\theta$$

$$F_{R_x} = -\gamma wR\left[-L\cos\theta + \frac{R}{2}\cos^2\theta\right]_0^{\pi/2} = -\gamma wR\left(L - \frac{R}{2}\right)$$

With $\gamma = 62.4$ lbf/ft^3, $w = 10$ ft, $R = 4$ ft, then

$$F_{R_x} = -62.4 \text{ lbf/ft}^3 \times 10 \text{ ft} \times 4 \text{ ft} \times (10 - \tfrac{4}{2}) \text{ ft}$$

$$F_{R_x} = -19,970 \text{ lbf}$$

$\left\{\begin{array}{l}\text{Horizontal component of} \\ F_R \text{ acts to the right}\end{array}\right\}$ F_{R_x}

To find the line of action of F_{R_x}, moment of F_{R_x} about 0 must be equal to sum of moments of dF_x about 0.

$$y'\hat{j} \times F_{R_x}\hat{i} = \int y\hat{j} \times dF_x\hat{i} = \int_A y\hat{j} \times (-p\,dA \sin\theta)\hat{i}$$

$$-y'F_{R_x}\hat{k} = \hat{k}\int_A yp\,dA \sin\theta$$

$$y' = -\frac{1}{F_{R_x}}\int_A yp\,dA \sin\theta = -\frac{1}{F_{R_x}}\int_0^{\pi/2} y\gamma hwR\,d\theta \sin\theta$$

$$= -\frac{1}{F_{R_x}}\int_0^{\pi/2} y\gamma(L - R\cos\theta)wR\,d\theta \sin\theta$$

Need $y = y(\theta)$. From the diagram $y = R \cos \theta$

$$\therefore \quad y' = -\frac{1}{F_{R_x}} \int_0^{\pi/2} R \cos \theta \gamma (L - R \cos \theta) wR \sin \theta \, d\theta$$

$$= -\frac{R^2 \gamma w}{F_{R_x}} \int_0^{\pi/2} (L \cos \theta \sin \theta - R \cos^2 \theta \sin \theta) \, d\theta$$

$$= -\frac{R^2 \gamma w}{F_{R_x}} \left[L \frac{\sin^2 \theta}{2} + \frac{R \cos^3 \theta}{3} \right]_0^{\pi/2}$$

$$= -\frac{R^2 \gamma w}{F_{R_x}} \left[\frac{L}{2} - \frac{R}{3} \right]$$

$$= \frac{-(4)^2 \, \text{ft}^2}{-19,970 \, \text{lbf}} \times \frac{62.4 \, \text{lbf}}{\text{ft}^3} \times 10 \, \text{ft} \left[\frac{10}{2} - \frac{4}{3} \right] \text{ft}$$

$$y' = 1.83 \, \text{ft} \qquad \text{KIM} \rightarrow \text{MARTHA} \rightarrow \text{SHERYL} \rightarrow \text{POLLY}$$

$$F_{R_y} = \int dF_y = \vec{F}_R \cdot \hat{j} = \int d\vec{F} \cdot \hat{j} = -\int p \, d\vec{A} \cdot \hat{j} = -\int_A p \, dA(1) \cos \theta$$

$$= -\int_0^{\pi/2} pwR \, d\theta \cos \theta = -\int_0^{\pi/2} \gamma hwR \cos \theta \, d\theta$$

$$= -\int_0^{\pi/2} \gamma (L - R \cos \theta) wR \cos \theta \, d\theta = -\gamma Rw \int_0^{\pi/2} (L \cos \theta - R \cos^2 \theta) \, d\theta$$

$$= -\gamma Rw \left[L \sin \theta - R \left(\frac{\theta}{2} + \frac{\sin 2\theta}{4} \right) \right]_0^{\pi/2}$$

$$= -\gamma Rw \left(L - R \frac{\pi}{4} \right)$$

$$= -(62.4) \, \text{lbf/ft}^3 \times 4 \, \text{ft} \times 10 \, \text{ft} \times (10 - \pi) \, \text{ft}$$

$$F_{R_y} = -17,100 \, \text{lbf} \qquad \qquad \{\text{acts downward}\} \, F_{R_y}$$

To find the line of action of F_{R_y}, moment of F_{R_y} about 0 must be equal to sum of moments of dF_y about 0.

$$x'\hat{i} \times F_{R_y}\hat{j} = \int x\hat{i} \times dF_y\hat{j} = \int_A x\hat{i} \times (-p \, dA \cos \theta)\hat{j}$$

$$x'F_{R_y}\hat{k} = -\hat{k} \int_A xp \, dA \cos \theta$$

hydrostatic forces on submerged surfaces

$$x' = -\frac{1}{F_{R_y}} \int_A xp \, dA \cos \theta = -\frac{1}{F_{R_y}} \int_0^{\pi/2} x\gamma hwR \, d\theta \cos \theta$$

$$x' = -\frac{1}{F_{R_y}} \int_0^{\pi/2} x\gamma(L - R \cos \theta)wR \cos \theta \, d\theta$$

Along the curved surface $x = R \sin \theta$

$$\therefore \quad x' = -\frac{1}{F_{R_y}} \int_0^{\pi/2} R \sin \theta\gamma(L - R \cos \theta)wR \cos \theta \, d\theta$$

$$= -\frac{wR^2\gamma}{F_{R_y}} \int_0^{\pi/2} (L \sin \theta \cos \theta - R \cos^2 \theta \sin \theta) \, d\theta$$

$$= -\frac{wR^2\gamma}{F_{R_y}} \left[\frac{L}{2} \sin^2 \theta + R \frac{\cos^3 \theta}{3} \right]_0^{\pi/2} = -\frac{wR^2\gamma}{F_{R_y}} \left[\frac{L}{2} - \frac{R}{3} \right]$$

$$x' = -\frac{10 \, \text{ft} \times 16 \, \text{ft}^2}{-17,100 \, \text{lbf}} \times \frac{62.4 \, \text{lbf}}{\text{ft}^3} \times \left(5 - \frac{4}{3} \right) \text{ft} = 2.14 \, \text{ft} \quad \longleftarrow \qquad x'$$

$\left\{ \begin{array}{l} \text{This problem illustrates application of the basic equations for hydrostatic forces to a} \\ \text{surface with constant radius of curvature.} \end{array} \right\}$

3–7 Buoyancy and Stability

If an object is completely immersed or floating on the surface of a liquid, the force acting on it due to liquid pressure is termed buoyancy. Consider the object shown in Fig. 3.9, immersed in static liquid.

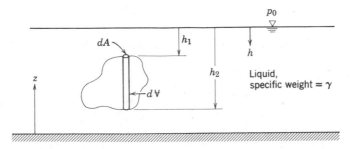

Fig. 3.9 *Immersed body in static liquid.*

90   **3/fluid statics**

The vertical force on the body due to hydrostatic pressure may be found most easily by considering cylindrical volume elements similar to the one shown in Fig. 3.9. For a static fluid

$$\frac{dp}{dh} = \gamma$$

Integrating for constant γ

$$p = p_0 + \gamma h$$

The net vertical force on the element is therefore

$$dF_z = (p_0 + \gamma h_2)\, dA - (p_0 + \gamma h_1)\, dA = \gamma(h_2 - h_1)\, dA$$

But $(h_2 - h_1)\, dA = d\forall$, the volume of the element. Thus

$$F_z = \int dF_z = \int_\forall \gamma\, d\forall = \gamma\forall \tag{3.18}$$

where $\forall$ is the volume of the object. Thus the net vertical pressure force, or buoyancy of the object, equals the weight of liquid displaced by the object. This relation was first put to practical use by Archimedes who used it in 220 B.C. to determine the gold content in the crown of King Hiero II (Example 3.8). Consequently it is often called "Archimedes' Principle." In more current technical applications, Eq. 3.18 is used to design displacement vessels, flotation gear, and bathyscaphes.

The line of action of the buoyancy force may be determined using the methods of Section 3–6.2. Since floating bodies are in equilibrium under body and buoyancy forces, the location of the line of action of the buoyancy force determines stability, as shown in Fig. 3.10.

As shown in Fig. 3.10, the body force, or weight, of an object acts through its center of gravity, CG. In Fig. 3.10a, the buoyant force is offset and produces a couple that tends to right the craft. In Fig. 3.10b, the couple tends to capsize the

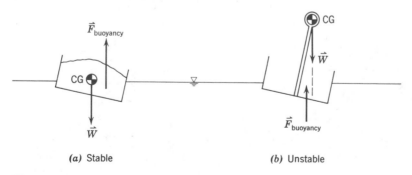

(a) Stable   (b) Unstable

Fig. 3.10 *Stability of floating bodies.*

bouyancy and stability

craft. In sailing, wind loads bring additional forces onto a boat that must be considered in analyzing stability.

Example 3.8

King Hiero ordered a new crown to be made from pure gold. When he received the crown, he suspected that other metals had been used in its construction. Archimedes discovered that the crown weighed 4.7 lbf when immersed in water, and that it displaced 18.9 in.³ of water. He concluded that the crown could not be pure gold. Do you agree?

EXAMPLE PROBLEM 3.8

GIVEN:

Crown volume $= \forall = 18.9$ in.^{3}
Weight in water $= F = 4.7$ lbf

FIND:

Average specific weight of material in crown.

SOLUTION:

Apply $\vec{F} = m\vec{a}$ to immersed crown.

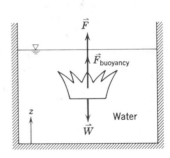

Basic equations: $\Sigma \vec{F} = m\vec{a}$

$$\vec{F}_{buoyancy} = \gamma_{H_2O} \forall \hat{k}$$

Assume: $\vec{a} = 0$ for crown.

Then

$$\Sigma \vec{F} = \vec{F} + \vec{W} + \vec{F}_{buoyancy} = m\vec{a} = 0$$

or

$$(F - W + \gamma_{H_2O}\forall)\hat{k} = 0$$

EXAMPLE PROBLEM 3.8 (continued)

But $W = \gamma_c \forall$, where γ_g is the specific weight of the crown, so

$$W = F + \gamma_{H_2O}\forall = \gamma_c \forall$$

and

$$\gamma_c = \gamma_{H_2O} + \frac{F}{\forall} = \frac{62.4 \text{ lbf}}{\text{ft}^3} \times \frac{\text{ft}^3}{1728 \text{ in.}^3} + \frac{4.7 \text{ lbf}}{18.9 \text{ in.}^3} = 0.036 + 0.249$$

$$\gamma_c = 0.285 \frac{\text{lbf}}{\text{in.}^3} \qquad\qquad\qquad\qquad\qquad \gamma_c$$

$\left\{ \begin{array}{l} \text{This is about the same specific weight as steel. Since gold} \\ \text{is heavier than steel, the crown could not be pure gold.} \end{array} \right\}$

problems

3.1 A cubical box, 2 ft on a side, half filled with oil (specific gravity = 0.80), is given a constant horizontal acceleration of 6.44 ft/sec². Determine the slope of the free surface and the pressure along the bottom of the box.

3.2 A rectangular container of water undergoes constant acceleration down an incline as shown in Fig. 3.11. Determine the slope of the free surface using the coordinate system shown.

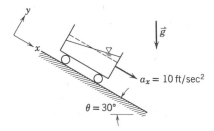

Fig. 3.11

3.3 A pail, 1 ft in diameter and 1 ft deep, weighs 3 lbf, and contains 8 in. of water. The pail is swung in a vertical circle of 3 ft radius at a speed of 10 ft/sec. The water may be assumed to move as a solid body.

At the instant when the pail is at the top of its trajectory, compute the tension in the string and the pressure on the bottom of the pail from the water.

3.4 A sealed chamber, which contains manometer oil (specific gravity = 0.8), rotates about its axis with angular velocity, ω. Derive an expression for the radial pressure gradient in the oil, $\partial p/\partial r$, in terms of radius, r, and angular velocity, ω.

3.5   A cylindrical container, similar to that analyzed in Example Problem 3.2, is rotated at constant angular velocity about its axis. The cylinder is 1 ft in diameter, and initially contains water that is 4 in. deep. Determine the maximum rate at which the container can be rotated before the liquid free surface just touches the bottom of the tank. Is your answer dependent on the density of the liquid? Explain.

3.6   Evaluate dp/dz for standard air on a windless day at sea level and at 10,000 ft. (The coordinate z is measured positive upward.)

3.7   The variation of pressure with height in still air is found to be $dp/dz = -0.068$ lbf/ft^3 at a location where the pressure is 14.7 psia. The coordinate, z, is measured vertically upward. Determine the temperature of the air at this location, assuming ideal gas behavior.

3.8   A sample of salt water has been obtained from the Great Salt Lake. Measurements show that the pressure gradient, $dp/dz = -75$ lbf/ft^3. Determine the concentration of salt in the water by weight, assuming that all salt is completely dissolved.

3.9   On integrating the static pressure-depth equation for an incompressible fluid, it is assumed that the gravitational "constant," g, is a constant. The Law of Gravitational Attraction is

$$g = g_0 \left(\frac{R}{R + h} \right)^2$$

where R is the radius of the earth, and h is altitude above the surface. Find the percent variation in g for the following two cases (take $R = 4000$ miles):

(a) $h = 6$ mile altitude
(b) $h = -4$ mile altitude

3.10   A model has been proposed that accounts for density variations in the earth's atmosphere according to the formula:

$$\gamma = \gamma_0 - kz$$

where $k = 1.93 \times 10^{-6}$ lbf/ft^4, $z =$ altitude above earth's surface, $\gamma_0 =$ specific weight of air at sea level (0.0765 lbf/ft^3). Assuming a sea level pressure of 14.7 psia, compute the pressure at an altitude of 20,000 ft. Compare with the Standard Atmosphere.

3.11   If the variation in specific weight of atmospheric air between sea level and an altitude of 3600 ft were given by $\gamma = \gamma_0 - k\sqrt{z}$, where γ_0 is the specific weight of air at sea level, z is the altitude above sea level, and $k = 0.00020$ lbf/ft^3-ft$^{1/2}$, determine the pressure in psia at an altitude of 3600 ft when sea level conditions are 14.7 psia, 59 F.

3.12   As a result of changes in temperature, salinity, and pressure, the density of sea water increases with increasing depth in accordance with the equation

$$\rho = \rho_s + bh$$

where ρ_s is the density at the surface, h is depth below the surface, and b is a positive constant. Develop an algebraic equation for the pressure as a function of depth.

3.13   The container shown in Fig. 3.12 contains water at a 5 ft depth. The pressure above the water surface is 5 psia. Calculate the absolute pressure on the inside of the bottom surface of the container.

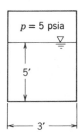

Fig. 3.12

$p = 5$ psia

5′

3′

3.14   Because the pressure falls, water boils at a lower temperature with increasing altitude. Consequently, cake mixes and boiled eggs, among other foods, must be cooked different lengths of time. Determine the boiling temperature of water at 4000 and 8000 ft elevation on a standard day, and compare with the sea level value.

3.15   Automobiles suffer a power loss with altitude as a result of the decrease in air density. If the volumetric efficiency of an engine remains constant, and the carburetion is adjusted to maintain the same air-fuel ratio, determine the percentage loss in power for an engine at 10,000 ft elevation, compared to sea level on a standard day.

3.16   A mercury barometer is read at the same location on different days. Each time, the pressure reading is 29.5 in. of mercury, but the ambient temperatures are 70 F and 95 F on the two days. Determine the actual atmospheric pressures on the two days, in lbf/ft², and the difference between them, in psia.

3.17   Determine the gage pressure in psig at point a in Fig. 3.13, if fluid A has a specific gravity of 0.75 and fluid B has a specific gravity of 1.20. The fluid surrounding point a is water, and the tank on the left is open to the atmosphere.

3.18   A reservoir manometer is calibrated for use with a fluid of specific gravity 0.827. The reservoir diameter is $\frac{5}{8}$ in. and the (vertical) tube diameter is $\frac{3}{16}$ in. Calculate the required distance between marks on the vertical scale for one in. of water pressure difference.

95   **problems**

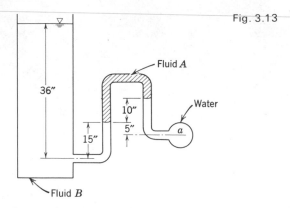

Fig. 3.13

Fluid A

Water

36"

10"

5"

15"

a

Fluid B

3.19   The inclined manometer shown in Fig. 3.14 is filled with gage oil (SG = 0.897). Compute the angle θ which will give a 5 in. oil deflection along the inclined tube for an applied pressure of 1 in. of water (gage).

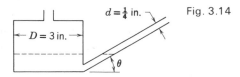

$d = \frac{1}{4}$ in.

Fig. 3.14

$D = 3$ in.

θ

3.20   A crude accelerometer can be made from a liquid-filled U-tube as shown in Fig. 3.15. Derive an expression for the acceleration, $\vec{a}$, in terms of liquid level variation, h, tube geometry, and fluid properties.

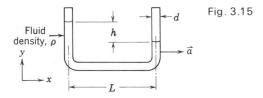

Fig. 3.15

Fluid density, ρ

h

d

y

$\vec{a}$

x

L

3.21   Water is usually assumed to be an incompressible fluid when evaluating static pressure variations. Actually, it is about 100 times more compressible than steel. The bulk modulus is defined as $E_v = dp/(d\rho/\rho)$ and it may be assumed constant (see Appendix A for values). Compute the percent change in density for water

3/fluid statics

raised to a pressure of 2000 psia if the original density at atmospheric pressure is 1.94 slug/ft³.

3.22   Water is usually assumed to be incompressible when evaluating static pressure variations. Actually its compressibility can be important in the design of submersible vehicles. Assume that the bulk modulus of water is constant. Compute the pressure and water density at a depth of 4 miles in sea water. The density at the sea water surface is 64 lbm/ft³.

Fig. 3.16

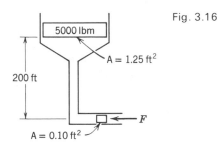

3.23   A hypothetical dense compressible fluid is used in the system shown in Fig. 3.16. Both pistons are frictionless and their outer faces are exposed to atmospheric pressure. The equation of state for the fluid is assumed to be:

$$\gamma = Kp^{1/2} \quad \text{where} \quad K = 0.10\frac{\text{lbf}^{1/2}}{\text{ft}^2}$$

Find the pressure on the inner face of the lower piston, and the force, F, required for equilibrium.

3.24   A door 2 ft wide and 3 ft high is located in a plane vertical wall of a water tank. The door is hinged along its upper edge, which is 2 ft below the water surface. Atmospheric pressure acts on the outer surface of the door and at the water surface. Determine the total resultant force due to all fluids acting on the door.

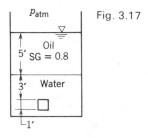

Fig. 3.17

97   **problems**

3.25 If, in Problem 3.24, the water surface pressure is raised to 20 psia, determine the total resultant force from all fluids acting on the door.

3.26 A one foot cube is submerged as shown in Fig. 3.17. Calculate the actual force of the water on the bottom surface, and the net vertical force on the cube.

3.27 The door shown in Fig. 3.18 is 5 ft wide and 10 ft high. Find the resultant force from all fluids acting on the door.

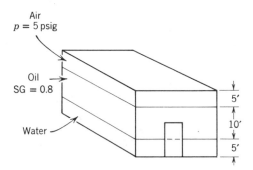

Fig. 3.18

3.28 Most large dams contain flood gates that can be raised to release stored water (Fig. 3.19). The gate shown slides against a plate on each side. The gate weighs 10,000 lbf.

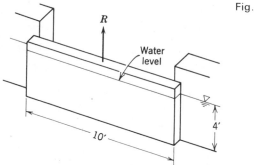

Fig. 3.19

(a) Find the normal force on the gate due to the water.
(b) If μ_s (coefficient of static friction) $= 0.4$ between the gate and the supports, determine the magnitude of the force, R, required to start the gate in motion.

3.29 An open tank with a rectangular vertical side 2 ft wide and 6 ft high is filled with a liquid of variable specific weight, γ, with $\gamma = 50 + 2y$ (lbf/ft^3), where y is measured

98   **3/fluid statics**

vertically downward from the free surface. Find the magnitude of the force on the side of the tank.

3.30 A vertical plane surface is submerged in water as shown in Fig. 3.20. The width of the surface is w. The specific weight of the incompressible liquid is γ. Find:

(a) A general expression for F.
(b) A general expression for a.

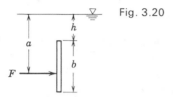

Fig. 3.20

3.31 A door 2 ft wide and 3 ft high is located in a plane vertical wall of a water tank. The door is hinged along its upper edge, which is 2 ft below the water surface. Atmospheric pressure acts on the outer surface of the door. If the pressure at the water surface is atmospheric, what force must be applied at the lower edge of the door in order to keep the door from opening?

3.32 If in Problem 3.31 the pressure at the water surface is 20 psia, what force must be applied at the lower edge of the door in order to keep the door from opening?

3.33 The door shown in Fig. 3.21 is hinged along its bottom. A pressure of 100 psfg is applied to the liquid free surface. Find the force, F, required to hold the door shut.

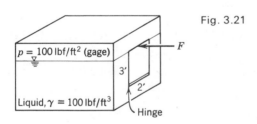

Fig. 3.21

3.34 An aquarium at Marineland has a window located as shown in Fig. 3.22. The resultant force from sea water ($\gamma = 64$ lbf/ft^3) on the window is 1280 lbf. Determine the point of application of the resultant force, in feet below the top of the window.

3.35 The gate $A0C$ shown in Fig. 3.23 is 6 ft wide and is hinged at point 0. Neglecting the weight of the gate, determine the force in the bar AB.

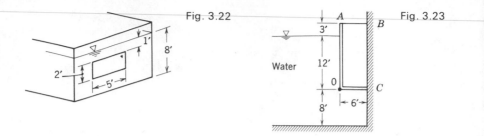

Fig. 3.22

Fig. 3.23

3.36 As water rises on the left-hand side of the rectangular gate, the gate will open automatically (Fig. 3.24). At what depth above the hinge will this occur? Neglect the weight of the gate.

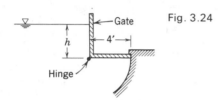

Fig. 3.24

3.37 A tank with a center partition has a small 1.5 ft wide by 2 ft high "door" at the bottom. This door is hinged along the top edge. The left tank has 2 ft of water while the right side contains 3 ft of nitric acid (specific gravity 1.5). What force (magnitude and direction) is required at the lower edge of the door to hold it closed?

3.38 The gate shown is hinged at H (Fig. 3.25). The gate is 5 ft wide normal to the plane of the diagram. Calculate the force required at A to hold the gate closed.

3.39 A submarine is 100 ft below the surface as shown in Fig. 3.26. Find the net force, F, required to open the circular hatch when applied as shown. The pressure inside the submarine is equal to atmospheric pressure.

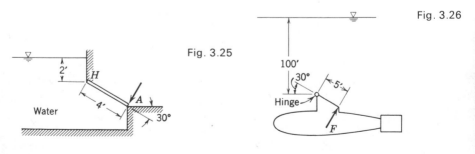

Fig. 3.25

Fig. 3.26

3.40 The gate shown in Fig. 3.27 is 8 ft wide and for the purpose of analysis can be considered weightless. For what depth of water will this rectangular gate be in equilibrium at an angle of 60° as shown?

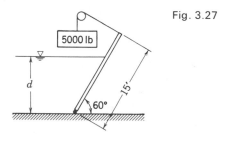

Fig. 3.27

3.41 A plane gate is held in equilibrium by the force, F, as shown in Fig. 3.28. The gate weighs 600 lbf/ft of gate width and its center of gravity is 6 ft from the hinge at 0. Find the unknown force, F, when $h = 5$ ft and $\theta = 30°$.

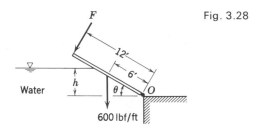

Fig. 3.28

3.42 A gate of mass 4160 lbm is mounted on a frictionless hinge along the lower edge. The length of the reservoir and gate (perpendicular to the plane of view) is 25 ft. For the equilibrium conditions shown in Fig. 3.29, compute the width, w, of the gate.

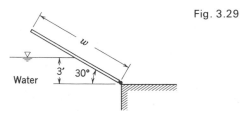

Fig. 3.29

problems

3.43 Gate *AB* is 3 ft wide and 2 ft long. It is inclined at $\alpha = 60°$ when closed. Find the moment about hinge *A* exerted by the water (Fig. 3.30).

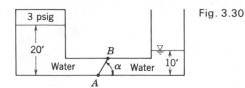

Fig. 3.30

3.44 The rectangular gate *AB*, as shown in Fig. 3.31, is 5 ft wide. Find the force exerted against the stop at *A*. Assume the gate weight is negligible.

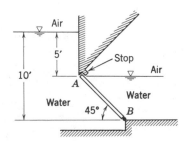

Fig. 3.31

3.45 Due to temperature gradients in a liquid in a large tank, the density is not constant. The density variation is given by

$$\rho = \rho_0(1 + h^{1/5})$$

Find:

(a) The resultant force on the area indicated in Fig. 3.32 due only to the liquid. Your answer should be simplified but may be given in integral form (i.e. the integrand should be expressed as a function of y or h).

(b) The integral expression for the position y' at which the resultant force acts. Again, simplify the expression but do not perform the indicated integrations.

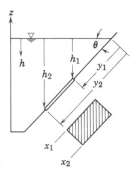

Fig. 3.32

3/fluid statics

3.46   A parabolic gate is shown in Fig. 3.33. The gate is 5 ft wide, and the fluid is water. Determine the magnitude and line of action of the horizontal force on the gate due to the water.

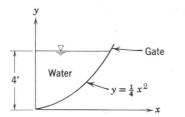

Fig. 3.33

3.47   The gate shown in Fig. 3.34 is 5 ft wide. Determine the magnitude and moment of the vertical component of the force about 0. The fluid is water.

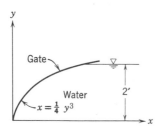

Fig. 3.34

3.48   Determine the vertical force, and the line of action of the vertical force on the curved section AB (Fig. 3.35). The fluid is water and the section AB is 1 ft wide. Atmospheric pressure acts at the free surface.

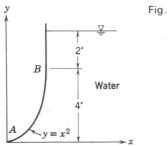

Fig. 3.35

103   **problems**

3.49 An open tank is filled with water to the depth indicated in Fig. 3.36. Atmospheric pressure acts on all outer surfaces of the tank. Determine the magnitude and line of action of the vertical component of the force of the water on the curved part of the tank bottom.

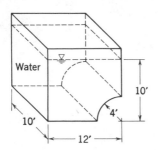

Fig. 3.36

3.50 A spillway gate formed in the shape of a circular arc is w feet wide (Fig. 3.37).

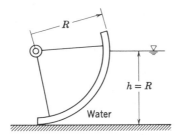

Fig. 3.37

Find:

(a) The magnitude and direction of the vertical component of the force due to all fluids acting on the gate.

(b) A point on the line of action of the total resultant force on the gate due to all fluids.

3.51 A gate, which is in the form of a quarter-cylinder, is hinged at A and is 6 ft wide

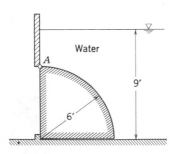

Fig. 3.38

normal to the paper (Fig. 3.38). The bottom of the gate is 9 ft below the water surface. Determine:

(a) The magnitude of the horizontal force.
(b) The line of action of the horizontal force.
(c) The magnitude of the vertical force.
(d) The line of action of the resultant force.

3.52   The tank shown in Fig. 3.39 is 2 ft wide (i.e. 2 ft perpendicular to the xz plane). It is filled with water to a depth of 8 ft. The curved portion of the tank is specified

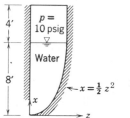

Fig. 3.39

by the equation

$$x = \tfrac{1}{2}z^2$$

The air between the top of the tank and the water is pressurized to 10 psig. The net vertical force on the curved portion of the tank is 14,180 lbf. Determine the line of action of the vertical force on the curved portion of the tank.

3.53   A cylindrical weir has a diameter of 10 ft and a length of 18 ft (Fig. 3.40). Give the magnitude and direction of the resultant force acting on the weir from the water.

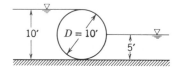

Fig. 3.40

3.54   Find the weight of the cylinder shown in Fig. 3.41. It is 3 ft long, and is supported by the liquid (water). Assume no friction between the cylinder and solid wall, and do not use the buoyancy force technique, except to check your results.

105   **problems**

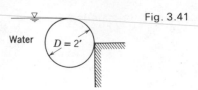

Fig. 3.41

Water $D = 2'$

3.55 Two different liquids are separated by a weightless curved divider with a constant radius of curvature. Each has a different pressure at its free surface designated by p_{0_1} and p_{0_2} in Fig. 3.42. The hinge is frictionless. The densities, p_1 and p_2 are 33 lbm/ft³ and 100 lbm/ft³ respectively. Find the difference in chamber pressures that is just sufficient to keep the gate closed.

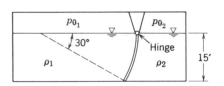

Fig. 3.42

p_{0_1} p_{0_2}

30°

Hinge

p_1 p_2 15′

3.56 A cube is submerged in a liquid at the depth shown (Fig. 3.43). The mass of the cube is 0.5 slug and the tension in the cord is 10 lbf. Find the specific weight of the liquid.

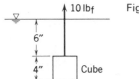

10 lbf   Fig. 3.43

6″

4″   Cube

3.57 A hydrometer is a specific gravity indicator, the value being indicated by the level at which the free surface intersects the stem when floating in a liquid. The 1.0 mark is the level when in distilled water. For the unit shown in Fig. 3.44 the

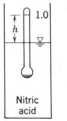

Fig. 3.44

1.0

h

Nitric
acid

immersed volume in distilled water is 1 in.³ The stem is 0.25 in. in diameter. Find the distance, h, from the 1.0 mark to the surface when the hydrometer is placed in a nitric acid solution of specific gravity 1.5.

3.58   Find the specific weight of the sphere shown in Fig. 3.45 if its volume is 1 ft³. State all assumptions. Is the weight necessary to float the sphere?

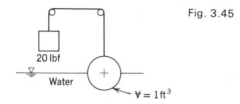

Fig. 3.45

20 lbf

Water

$V = 1 ft^3$

3.59   One ft³ of material weighing 67 lbf is allowed to sink in the water as shown in Fig. 3.46. A circular wooden rod 10 ft long and 3 in.² in cross section is attached to the weight and also to the wall. If the rod weighs 3 lbf, what will be the angle θ for equilibrium?

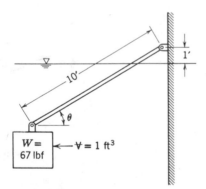

Fig. 3.46

10′

1′

θ

$W =$ 67 lbf

$V = 1 ft^3$

basic equations in integral form for a control volume

We shall begin our study of fluids in motion by developing the basic equations in integral form for application to control volumes. Why the control volume formulation rather than the system formulation? There are perhaps two basic reasons. First, since fluid media are capable of continuous distortion and deformation, it is often extremely difficult to identify and follow the same mass of fluid at all times (as must be done in applying the system formulation). Second, we are often interested, not in the motion of a given mass of fluid, but rather in the effect of the overall fluid motion on some device or structure. Thus it is more convenient to apply the basic laws to a fixed volume in space, that is, to employ a control volume analysis.

There are two possible approaches to formulating the basic laws for a control volume. We could start with a word statement of the basic laws for a control volume and proceed to develop the corresponding mathematical formulation for each basic equation. For example, the conservation of mass for a control volume would be stated as: the net rate of mass efflux from a control volume (rate of outflow minus rate of inflow) plus the rate of change of mass within the control volume is equal to zero. From this word statement we could develop the mathematical formulation. We could proceed in a similar fashion for each of the other four basic laws. In doing this we would repeat the same steps five times using the different quantities: mass, momentum, angular momentum, energy, and entropy. It seems more desirable (and certainly economical in both space and time) to develop a general formulation that is universally applicable to all five of the basic equations. It is this second approach that we will follow. The logical question at this point is, "How shall we do it?" The answer is that we shall go back to what we already know—the basic equations for a system and the characteristics of systems and control volumes—and proceed from there.

4–1 Basic Laws for a System

The basic laws for a system are familiar from your earlier work in mechanics and thermodynamics. We shall summarize them briefly here and for reasons that will become apparent in the next section, we shall write each of the basic equations for a system as a rate equation.

4–1.1 CONSERVATION OF MASS

Since a system is, by definition, an arbitrary collection of matter of fixed identity, that is, composed of the same quantity of matter at all times, the conservation of mass simply states that the mass, M, of the system is constant. On a rate basis, we have

$$\left.\frac{dM}{dt}\right)_{\text{system}} = 0 \tag{4.1a}$$

where,

$$M_{\text{system}} = \int_{\text{Mass (system)}} dm = \int_{\forall\,(\text{system})} \rho\,d\forall \tag{4.1b}$$

4–1.2 NEWTON'S SECOND LAW

Newton's second law states that for a system moving relative to an inertial reference frame, the sum of all external forces acting on the system is equal to the time rate of change of the linear momentum of the system, that is

$$\vec{F} = \left.\frac{d\vec{P}}{dt}\right)_{\text{system}} \tag{4.2a}$$

where the linear momentum, $\vec{P}$, of the system is given by

$$\vec{P}_{\text{system}} = \int_{\text{Mass (system)}} \vec{V}\,dm = \int_{\forall\,(\text{system})} \vec{V}\,\rho\,d\forall \tag{4.2b}$$

4–1.3 CONSERVATION OF ANGULAR MOMENTUM*

Conservation of angular momentum, or moment of momentum, states that for a system, the rate of change of angular momentum equals the sum of all torques acting on the system, that is,

$$\vec{T} = \frac{d\vec{H}}{dt}\bigg)_{\text{system}} \qquad (4.3a)$$

where the angular momentum of the system is given by

$$\vec{H}_{\text{system}} = \int_{\text{Mass (system)}} \vec{r} \times \vec{V} \, dm = \int_{\forall \text{(system)}} \vec{r} \times \vec{V} \, \rho \, d\forall \qquad (4.3b)$$

Torque can be produced by surface and body forces, and also by shafts which cross the system boundary, that is,

$$\vec{T} = \vec{r} \times \vec{F}_s + \int_{\text{Mass (system)}} \vec{r} \times \vec{g} \, dm + \vec{T}_{\text{shaft}} \qquad (4.3c)$$

4–1.4 THE FIRST LAW OF THERMODYNAMICS

The first law of thermodynamics is a statement of conservation of energy for a system, that is,

$$\delta Q - \delta W = \Delta E$$

In rate form the equation can be written as

$$\dot{Q} - \dot{W} = \frac{dE}{dt}\bigg)_{\text{system}} \qquad (4.4a)$$

where the total energy of the system is given by

$$E_{\text{system}} = \int_{\text{Mass (system)}} e \, dm = \int_{\forall \text{(system)}} e\rho \, d\forall \qquad (4.4b)$$

In Eq. 4.4a the rate of heat transfer, $\dot{Q}$, is taken as positive when heat is added to the system from the surroundings; the rate of work done, $\dot{W}$, is taken as positive when work is done by the system on its surroundings.

4–1.5 THE SECOND LAW OF THERMODYNAMICS

If an amount of heat, δQ, is transferred to a system at temperature, T, the second law of thermodynamics states that the change in entropy, dS, of the system is

* Sections marked ** may be omitted without loss of continuity in the text material.

basic laws for a system

given by

$$dS \geq \frac{\delta Q}{T}$$

On a rate basis we can write

$$\left. \frac{dS}{dt} \right)_{system} \geq \frac{1}{T} \dot{Q} \tag{4.5a}$$

where the total entropy of the system is given by

$$S_{system} = \int_{Mass \, (system)} s \, dm = \int_{\forall \, (system)} s\rho \, d\forall \tag{4.5b}$$

4–2 *Relation of System Derivatives to Control Volume Formulation*

In the previous section we summarized the basic equations for a system. We found that, when written on a rate basis, each equation involved the time derivative of an extensive property of the system (the total mass, momentum, angular momentum, energy, or entropy of the system). In developing the control volume formulation of each basic law from the system formulation, we shall use the symbol, N, to designate any arbitrary extensive property of the system. The corresponding intensive property (extensive property per unit mass) will be designated by η. Thus

$$N_{system} = \int_{Mass \, (system)} \eta \, dm = \iiint_{\forall \, (system)} \eta\rho \, d\forall \tag{4.6}$$

By comparing Eq. 4.6 with Eqs. 4.1b, 4.2b, 4.3b, 4.4b and 4.5b, we see that if:

$$N = M, \quad \text{then } \eta = 1$$
$$N = \vec{P}, \quad \text{then } \eta = \vec{V}$$
$$N = \vec{H}, \quad \text{then } \eta = \vec{r} \times \vec{V}$$
$$N = E, \quad \text{then } \eta = e$$
$$N = S, \quad \text{then } \eta = s$$

The basic problem in going from the system to the control volume formulation of the basic laws is to relate the rate of change of an arbitrary extensive property, N, for a system, to the time variation of this property associated with a control volume. Since mass crosses the boundaries of a control volume, time variations of the property N associated with the control volume involve the mass flux and the associated properties convected with it. A convenient way to account for mass flux is to use a limiting process involving a system and a control volume that coincide at a certain instant of time. Flux quantities in regions of overlap and

4/basic equations in integral form for a control volume

regions surrounding the control volume are then formulated approximately, and the limiting process is applied to obtain exact results. The final equation relates the rate of change of an arbitrary extensive property, N, for a system to the time variation of this property associated with a control volume.

One final comment before beginning the derivation: the details of the derivation are involved algebraically, but the concept is extremely simple and straightforward.

4–2.1 DERIVATION

The configuration of the system and control volume to be used in the analysis is shown in Fig. 4.1. The flow field, $\vec{V}(x, y, z, t)$, is arbitrary relative to the coordinates x, y, z. The control volume is fixed in space; by definition, the system must always

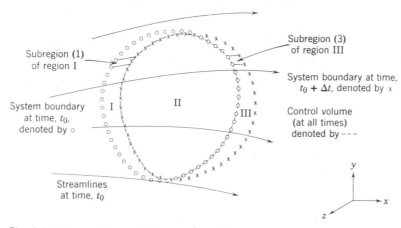

Fig. 4.1 *System and control volume configuration.*

consist of the same fluid particles, and consequently it must move with the flow field. In Fig. 4.1 the boundaries of the system are shown at two different instants of time, t_0 and $t_0 + \Delta t$. At time t_0, the system occupies regions I and II; at time $t_0 + \Delta t$, the system occupies regions II and III. The control volume always consists of region II, which is part of the system for $t_0 \leq t \leq t_0 + \Delta t$.

The system has been chosen so that the mass within region I enters the control volume in the time interval Δt and the mass in region III leaves the control volume in the same time interval. Thus as the time increment, Δt, is made smaller and smaller, the boundaries of the system and the control volume approach each other. In the limit as $\Delta t \to 0$, the system boundaries and the control surface (boundary of the control volume) coincide; that is, as $\Delta t \to 0$, the system and the control volume occupy the same volume, and have the same boundaries.

113   **relation of system derivatives to control volume formulation**

Recall that our objective here is to relate the rate of change of any arbitrary extensive property, N, of the system to the time variations of the property associated with the control volume. If η is the distribution of N per unit mass, then

$$N_{\text{system}} = \int_{\text{Mass (system)}} \eta \, dm = \iiint_{\forall \, (\text{system})} \eta \rho \, d\forall \qquad (4.6)$$

Then the rate of change of N_{system} is given by

$$\left. \frac{dN}{dt} \right)_{\text{system}} \equiv \lim_{\Delta t \to 0} \frac{(N_s)_{t_0 + \Delta t} - (N_s)_{t_0}}{\Delta t} \qquad (4.7)$$

For convenience, the subscript, s, has been used to denote the system in the definition of a derivative employed in Eq. 4.7.

At time $t_0 + \Delta t$, the system occupies regions II and III; at time t_0, the system occupies regions I and II. Consequently we can write

$$N_s)_{t_0 + \Delta t} = (N_{\text{II}} + N_{\text{III}})_{t_0 + \Delta t} = \left[\int_{\forall_{\text{II}}} \eta \rho \, d\forall \right]_{t_0 + \Delta t} + \left[\int_{\forall_{\text{III}}} \eta \rho \, d\forall \right]_{t_0 + \Delta t}$$

and

$$N_s)_{t_0} = (N_{\text{I}} + N_{\text{II}})_{t_0} = \left[\int_{\forall_{\text{I}}} \eta \rho \, d\forall \right]_{t_0} + \left[\int_{\forall_{\text{II}}} \eta \rho \, d\forall \right]_{t_0}$$

Upon substituting these expressions into the definition of the system derivative, Eq. 4.7, we obtain

$$\left. \frac{dN}{dt} \right]_{\text{system}} =$$

$$\lim_{\Delta t \to 0} \frac{\left[\int_{\text{II}} \eta \rho \, d\forall \right]_{t_0 + \Delta t} + \left[\int_{\text{III}} \eta \rho \, d\forall \right]_{t_0 + \Delta t} - \left[\int_{\text{I}} \eta \rho \, d\forall \right]_{t_0} - \left[\int_{\text{II}} \eta \rho \, d\forall \right]_{t_0}}{\Delta t}$$

$$(4.8)$$

Since the limit of a sum is equal to the sum of the limits, we can write

$$\left. \frac{dN}{dt} \right]_{\text{system}} = \lim_{\Delta t \to 0} \frac{\left[\int_{\text{II}} \eta \rho \, d\forall \right]_{t_0 + \Delta t} - \left[\int_{\text{II}} \eta \rho \, d\forall \right]_{t_0}}{\Delta t}$$

$$①$$

$$+ \lim_{\Delta t \to 0} \frac{\left[\int_{\text{III}} \eta \rho \, d\forall \right]_{t_0 + \Delta t}}{\Delta t} - \lim_{\Delta t \to 0} \frac{\left[\int_{\text{I}} \eta \rho \, d\forall \right]_{t_0}}{\Delta t} \qquad (4.9)$$

$$② \qquad\qquad\qquad ③$$

Our task now is to evaluate each of the three terms in Eq. 4.9.

Term ① in Eq. 4.9

$$\lim_{\Delta t \to 0} \frac{\left[\int_{II} \eta\rho\, d\forall\right]_{t_0 + \Delta t} - \left[\int_{II} \eta\rho\, d\forall\right]_{t_0}}{\Delta t}$$

$$= \lim_{\Delta t \to 0} \frac{N_{II})_{t_0 + \Delta t} - N_{II})_{t_0}}{\Delta t} = \frac{\partial N_{II}}{\partial t} = \left. \frac{\partial N}{\partial t}\right]_{\substack{\text{Control} \\ \text{Volume}}}$$

$$= \frac{\partial}{\partial t}\int_{CV} \eta\rho\, d\forall$$

since region II is the control volume. The last equality follows directly from the definition of N, Eq. 4.6.

Term ② in Eq. 4.9

$$\lim_{\Delta t \to 0} \frac{\left[\int_{III} \eta\rho\, d\forall\right]_{t_0 + \Delta t}}{\Delta t} = \lim_{\Delta t \to 0} \frac{N_{III})_{t_0 + \Delta t}}{\Delta t}$$

To evaluate $N_{III})_{t_0 + \Delta t}$, let us look at an enlarged view of a typical subregion of region III as shown in Fig. 4.2. The vector $d\vec{A}$ has a magnitude equal to the element of area dA of the control surface; the direction of the vector $d\vec{A}$ is that of the

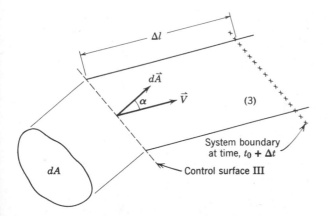

Fig. 4.2 *Enlarged view of subregion (3) from Fig. 4.1.*

115   **relation of system derivatives to control volume formulation**

outward drawn normal to the element of the control surface area. The angle α is the angle between $d\vec{A}$ and the velocity vector, $\vec{V}$. Since the mass in region III is that which flows out of the control volume in the time increment, Δt, the angle α will always be less than $\pi/2$ over the entire area of the control surface bounding region III.

For subregion (3) we can write

$$dN_3)_{t_0+\Delta t} = (\eta \rho \, d\forall)_{t_0+\Delta t} = [\eta \rho (\Delta l \cos \alpha \, dA)]_{t_0+\Delta t}$$

since $d\forall = \Delta l \cos \alpha \, dA$. Then for the entire region III,

$$N_{\text{III}})_{t_0+\Delta t} = \left[\int_{\text{CS}_{\text{III}}} \eta \rho \, \Delta l \cos \alpha \, dA \right]_{t_0+\Delta t}$$

where CS_{III} is the surface common to region III and the control volume. In this expression for $N_{\text{III}})_{t_0+\Delta t}$, Δl represents the distance traveled by a particle on the system surface in time Δt, along a streamline that existed at time t_0.

Now that we have an expression for $N_{\text{III}})_{t_0+\Delta t}$, we can evaluate term ② in Eq. 4.9:

$$\lim_{\Delta t \to 0} \frac{\left[\int_{\text{III}} \eta \rho \, d\forall \right]_{t_0+\Delta t}}{\Delta t} = \lim_{\Delta t \to 0} \frac{N_{\text{III}})_{t_0+\Delta t}}{\Delta t}$$

$$= \lim_{\Delta t \to 0} \frac{\int_{\text{CS}_{\text{III}}} \eta \rho \, \Delta l \cos \alpha \, dA}{\Delta t}$$

$$= \lim_{\Delta t \to 0} \int_{\text{CS}_{\text{III}}} \eta \rho \frac{\Delta l}{\Delta t} \cos \alpha \, dA$$

$$= \int_{\text{CS}_{\text{III}}} \eta \rho |\vec{V}| \cos \alpha |d\vec{A}|$$

The last equality follows from the fact that

$$\lim_{\Delta t \to 0} \frac{\Delta l}{\Delta t} = |\vec{V}| \quad \text{and} \quad dA = |d\vec{A}|$$

Term ③ in Eq. 4.9

$$-\lim_{\Delta t \to 0} \frac{\left[\int_{\text{I}} \eta \rho \, d\forall \right]_{t_0}}{\Delta t} = -\lim_{\Delta t \to 0} \frac{N_{\text{I}})_{t_0}}{\Delta t}$$

To evaluate $N_1)_{t_0}$ look at an enlarged view of a typical subregion of region I as shown in Fig. 4.3.

The vector $d\vec{A}$ has a magnitude equal to the element of area dA of the control surface; the direction of the vector $d\vec{A}$ is that of the outward drawn normal from

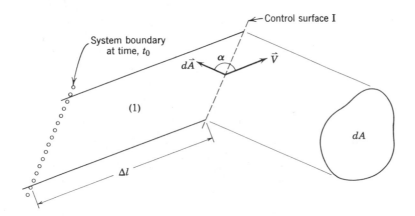

Fig. 4.3 *Enlarged view of subregion (1) from Fig. 4.1.*

the element of the control surface area. The angle α is the angle between $d\vec{A}$ and the velocity vector, $\vec{V}$. Since the mass in region I flows into the control volume in the time increment, Δt, the angle α will always be greater than $\pi/2$ over the entire area of the control surface bounding region I.

For subregion (1) we can write

$$dN_1)_{t_0} = (\eta\rho\,d\forall)_{t_0} = [\eta\rho\,\Delta l(-\cos\alpha)\,dA]_{t_0}$$

since $d\forall = \Delta l(-\cos\alpha)\,dA$. Why the minus sign you ask? Recall that volume is a scalar quantity that must have a positive numerical value. Since $\alpha > \pi/2$, then $\cos\alpha$ will be negative. Hence the need for the minus sign.

Then for the entire region I

$$N_1)_{t_0} = \left[\int_{CS_1} -\eta\rho\,\Delta l\cos\alpha\,dA\right]_{t_0}$$

where CS_1 is the surface common to region I and the control volume. In this expression for $N_1)_{t_0}$, Δl represents the distance traveled by a particle on the system surface in time Δt, along a streamline that existed at time t_0.

117   **relation of system derivatives to control volume formulation**

Now that we have an expression for $N_1)_{t_0}$, we can evaluate term ③ in Eq. 4.9:

$$- \lim_{\Delta t \to 0} \frac{\left[\int_I \eta \rho \, d\forall \right]_{t_0}}{\Delta t} = - \lim_{\Delta t \to 0} \frac{N_1)_{t_0}}{\Delta t}$$

$$= - \lim_{\Delta t \to 0} \frac{\int_{CS_I} -\eta \rho \, \Delta l \cos \alpha \, dA}{\Delta t}$$

$$= \lim_{\Delta t \to 0} \int_{CS_I} \eta \rho \frac{\Delta l}{\Delta t} \cos \alpha \, dA$$

$$= \int_{CS_I} \eta \rho |\vec{V}| \cos \alpha |d\vec{A}|$$

The last equality follows from the fact that

$$\lim_{\Delta t \to 0} \frac{\Delta l}{\Delta t} = |\vec{V}| \quad \text{and} \quad dA = |d\vec{A}|$$

Now that we have obtained expressions for each of the three terms on the right-hand side of Eq. 4.9, let us substitute these expressions back into the equation. Equation 4.9 can then be written as

$$\left. \frac{dN}{dt} \right)_{\text{system}} = \frac{\partial}{\partial t} \int_{CV} \eta \rho \, d\forall + \int_{CS_I} \eta \rho |\vec{V}| \cos \alpha |d\vec{A}|$$

$$+ \int_{CS_{III}} \eta \rho |\vec{V}| \cos \alpha |d\vec{A}|$$

Referring to Fig. 4.1, we see that the entire control surface, CS, consists of three surfaces, that is,

$$\text{CS} = \text{CS}_I + \text{CS}_{III} + \text{CS}_p$$

where CS_p is characterized by either $\alpha = \pi/2$ or $\vec{V} = 0$, that is, by no flow across the surface.

Consequently we can write

$$\left. \frac{dN}{dt} \right)_{\text{system}} = \frac{\partial}{\partial t} \int_{CV} \eta \rho \, d\forall + \int_{CS} \eta \rho |\vec{V}| \cos \alpha |d\vec{A}| \qquad (4.10)$$

Recognizing that $|\vec{V}| \cos \alpha |d\vec{A}| = \vec{V} \cdot d\vec{A}$, Eq. 4.10 becomes

$$\left. \frac{dN}{dt} \right)_{\text{system}} = \frac{\partial}{\partial t} \int_{CV} \eta \rho \, d\forall + \int_{CS} \eta \rho \vec{V} \cdot d\vec{A} \qquad (4.11)$$

Equation 4.11 is the relation we set out to obtain, that is, a relation between

$dN/dt)_{\text{system}}$ and a control volume formulation. Since we shall apply it five times using different physical quantities for N, we shall refer to it as a "recipe."

4–2.2 PHYSICAL MEANING OF THE RECIPE

We have taken several pages to derive Eq. 4.11. Recall that our objective in deriving this equation was to obtain a general relation between the rate of change of any arbitrary extensive property, N, of a system and the time variations of the property associated with the control volume. The main reason for deriving it was to reduce the amount of algebra required in obtaining the control volume formulations of the basic equations. Prior to examining Eq. 4.11, let us rewrite it using a triple integral to indicate integration over a volume and a double integral to indicate integration over an area.

$$\frac{dN}{dt}\bigg)_{\text{system}} = \frac{\partial}{\partial t} \iiint_{CV} \eta\rho \, d\forall + \iint_{CS} \eta\rho \vec{V} \cdot d\vec{A} \qquad (4.12)$$

It is important to recall that in deriving Eq. 4.12, the limiting process employed (i.e. taking the limit as $\Delta t \to 0$) insured that the relation is valid at the instant when the system and the control volume coincide. In using Eq. 4.12 to go from the system formulations of the basic laws to the control volume formulations, we recognize that Eq. 4.12 relates the rate of change of any arbitrary extensive property, N, of a system to the time variation of the property associated with a control volume at the instant when the system and the control volume coincide; this is true since, as $\Delta t \to 0$, the system and the control volume occupy the same volume and have the same boundaries.

Before proceeding to use Eq. 4.12 to develop control volume formulations of the basic laws, let us make sure we understand the meaning of each of the terms and symbols in the equation:

$\dfrac{dN}{dt}\bigg)_{\text{system}}$ is the total rate of change of any arbitrary extensive property of the system.

$\dfrac{\partial}{\partial t}\iiint_{CV} \eta\rho \, d\forall$ is the time rate of change of the arbitrary extensive property N within the control volume.

    : η is the intensive property corresponding to N, that is, $\eta = N$ per unit mass.

    : $\rho \, d\forall$ is an element of mass contained in the control volume.

    : $\iiint_{CV} \eta\rho \, d\forall$ is the total amount of the extensive property N contained within the control volume.

$$\iint_{cs} \eta \rho \vec{V} \cdot d\vec{A}$$

is the net rate of efflux of the extensive property N through the control surface.

: $\rho \vec{V} \cdot d\vec{A}$ is the rate of mass efflux through the element of area $d\vec{A}$ per unit time (we recognize that the dot product is a scalar product; the sign of $\rho \vec{V} \cdot d\vec{A}$ depends on the direction of the velocity vector, $\vec{V}$, relative to the area vector, $d\vec{A}$).

: $\eta \rho \vec{V} \cdot d\vec{A}$ is the net rate of efflux of the extensive property N through the area $d\vec{A}$.

Two additional points about Eq. 4.12 should be made. It should be clear from the derivation of the equation, that the velocity, $\vec{V}$, in the equation is measured relative to the control volume. In developing Eq. 4.12, we considered a control volume fixed in space, that is, fixed relative to the reference coordinates x, y, z. The velocity field was specified relative to the same reference coordinates. Since, in our development, the system moved in the specified velocity field, then the time rate of change of the arbitrary extensive property, N, within the control volume must also be evaluated by an observer fixed in the reference coordinates.

We shall further emphasize these points in deriving the control volume formulation of each of the basic laws. In each case we begin with the familiar system formulation and employ Eq. 4.12 to derive the control volume formulation.

4–3 Conservation of Mass

In considering the flow through a control volume, we recognize that if, at a given instant of time, the amount of mass flowing into the control volume is not the same as the amount of mass flowing out of the control volume, there will be a change in the amount of mass within the control volume. If the outflow is greater than the inflow, there will be a decrease in the amount of mass in the control volume; conversely, there will be an increase in the amount of mass within the control volume if the inflow is greater than the outflow. Thus on purely physical grounds we can provide a word statement of the conservation of mass:

$$0 = \begin{bmatrix} \text{Rate of mass} \\ \text{flow out of the} \\ \text{control volume} \end{bmatrix} - \begin{bmatrix} \text{Rate of mass} \\ \text{flow into the} \\ \text{control volume} \end{bmatrix} + \begin{bmatrix} \text{Rate of change} \\ \text{of mass inside the} \\ \text{control volume} \end{bmatrix}$$

Since the difference between the rate of mass outflow and the rate of mass inflow is equal to the net rate of efflux (outflow) of mass through the control volume, the above word statement of the conservation of mass could be written as

$$0 = \begin{bmatrix} \text{Net rate of mass efflux (outflow)} \\ \text{through the control surface} \end{bmatrix} + \begin{bmatrix} \text{Rate of change of mass} \\ \text{inside the control volume} \end{bmatrix}$$

Now that we have a physical formulation of conservation of mass for a control volume, let us obtain a mathematical formulation. We begin with the mathematical formulation for a system and employ Eq. 4.12 to transform it to the control volume formulation.

4–3.1 CONTROL VOLUME EQUATION

Recall that conservation of mass simply states that the mass of a system is constant, that is

$$\frac{dM}{dt}\bigg)_{\text{system}} = 0 \tag{4.1a}$$

where

$$M_{\text{system}} = \int_{\text{Mass (system)}} dm = \int_{\forall\,(\text{system})} \rho\, d\forall \tag{4.1b}$$

The recipe relating the system and control volume formulations is given by Eq. 4.12

$$\frac{dN}{dt}\bigg)_{\text{system}} = \frac{\partial}{\partial t}\iiint_{\text{CV}} \eta\rho\, d\forall + \iint_{\text{CS}} \eta\rho\vec{V}\cdot d\vec{A} \tag{4.12}$$

where

$$N_{\text{system}} = \int_{\text{Mass (system)}} \eta\, dm = \iiint_{\forall\,(\text{system})} \eta\rho\, d\forall \tag{4.6}$$

Then, in using the recipe to derive the control volume formulation of the conservation of mass, we set

$$N = M \quad \text{and} \quad \eta = 1$$

With this substitution, we obtain

$$\frac{dM}{dt}\bigg)_{\text{system}} = \frac{\partial}{\partial t}\iiint_{\text{CV}} \rho\, d\forall + \iint_{\text{CS}} \rho\vec{V}\cdot d\vec{A} \tag{4.13}$$

The conservation of mass for a system is given by Eq. 4.1a

$$\frac{dM}{dt}\bigg)_{\text{system}} = 0 \tag{4.1a}$$

Comparing Eqs. 4.1a and 4.13 we arrive at the control volume formulation of the conservation of mass:

$$0 = \frac{\partial}{\partial t} \iiint_{CV} \rho \, d\forall + \iint_{CS} \rho \vec{V} \cdot d\vec{A} \qquad (4.14)$$

In Eq. 4.14 the first term represents the rate of change of mass within the control volume; the second term represents the net rate of mass efflux through the control surface. It is comforting to note that the mathematical formulation is consistent with the physical formulation that we developed at the outset.

It is worth reiterating at this point that the velocity, $\vec{V}$, in Eq. 4.14 is measured relative to the control surface. Furthermore the dot product $\rho \vec{V} \cdot d\vec{A}$ is a scalar product. The sign depends on the direction of the velocity vector, $\vec{V}$, relative to the area vector, $d\vec{A}$. Referring back to the derivation of the recipe (Section 4–2.1) we see that the dot product, $\rho \vec{V} \cdot d\vec{A}$, is positive where flow is out through the control surface, negative where flow is in through the control surface, and zero where the flow is tangent to the control surface.

4–3.2 SPECIAL CASES

In special cases it is possible to simplify Eq. 4.14. Two such cases are worth noting.

Consider first the case of incompressible flow—that is, flow in which the density remains constant. Since ρ is a constant, it may be taken out from under the integral signs in Eq. 4.14. For this case we write Eq. 4.14 as

$$0 = \frac{\partial}{\partial t} \rho \iiint_{CV} d\forall + \rho \iint_{CS} \vec{V} \cdot d\vec{A} \qquad (4.15a)$$

The integral of $d\forall$ over the control volume is simply the volume of the control volume. Thus we write Eq. 4.15a as

$$0 = \frac{\partial}{\partial t} [\rho \forall] + \rho \iint_{CS} \vec{V} \cdot d\vec{A} \qquad (4.15b)$$

Since ρ is a constant and the size of the control volume is fixed, we can write the conservation of mass for incompressible flow as

$$0 = \iint_{CS} \vec{V} \cdot d\vec{A} \qquad (4.15c)$$

Note that we have not assumed the flow to be steady in reducing Eq. 4.14 to the form 4.15c. We have only imposed the restriction of incompressible flow. Thus Eq. 4.15c is a statement of the conservation of mass for an incompressible flow that may be steady or unsteady.

Consider now the general case of steady flow in which the flow is not incompressible. Since the flow is steady, this means that $\rho = \rho(x, y, z)$. By definition a steady flow is one in which none of the fluid properties varies with time. Consequently the first term of Eq. 4.14 must be zero and, hence, for any steady flow the statement of the conservation of mass reduces to

$$0 = \iint_{CS} \rho \vec{V} \cdot d\vec{A} \qquad (4.16)$$

Note that when $\rho \vec{V} \cdot d\vec{A}$ is negative, mass flows in through the control surface. Mass flows out through the control surface in regions where $\rho \vec{V} \cdot d\vec{A}$ is positive. This fact provides a quick check of the signs on the various flux terms in an analysis.

Example 4.1

Consider the steady flow of water ($\rho = 1.94 \text{ slug/ft}^3$) through the device shown in the diagram. The following conditions are known:

$$A_1 = 0.2 \text{ ft}^2, \qquad A_2 = 0.5 \text{ ft}^2, \qquad A_3 = A_4 = 0.4 \text{ ft}^2$$

Mass flow out through section ③ is given as 3.88 slug/sec. Volumetric flow in through section ④ is given as 1 ft³/sec, and $\vec{V}_1 = 10\,\hat{\imath}$ ft/sec. If all properties can be considered uniform across all inlet and outlet flow sections, determine the velocity at section ②.

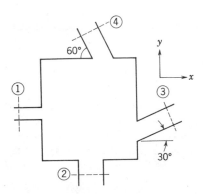

conservation of mass

GIVEN:

Steady flow of water through the device.
Properties uniform at all ports.

$$A_1 = 0.2 \text{ ft}^2 \qquad A_2 = 0.5 \text{ ft}^2$$

$$A_3 = A_4 = 0.4 \text{ ft}^2 \qquad \rho = 1.94 \text{ slug/ft}^3 \quad ①$$

$$\dot{m}_3 = 3.88 \text{ slug/sec (outflow)}$$

Volumetric flow in at ④ $= 1.0 \text{ ft}^3/\text{sec}$

$$\vec{V}_1 = 10\hat{\imath} \text{ ft/sec}$$

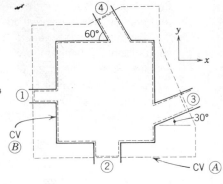

FIND:

Velocity at section ②.

SOLUTION:

Choose a control volume. Two possibilities are shown by dashed lines.

Basic equation: $\qquad 0 = \dfrac{\partial}{\partial t} \displaystyle\iiint_{CV} \rho \, d\forall + \iint_{CS} \rho \vec{V} \cdot d\vec{A}$

For steady flow,

$$\frac{\partial}{\partial t} \iiint_{CV} \rho \, d\forall = 0 \quad \left\{ \begin{array}{c} \text{by definition of steady} \\ \text{flow} \end{array} \right\}$$

$$\therefore \quad \iint_{CS} \rho \vec{V} \cdot d\vec{A} = 0$$

In looking at either control volume we see that there are four sections where mass flows across the control surface. Thus we write

$$\iint_{CS} \rho \vec{V} \cdot d\vec{A} = \iint_{A_1} \rho \vec{V} \cdot d\vec{A} + \iint_{A_2} \rho \vec{V} \cdot d\vec{A} + \iint_{A_3} \rho \vec{V} \cdot d\vec{A} + \iint_{A_4} \rho \vec{V} \cdot d\vec{A} = 0 \qquad (1)$$

Let us look at these integrals one at a time, recognizing properties are uniform over each area, and $\rho = $ constant.

$$\iint_{A_1} \rho \vec{V} \cdot d\vec{A} = -\iint_{A_1} |\rho V \, dA| = -|\rho V_1 A_1| \quad \xrightarrow{\vec{V}_1}_{\overleftarrow{\vec{A}_1} \;①} \quad \left\{ \begin{array}{l} \text{sign of } \vec{V} \cdot d\vec{A} \text{ is negative at} \\ \text{surface } ①. \end{array} \right\}$$

$\left\{ \text{With the absolute value signs indicated, we have accounted for the direction of } \vec{V} \text{ and } d\vec{A} \right.$
$\left. \text{in taking the dot product.} \right\}$

Since we do not know the direction of $\vec{V}_2$, we will leave it for the moment.

$$\iint_{A_3} \rho \vec{V} \cdot d\vec{A} = \iint_{A_3} |\rho V \, dA| = |\rho V_3 A_3| = \dot{m}_3$$

$\left\{\begin{array}{l}\text{sign of } \vec{V} \cdot d\vec{A} \text{ is positive} \\ \text{at surface ③ since} \\ \text{flow is out.}\end{array}\right\}$

$$\iint_{A_4} \rho \vec{V} \cdot d\vec{A} = -\iint_{A_4} |\rho V \, dA| = -|\rho V_4 A_4| = -\rho |V_4 A_4|$$

$$= -\rho|\dot{q}| \quad \text{where } \dot{q} = \text{volumetric flow rate}$$

$\left\{\begin{array}{l}\text{sign of } \vec{V} \cdot d\vec{A} \text{ is negative} \\ \text{at surface ④.}\end{array}\right\}$

From Eq. 1 above,

$$\iint_{A_2} \rho \vec{V} \cdot d\vec{A} = -\iint_{A_1} \rho \vec{V} \cdot d\vec{A} - \iint_{A_3} \rho \vec{V} \cdot d\vec{A} - \iint_{A_4} \rho \vec{V} \cdot d\vec{A}$$

$$= +|\rho V_1 A_1| - \dot{m}_3 + \rho|\dot{q}_4|$$

$$= |1.94 \text{ slug/ft}^3 \times 10 \text{ ft/sec} \times 0.2 \text{ ft}^2| - 3.88 \text{ slug/sec}$$

$$+ 1.94 \text{ slug/ft}^3 \times |1.0 \text{ ft}^3/\text{sec}|$$

$$\iint_{A_2} \rho \vec{V} \cdot d\vec{A} = 1.94 \text{ slug/sec}$$

Since this is positive, $\vec{V} \cdot d\vec{A}$ at ② is positive. Flow is out as shown below:

$$\iint_{A_2} \rho \vec{V} \cdot d\vec{A} = \iint_{A_2} |\rho V \, dA| = |\rho V_2 A_2| = 1.94 \text{ slug/sec}$$

$$|V_2| = \frac{1.94 \text{ slug/sec}}{\rho A_2} = 1.94 \text{ slug/sec} \times \frac{1}{1.94} \text{ ft}^3/\text{slug} \times \frac{1}{0.5 \text{ ft}^2} = 2 \text{ ft/sec}$$

$$\therefore \quad \vec{V}_2 = -2\hat{j} \text{ ft/sec}$$

$\left\{\text{This problem illustrates the procedure for evaluating } \iint_{CS} \rho \vec{V} \cdot d\vec{A}\right\}$

Example 4.2

A 2 cubic foot tank contains air at 100 psig, 60 F. At time, $t = 0$, air escapes from the tank through a valve with a flow area of 0.1 in.^{2} The air passing through the valve has a velocity of 1020 ft/sec and a density of 0.0117 slug/ft^3. Properties in

conservation of mass

the rest of the tank may be assumed to be uniform. Determine the instantaneous
rate of change of density in the tank at $t = 0$.

EXAMPLE PROBLEM 4.2

GIVEN:

Tank of volume $\forall = 2 \text{ ft}^3$ contains air at $p = 100 \text{ psig}$, $T = 60 \text{ F}$. At $t = 0$, air escapes
through a valve. Air leaves with $V = 1020 \text{ ft/sec}$, $\rho = 0.0117 \text{ slug/ft}^3$ through an area,
$A = 0.1 \text{ in.}^2$
Properties in tank are uniform at any instant.

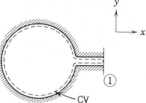

FIND:

Rate of change of air density in the tank at $t = 0$.

SOLUTION:

Choose a control volume as shown by the dashed line.

Basic equation:
$$0 = \frac{\partial}{\partial t} \iiint_{CV} \rho \, d\forall + \iint_{CS} \rho \vec{V} \cdot d\vec{A}$$

Since properties are assumed uniform in the tank at any instant, the density is not a
function of the volume, that is, $\rho \neq \rho(\forall)$. Consequently we can take ρ out from inside the
integral of the first term.

$$\therefore \quad \frac{\partial}{\partial t} \left[\rho \iiint_{CV} d\forall \right] + \iint_{CS} \rho \vec{V} \cdot d\vec{A} = 0$$

Now, $\displaystyle\iiint_{CV} d\forall = \forall$ and hence

$$\frac{\partial}{\partial t}(\rho \forall) + \iint_{CS} \rho \vec{V} \cdot d\vec{A} = 0$$

The only place where mass crosses the boundary of the control volume is at surface ①.
Hence

$$\iint_{CS} \rho \vec{V} \cdot d\vec{A} = \iint_{A_1} \rho \vec{V} \cdot d\vec{A} \quad \text{and} \quad \frac{\partial}{\partial t}(\rho \forall) + \iint_{A_1} \rho \vec{V} \cdot d\vec{A} = 0$$

At surface ① the sign of $\rho \vec{V} \cdot d\vec{A}$ is positive

$$\vec{A}_1$$

$$\therefore \quad \frac{\partial}{\partial t}(\rho \forall) + \int\int_{A_1} |\rho V \, dA| = 0$$

Assuming properties are uniform over surface ①, then

$$\frac{\partial}{\partial t}(\rho \forall) + |\rho_1 V_1 A_1| = 0$$

and

$$\frac{\partial}{\partial t}(\rho \forall) = -|\rho_1 V_1 A_1|$$

Since the volume, $\forall$, of the tank is not a function of time

$$\forall \frac{\partial \rho}{\partial t} = -|\rho_1 V_1 A_1|$$

and

$$\frac{\partial \rho}{\partial t} = -\frac{|\rho_1 V_1 A_1|}{\forall}$$

At time $t = 0$,

$$\frac{\partial \rho}{\partial t} = -0.0117 \frac{\text{slug}}{\text{ft}^3} \times 1020 \frac{\text{ft}}{\text{sec}} \times 0.1 \text{ in.}^2 \times \frac{1}{2 \text{ ft}^3} \times \frac{\text{ft}^2}{144 \text{ in.}^2}$$

$$\frac{\partial \rho}{\partial t} = -0.0041 \frac{\text{slug/ft}^3}{\text{sec}} \longleftarrow \qquad \qquad \{\text{the density is decreasing}\}$$

$\left\{\begin{array}{l}\text{This problem illustrates the application of the control volume formulation of the} \\ \text{conservation of mass to an unsteady flow.}\end{array}\right\}$

Example 4.3

The fluid in direct contact with a stationary solid boundary has zero velocity, that is, there is no slip at the boundary. Thus the flow over a flat plate adheres to the plate surface and forms a boundary layer, as depicted in the figure below. The flow ahead of the plate is uniform with velocity, $\vec{V} = U_0\hat{\imath}$; $U_0 = 100$ ft/sec. The velocity distribution at CD is approximated as being linear $(\vec{V}_C = U_0\hat{\imath}, \vec{V}_D = 0)$. The

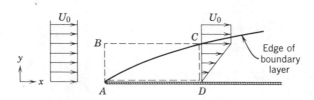

boundary layer thickness at this location is 0.2 in. The fluid is air with density $\rho = 0.0024$ slug/ft³. Assuming the plate width to be 2 ft, calculate the mass flow across the surface BC of control volume $ABCD$.

EXAMPLE PROBLEM 4.3

GIVEN:

Steady, incompressible flow over a flat plate, $\rho = 0.0024$ slug/ft³.
Velocity ahead of plate is uniform

$$\vec{V} = \hat{\imath}U_0 \qquad U_0 = 100 \text{ ft/sec}$$

At $x = x_D$: the x component of velocity varies linearly with y.

$$\text{at} \quad y = y_C = \delta, \qquad u = U_0$$

$$\text{at} \quad y = 0, \qquad u = 0$$

: boundary layer thickness, $\delta = 0.2$ in.
Width of plate, $w = 2$ ft.

FIND:

The mass flow rate across the surface BC.

SOLUTION:

The control volume is selected as shown by the dashed lines

Basic equation: $\qquad 0 = \dfrac{\partial}{\partial t}\iiint_{CV} \rho\, d\forall + \iint_{CS} \rho\vec{V}\cdot d\vec{A}$

For steady flow,

$$\frac{\partial}{\partial t}\iiint_{CV} \rho\, d\forall = 0 \quad \text{and, hence,} \quad \iint_{CS} \rho\vec{V}\cdot d\vec{A} = 0$$

Assuming there is no flow in the z direction, then

$$\iint_{CS} \rho\vec{V}\cdot d\vec{A} = 0 = \iint_{A_{AB}} \rho\vec{V}\cdot d\vec{A} + \iint_{A_{BC}} \rho\vec{V}\cdot d\vec{A} + \iint_{A_{CD}} \rho\vec{V}\cdot d\vec{A} + \overset{0\left(\substack{\text{no flow} \\ \text{across } DA}\right)}{\iint_{A_{DA}} \rho\vec{V}\cdot d\vec{A}}$$

$$\therefore \iint_{A_{BC}} \rho\vec{V}\cdot d\vec{A} = -\iint_{A_{AB}} \rho\vec{V}\cdot d\vec{A} - \iint_{A_{CD}} \rho\vec{V}\cdot d\vec{A} \quad \left\{\begin{array}{l}\text{We need to evaluate the integrals on} \\ \text{the right side of the equation.}\end{array}\right\}$$

$$\iint_{A_{AB}} \rho \vec{V} \cdot d\vec{A} = -\iint_{A_{AB}} |\rho u \, dA| = -\int_{y_A}^{y_B} |\rho u w \, dy| \qquad \begin{aligned}\vec{V} \cdot d\vec{A} \text{ is negative} \\ dA = w \, dy\end{aligned}$$

$$= -\int_0^\delta |\rho u w \, dy| = -\left|\int_0^\delta \rho U_0 w \, dy\right| \qquad \begin{aligned}u = U_0 \text{ over} \\ \text{area } AB\end{aligned}$$

$$= -|[\rho U_0 w y]_0^\delta| = -|\rho U_0 w \delta|$$

$$\iint_{A_{CD}} \rho \vec{V} \cdot d\vec{A} = \iint_{A_{CD}} |\rho u \, dA| = \int_{y_D}^{y_C} |\rho u w \, dy| \qquad \begin{aligned}\vec{V} \cdot d\vec{A} \text{ is positive} \\ dA = w \, dy\end{aligned}$$

$$= \int_0^\delta |\rho u w \, dy| = \left|\int_0^\delta \rho \frac{U_0 y}{\delta} w \, dy\right| \qquad \left\{\text{across surface } CD, u = U_0 \frac{y}{\delta}\right\}$$

$$= \left|\left[\frac{\rho U_0 w}{\delta} \frac{y^2}{2}\right]_0^\delta\right| = \left|\frac{\rho U_0 w \delta}{2}\right|$$

$$\therefore \quad \iint_{A_{BC}} \rho \vec{V} \cdot d\vec{A} = -\iint_{A_{AB}} \rho \vec{V} \cdot d\vec{A} - \iint_{A_{CD}} \rho \vec{V} \cdot d\vec{A} = \rho U_0 w \delta - \frac{\rho U_0 w \delta}{2}$$

$$= \frac{\rho U_0 w \delta}{2} = \frac{1}{2} \times 0.0024 \frac{\text{slug}}{\text{ft}^3} \times 100 \frac{\text{ft}}{\text{sec}} \times 2 \, \text{ft} \times 0.2 \, \text{in.} \times \frac{\text{ft}}{12 \, \text{in.}}$$

$$\iint_{A_{BC}} \rho \vec{V} \cdot d\vec{A} = 0.004 \, \text{slug/sec} \qquad \begin{aligned}\text{positive sign indicates flow out} \\ \text{across surface } BC.\end{aligned} \qquad \dot{m}_{BC}$$

{ This problem illustrates the application of the control volume formulation of the conservation of mass to the case of nonuniform flow at a section. }

4-4 Momentum Equation for Inertial Control Volume

We are interested in developing a mathematical formulation of Newton's second law suitable for application to a control volume analysis. In this section our derivation will be restricted to an inertial control volume, that is a control volume that is not accelerating relative to a stationary frame of reference (inertial coordinate system).

In deriving the control volume formulation the procedure is completely analogous to the procedure followed in deriving the mathematical formulation for the conservation of mass applied to a control volume. We begin with the mathematical formulation for a system and then employ the recipe of Eq. 4.12 to go from the system to the control volume formulation.

4-4.1 CONTROL VOLUME EQUATION

Recall that Newton's second law for a system moving relative to an inertial coordinate system was given by Eq. 4.2a

$$\vec{F} = \left.\frac{d\vec{P}}{dt}\right)_{system} \tag{4.2a}$$

where the linear momentum, $\vec{P}$, of the system is given by

$$\vec{P}_{system} = \int_{Mass\,(system)} \vec{V}\,dm = \int_{\forall\,(system)} \vec{V}\rho\,d\forall \tag{4.2b}$$

and the resultant force, $\vec{F}$, includes all surface and body forces acting on the system. That is

$$\vec{F} = \vec{F}_S + \vec{F}_B$$

The recipe for relating the system and control volume formulations is given by Eq. 4.12

$$\left.\frac{dN}{dt}\right)_{system} = \frac{\partial}{\partial t}\iiint_{CV} \eta\rho\,d\forall + \iint_{CS} \eta\rho\vec{V}\cdot d\vec{A} \tag{4.12}$$

where

$$N_{system} = \int_{Mass\,(system)} \eta\,dm = \iiint_{\forall\,(system)} \eta\rho\,d\forall \tag{4.6}$$

Then in using the recipe to derive the control volume formulation of Newton's second law, we set

$$N = \vec{P} \quad\text{and}\quad \eta = \vec{V}$$

From Eq. 4.12, with this substitution, we obtain

$$\left.\frac{d\vec{P}}{dt}\right)_{system} = \frac{\partial}{\partial t}\iiint_{CV} \vec{V}\rho\,d\forall + \iint_{CS} \vec{V}\rho\vec{V}\cdot d\vec{A} \tag{4.17}$$

From Eq. 4.2a

$$\left.\frac{d\vec{P}}{dt}\right)_{system} = \vec{F}\Big)_{on\,system} \tag{4.2a}$$

4/basic equations in integral form for a control volume

Since, in deriving the recipe, the system and the control volume coincided at time t,

$$\vec{F}]_{\text{on system}} = \vec{F}]_{\text{on control volume}}$$

In light of this, Eqs. 4.2a and 4.17 yield the control volume formulation of Newton's second law for a nonaccelerating control volume

$$\vec{F} = \vec{F}_S + \vec{F}_B = \frac{\partial}{\partial t} \iiint_{CV} \vec{V}\rho\, d\forall + \iint_{CS} \vec{V}\rho\vec{V}\cdot d\vec{A} \qquad (4.18)$$

This equation states that the sum of all forces (surface and body forces) acting on a nonaccelerating control volume is equal to the sum of the rate of change of momentum inside the control volume and the net rate of efflux of momentum through the control surface.

The derivation of the momentum equation for a control volume was straight-forward. Application of this basic equation to the solution of problems will not be difficult if you exercise care in using the equation.

In applying any basic equation to a control volume analysis the first step should be to draw the boundaries of the control volume and label appropriate coordinate directions. In Eq. 4.18, the force, $\vec{F}$, represents all forces acting on the control volume. It includes both surface forces and body forces. If we denote the body force per unit mass as $\vec{B}$, then

$$\vec{F}_B = \int \vec{B}\, dm = \iiint_{CV} \vec{B}\rho\, d\forall$$

The nature of the forces acting on the control volume will undoubtedly influence the choice of the control volume boundaries.

All velocities, $\vec{V}$, in Eq. 4.18 are measured relative to the control volume. The momentum flux, $\vec{V}\rho\vec{V}\cdot d\vec{A}$, through an element of the control surface area, $d\vec{A}$, is a vector. The sign of the scalar product $\rho\vec{V}\cdot d\vec{A}$ depends on the direction of the velocity vector, $\vec{V}$, relative to the area vector, $d\vec{A}$. The sign of the vector velocity, $\vec{V}$, is dependent on the choice of the coordinate system.

The momentum equation is a vector equation. As with all vector equations, it may be written in scalar component equations. Relative to an x, y, z coordinate system, the scalar components of Eq. 4.18 may be written as

$$F_x = F_{S_x} + F_{B_x} = \frac{\partial}{\partial t} \iiint_{CV} u\rho\, d\forall + \iint_{CS} u\rho\vec{V}\cdot d\vec{A} \qquad (4.19a)$$

131   **momentum equation for inertial control volume**

$$F_y = F_{S_y} + F_{B_y} = \frac{\partial}{\partial t} \iiint_{CV} v\rho \, d\forall + \iint_{CS} v\rho \vec{V} \cdot d\vec{A} \qquad (4.19b)$$

$$F_z = F_{S_z} + F_{B_z} = \frac{\partial}{\partial t} \iiint_{CV} w\rho \, d\forall + \iint_{CS} w\rho \vec{V} \cdot d\vec{A} \qquad (4.19c)$$

In using the scalar equations, it is again necessary to select a coordinate system at the outset. The positive directions of the velocity components, u, v, and w and the force components F_x, F_y, and F_z are then established relative to the selected coordinate system. As we have previously pointed out, the sign of the scalar product $\rho\vec{V} \cdot d\vec{A}$ depends on the direction of the velocity vector, $\vec{V}$, relative to the area vector, $d\vec{A}$. Thus the flux term in either Eq. 4.18 or Eqs. 4.19 is a product of two quantities, both of which have signs associated with them. It is suggested that you proceed in two steps in determining the momentum flux through any portion of a control surface. The first step should be to determine the sign of $\rho\vec{V} \cdot d\vec{A}$, that is

$$\rho\vec{V} \cdot d\vec{A} = \rho|V \, dA| \cos \alpha = \pm|\rho V \, dA \cos \alpha|$$

Since the sign for each velocity component, u, v, and w, depends on the choice of coordinate system, the sign should be accounted for when substituting numerical values into the terms $u\rho\vec{V} \cdot d\vec{A} = u\{\pm|\rho V \, dA \cos \alpha|\}$, and so on.

Example 4.4

During a demonstration the police decide to use fire hoses to knock over a barricade built by the demonstrators. The barricade is built in the form of a flat plate as shown in the diagram. The velocity of the water leaving the nozzle is 45 ft/sec relative to the nozzle; the nozzle area is 0.1 ft². Assuming the water is directed normal to the plate, determine the horizontal force on the barricade.

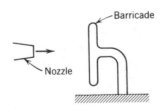

GIVEN:

Water from a stationary nozzle strikes a stationary flat plate as shown (water directed normal to the plate).

$$\vec{V}_j = 45\hat{\imath} \text{ ft/sec}$$

Area of nozzle $= 0.1 \text{ ft}^2$

FIND:

Horizontal force on the support.

SOLUTION:

Note that we have chosen a coordinate system in defining the problem. We must now choose a suitable control volume. A number of possible choices are shown by the dashed lines.

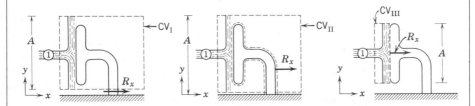

In all three cases the water from the nozzle crosses the control surface through area A_1 (assumed equal to the nozzle area) and the water is assumed to leave the control volume tangent to the plate surface, that is, in the $+y$ or $-y$ direction. Before trying to decide which is the "best" control volume to use, let us write the basic equations.

$$\vec{F} = \vec{F}_S + \vec{F}_B = \frac{\partial}{\partial t} \iiint_{CV} \vec{V}\rho \, d\forall + \iint_{CS} \vec{V}\rho\vec{V} \cdot d\vec{A} \quad \text{and} \quad \frac{\partial}{\partial t} \iiint_{CV} \rho \, d\forall + \iint_{CS} \rho\vec{V} \cdot d\vec{A} = 0$$

Regardless of our choice of control volume, the flow is steady and, hence, the basic equations become

$$\vec{F} = \vec{F}_S + \vec{F}_B = \iint_{CS} \vec{V}\rho\vec{V} \cdot d\vec{A} \quad \text{and} \quad \iint_{CS} \rho\vec{V} \cdot d\vec{A} = 0$$

Evaluating the momentum flux term will lead to the same result for all of the control volumes. We should choose the control volume that allows the most straightforward evaluation of the forces.

Remember in applying the momentum equation the force $\vec{F}$ represents all forces acting *on* the control volume.

Let us solve the problem using each of the three control volumes.

CV_1

The control volume has been selected so that the area of the left surface and the right surface are equal. Denote this area by A.

The control volume cuts through the support. We denote the force of the support on the control volume as R_x and assume it to be positive. (The force of the control volume on the support is equal and opposite to R_x.)

Since we are looking for the horizontal force, we write the x component of the steady flow momentum equation

$$F_{S_x} + F_{B_x} = \iint_{CS} u\rho\vec{V} \cdot d\vec{A}$$

There are no body forces in the x direction. $F_{B_x} = 0$ and

$$F_{S_x} = \iint_{CS} u\rho\vec{V} \cdot d\vec{A}$$

To evaluate F_{S_x} we must include all surface forces acting on the control volume

$$F_{S_x} = \qquad p_a A \qquad - \qquad p_a A \qquad + \qquad R_x$$

force due to atmospheric pressure acts to right (positive direction) on left surface	force due to atmospheric pressure acts to left (negative direction) on right surface	force of support on control volume (assumed positive)

Consequently, $F_{S_x} = R_x$

$$\therefore \quad R_x = \iint_{CS} u\rho\vec{V} \cdot d\vec{A} = \iint_{A_1} u\rho\vec{V} \cdot d\vec{A} \qquad \left\{ \begin{array}{l} \text{For mass crossing top and} \\ \text{bottom surfaces, } u = 0. \end{array} \right\}$$

$$= -\iint_{A_1} u|\rho V_1 \, dA| \qquad \left\{ \begin{array}{l} \text{at } ① \; \rho\vec{V} \cdot d\vec{A} = -|\rho V_1 \, dA| \text{ since direction} \\ \text{of } \vec{V}_1 \text{ and } d\vec{A}_1 \text{ are } 180° \text{ apart.} \end{array} \right\}$$

$$= -u_1|\rho V_1 A_1| \qquad \qquad \{\text{properties uniform over } A_1\}$$

$$= -45 \text{ ft/sec} \times |1.94 \text{ slug/ft}^3 \times 45 \text{ ft/sec} \times 0.1 \text{ ft}^2| \qquad \{u_1 = 45 \text{ ft/sec}\}$$

$$= -392 \text{ slug-ft/sec}^2 \times \text{lbf-sec}^2/\text{slug-ft}$$

$$R_x = -392 \text{ lbf} \qquad \left\{ \begin{array}{l} R_x \text{ acts opposite to positive} \\ \text{direction assumed.} \end{array} \right\}$$

The force on the support, $K_x = -R_x = 392 \text{ lbf}$ $\qquad \{\text{force on support acts to the right}\}$

CV_{II}

The control volume has been selected so that the area of the left surface and the right surface are equal. Denote this area by A.

The control volume does not cut through the support. However, the control volume is in contact with the support over several portions of the area of the control surface. There is a force exerted by the support on the control surface. We denote the x component of this force as R_x.

Then for this control volume the x momentum equation

$$F_{S_x} = \iint_{CS} u\rho \vec{V} \cdot d\vec{A}$$

leads directly (in the same step by step manner) to the same solution as for CV_I above.

CV_{III}

The control volume has been selected so that the area of the left surface and the right surface are equal to the area of the plate. Denote this area by A.

As in the case of CV_{II} the x component of the force of the support on the CV is denoted by R_x.

Then the x momentum equation,

$$F_{S_x} = \iint_{CS} u\rho \vec{V} \cdot d\vec{A}$$

yields

$$F_{S_x} = p_a A + R_x = \iint_{A_1} u\rho \vec{V} \cdot d\vec{A} = -\iint_{A_1} u|\rho V_1\, dA| = -392 \text{ lbf}$$

$$R_x = -p_a A - 392 \text{ lbf} \qquad K_x = -R_x = p_a A + 392 \text{ lbf}$$

To determine the net force on the plate we need a free-body diagram of the plate.

$$\Sigma F_x = K_x - p_a A \qquad \text{\{since atmospheric pressure acts on the back of the plate\}}$$

$$\Sigma F_x = p_a A + 392 \text{ lbf} - p_a A = 392 \text{ lbf}$$

{ This problem illustrates the application of the momentum equation to an inertial control volume, with emphasis on choosing a suitable control volume. }

Example 4.5

A metal container 2 ft high with an inside cross-section area of 1 ft^2 weighs 5 lbf when empty. The container is placed on a scale and water flows in through an opening in the top and out through the two equal area openings in the sides as shown in the diagram. Under steady flow conditions the height of the water in the tank is 1.9 ft. The pressure is atmospheric across all openings. Determine the reading on the scale.

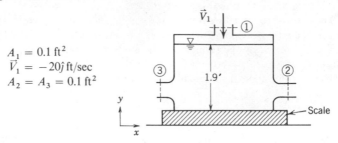

$A_1 = 0.1 \text{ ft}^2$
$\vec{V}_1 = -20\hat{j} \text{ ft/sec}$
$A_2 = A_3 = 0.1 \text{ ft}^2$

EXAMPLE PROBLEM 4.5

GIVEN:

Metal container of height 2 ft and cross-section area, $A = 1 \text{ ft}^2$, weighs 5 lbf when empty. Container is placed on scale. Under steady flow conditions water depth is 1.9 ft.
   Water enters vertically at section ① and leaves horizontally through sections ② and ③.

$A_1 = 0.1 \text{ ft}^2$

$\vec{V}_1 = -20\hat{j} \text{ ft/sec}$

$A_2 = A_3 = 0.1 \text{ ft}^2$

The pressure is atmospheric at all openings.

FIND:

The reading on the scale.

SOLUTION:

Choose a control volume as shown. R_y is the force of the scale on the control volume and is assumed positive.

Basic equations:

$$\vec{F}_S + \vec{F}_B = \overset{= \, 0 \,(\text{steady flow})}{\cancel{\frac{\partial}{\partial t} \iiint_{CV} \vec{V}\rho \, d\forall}} + \iint_{CS} \vec{V}\rho \vec{V} \cdot d\vec{A}$$

$$\overset{= \, 0 \,(\text{sf})}{\cancel{\frac{\partial}{\partial t} \iiint_{CV} \rho \, d\forall}} + \iint_{CS} \rho \vec{V} \cdot d\vec{A} = 0$$

136   4/basic equations in integral form for a control volume

EXAMPLE PROBLEM 4.5 (continued)

Writing the y component of the momentum equation

$$F_{S_y} + F_{B_y} = \iint_{CS} v\rho \vec{V} \cdot d\vec{A} \tag{1}$$

$F_{S_y} = -p_{atm}A + R_y$ \hspace{2cm} {net force due to atmospheric pressure acts down.}

$F_{B_y} = -W_{tank} - W_{H_2O}$ \hspace{1.5cm} {both body forces act in negative y direction.}

$$W_{H_2O} = \gamma \forall = \gamma Ah$$

$$\iint_{CS} v\rho \vec{V} \cdot d\vec{A} = \iint_{A_1} v\rho \vec{V} \cdot d\vec{A} = -\iint_{A_1} v|\rho_1 V_1 \, dA| \qquad \left\{ \begin{array}{l} \vec{V} \cdot d\vec{A} \text{ is negative at } ① \\ v = 0 \text{ at stations } ② \text{ and } ③ \end{array} \right\}$$

$$= -v_1|\rho_1 V_1 A_1| \qquad \text{assuming uniform properties at } ①.$$

Substituting into equation (1)

$$-p_{atm}A + R_y - W_{tank} - \gamma Ah = -v_1|\rho_1 V_1 A_1|$$

or

$R_y = p_{atm}A + W_{tank} + \gamma Ah - v_1|\rho_1 V_1 A_1|$

Substituting numbers with $v_1 = -20$ ft/sec

$R_y = p_a A + 5 \text{ lbf} + 62.4 \text{ lbf/ft}^3 \times 1 \text{ ft}^2 \times 1.9 \text{ ft}$

$\qquad -(-20 \text{ ft/sec})|1.94 \text{ slug/ft}^3 \times (-20 \text{ ft/sec}) \times 0.1 \text{ ft}^2|$

$\qquad = p_a A + 5 \text{ lbf} + 118.5 \text{ lbf} + 77.6 \text{ lbf}$

$R_y = p_a A + 201.1 \text{ lbf}$ \hspace{3cm} {force of scale on CV is upward.}

The force of the control volume on the scale is $K_y = -R_y = -p_a A - 201.1$ lbf.
  Since atmospheric pressure acts on the scale, the net force on the scale is -201.1 lbf.
  The minus sign indicates that the force on the scale is downward in our coordinate
system.
∴ The scale reading is 201 lbf.

$\left\{ \begin{array}{l} \text{This problem illustrates the application of the momentum equation} \\ \text{to an inertial control volume wherein body forces are included.} \end{array} \right\}$

Example 4.6

Water in an open channel flows under a sluice gate as shown in the sketch. The
flow is incompressible, and is uniform at sections ① and ②. Hydrostatic pressure
distributions at sections ① and ② may be assumed because the flow streamlines

are essentially straight there. Determine the magnitude and direction of the force exerted on the gate by the flow.

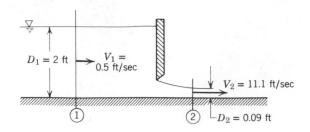

EXAMPLE PROBLEM 4.6

GIVEN:

Flow under sluice gate. Width = w.

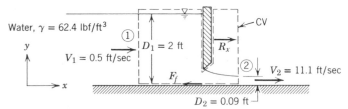

FIND:

Force exerted (per unit width) on the gate.

SOLUTION:

Choose CV and coordinate system shown for analysis. Apply x component of momentum equation.

Basic equation:

$$F_{S_x} + \overset{=\,0(2)}{\cancel{F_{B_x}}} = \overset{=\,0(3)}{\cancel{\frac{\partial}{\partial t} \iiint_{CV} u\rho\, d\forall}} + \iint_{CS} u\rho \vec{V} \cdot d\vec{A}$$

Assumptions: (1) F_f negligible (i.e. neglect friction on channel bottom)
(2) $F_{B_x} = 0$
(3) Steady flow
(4) Incompressible flow
(5) Uniform flow at each section
(6) Hydrostatic pressure distribution at ① and ②

Then

$$F_{S_x} = u_1\{-|\rho V_1 w D_1|\} + u_2\{|\rho V_2 w D_2|\}$$

The surface forces acting on the CV are due to pressure and the unknown force, R_x.

From assumption (6),

$$\frac{dp}{dy} = -\gamma; \qquad p = p_0 + \gamma(y_0 - y) = p_{atm} + \gamma(D - y)$$

Evaluating F_{S_x},

$$F_{S_x} = \int_0^{D_1} p_1 \, dA_1 - \int_0^{D_2} p_2 \, dA_2 - p_{atm}(D_1 - D_2)w + R_x$$

$$= \int_0^{D_1} [p_{atm} + \gamma(D_1 - y)]w \, dy - \int_0^{D_2} [p_{atm} + \gamma(D_2 - y)]w \, dy$$

$$- p_{atm}(D_1 - D_2)w + R_x$$

$$= p_{atm}D_1 w + \frac{\gamma D_1^2}{2}w - p_{atm}D_2 w - \frac{\gamma D_2^2}{2}w - p_{atm}D_1 w + p_{atm}D_2 w + R_x$$

or

$$F_{S_x} = R_x + \frac{\gamma w}{2}(D_1^2 - D_2^2)$$

Substituting in the momentum equation, with $u_1 = V_1$ and $u_2 = V_2$, then

$$R_x + \frac{\gamma w}{2}(D_1^2 - D_2^2) = V_1(-\rho V_1 w D_1) + V_2(\rho V_2 w D_2)$$

or

$$R_x = \rho w(V_2^2 D_2 - V_1^2 D_1) - \frac{\gamma w}{2}(D_1^2 - D_2^2)$$

and

$$\frac{R_x}{w} = \rho(V_2^2 D_2 - V_1^2 D_1) - \frac{\gamma}{2}(D_1^2 - D_2^2)$$

$$= 1.94 \frac{\text{slug}}{\text{ft}^3}[(11.1)^2(0.09) - (0.5)^2(2)]\frac{\text{ft}^2}{\text{sec}^2}\text{ft}$$

$$- \frac{1}{2}\frac{62.4 \text{ lbf}}{\text{ft}^3}[(2)^2 - (0.09)^2] \text{ ft}^2$$

$$= (1.94)(11.1 - 0.5) \text{ lbf/ft} - (31.2)(4 - 0.01) \text{ lbf/ft}$$

$$\frac{R_x}{w} = -104 \text{ lbf/ft}$$

R_x is the unknown external force acting *on* the control volume. It is applied to the CV by the gate. Therefore the force from all fluids *on* the gate is K_x, where $K_x = -R_x$. Thus

$$\frac{K_x}{w} = -\frac{R_x}{w} = 104\,\text{lbf/ft}$$

K_x/w

{Thus the resulting force *on* the gate is 104 lbf/ft, applied to the right.}

Example 4.7

Water is flowing steadily through the 90° reducing elbow shown in the diagram. At the inlet to the elbow the pressure is 30 psia and the cross-section area is 0.1 ft². At the outlet the cross-section area is 0.025 ft² and the velocity is 48 ft/sec. The pressure at the outlet is atmospheric. Determine the force required to hold the elbow in place.

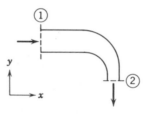

EXAMPLE PROBLEM 4.7

GIVEN:

Steady flow of water through 90° reducing elbow

$p_1 = 30\,\text{psia}$ $A_1 = 0.1\,\text{ft}^2$

$\vec{V}_2 = -48\hat{j}\,\text{ft/sec}$ $A_2 = 0.025\,\text{ft}^2$

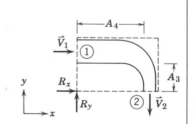

FIND:

Force required to hold elbow in place.

SOLUTION:

Choose control volume as shown by the dashed line.

R_x, R_y are components of force required to hold the elbow in place. (They are assumed positive and are forces acting on CV.)

A_3 is area of vertical sides of CV excluding A_1, that is, $A_{\text{vertical sides}} = A_1 + A_3$.

A_4 is area of horizontal sides of CV excluding A_2, that is, $A_{\text{horizontal sides}} = A_2 + A_4$.

Basic equations:

$$\vec{F} = \vec{F}_S + \vec{F}_B = \overset{0\,(\text{steady flow})}{\cancel{\frac{\partial}{\partial t} \iiint_{CV} \vec{V}\rho\, d\forall}} + \iint_{CS} \vec{V}\rho\vec{V}\cdot d\vec{A}$$

$$0 = \overset{0\,(\text{sf})}{\cancel{\frac{\partial}{\partial t} \iiint_{CV} \rho\, d\forall}} + \iint_{CS} \rho\vec{V}\cdot d\vec{A}$$

Assumptions: (1) Uniform flow over each section.
(2) Atmospheric pressure, $p_a = 14.7$ psia, $p_2 = 14.7$ psia.
(3) Incompressible flow.

Writing the x component of the momentum equation

$$F_{S_x} = \iint_{CS} u\rho\vec{V}\cdot d\vec{A} = \iint_{A_1} u\rho\vec{V}\cdot d\vec{A} \qquad \{F_{B_x} = 0 \text{ and } u_2 = 0\}$$

$$p_1 A_1 + p_a A_3 - p_a(A_1 + A_3) + R_x = \iint_{A_1} u\rho\vec{V}\cdot d\vec{A} \qquad \left\{\begin{array}{l} \text{Pressure over right side is } p_a. \\ \text{Pressure over left side is } p_1 \\ \text{on } A_1 \text{ and } p_a \text{ on } A_3. \end{array}\right\}$$

$$(p_1 - p_a)A_1 + R_x = -\iint_{A_1} u|\rho V_1\, dA| \qquad \{\vec{V}\cdot d\vec{A} \text{ is negative at } A_1\}$$

$$R_x = -p_{1_g}A_1 - u_1|\rho V_1 A_1| \qquad \{p_1 - p_a = p_{1_{\text{gage}}}\}$$

To find V_1 use continuity equation:

$$\iint_{CS} \rho\vec{V}\cdot d\vec{A} = 0 = \iint_{A_1} \rho\vec{V}\cdot d\vec{A} + \iint_{A_2} \rho\vec{V}\cdot d\vec{A}$$

$$\therefore \quad 0 = -\iint_{A_1} |\rho V dA| + \iint_{A_2} |\rho V dA| = -|\rho V_1 A_1| + |\rho V_2 A_2|$$

and

$$|V_1| = |V_2|\frac{A_2}{A_1} = 48 \text{ ft/sec} \times \frac{0.025}{0.10} = 12 \text{ ft/sec} \qquad \therefore \quad \vec{V}_1 = 12\hat{\imath} \text{ ft/sec}$$

$$R_x = -p_{1_g}A_1 - u_1|\rho V_1 A_1|$$

$$= -15.3 \text{ lbf/in.}^2 \times 0.1 \text{ ft}^2 \times 144 \text{ in.}^2/\text{ft}^2$$

$$-12 \text{ ft/sec} |1.94 \text{ slug/ft}^3 \times 12 \text{ ft/sec} \times 0.1 \text{ ft}^2 \times \text{lbf-sec}^2/\text{slug-ft}|$$

$$R_x = -220 \text{ lbf} - 28 \text{ lbf} = -248 \text{ lbf} \qquad\qquad \{R_x \text{ to hold elbow acts to left}\} \quad R_x$$

Writing the y component of the momentum equation

$$F_{S_y} + F_{B_y} = \iint_{\text{CS}} v\rho \vec{V} \cdot d\vec{A} = \iint_{A_2} v\rho \vec{V} \cdot d\vec{A} \qquad\qquad \{v_1 = 0\}$$

$$p_a A_4 + p_a A_2 - p_a A_4 - p_a A_2 + F_{B_y} + R_y = \iint_{A_2} v|\rho V \, dA| \quad \left\{ \begin{array}{l} \text{Pressure is } p_a \text{ over} \\ \text{top and bottom of CV.} \\ \vec{V} \cdot d\vec{A} \text{ is positive at } \textcircled{2}. \end{array} \right\}$$

$$F_{B_y} + R_y = v_2|\rho V_2 A_2|$$

$$R_y = -F_{B_y} + v_2|\rho V_2 A_2| \quad \left\{ \begin{array}{l} \text{Since we do not know the volume of} \\ \text{the elbow, we cannot evaluate } F_{B_y}. \end{array} \right\}$$

Putting in numbers, recognizing $\vec{V}_2 = -48\hat{\jmath}$ ft/sec, so $v_2 = -48$ ft/sec

$$R_y = -F_{B_y} + (-48 \text{ ft/sec}) \times |1.94 \text{ slug/ft}^3 \times (-48 \text{ ft/sec}) \times 0.025 \text{ ft}^2 \times \text{lbf-sec}^2/\text{slug-ft}|$$

$$R_y = -F_{B_y} - 112 \text{ lbf}$$

Neglecting F_{B_y},

$$R_y = -112 \text{ lbf} \qquad\qquad\qquad \{R_y \text{ to hold elbow acts down.}\} \quad R_y$$

$$\left\{ \begin{array}{l} \text{This problem illustrates the application of the momentum equation to an inertial} \\ \text{control volume in which the pressure is not atmospheric across the entire control surface.} \end{array} \right\}$$

Example 4.8

A horizontal conveyor belt moving at 3 ft/sec receives sand from a hopper. The sand falls vertically from the hopper to the belt at speed of 5 ft/sec and a flowrate of 500 lbm/sec (the density of sand is approximately 2700 lbm/cubic yard). The conveyor belt is initially empty, but begins to fill with sand. If friction in the drive system and rollers is negligible, find the tension required to pull the belt.

GIVEN:

Conveyor and hopper shown
in sketch.

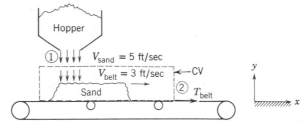

FIND:

T_{belt}.

SOLUTION:

Use control volume and coordinates shown.
Apply x component of momentum equation.

Basic equations:

$$F_{S_x} + \cancelto{0(2)}{F_{B_x}} = \frac{\partial}{\partial t}\iiint_{CV} u\rho\, d\forall + \iint_{CS} u\rho\vec{V}\cdot d\vec{A} \; : \quad 0 = \frac{\partial}{\partial t}\iiint_{CV}\rho\, d\forall + \iint_{CS}\rho\vec{V}\cdot d\vec{A}$$

Assumptions: (1) $F_{S_x} = T_{belt} = T$
(2) $F_{B_x} = 0$
(3) Uniform flow at ①
(4) All sand on belt moves with $V_{belt} = V_b$

Then

$$T = \frac{\partial}{\partial t}\iiint_{CV} u\rho\, d\forall + u_1\{-|\rho V_1 A_1|\} + u_2\{|\rho V_2 A_2|\}$$

or with $u = V_b$

$$T = \frac{\partial}{\partial t}[V_b M_s] + \cancelto{0}{u_1}\{-\dot{m}_s\} + 0 = M_s\cancelto{0}{\frac{\partial V_b}{\partial t}} + V_b\frac{\partial M_s}{\partial t} \qquad \{|\rho V_1 A_1| = \dot{m}_s\}$$

where M_s is the mass of sand on the belt. From the continuity equation,

$$\frac{\partial}{\partial t}\iiint_{CV}\rho\, d\forall = \frac{\partial}{\partial t}M_s = -\iint_{CS}\rho\vec{V}\cdot d\vec{A} = \dot{m}_s = 500\frac{lbm}{sec}$$

Then

$$T = V_b\dot{m}_s = 3\frac{ft}{sec} \times \frac{500\ lbm}{sec} \times \frac{slug}{32.2\ lbm} \times \frac{lbf\text{-}sec^2}{slug\text{-}ft}$$

$$T = 46.6\ lbf \qquad\qquad\qquad\qquad\qquad T$$

{ The purpose of this problem is to illustrate an application of the momentum equation
to a problem in which the rate of change of momentum within the control volume is not
equal to zero. }

4-4.2 CONTROL VOLUME MOVING WITH CONSTANT VELOCITY

In the preceding problems, which illustrate the application of the momentum equation to inertial control volumes, we have considered only stationary control volumes. It should be emphasized that a control volume (fixed relative to reference frame xyz) moving with a constant velocity, $\vec{V}_{rf}$, relative to a fixed (inertial) reference frame XYZ, is also inertial, since it has no acceleration with respect to XYZ.

Equation 4.12, relating system derivatives to the control volume formulation, is valid for any motion of the coordinate system xyz (fixed to the control volume) provided:

1. All velocities are measured relative to the control volume.
2. All time derivatives are measured relative to the control volume.

To emphasize this point, we rewrite Eq. 4.12 as

$$\left. \frac{dN}{dt} \right|_{\text{system}} = \frac{\partial}{\partial t} \iiint_{CV} \eta \rho \, d\forall + \iint_{CS} \eta \rho \vec{V}_{xyz} \cdot d\vec{A} \tag{4.20}$$

Since all time derivatives must be measured relative to the control volume, in using this equation to obtain the momentum equation for an inertial control volume from the system formulation, we really set

$$N = \vec{P}_{xyz} \quad \text{and} \quad \eta = \vec{V}_{xyz}$$

to obtain the equation

$$\vec{F} = \vec{F}_S + \vec{F}_B = \frac{\partial}{\partial t} \iiint_{CV} \vec{V}_{xyz} \rho \, d\forall + \iint_{CS} \vec{V}_{xyz} \rho \vec{V}_{xyz} \cdot d\vec{A} \tag{4.21}$$

Equation 4.21 is the formulation of Newton's second law applied to any inertial control volume (stationary or moving with a constant velocity). It is identical to Eq. 4.18 except that we have included the subscript xyz to emphasize that quantities must be measured relative to the control volume. The momentum equation is applied to an inertial control volume moving with a constant velocity in Example 4.9.

Example 4.9

The sketch shows a vane with a turning angle of 60°. The vane moves with a constant velocity $U = 20$ ft/sec and receives a jet of water, leaving a stationary nozzle with velocity $\vec{V}_j = 100\,\hat{i}$ ft/sec. The nozzle has an exit area of 0.03 ft². Determine the total force on the moving vane.

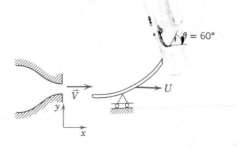

EXAMPLE PROBLEM 4.9

GIVEN:

Vane having a turning angle $\theta = 60°$, moves with a constant velocity $\vec{U} = 20\hat{\imath}$ ft/sec. Water flowing from a constant area nozzle, $A = 0.03$ ft^2, with a velocity $\vec{V}_j = 100\hat{\imath}$ ft/sec flows over the vane as shown.

FIND:

The force on the vane.

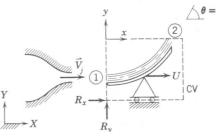

SOLUTION:

Select a control volume moving with the vane at velocity U, as shown by the dashed line. R_x, R_y are the components of the force required to maintain the velocity of the control volume at $20\hat{\imath}$ ft/sec.

The control volume is inertial since it is not accelerating (U = constant). Remember all velocities must be measured relative to the control volume in applying the basic equations.

Basic equations: $\vec{F}_S + \vec{F}_B = \dfrac{\partial}{\partial t} \displaystyle\iiint_{CV} \vec{V}_{xyz}\rho \, d\forall + \iint_{CS} \vec{V}_{xyz}\rho\vec{V}_{xyz} \cdot d\vec{A}$

$0 = \dfrac{\partial}{\partial t} \displaystyle\iiint_{CV} \rho \, d\forall + \iint_{CS} \rho\vec{V}_{xyz} \cdot d\vec{A}$

Assumptions: (1) Flow is steady relative to vane
(2) Velocity magnitude is constant along the vane, that is, $|\vec{V}_1| = |\vec{V}_2|$ $= V_j - U$
(3) Properties are uniform at sections ① and ②
(4) $F_{B_x} = 0$

(5) $\rho = $ constant $= 1.94 \dfrac{\text{slug}}{\text{ft}^3}$

---

momentum equation for inertial control volume

Writing the x component of the momentum equation

$$F_{S_x} + \cancel{F_{B_x}}^{= 0(4)} = \cancel{\frac{\partial}{\partial t} \int\int\int_{CV} u\rho \, d\forall}^{= 0(1)} + \int\int_{CS} u\rho \vec{V} \cdot d\vec{A}$$

$$R_x = \int\int_{CS} u\rho \vec{V} \cdot d\vec{A} = \int\int_{A_1} u\rho \vec{V} \cdot d\vec{A} + \int\int_{A_2} u\rho \vec{V} \cdot d\vec{A} \qquad \left\{ \begin{array}{l} \text{No net pressure force:} \\ \text{pressure} = p_a \text{ on all sides.} \end{array} \right\}$$

$$= -\int\int_{A_1} u|\rho V_1 \, dA| + \int\int_{A_2} u|\rho V_2 \, dA|$$

$$R_x = -u_1|\rho V_1 A_1| + u_2|\rho V_2 A_2| \qquad \left\{ \begin{array}{l} \text{Remember all velocities are relative} \\ \text{to the control volume.} \end{array} \right\}$$

From the continuity equation

$$0 = \int\int_{CS} \rho \vec{V} \cdot d\vec{A} = \int\int_{A_1} \rho \vec{V} \cdot d\vec{A} + \int\int_{A_2} \rho \vec{V} \cdot d\vec{A} = -\int\int_{A_1} |\rho V \, dA| + \int\int_{A_2} |\rho V \, dA|$$

$$\therefore \quad -|\rho V_1 A_1| + |\rho V_2 A_2| = 0 \quad \text{and} \quad |\rho V_1 A_1| = |\rho V_2 A_2|$$

$$\therefore \quad R_x = -u_1|\rho V_1 A_1| + u_2|\rho V_1 A_1| = [u_2 - u_1]|\rho V_1 A_1|$$

$$= [(V_j - U)\cos\theta - (V_j - U)]|\rho V_1 A_1| = (V_j - U)(\cos\theta - 1)|\rho(V_j - U)A_1|$$

$$R_x = (100 - 20) \, \text{ft/sec} \times (-0.5)|1.94 \, \text{slug/ft}^3 \times (100 - 20) \, \text{ft/sec} \times 0.03 \, \text{ft}^2| = -186 \, \text{lbf}$$

R_x is force required to maintain the vane velocity of U.

$$\therefore \quad \text{Force of water on vane } K_x = -R_x = 186 \, \text{lbf} \qquad\qquad \{\text{to right}\} \qquad\qquad K_x$$

Writing the y component of the momentum equation,

$$F_{S_y} + \cancel{F_{B_y}}^{\simeq 0 \text{ (neglect weight)}} = \cancel{\frac{\partial}{\partial t} \int\int\int_{CV} v\rho \, d\forall}^{= 0(1)} + \int\int_{CS} v\rho \vec{V} \cdot d\vec{A}$$

$$\therefore \quad R_y = \int\int_{CS} v\rho \vec{V} \cdot d\vec{A} = \int\int_{A_2} v\rho \vec{V} \cdot d\vec{A} \qquad\qquad \{v_1 = 0\}$$

$$= \int\int_{A_2} v|\rho V \, dA| = v_2|\rho V_2 A_2| = v_2|\rho V_1 A_1| \qquad\qquad \{\text{Recall } |\rho V_2 A_2| = |\rho V_1 A_1|\}$$

$$= (V_j - U)\sin\theta|\rho(V_j - U)A_1|$$

$$= (100 - 20) \, \text{ft/sec} \times (0.866)|1.94 \, \text{slug/ft}^3 \times (100 - 20) \, \text{ft/sec} \times 0.03 \, \text{ft}^2|$$

$$R_y = 323 \, \text{lbf} \quad \{\text{upward}\}$$

EXAMPLE PROBLEM 4.9 (continued)

Force of water on vane,

$$K_y = -R_y = -323 \, \text{lbf} \qquad \longleftarrow \qquad \{\text{downward}\} \, K_y$$

$\left\{ \begin{array}{l} \text{This problem illustrates the point that in applying the momentum equation to an inertial} \\ \text{control volume all velocities are measured relative to the control volume.} \end{array} \right\}$

4–5 Momentum Equation for Linearly Accelerating Control Volume

For an inertial control volume, that is, a control volume that is not accelerating relative to a stationary frame of reference, we developed Eq. 4.21 (same as Eq. 4.18) as the control volume formulation of Newton's second law.

$$\vec{F} = \vec{F}_S + \vec{F}_B = \frac{\partial}{\partial t} \iiint_{CV} \vec{V}_{xyz} \rho \, d\forall + \iint_{CS} \vec{V}_{xyz} \rho \vec{V}_{xyz} \cdot d\vec{A} \qquad (4.21)$$

Since we are interested in analyzing control volumes that may accelerate relative to an inertial coordinate system (a rocket must accelerate if it is to get off the ground), it is logical to ask whether Eq. 4.21 can be used for accelerating control volumes. In arriving at an answer we should look at the items we employed in deriving Eq. 4.21. In the derivation of Eq. 4.21 we used the recipe (Eq. 4.12) and the system equation (Eq. 4.2).

In deriving the recipe (Section 4–2.1) the control volume was fixed relative to x, y, z; the flow field, $\vec{V}(x, y, z, t)$, was specified relative to the coordinates x, y, z. No restriction was placed on the motion of the x, y, z reference frame. Therefore, Eq. 4.12 is valid at any instant for any arbitrary motion of the coordinates x, y, z provided that all time derivatives and velocities in the equation are measured relative to the control volume.

4–5.1 CONTROL VOLUME EQUATION

The system equation (Eq. 4.2) which states that

$$\vec{F} = \frac{d\vec{P}}{dt}\bigg)_{\text{system}} \qquad (4.2a)$$

where the linear momentum, $\vec{P}$, of the system is given by

$$\vec{P}_{\text{system}} = \int_{\text{Mass (system)}} \vec{V} \, dm = \int_{\forall \text{(system)}} \vec{V} \rho \, d\forall \qquad (4.2b)$$

is valid only for velocities measured relative to an inertial reference frame. Thus, if we denote the inertial reference frame by *XYZ*, then Newton's second law states that

$$\vec{F} = \frac{d\vec{P}_{XYZ}}{dt}\bigg)_{system} \quad (4.22)$$

Consider a pinball machine on a train. Here the inertial coordinate system *XYZ* can be taken fixed to the ground. The motion of a pinball is described most easily relative to a coordinate system *x, y, z* fixed on the train. Clearly the description of the motion of the ball relative to the train (coordinates *x, y, z*) is different from the description relative to the ground (coordinates *X, Y, Z*) if the train is in motion.

In order to obtain the momentum equation for a linearly accelerating control volume, that is, one which is undergoing translation without rotation relative to an inertial reference frame (*X, Y, Z*), we shall employ Eq. 4.12 and the appropriate system equation, Eq. 4.22. In Eq. 4.12, all time derivatives and velocities are measured relative to the control volume; thus it is necessary to relate $\vec{P}_{XYZ}$ of the system to $\vec{P}_{xyz}$ of the system.

To employ our knowledge from our earlier course in dynamics to advantage, let us write Newton's second law for a system in terms of the acceleration of the system, remembering that the acceleration must be measured relative to an inertial reference frame that we have designated as *X, Y, Z*. We can write

$$\vec{F} = \frac{d\vec{P}_{XYZ}}{dt}\bigg]_{system} = \frac{d}{dt}\int_{Mass\,(system)} \vec{V}_{XYZ}\,dm = \int_{Mass\,(system)} \frac{d\vec{V}_{XYZ}}{dt}\,dm$$

$$\vec{F} = \int_{Mass\,(system)} \vec{a}_{XYZ}\,dm \quad (4.23)$$

The only problem now is to obtain a suitable expression for $\vec{a}_{XYZ}$, for the special case in which the coordinates *x, y, z* are undergoing pure translation, without rotation, relative to the inertial frame *X, Y, Z*.

Since the motion of *x, y, z* is pure translation, without rotation, relative to an inertial reference frame *X, Y, Z*, then

$$\vec{a}_{XYZ} = \vec{a}_{xyz} + \vec{a}_{rf} \quad (4.24)$$

where

$\vec{a}_{XYZ}$ is the linear acceleration of the system relative to the inertial reference frame *X, Y, Z,*

$\vec{a}_{xyz}$ is the linear acceleration of the system relative to the noninertial reference frame *x, y, z,* and

$\vec{a}_{rf}$ is the linear acceleration of the noninertial reference frame *x, y, z* relative to the inertial frame *X, Y, Z.*

For this case we write the system equation as

$$\vec{F} = \int_{\text{Mass (system)}} \vec{a}_{XYZ}\, dm = \int_{\text{Mass (system)}} (\vec{a}_{xyz} + \vec{a}_{rf})\, dm$$

Alternately,

$$\vec{F} - \int_{\text{Mass (system)}} \vec{a}_{rf}\, dm = \int_{\text{Mass (system)}} \vec{a}_{xyz}\, dm$$

Since

$$\vec{a}_{xyz} = \frac{d\vec{V}_{xyz}}{dt}$$

then

$$\vec{F} - \int_{\text{Mass (system)}} \vec{a}_{rf}\, dm = \int_{\text{Mass (system)}} \frac{d\vec{V}_{xyz}}{dt}\, dm = \frac{d}{dt}\int_{\text{Mass (system)}} \vec{V}_{xyz}\, dm = \frac{d\vec{P}_{xyz}}{dt}\bigg)_{\text{system}}$$

Since $dm = \rho\, d\forall$, the system equation becomes

$$\vec{F} - \iiint_{\forall\,(\text{system})} \vec{a}_{rf}\rho\, d\forall = \frac{d\vec{P}_{xyz}}{dt}\bigg)_{\text{system}} \tag{4.25a}$$

where the linear momentum, $\vec{P}_{xyz}$, of the system is given by

$$\vec{P}_{xyz_{\text{system}}} = \int_{\text{Mass (system)}} \vec{V}_{xyz}\, dm = \int_{\forall\,(\text{system})} \vec{V}_{xyz}\rho\, d\forall \tag{4.25b}$$

and the force, $\vec{F}$, includes all surface and body forces acting on the system.

The recipe for relating the system formulation to the formulation for a moving control volume is given by Eq. 4.20

$$\frac{dN}{dt}\bigg)_{\text{system}} = \frac{\partial}{\partial t}\iiint_{\text{CV}} \eta\rho\, d\forall + \iint_{\text{CS}} \eta\rho\vec{V}_{xyz}\cdot d\vec{A} \tag{4.20}$$

where

$$N_{\text{system}} = \int_{\text{Mass (system)}} \eta\, dm = \iiint_{\forall\,(\text{system})} \eta\rho\, d\forall \tag{4.6}$$

Then in using Eq. 4.6 to derive the control volume formulation of Newton's second law, we set

$$N = \vec{P}_{xyz} \quad \text{and} \quad \eta = \vec{V}_{xyz}$$

From Eq. 4.20, with this substitution, we obtain

$$\left. \frac{d\vec{P}_{xyz}}{dt} \right)_{\text{system}} = \frac{\partial}{\partial t} \iiint_{CV} \vec{V}_{xyz}\rho \, d\forall + \iint_{CS} \vec{V}_{xyz}\rho \vec{V}_{xyz} \cdot d\vec{A} \qquad (4.26)$$

From the system equation (Eq. 4.25a),

$$\left. \frac{d\vec{P}_{xyz}}{dt} \right)_{\text{system}} = \vec{F}_{\text{on system}} - \iiint_{\forall\,(\text{system})} \vec{a}_{rf}\rho \, d\forall \qquad (4.25a)$$

Since the system and the control volume coincided at time t, then,

$$\vec{F}_{\text{on system}} - \iiint_{\forall\,(\text{system})} \vec{a}_{rf}\rho \, d\forall = \vec{F}_{\text{on CV}} - \iiint_{CV} \vec{a}_{rf}\rho \, d\forall$$

In light of this, Eqs. 4.25a and 4.26 may be combined to yield the control volume formulation of Newton's second law for a control volume translating, without rotation, relative to an inertial reference frame:

$$\vec{F} - \iiint_{CV} \vec{a}_{rf}\rho \, d\forall = \frac{\partial}{\partial t} \iiint_{CV} \vec{V}_{xyz}\rho \, d\forall + \iint_{CS} \vec{V}_{xyz}\rho \vec{V}_{xyz} \cdot d\vec{A} \qquad (4.27)$$

Since, $\vec{F} = \vec{F}_S + \vec{F}_B$ and the body force, $\vec{F}_B$, can be written

$$\vec{F}_B = \iiint_{CV} \vec{B}\rho \, d\forall$$

then Eq. 4.27 becomes

$$\vec{F}_S + \iiint_{CV} \vec{B}\rho \, d\forall - \iiint_{CV} \vec{a}_{rf}\rho \, d\forall = \frac{\partial}{\partial t} \iiint_{CV} \vec{V}_{xyz}\rho \, d\forall + \iint_{CS} \vec{V}_{xyz}\rho \vec{V}_{xyz} \cdot d\vec{A}$$

$$(4.28)$$

In comparing the momentum equation for a linearly accelerating control volume, Eq. 4.28, to that for a nonaccelerating control volume, Eq. 4.21, we see that the only difference is the presence of one additional term in Eq. 4.28. When the control volume is not accelerating relative to an inertial reference frame (X, Y, Z), that is, when $\vec{a}_{rf} = 0$, then Eq. 4.28 reduces to Eq. 4.21.

The precautions concerning the use of Eq. 4.21 also apply to the application of Eq. 4.28. Before attempting to apply either equation, one should draw the boundaries of the control volume and label appropriate coordinate directions. For an accelerating control volume, one must label a coordinate system (x, y, z) on the control volume and an inertial reference frame (X, Y, Z).

4/basic equations in integral form for a control volume

In Eq. 4.28, $\vec{F}_S$ represents all surface forces acting on the control volume. One should recognize that both the remaining terms on the lefthand side of the equation may be functions of time, since the mass within the control volume may vary with time. Furthermore, the acceleration, $\vec{a}_{rf}$, of the reference frame x, y, z relative to an inertial frame in general will be a function of time.

All velocities in Eq. 4.28 are measured relative to the control volume. The momentum flux, $\vec{V}_{xyz}\rho\vec{V}_{xyz} \cdot d\vec{A}$, through an element of the control surface area, $d\vec{A}$, is a vector. The sign of the scalar product $\rho\vec{V}_{xyz} \cdot d\vec{A}$ depends on the direction of the velocity vector, $\vec{V}_{xyz}$, relative to the area vector, $d\vec{A}$. The sign of the vector velocity, $\vec{V}_{xyz}$, is dependent on the choice of the coordinate system.

The momentum equation is a vector equation. As with all vector equations, it may be written in scalar component equations. The scalar components of Eq. 4.28 are

$$F_{S_x} + \iiint_{CV} B_x\rho\, d\forall - \iiint_{CV} a_{rf_x}\rho\, d\forall = \frac{\partial}{\partial t}\iiint_{CV} u_{xyz}\rho\, d\forall + \iint_{CS} u_{xyz}\rho\vec{V}_{xyz}\cdot d\vec{A}$$

(4.29a)

$$F_{S_y} + \iiint_{CV} B_y\rho\, d\forall - \iiint_{CV} a_{rf_y}\rho\, d\forall = \frac{\partial}{\partial t}\iiint_{CV} v_{xyz}\rho\, d\forall + \iint_{CS} v_{xyz}\rho\vec{V}_{xyz}\cdot d\vec{A}$$

(4.29b)

$$F_{S_z} + \iiint_{CV} B_z\rho\, d\forall - \iiint_{CV} a_{rf_z}\rho\, d\forall = \frac{\partial}{\partial t}\iiint_{CV} w_{xyz}\rho\, d\forall + \iint_{CS} w_{xyz}\rho\vec{V}_{xyz}\cdot d\vec{A}$$

(4.29c)

Example 4.10

A vane with turning angle $\theta = 60°$ is attached to a cart. The cart and vane weigh 150 lbf, and roll on a level track. Friction and air resistance may be neglected. The vane receives a jet of water, which leaves a stationary nozzle horizontally at 100 ft/sec. The nozzle exit area is 0.03 ft². Determine the velocity of the cart as a function of time and plot the results.

EXAMPLE PROBLEM 4.10

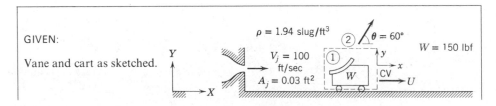

GIVEN:

Vane and cart as sketched.

$\rho = 1.94$ slug/ft³
$\theta = 60°$
$W = 150$ lbf
$V_j = 100$ ft/sec
$A_j = 0.03$ ft²

FIND:

(a) $U(t)$
(b) Plot results

SOLUTION:

Choose the control volume and coordinate systems shown for the analysis. Note that X, Y is a fixed frame, while frame x, y moves with the cart. Apply the x component of the momentum equation.

$$= 0(1) \quad = 0(2) \qquad \simeq 0(3)$$

$$\text{Basic equation}: \cancel{F_{S_x}} + \cancel{F_{B_x}} - \iiint_{CV} a_{rf_x}\rho \, d\forall = \frac{\partial}{\partial t}\iiint_{CV} u_{xyz}\rho \, d\forall + \iint_{CS} u_{xyz}\rho \vec{V}_{xyz} \cdot d\vec{A}$$

Assumptions: (1) $F_{S_x} = 0$, since no resistance is present
(2) $F_{B_x} = 0$
(3) Neglect rate of change of u_{xyz} for water in contact with vane:

$$\frac{\partial}{\partial t}\iiint_{CV} u_{xyz}\rho \, d\forall \simeq 0$$

(4) Uniform flow at sections ① and ②, $\rho = 1.94 \text{ slug/ft}^3$
(5) $|\vec{V}_{xyz_1}| = |\vec{V}_{xyz_2}|$, that is, water stream is not slowed by friction on vane
(6) $A_2 = A_1 = A_j$

Then

$$-\iiint_{CV} a_{rf_x}\rho \, d\forall = u_{xyz_1}\{-|\rho V_{xyz_1} A_1|\} + u_{xyz_2}\{|\rho V_{xyz_2} A_2|\}$$

where all quantities are measured relative to the xyz frame. Dropping subscripts rf and xyz,

$$-\iiint_{CV} a_x\rho \, d\forall = u_1\{-|\rho V_1 A_1|\} + u_2\{|\rho V_2 A_2|\} \tag{1}$$

Let us evaluate these terms separately:

$$-\iiint_{CV} a_x\rho \, d\forall = -a_x M_{CV} = -a_x\frac{W}{g} = -\frac{dU}{dt}\frac{W}{g}$$

$$u_1\{-|\rho V_1 A_1|\} = (V_j - U)\{-|\rho(V_j - U)A_j|\} = -\rho(V_j - U)^2 A_j$$

$$u_2\{|\rho V_2 A_2|\} = (V_j - U)\cos\theta\{|\rho(V_j - U)A_j|\} \qquad \left\{\begin{matrix}\text{From assumptions (4)}\\ \text{and (5), } A_2 = A_1 = A_j\end{matrix}\right\}$$

$$= \rho(V_j - U)^2 A_j\cos\theta$$

Substituting in Eq. 1,

$$-\frac{W}{g}\frac{dU}{dt} = -\rho(V_j - U)^2 A_j + \rho(V_j - U)^2 A_j \cos\theta$$

$$= (\cos\theta - 1)\rho(V_j - U)^2 A_j$$

Separating variables

$$\frac{dU}{(V_j - U)^2} = \frac{(1 - \cos\theta)\rho A_j g}{W}dt = b\,dt$$

Integrating between limits $U = 0$ at $t = 0$, and $U = U$ at $t = t$

$$\int_0^U \frac{dU}{(V_j - U)^2} = \frac{1}{(V_j - U)}\bigg]_0^U = \int_0^t b\,dt = bt$$

or

$$\frac{1}{(V_j - U)} - \frac{1}{V_j} = bt$$

which becomes

$$\frac{U}{V_j(V_j - U)} = bt$$

Solving for U

$$\frac{U}{V_j} = \frac{V_j bt}{1 + V_j bt}$$

For the given data

$$V_j b = V_j\frac{(1 - \cos\theta)\rho A_j g}{W} = (1 - \tfrac{1}{2}) \times \frac{100\text{ ft}}{\text{sec}} \times \frac{1.94\text{ slug}}{\text{ft}^3} \times 0.03\text{ ft}^2 \times \frac{32.2\text{ ft}}{\text{sec}^2}$$

$$\times \frac{1}{150\text{ lbf}} \times \frac{\text{lbf-sec}^2}{\text{slug-ft}}$$

$$= \frac{(0.5)(1.94)(3)(3.22)10}{(1.5)10^2} = 0.625\text{ sec}^{-1}$$

Thus

$$\frac{U}{V_j} = \frac{0.625t}{1 + 0.625t}(t\text{ in sec}) \qquad\qquad U(t)$$

Plot:

$$\frac{U}{V_j}$$

Plot

momentum equation for linearly accelerating control volume

Example 4.11

A small rocket, with an initial weight of 900 lbf is to be launched vertically. Upon ignition the rocket consumes fuel at the rate of 10 lbm/sec and ejects gas at atmospheric pressure with a velocity of 3220 ft/sec relative to the rocket. Determine

(a) The initial acceleration of the rocket.
(b) The rocket velocity after 10 sec if air resistance is neglected.

EXAMPLE PROBLEM 4.11

GIVEN:

Small rocket accelerates vertically from rest.
Air resistance may be neglected.
Initial weight, $W_0 = 900$ lbf.
Fuel consumption, $\dot{m}_{out} = 10$ lbm/sec.
Exhaust velocity, $V_j = 3220$ ft/sec relative to rocket.

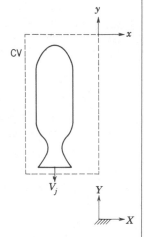

FIND:

(a) Initial acceleration of the rocket.
(b) Rocket velocity after 10 sec.

SOLUTION:

Choose a control volume as shown by dashed lines. Because the control volume is accelerating we need to define an inertial coordinate system XY. The coordinate system xy is attached to the CV.

Basic equation: $\vec{F}_S + \iiint_{CV} \vec{B}\rho \, d\forall - \iiint_{CV} \vec{a}_{rf}\rho \, d\forall = \frac{\partial}{\partial t} \iiint_{CV} \vec{V}_{xyz}\rho \, d\forall + \iint_{CS} \vec{V}_{xyz}\rho\vec{V}_{xyz} \cdot d\vec{A}$

Writing the y component of the equation

$$F_{S_y} + \iiint_{CV} B_y\rho \, d\forall - \iiint_{CV} a_{rf_y}\rho \, d\forall = \frac{\partial}{\partial t} \iiint_{CV} v_{xyz}\rho \, d\forall + \iint_{CS} v_{xyz}\rho\vec{V}_{xyz} \cdot d\vec{A}$$

Assumptions: (1) Atmospheric pressure acts on all surfaces of the CV. Since air resistance is neglected then $F_{S_y} = 0$.
(2) Only body force is that due to gravity and hence $B_y = -g$.
(3) Flow leaving the rocket is uniform and V_j is constant.

Under these assumptions the y component of the momentum equation reduces to

$$-\iiint_{CV} g\rho \, d\forall - \iiint_{CV} a_{rf_y}\rho \, d\forall = \frac{\partial}{\partial t} \iiint_{CV} v_{xyz}\rho \, d\forall + \iint_{CS} v_{xyz}\rho \vec{V}_{xyz} \cdot d\vec{A} \qquad (1)$$

$\underbrace{}_{\textstyle ⒶA}\qquad\qquad \underbrace{}_{\textstyle ⒷB}\qquad\qquad\quad \underbrace{}_{\textstyle ⒸC}\qquad\quad\; \underbrace{}_{\textstyle ⒹD}$

Ⓐ $\;$ Ⓑ $\;$ Ⓒ $\;$ Ⓓ

Let us look at the equation term by term:

$$Ⓐ \; - \iiint_{CV} g\rho \, d\forall = -g \iiint_{CV} \rho \, d\forall = -gM_{CV}, \quad \text{since } g \text{ is constant}$$

We recognize that the mass of the CV will be a function of time because mass is leaving the CV at a rate, $\dot{m}$.

To determine M_{CV} as a function of time, we need to look at the conservation of mass equation

$$\frac{\partial}{\partial t} \iiint_{CV} \rho \, d\forall + \iint_{CS} \rho \vec{V} \cdot d\vec{A} = 0$$

$$\therefore \quad \frac{\partial}{\partial t} \iiint_{CV} \rho \, d\forall = -\iint_{CS} \rho \vec{V} \cdot d\vec{A} = -\iint_{A_j} \rho \vec{V} \cdot d\vec{A} = -\iint_{A_j} |\rho V \, dA| = -|\dot{m}|$$

or

$$\frac{\partial}{\partial t} M_{CV} = -|\dot{m}| \qquad\qquad \{\text{the mass of the CV is decreasing with time}\}$$

Since the mass of the CV is only a function of time, we can write

$$\frac{dM_{CV}}{dt} = -|\dot{m}|$$

To find the mass of the CV at any time, t, we integrate

$$\int_{M_0}^{M} dM_{CV} = -\int_{0}^{t} |\dot{m}| \, dt \qquad \text{where at } t = 0, M_{CV} = M_0, \text{and at } t = t, M_{CV} = M$$

$$M - M_0 = -|\dot{m}|t$$

or

$$M = M_0 - \dot{m}t \qquad\qquad\qquad \left\{\begin{matrix}\text{since } \dot{m} \text{ is positive we can} \\ \text{drop the absolute value sign.}\end{matrix}\right\}$$

Substituting the expression for M into the expression for $-\iiint_{CV} g\rho \, d\forall$, we obtain

$$-\iiint_{CV} g\rho \, d\forall = -gM_{CV} = -g(M_0 - \dot{m}t)$$

$$Ⓑ \; -\iiint_{CV} a_{rf_y}\rho \, d\forall$$

Since all elements of the CV are accelerating with the same a_{rf_y}, that is, a_{rf_y} is not a function of xyz (it is a function of Y), then

$$-\iiint_{CV} a_{rf_y}\rho\,d\forall = -a_{rf_y}\iiint_{CV}\rho\,d\forall = -a_{rf_y}M_{CV} = -a_{rf_y}(M_0 - \dot{m}t)$$

Ⓒ $\quad \dfrac{\partial}{\partial t}\iiint_{CV} v_{xyz}\rho\,d\forall$

is the time rate of change of the y momentum of the fluid in the control volume measured relative to the control volume.

While the y momentum of the fluid inside the CV, measured relative to the CV is a large number, it does not change appreciably with time. To see this we must recognize that:

(1) The unburned fuel and the rocket structure have zero momentum relative to the rocket.

(2) The velocity of the gas at the nozzle exit remains constant with time as does the velocity at various points in the nozzle.

Consequently it is reasonable to assume

$$\frac{\partial}{\partial t}\iiint_{CV} v_{xyz}\rho\,d\forall \approx 0$$

Ⓓ $\quad \displaystyle\iint_{CS} v_{xyz}\rho\vec{V}_{xyz}\cdot d\vec{A} = \iint_{A_j} v_{xyz}\rho\vec{V}_{xyz}\cdot d\vec{A} = \iint_{A_j} v_{xyz}|\rho V_{xyz}\,dA| = v_{xyz}|\rho V_{xyz}A|$

$$= -V_j|\rho V_j A_j| = -V_j|\dot{m}|,$$

since $\vec{V}_j = -V_j\hat{j} = -3220\hat{j}$ ft/sec.

Substituting back into equation (1) we have

$$-g(M_0 - \dot{m}t) - a_{rf_y}(M_0 - \dot{m}t) = -V_j\dot{m}$$

and

$$a_{rf_y} = \frac{V_j\dot{m}}{M_0 - \dot{m}t} - g \tag{2}$$

At time, $t = 0$

$$\left.a_{rf_y}\right|_{t=0} = \frac{V_j\dot{m}}{M_0} - g = \frac{V_j\dot{m}g}{W_0} - g$$

$$= \frac{3220\text{ ft}}{\text{sec}} \times \frac{10\text{ lbm}}{\text{sec}} \times \frac{32.2\text{ ft}}{\text{sec}^2} \times \frac{1}{900\text{ lbf}} \times \frac{\text{slug}}{32.2\text{ lbm}} \times \frac{\text{lbf-sec}^2}{\text{slug-ft}} - \frac{32.2\text{ ft}}{\text{sec}^2}$$

$$a_{rf_y})_{t=0} = 35.8\text{ ft/sec}^2 - 32.2\text{ ft/sec}^2 = 3.6\text{ ft/sec}^2 \quad\underleftarrow{\qquad\qquad} a_{rf_y})_{t=0}$$

EXAMPLE PROBLEM 4.11 (continued)

Now a_{rf_y} is the acceleration of the CV

$$a_{rf_y} = \frac{dV_{ycv}}{dt}$$

$$\therefore \quad \frac{dV_{ycv}}{dt} = \frac{V_j \dot{m}}{M_0 - \dot{m}t} - g$$

Separating variables and integrating

$$\int_0^{V_{ycv}} dV_{ycv} = \int_0^t \frac{V_j \dot{m}}{M_0 - \dot{m}t} - \int_0^t g \, dt$$

$$V_{ycv} = -V_j \ln \left[\frac{M_0 - \dot{m}t}{M_0} \right] - gt$$

At $t = 10$ sec

$$V_{ycv} = -3220 \, \text{ft/sec} \times \ln \left[\frac{800}{900} \right] - 32.2 \, \text{ft/sec}^2 \times 10 \, \text{sec} = 380 - 322 = 58 \, \text{ft/sec} \quad \underleftarrow{V_{ycv})_{t=10}}$$

{ This problem illustrates the application of the momentum equation to a linearly accelerating control volume. }

**4–6 Momentum Equation for Control Volume with Arbitrary Acceleration

In Section 4–5 we formulated the momentum equation for a linearly accelerating control volume. The purpose of this section is to extend the formulation to include rotational and angular acceleration of the control volume in addition to linear acceleration.

First we develop an expression for Newton's second law in an arbitrary, non-inertial coordinate system. Then we apply the recipe to complete the formulation for a control volume.

4–6.1 CONTROL VOLUME EQUATION

Newton's second law for a system moving relative to an inertial coordinate system is given by

$$\vec{F} = \frac{d\vec{P}_{XYZ}}{dt} \bigg)_{\text{system}}$$

** This section may be omitted without loss of continuity in the text material.

    momentum equation for control volume with arbitrary acceleration

But

$$\vec{F} = \frac{d}{dt} \int_{M \,(\text{system})} \vec{V}_{XYZ} \, dm = \int_{M \,(\text{system})} \frac{d\vec{V}_{XYZ}}{dt} \, dm$$

or

$$\vec{F} = \int_{M \,(\text{system})} \vec{a}_{XYZ} \, dm \qquad (4.30)$$

The basic problem is to obtain a formulation of $\vec{a}_{XYZ}$ in terms of $\vec{a}_{xyz}$, the acceleration measured relative to a noninertial coordinate system. For this purpose, consider the noninertial reference frame, xyz, shown in Fig. 4.4. The position vector of a particle is denoted by position vector $\vec{X}$ relative to the inertial reference frame XYZ. The noninertial frame, xyz, is located by position vector $\vec{R}$ relative to the fixed frame. The noninertial frame rotates with angular velocity $\vec{\omega}$.[1] Finally, the particle is located relative to the moving frame by position vector $\vec{r} = \hat{i}x + \hat{j}y + \hat{k}z$. From the geometry of the figure, $\vec{X} = \vec{R} + \vec{r}$.

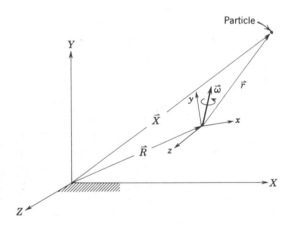

Fig. 4.4 *Inertial (XYZ) and noninertial (xyz) reference frames.*

The velocity of the particle relative to an observer in the XYZ system is

$$\vec{V}_{XYZ} = \frac{d\vec{X}}{dt} = \frac{d\vec{R}}{dt} + \frac{d\vec{r}}{dt} = \vec{V}_{rf} + \frac{d\vec{r}}{dt} \qquad (4.31)$$

[1] Note that any arbitrary motion can be decomposed into a translation plus a rotation.

4/basic equations in integral form for a control volume

However, we must be careful in evaluating $d\vec{r}/dt$, because both the magnitude $|\vec{r}|$ and the orientation of the unit vectors, $\hat{\imath}, \hat{\jmath}$, and $\hat{k}$ are functions of time. Thus

$$\frac{d\vec{r}}{dt} = \frac{d}{dt}(x\hat{\imath} + y\hat{\jmath} + z\hat{k}) = \hat{\imath}\frac{dx}{dt} + x\frac{d\hat{\imath}}{dt} + \hat{\jmath}\frac{dy}{dt} + y\frac{d\hat{\jmath}}{dt} + \hat{k}\frac{dz}{dt} + z\frac{d\hat{k}}{dt} \quad (4.32a)$$

The terms dx/dt, dy/dt, and dz/dt are the velocity components of the particle relative to xyz. Thus

$$\vec{V}_{xyz} = \hat{\imath}\frac{dx}{dt} + \hat{\jmath}\frac{dy}{dt} + \hat{k}\frac{dz}{dt} \quad (4.32b)$$

You probably recall from dynamics that for a rotating coordinate system,

$$\vec{\omega} \times \vec{r} = x\frac{d\hat{\imath}}{dt} + y\frac{d\hat{\jmath}}{dt} + z\frac{d\hat{k}}{dt} \quad (4.32c)$$

This result is also derived in Example Problem 4.12. Combining Eqs. 4.32a, b and c, we obtain

$$\frac{d\vec{r}}{dt} = \vec{V}_{xyz} + \vec{\omega} \times \vec{r} \quad (4.32d)$$

Substituting into Eq. 4.31,

$$\vec{V}_{XYZ} = \vec{V}_{rf} + \vec{V}_{xyz} + \vec{\omega} \times \vec{r} \quad (4.33)$$

The acceleration of the particle relative to an observer in the XYZ system is

$$\vec{a}_{XYZ} = \frac{d\vec{V}_{XYZ}}{dt} = \frac{d\vec{V}_{rf}}{dt} + \frac{d\vec{V}_{xyz}}{dt} + \frac{d}{dt}(\vec{\omega} \times \vec{r})$$

or

$$\vec{a}_{XYZ} = \vec{a}_{rf} + \frac{d\vec{V}_{xyz}}{dt} + \frac{d}{dt}(\vec{\omega} \times \vec{r}) \quad (4.34)$$

Both $\vec{V}_{xyz}$ and $\vec{r}$ are measured relative to xyz, so the same caution observed in developing Eq. 4.32d applies. Thus

$$\frac{d\vec{V}_{xyz}}{dt} = \vec{a}_{xyz} + \vec{\omega} \times \vec{V}_{xyz} \quad (4.35a)$$

and

$$\frac{d}{dt}(\vec{\omega} \times \vec{r}) = \frac{d\vec{\omega}}{dt} \times \vec{r} + \vec{\omega} \times \frac{d\vec{r}}{dt}$$

$$= \dot{\vec{\omega}} \times \vec{r} + \vec{\omega} \times (\vec{V}_{xyz} + \vec{\omega} \times \vec{r})$$

or

$$\frac{d}{dt}(\vec{\omega} \times \vec{r}) = \dot{\vec{\omega}} \times \vec{r} + \vec{\omega} \times \vec{V}_{xyz} + \vec{\omega} \times (\vec{\omega} \times \vec{r}) \quad (4.35b)$$

159   **momentum equation for control volume with arbitrary acceleration**

Substituting Eqs. 4.35a and b into Eq. 4.34, we obtain

$$\vec{a}_{XYZ} = \vec{a}_{rf} + \vec{a}_{xyz} + 2\vec{\omega} \times \vec{V}_{xyz} + \vec{\omega} \times (\vec{\omega} \times \vec{r}) + \dot{\vec{\omega}} \times \vec{r} \qquad (4.36)$$

The physical meaning of each term in Eq. 4.36 is

$\vec{a}_{XYZ}$:   Absolute linear acceleration of a particle relative to a fixed reference frame.

$\vec{a}_{rf}$:   Absolute linear acceleration of a moving reference frame relative to a fixed frame.

$\vec{a}_{xyz}$:   Linear acceleration of a particle *relative* to a moving reference frame (this acceleration would be that seen by an observer on the moving frame).

$2\vec{\omega} \times \vec{V}_{xyz}$:   Coriolis acceleration due to motion of the particle *within* the moving frame.

$\vec{\omega} \times (\vec{\omega} \times \vec{r})$:   Centripetal acceleration due to rotation of the moving frame.

$\dot{\vec{\omega}} \times \vec{r}$:   Tangential acceleration due to angular *acceleration* of the moving reference frame.

Substituting $\vec{a}_{XYZ}$, as given by Eq. 4.36, into Eq. 4.30, we obtain

$$\vec{F}_{system} = \int_{M\,(system)} [\vec{a}_{rf} + \vec{a}_{xyz} + 2\vec{\omega} \times \vec{V}_{xyz} + \vec{\omega} \times (\vec{\omega} \times \vec{r}) + \dot{\vec{\omega}} \times \vec{r}]\, dm$$

or

$$\int_{M\,(system)} \vec{a}_{xyz}\, dm = \vec{F} - \int_{M\,(system)} [\vec{a}_{rf} + 2\vec{\omega} \times \vec{V}_{xyz} + \vec{\omega} \times (\vec{\omega} \times \vec{r}) + \dot{\vec{\omega}} \times \vec{r}]\, dm \quad (4.37a)$$

But

$$\int_{M\,(system)} \vec{a}_{xyz}\, dm = \int_{M\,(system)} \frac{d\vec{V}_{xyz}}{dt}\, dm = \frac{d}{dt} \int_{M\,(system)} \vec{V}_{xyz}\, dm = \frac{d\vec{P}_{xyz}}{dt}\Bigg)_{system} \qquad (4.37b)$$

where all time derivatives are those seen by an observer fixed in the noninertial frame, *xyz*. Combining Eqs. 4.37a and 4.37b, we obtain

$$\frac{d\vec{P}_{xyz}}{dt}\Bigg)_{system} = \vec{F} - \int_{M\,(system)} [\vec{a}_{rf} + 2\vec{\omega} \times \vec{V}_{xyz} + \vec{\omega} \times (\vec{\omega} \times \vec{r}) + \dot{\vec{\omega}} \times \vec{r}]\, dm$$

or

$$\frac{d\vec{P}_{xyz}}{dt}\Bigg)_{system} = \vec{F}_S + \vec{F}_B - \int_{M\,(system)} [\vec{a}_{rf} + 2\vec{\omega} \times \vec{V}_{xyz}$$

$$+ \vec{\omega} \times (\vec{\omega} \times \vec{r}) + \dot{\vec{\omega}} \times \vec{r}]\rho\, d\forall \qquad (4.38)$$

Equation 4.38 expresses the conservation of momentum, $\vec{P}_{xyz}$, measured relative to xyz, as seen by an observer in xyz. Since the coordinate frame, xyz, moves, so does the control volume, to which the observer is fixed. The control volume formulation may be completed by combining Eq. 4.38 with Eq. 4.20,

$$\left.\frac{dN}{dt}\right)_{\text{system}} = \frac{\partial}{\partial t} \iiint_{CV} \eta\rho \, d\forall + \iint_{CS} \eta\rho \vec{V}_{xyz} \cdot d\vec{A} \qquad (4.20)$$

where

$$N = \vec{P}_{xyz} \quad \text{and} \quad \eta = \vec{V}_{xyz}$$

Then, since the system and control volume coincide at time t,

$$\vec{F}_S + \vec{F}_B - \iiint_{CV} [\vec{a}_{rf} + 2\vec{\omega} \times \vec{V}_{xyz} + \vec{\omega} \times (\vec{\omega} \times \vec{r}) + \dot{\vec{\omega}} \times \vec{r}]\rho \, d\forall$$

$$= \frac{\partial}{\partial t} \iiint_{CV} \vec{V}_{xyz}\rho \, d\forall + \iint_{CS} \vec{V}_{xyz}\rho \vec{V}_{xyz} \cdot d\vec{A} \qquad (4.39)$$

Equation 4.39 expresses conservation of linear momentum for a control volume moving with arbitrary acceleration, fixed within a noninertial reference frame, xyz. All time derivatives in Eq. 4.39 are evaluated as seen by an observer fixed in xyz, that is, they are *relative* to xyz. An example illustrating the use of Eq. 4.39 is given in Example Problem 4.13.

Note that Eq. 4.39 reduces to Eq. 4.28 when the angular terms are zero, and to Eq. 4.21 for an inertial control volume ($\vec{a}_{rf} = 0$), with zero angular velocity.

Example 4.12

A reference frame, xyz, moves arbitrarily with respect to a fixed frame, XYZ. A particle moves with velocity, $\vec{V}_{xyz} = (dx/dt)\hat{i} + (dy/dt)\hat{j} + (dz/dt)\hat{k}$, where time derivatives are those seen by an observer in xyz. Show that

$$\vec{V}_{XYZ} = \vec{V}_{rf} + \vec{V}_{xyz} + \vec{\omega} \times \vec{r}$$

as seen by an observer fixed in XYZ.

GIVEN:

Fixed and noninertial frames as shown.

FIND:

$\vec{V}_{XYZ}$ in terms of $\vec{V}_{xyz}$, $\vec{\omega}$, $\vec{r}$, and $\vec{V}_{rf}$.

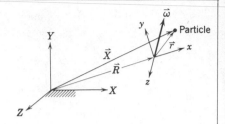

SOLUTION:

From the geometry of the sketch

$$\vec{X} = \vec{R} + \vec{r}$$

so

$$\vec{V}_{XYZ} = \frac{d\vec{X}}{dt} = \frac{d\vec{R}}{dt} + \frac{d\vec{r}}{dt} = \vec{V}_{rf} + \frac{d\vec{r}}{dt}$$

But

$$\vec{r} = x\hat{i} + y\hat{j} + z\hat{k}, \quad \text{so} \quad \frac{d\vec{r}}{dt} = \frac{dx}{dt}\hat{i} + \frac{dy}{dt}\hat{j} + \frac{dz}{dt}\hat{k} + x\frac{d\hat{i}}{dt} + y\frac{d\hat{j}}{dt} + z\frac{d\hat{k}}{dt}$$

or

$$\frac{d\vec{r}}{dt} = \vec{V}_{xyz} + x\frac{d\hat{i}}{dt} + y\frac{d\hat{j}}{dt} + z\frac{d\hat{k}}{dt}$$

Therefore the problem is to evaluate $x(d\hat{i}/dt) + y(d\hat{j}/dt) + z(d\hat{k}/dt)$, where d/dt is the time derivative seen by the observer fixed in frame xyz.

Consider the unit vector $\hat{i}$ first. It will rotate in the xy plane due to ω_z, as shown below:

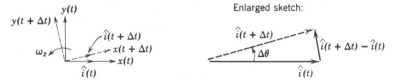

Enlarged sketch:

Now

$$\frac{d\hat{i}}{dt}\bigg]_{\substack{\text{due} \\ \text{to } \omega_z}} = \lim_{\Delta t \to 0}\left[\frac{\hat{i}(t + \Delta t) - \hat{i}(t)}{\Delta t}\right] = \lim_{\Delta t \to 0}\left[\frac{(1)\Delta\theta\hat{j}}{\Delta t}\right] = \lim_{\Delta t \to 0}\left[\frac{(1)\omega_z\Delta t\hat{j}}{\Delta t}\right]$$

$$\frac{d\hat{i}}{dt}\bigg]_{\substack{\text{due} \\ \text{to } \omega_z}} = \hat{j}\omega_z$$

Similarly, $\hat{i}$ will rotate in the xz plane due to ω_y

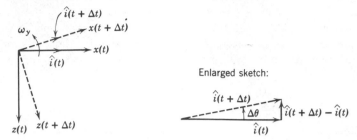

Enlarged sketch:

Then

$$\frac{d\hat{i}}{dt}\bigg]_{\substack{\text{due} \\ \text{to } \omega_y}} = \lim_{\Delta t \to 0} \left[\frac{\hat{i}(t + \Delta t) - \hat{i}(t)}{\Delta t} \right] = \lim_{\Delta t \to 0} \left[\frac{(1)\Delta\theta(-\hat{k})}{\Delta t} \right] = \lim_{\Delta t \to 0} \left[\frac{(1)\omega_y\Delta t(-\hat{k})}{\Delta t} \right]$$

$$\frac{d\hat{i}}{dt}\bigg]_{\substack{\text{due} \\ \text{to } \omega_y}} = -\hat{k}\omega_y$$

Rotation about x due to ω_x causes no change in $\hat{i}$. Combining terms,

$$\frac{d\hat{i}}{dt} = \omega_z\hat{j} - \omega_y\hat{k}$$

By similar reasoning,

$$\frac{d\hat{j}}{dt} = \omega_x\hat{k} - \omega_z\hat{i}$$

and

$$\frac{d\hat{k}}{dt} = \omega_y\hat{i} - \omega_x\hat{j}$$

Thus

$$x\frac{d\hat{i}}{dt} + y\frac{d\hat{j}}{dt} + z\frac{d\hat{k}}{dt} = (z\omega_y - y\omega_z)\hat{i} + (x\omega_z - z\omega_x)\hat{j} + (y\omega_x - x\omega_y)\hat{k}$$

But

$$\vec{\omega} \times \vec{r} = \begin{vmatrix} \hat{i} & \hat{j} & \hat{k} \\ \omega_x & \omega_y & \omega_z \\ x & y & z \end{vmatrix} = (z\omega_y - y\omega_z)\hat{i} + (x\omega_z - z\omega_x)\hat{j} + (y\omega_x - x\omega_y)\hat{k}$$

Combining these results,

$$\vec{V}_{XYZ} = \vec{V}_{rf} + \vec{V}_{xyz} + \vec{\omega} \times \vec{r} \qquad\qquad \vec{V}_{XYZ}$$

momentum equation for control volume with arbitrary acceleration

Example 4.13

A small "pod" is propelled by a jet. The pod is mounted on a strut at a radius of 3 ft from a fixed center. The pod has a mass of 5 lbm; the jet velocity, density, and area are 200 ft/sec, 0.1 lbm/ft³ and 0.1 in.², respectively. At a particular instant, the pod moves at a speed of 150 ft/sec, and the air drag is 2 lbf. Friction and the mass of the strut may be neglected. Determine the angular acceleration of the pod at the given instant.

EXAMPLE PROBLEM 4.13

GIVEN:

System shown in sketch.

FIND:

$\dot{\vec{\omega}}$.

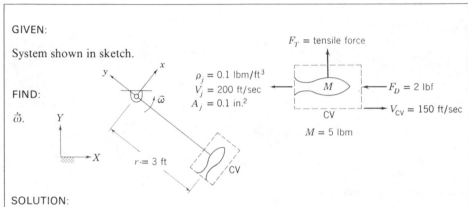

SOLUTION:

Apply momentum equation

Basic equation:
$$\vec{F}_S + \overset{= \, 0(2)}{\cancel{\vec{F}_B}} - \iint_{CV} [\overset{= \, 0(3)}{\cancel{\vec{a}_{rf}}} + 2\vec{\omega} \times \overset{\simeq \, 0(4)}{\cancel{\vec{V}_{xyz}}} + \vec{\omega} \times (\vec{\omega} \times \vec{r}) + \dot{\vec{\omega}} \times \vec{r}]\rho \, d\forall$$

$$= \frac{\partial}{\partial t} \iint_{CV} \overset{\simeq \, 0(4)}{\cancel{\vec{V}_{xyz}}} \rho \, d\forall + \iint_{CS} \vec{V}_{xyz} \rho \vec{V}_{xyz} \cdot d\vec{A}$$

Assumptions: (1) $\vec{F}_S = -F_D \hat{i} + F_T \hat{j}$
　　　　　　 (2) $\vec{F}_B = 0$
　　　　　　 (3) $\vec{a}_{rf} = 0$
　　　　　　 (4) Neglect $\vec{V}_{xyz}$ within CV
　　　　　　 (5) Uniform flow from CV

Then

$$F_T \hat{j} - F_D \hat{i} - \iint_{CV} [\vec{\omega} \times (\vec{\omega} \times \vec{r}) + \dot{\vec{\omega}} \times \vec{r}]\rho \, d\forall = -V_j \hat{i} \{|\rho_j V_j A_j|\}$$

EXAMPLE PROBLEM 4.13 (continued)

Now

$$\vec{\omega} = \frac{V_{CV}}{r}\hat{k} = \frac{150 \text{ ft}}{\text{sec}} \times \frac{1}{3 \text{ ft}}\hat{k} = 50 \text{ sec}^{-1}\hat{k}$$

$$\vec{r} = -3 \text{ ft}\,\hat{j}$$

$$\vec{\omega} \times \vec{r} = 50\hat{k} \times (-3\hat{j}) = 150 \text{ ft/sec}\,\hat{i}$$

$$\vec{\omega} \times (\vec{\omega} \times \vec{r}) = 50\hat{k} \times 150\hat{i} = 7500 \text{ ft/sec}^2\,\hat{j}$$

$$\dot{\vec{\omega}} \times \vec{r} = (\dot{\omega}\hat{k})\frac{1}{\text{sec}^2} \times (-3\hat{j}) \text{ ft} = 3\dot{\omega} \text{ ft/sec}^2\,\hat{i}$$

Substituting

$$F_T\hat{j} - F_D\hat{i} - (7500 \text{ ft/sec}^2\,\hat{j} + 3\dot{\omega} \text{ ft/sec}^2\,\hat{i})M = -\rho_j V_j^2 A_j\hat{i}$$

In components,

$$F_T = 7500 \text{ ft/sec}^2\,M$$

$$F_T = \frac{7500 \text{ ft}}{\text{sec}^2} \times 5 \text{ lbm} \times \frac{\text{slug}}{32.2 \text{ lbm}} \times \frac{\text{lbf-sec}^2}{\text{slug-ft}} = 1160 \text{ lbf}$$

and

$$-F_D - 3\dot{\omega}M = -\rho_j V_j^2 A_j$$

or

$$\dot{\omega} = \frac{\rho_j V_j^2 A_j}{3M} - \frac{F_D}{3M}$$

$$= \frac{0.1 \text{ lbm}}{\text{ft}^3} \times \frac{(200)^2 \text{ ft}^2}{\text{sec}^2} \times \frac{0.1 \text{ in.}^2}{3} \times \frac{1}{5 \text{ lbm}} \times \frac{\text{ft}^2}{144 \text{ in.}^2}$$

$$-\frac{2 \text{ lbf}}{3} \times \frac{1}{5 \text{ lbm}} \times \frac{32.2 \text{ lbm}}{\text{slug}} \times \frac{\text{slug-ft}}{\text{lbf-sec}^2}$$

$$\dot{\omega} = \frac{(4)10^2}{(3)(5)(144)} - \frac{(2)(32.2)}{(3)(5)} = 0.185 - 4.29 = -4.1 \text{ rad/sec}^2$$

Thus $\dot{\vec{\omega}} = \dot{\omega}\hat{k} = -4.1 \text{ rad/sec}^2\,\hat{k}$ ⟵ $\dot{\vec{\omega}}$

{Thus, the pod is decelerating, since $\dot{\omega} < 0$}

**4–7 Conservation of Angular Momentum

Next we wish to formulate the conservation of angular momentum for application to a control volume. We shall begin with the mathematical statement for a system and then employ our recipe to obtain the control volume equation.

** This section may be omitted without loss of continuity in the text material.

4-7.1 CONTROL VOLUME EQUATION

Conservation of angular momentum for a system was given by Eq. 4.3a

$$\vec{T} = \frac{d\vec{H}}{dt}\Bigg]_{\text{system}} \tag{4.3a}$$

where the angular momentum, H, of the system is given by

$$\vec{H}_{\text{system}} = \int_{M\,(\text{system})} \vec{r} \times \vec{V}\, dm \tag{4.3b}$$

The torque, $\vec{T}$, applied to a system is given by

$$\vec{T} = \vec{r} \times \vec{F}_s + \int_{M\,(\text{system})} \vec{r} \times \vec{g}\, dm + \vec{T}_{\text{shaft}} \tag{4.3c}$$

The recipe for relating the system and control volume formulations is given by Eq. 4.20

$$\frac{dN}{dt}\Bigg]_{\text{system}} = \frac{\partial}{\partial t} \iiint_{CV} \eta\rho\, d\forall + \iint_{CS} \eta\rho \vec{V}_{xyz} \cdot d\vec{A}$$

where

$$N_{\text{system}} = \int_{M\,(\text{system})} \eta\, dm$$

Setting

$$N = \vec{H} \quad \text{and} \quad \eta = \vec{r} \times \vec{V}$$

then

$$\frac{d\vec{H}}{dt}\Bigg]_{\text{system}} = \frac{\partial}{\partial t} \iiint_{CV} \vec{r} \times \vec{V}\rho\, d\forall + \iint_{CS} \vec{r} \times \vec{V}\, \rho \vec{V}_{xyz} \cdot d\vec{A} \tag{4.40}$$

Combining Eqs. 4.3a, 4.3c, and 4.40, we obtain

$$\vec{r} \times \vec{F}_s + \int_{M\,(\text{system})} \vec{r} \times \vec{g}\, dm + \vec{T}_{\text{shaft}}$$

$$= \frac{\partial}{\partial t} \iiint_{CV} \vec{r} \times \vec{V}\rho\, d\forall + \iint_{CS} \vec{r} \times \vec{V}\, \rho \vec{V}_{xyz} \cdot d\vec{A}$$

Since the system and control volume coincide at time t,

$$\vec{T}_{\text{system}} = \vec{T}_{CV}$$

4/basic equations in integral form for a control volume

and

$$\vec{r} \times \vec{F}_S + \iiint_{CV} \vec{r} \times \vec{g}\rho \, d\forall + \vec{T}_{\text{shaft}} = \frac{\partial}{\partial t} \iiint_{CV} \vec{r} \times \vec{V}\rho \, d\forall + \iint_{CS} \vec{r} \times \vec{V}\rho \vec{V}_{xyz} \cdot d\vec{A}$$

(4.41)

Equation 4.41 is a vector equation that expresses conservation of angular momentum for a control volume. For application to turbomachinery, it is usually used in scalar form by taking the dot product with $\vec{\omega}$, which is directed along the axis of rotation. Note that both *absolute* velocity, $\vec{V}$, and relative velocity, $\vec{V}_{xyz}$, appear in the equation. This is accepted practice in turbomachinery analysis, because the absolute velocities are often known from analysis of the upstream flow, and desired for design of downstream flow passages.

4–7.2 APPLICATION TO TURBOMACHINERY

Fluid handling devices that have purely rotary motion are termed *turbomachines*. Prime movers, which add energy to a fluid stream, are called pumps when the flow is liquid (or slurry), and fans, blowers, or compressors for gas and vapor handling units, depending on pressure rise. (Flow through fans is essentially incompressible, blowers have a pressure rise of a few psi, and compressors are units designed for a larger pressure rise.) Turbines extract energy from a fluid stream. They range from such simple devices as the windmill to sophisticated steam turbines with many stages of carefully designed blading. All of these devices can be analyzed by applying Eq. 4.41.

Flow through a turbomachine may be nearly axial, nearly radial, or mixed. In any event, it is convenient in general to consider a control volume enclosing the rotating member, the rotor or impeller, for analysis of torque reactions. If fluid viscosity is ignored, then surface forces on such a control volume are zero. The body force contribution may be neglected by symmetry. Then for steady flow, Eq. 4.41 becomes

$$\vec{T}_{\text{shaft}} = \iint_{CS} \vec{r} \times \vec{V} \, \rho \vec{V}_{xyz} \cdot d\vec{A}$$

Let us now simplify this equation to scalar form for axial and radial flow machines, and then illustrate its use with several examples.

a. Axial Flow Machines

Consider the axial flow rotor shown in Fig. 4.5. The flow relative to the rotor is assumed to follow the blades.

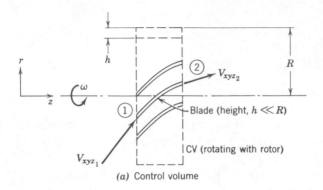

(a) Control volume

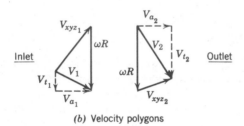

(b) Velocity polygons

Fig. 4.5 *Control volume and velocity polygons for analysis of an axial flow turbomachine.*

Basic equation:
$$\vec{T}_{\text{shaft}} = \iint_{CS} \vec{r} \times \vec{V} \rho \vec{V}_{xyz} \cdot d\vec{A}$$

Assumptions: (1) No surface forces produce torque
(2) No body forces produce torque
(3) Steady flow
(4) Uniform flow at each section
(5) $h \ll R$, $h_1 = h_2$
(6) Incompressible flow

Then

$$\vec{T}_{\text{shaft}} = \vec{r}_1 \times \vec{V}_1 \{ -|\rho V_{1a} A_1| \} + \vec{r}_2 \times \vec{V}_2 \{|\rho V_{2a} A_2|\}$$

$$T_{\text{shaft}} \hat{\imath}_z = \dot{m}(\vec{r}_2 \times \vec{V}_2 - \vec{r}_1 \times \vec{V}_1)$$

4/basic equations in integral form for a control volume

Using the notation of Fig. 4.5,

$$\vec{r} = R\hat{\imath}_r$$

$$\vec{V} = V_t\hat{\imath}_\theta + V_a\hat{\imath}_z$$

$$\vec{r} \times \vec{V} = R\hat{\imath}_r \times (V_t\hat{\imath}_\theta + V_a\hat{\imath}_z)$$

$$= RV_t(\hat{\imath}_r \times \hat{\imath}_\theta) + RV_a(\hat{\imath}_r \times \hat{\imath}_z)$$

$$= RV_t\hat{\imath}_z - RV_a\hat{\imath}_\theta$$

Substituting

$$T_{\text{shaft}}\hat{\imath}_z = \dot{m}[R(V_{2t} - V_{1t})\hat{\imath}_z + R(V_{1a} - V_{2a})\hat{\imath}_\theta]$$

In components,

$$T_{\text{shaft}} = \dot{m}R(V_{2t} - V_{1t})$$

$$V_{1a} = V_{2a}$$

If $V_{2t} > V_{1t}$, as shown in Fig. 4.5, then torque has the same sense as $\vec{\omega}$, and work is done *on* the control volume. The rate of work input is given by

$$\dot{W}_{\text{in}} = \vec{\omega} \cdot \vec{T}_{\text{shaft}} = \omega T_{\text{shaft}} = \omega\dot{m}R(V_{2t} - V_{1t})$$

From the second components, $V_{1a} = V_{2a}$, we obtained a result that is valid only for incompressible flow (we neglected any area change when we assumed $h_1 = h_2$).

Thus work input to an axial flow machine results in an increase in tangential velocity. Since ωR is the same for both inlet and outlet flows, then $V_2 > V_1$ also. Consequently, the kinetic energy per unit mass of flow leaving the machine is larger than that entering. Some of this kinetic energy can be converted to static pressure rise if a diffuser (see Section 8–9) is installed downstream. Alternatively, diffusion can be accomplished in a set of fixed vanes downstream from the rotor. The combination of a rotating and a fixed member is called a *stage*,

b. Radial Flow Machines

Consider the radial flow rotor shown in Fig. 4.6. Flow enters the rotor at radius R_1, and leaves at radius R_2. Any combination of blade angles is possible.

Basic equation:
$$\vec{T}_{\text{shaft}} = \int\!\!\int_{\text{CS}} \vec{r} \times \vec{V} \, \rho \vec{V}_{xyz} \cdot d\vec{A}$$

conservation of angular momentum

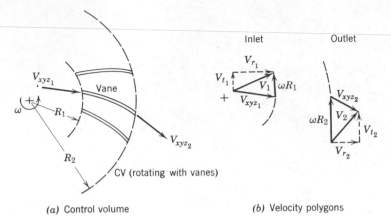

Inlet   Outlet

(a) Control volume   (b) Velocity polygons

Fig. 4.6 *Control volume and velocity polygons for analysis of a radial flow turbomachine.*

Assumptions: (1) No surface forces produce torque
              (2) No body forces produce torque
              (3) Steady flow
              (4) Uniform flow at each section
              (5) Incompressible flow
              (6) Flow enters and leaves tangent to vanes

Then

$$\vec{T}_{\text{shaft}} = \vec{r}_1 \times \vec{V}_1\{-|\rho V_{1r}A_1|\} + \vec{r}_2 \times \vec{V}_2\{|\rho V_{2r}A_2|\}$$

$$T_{\text{shaft}}\hat{i}_z = \dot{m}(\vec{r}_2 \times \vec{V}_2 - \vec{r}_1 \times \vec{V}_1)$$

Using the notation of Fig. 4.6,

$$\vec{r} = R\hat{i}_r$$

$$\vec{V} = V_r\hat{i}_r + V_t\hat{i}_\theta$$

$$\vec{r} \times \vec{V} = RV_r(\hat{i}_r \times \hat{i}_r) + RV_t(\hat{i}_r \times \hat{i}_\theta)$$

$$= RV_t\hat{i}_z$$

Substituting,

$$T_{\text{shaft}} = \dot{m}(R_2 V_{2t} - R_1 V_{1t})$$

If flow enters without swirl, then $V_{1t} = 0$, and

$$T_{\text{shaft}})_{\text{max}} = \dot{m}R_2 V_{2t}$$

Note that continuity shows for incompressible flow that $R_1 V_{1r} = R_2 V_{2r}$, so $V_{2r} = V_{1r}(R_1/R_2)$, provided there is no change in vane depth from inlet to outlet.

Work input to a radial flow machine produces an increase in tangential velocity and, hence, kinetic energy. A static pressure rise is produced by the rotating flow in the impeller (remember Example 3.2?). Additional static pressure may be regained by using a diffuser downstream from the impeller (see Section 8–9).

c. Mixed Flow Machines

Mixed flow machines may be analyzed also by applying Eq. 4.41, accounting correctly for the appropriate velocity and radius terms (Example Problem 4.14).

Example 4.14

Water at 150 gal/min enters a mixed flow pump impeller axially. The axial inlet velocity is uniform, and the tangential velocity varies linearly from zero at the centerline to 15 ft/sec at a diameter of 1 in. The outlet diameter of the impeller is 4 in. Flow leaves radially at 10 ft/sec relative velocity. The impeller speed is 3450 rpm.

Determine the impeller exit width, b, the torque input to the impeller, and the horsepower supplied.

EXAMPLE PROBLEM 4.14

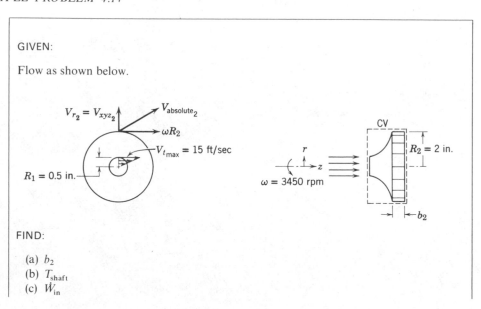

GIVEN:

Flow as shown below.

FIND:

(a) b_2
(b) T_{shaft}
(c) $\dot{W}_{in}$

conservation of angular momentum

SOLUTION:

Apply conservation of angular momentum.

Basic equations: $\vec{T}_{shaft} = \int\int_{CS} \vec{r} \times \vec{V} \rho \vec{V}_{xyz} \cdot d\vec{A}; \quad \dot{W}_{in} = \vec{\omega} \cdot \vec{T} = \omega T$

$$= \underset{\nearrow}{0}(3)$$

$$0 = \frac{\partial}{\partial t}\int\int\int_{CV} \rho \, d\forall + \int\int_{CS} \rho \vec{V} \cdot d\vec{A}$$

Assumptions: (1) No surface forces produce torque
(2) No body forces produce torque
(3) Steady flow
(4) Uniform axial flow at inlet, uniform radial flow at outlet
(5) Incompressible flow

Then from continuity

$$0 = \{-|\rho V_{1a}\pi R_1^2|\} + \{|\rho V_{2r}2\pi R_2 b_2|\}$$

or

$$Q = 2\pi R_2 b_2 V_{2r} = \text{volume flowrate}$$

$$b_2 = \frac{Q}{2\pi R_2 V_{2r}} = \frac{150 \text{ gal}}{\text{min}} \times \frac{\text{ft}^3}{7.48 \text{ gal}} \times \frac{1}{2\pi} \times \frac{1}{2 \text{ in.}} \times \frac{\text{sec}}{10 \text{ ft}} \times \frac{\text{min}}{60 \text{ sec}} \times \frac{12 \text{ in.}}{\text{ft}}$$

$$b_2 = 0.032 \text{ ft or } 0.384 \text{ in.} \qquad\qquad\qquad\qquad\qquad\qquad b_2$$

$$T_{shaft}\hat{i}_z = -\int_0^{R_1} \vec{r} \times \vec{V} \rho V_{1a} 2\pi r \, dr + \vec{r}_2 \times \vec{V}_2 \dot{m}$$

$$\vec{r} = r\hat{i}_r$$

$$\vec{V} = V_{1a}\hat{i}_z + V_{1t}\hat{i}_\theta$$

$$\vec{r} \times \vec{V} = rV_a(\hat{i}_r \times \hat{i}_z) + rV_t(\hat{i}_r \times \hat{i}_\theta)$$

$$= -rV_a(\hat{i}_\theta) + rV_t\hat{i}_z$$

$$V_{1t} = V_{t\,max}\frac{r}{R_1}$$

$$\vec{r}_2 = R_2\hat{i}_r$$

$$\vec{V}_2 = V_{2r}\hat{i}_r + V_{2t}\hat{i}_\theta$$

$$\vec{r}_2 \times \vec{V}_2 = R_2 V_{2r}(\hat{i}_r \overset{0}{\cancel{\times} \hat{i}_r}) + R_2 V_{2t}(\hat{i}_r \times \hat{i}_\theta) = R_2 V_{2t}\hat{i}_z$$

EXAMPLE PROBLEM 4.14 (*continued*)

Summing the z components,

$$T_{\text{shaft}} = -\int_0^{R_1} r\left(V_{t\max}\frac{r}{R_1}\right)\rho V_{1a}2\pi r\, dr + \dot{m}R_2V_{2t}$$

$$= -V_{t\max}\rho V_{1a}2\pi R_1^3 \int_0^1 \left(\frac{r}{R_1}\right)^3 d\left(\frac{r}{R_1}\right) + \dot{m}R_2V_{2t}$$

But $\dot{m} = \rho V_{1a}\pi R_1^2$, so

$$T_{\text{shaft}} = -2\dot{m}V_{t\max}(\tfrac{1}{4})R_1 + \dot{m}R_2V_{2t}$$

or

$$T_{\text{shaft}} = \rho Q(R_2 V_{2t} - \tfrac{1}{2}R_1 V_{t\max})$$

But $V_{2t} = \omega R_2$, so

$$T_{\text{shaft}} = \rho Q(\omega R_2^2 - \tfrac{1}{2}R_1 V_{t\max})$$

Thus

$$T_{\text{shaft}} = \frac{1.94\ \text{slug}}{\text{ft}^3} \times \frac{150\ \text{gal}}{\text{min}} \times \frac{\text{ft}^3}{7.48\ \text{gal}} \times \frac{\text{min}}{60\ \text{sec}} \times \left[\frac{3450\ \text{rev}}{\text{min}} \times \frac{2\pi\ \text{rad}}{\text{rev}} \times \frac{\text{min}}{60\ \text{sec}}\right.$$

$$\left. \times (2)^2\ \text{in.}^2 \times \frac{\text{ft}^2}{144\ \text{in.}^2} - \frac{1}{2}\left(\frac{1}{2}\right)\text{in.} \times \frac{15\ \text{ft}}{\text{sec}} \times \frac{\text{ft}}{12\ \text{in.}}\right]$$

$$T_{\text{shaft}} = 0.648\frac{\text{slug}}{\text{sec}}\left[(10 - 0.31)\frac{\text{ft}^2}{\text{sec}}\right] = 6.28\ \text{ft-lbf} \qquad\qquad\qquad \longleftarrow \quad T_{\text{shaft}}$$

$$\dot{W}_{\text{in}} = \vec{\omega}\cdot\vec{T}_{\text{shaft}} = \omega T_{\text{shaft}} = \frac{(3450)(2\pi)}{(60)}\frac{\text{rad}}{\text{sec}} \times 6.28\ \text{ft-lbf} \times \frac{\text{hp-sec}}{550\ \text{ft-lbf}}$$

$$\dot{W}_{\text{in}} = 4.1\ \text{hp} \qquad\qquad\qquad\qquad\qquad\qquad\qquad\qquad\qquad\qquad\qquad\qquad \longleftarrow \quad \dot{W}_{\text{in}}$$

4–8 The First Law of Thermodynamics

Development of the mathematical formulation of the first law of thermodynamics should not prove difficult for us at this point. We have used the recipe often enough to make the pattern of developing the control volume equation from the corresponding system equation reasonably clear. Furthermore the steady state form of the first law is familiar to you from your earlier work in thermodynamics.

Recall that the system formulation of the first law was given by Eq. 4.4a

$$\dot{Q} - \dot{W} = \frac{dE}{dt}\bigg)_{\text{system}} \qquad\qquad\qquad (4.4a)$$

where the total energy of the system is given by

$$E_{system} = \int_{M\,(system)} e\,dm = \int_{\forall\,(system)} e\rho\,d\forall \qquad (4.4b)$$

and

$$e = u + \frac{V^2}{2} + gz$$

In Eq. 4.4a the rate of heat transfer, $\dot{Q}$, is taken as positive when heat is added to the system from the surroundings; the rate of work done, $\dot{W}$, is taken as positive when work is done by the system on its surroundings.

The system and control volume formulations are related by Eq. 4.12.

$$\left.\frac{dN}{dt}\right)_{system} = \frac{\partial}{\partial t} \iiint_{CV} \eta\rho\,d\forall + \iint_{CS} \eta\rho\vec{V}\cdot d\vec{A} \qquad (4.12)$$

where

$$N_{system} = \int_{M\,(system)} \eta\,dm = \iiint_{\forall\,(system)} \eta\rho\,d\forall \qquad (4.6)$$

Then to derive the control volume formulation of the first law of thermodynamics, we set

$$N = E \quad \text{and} \quad \eta = e$$

From Eq. 4.12, with this substitution, we obtain

$$\left.\frac{dE}{dt}\right)_{system} = \frac{\partial}{\partial t} \iiint_{CV} e\rho\,d\forall + \iint_{CS} e\rho\vec{V}\cdot d\vec{A} \qquad (4.42)$$

In deriving Eq. 4.12, the system and the control volume coincided at time t, so

$$[\dot{Q} - \dot{W}]_{system} = [\dot{Q} - \dot{W}]_{control\ volume}$$

In light of this, Eqs. 4.4a and 4.42 yield the control volume formulation of the first law of thermodynamics.

$$\dot{Q} - \dot{W} = \frac{\partial}{\partial t} \iiint_{CV} e\rho\,d\forall + \iint_{CS} e\rho\vec{V}\cdot d\vec{A} \qquad (4.43)$$

where,

$$e = u + \frac{V^2}{2} + gz$$

Is Eq. 4.43 the form of the first law you used in your earlier thermodynamics course? (Note that for steady flow the first term on the right side of Eq. 4.43 is zero.) Even for steady flow, Eq. 4.43 is not quite the same form that you used previously in applying the first law to control volume problems. To obtain a formulation suitable and convenient for problem solutions, let us take a closer look at the work term, $\dot{W}$.

4–8.1 RATE OF WORK FOR A CONTROL VOLUME

The term $\dot{W}$ in Eq. 4.43 has a positive numerical value when work is done by the control volume on the surroundings. This interaction can only take place at the boundaries of the control volume.

The rate of work done on the control volume is of opposite sign to the work done by the control volume. The work done by the control volume is conveniently subdivided into four classifications.

$$\dot{W} = \dot{W}_s + \dot{W}_{normal} + \dot{W}_{shear} + \dot{W}_{other}$$

Let us consider these separately:

1. Shaft Work

We shall designate shaft work by W_s and, hence, the rate of work transferred out through the control surface by shaft work is designated by $\dot{W}_s$.

2. Work Done by Normal Stresses at the Control Surface

Recall that work requires a force to be moved through a distance. Thus, in moving a force, $\vec{F}$, through an infinitesimal distance, $d\vec{s}$, the work done is given by

$$\delta W = \vec{F} \cdot d\vec{s}$$

To obtain the rate at which work is done by the force, divide by the time increment Δt and take the limit as $\Delta t \to 0$. Thus, the rate of work done by the force, $\vec{F}$, is given by

$$\dot{W} = \lim_{\Delta t \to 0} \frac{\delta W}{\Delta t} = \lim_{\Delta t \to 0} \frac{\vec{F} \cdot d\vec{s}}{\Delta t}$$

that is

$$\dot{W} = \vec{F} \cdot \vec{V}$$

The rate of work done on an element of area, $d\vec{A}$, of the control surface by normal stresses is given by

$$d\vec{F} \cdot \vec{V} = \sigma_{nn} \, d\vec{A} \cdot \vec{V}$$

the first law of thermodynamics

The total rate of work done by the normal stresses acting over the entire control surface is given by

$$\iint_{CS} \sigma_{nn}\, d\vec{A} \cdot \vec{V} = \iint_{CS} \sigma_{nn}\vec{V} \cdot d\vec{A}$$

Since the work out across the boundaries of the control volume is the negative of the work done on the control volume, the rate of work out of the control volume due to normal stresses is given by

$$\dot{W}_{normal} = -\iint_{CS} \sigma_{nn}\vec{V} \cdot d\vec{A}$$

3. Work Done by Shear Stresses at the Control Surface

Just as there is work done by the normal stresses at the boundaries of the control volume, so may there be work done by the shear stresses.

The shear force acting on an element of area of the control surface is given by

$$d\vec{F} = \vec{\tau}\, dA$$

where the shear stress vector, $\vec{\tau}$, is the stress acting in the plane of dA.

The rate of work done by the shear stresses on the entire surface of the control volume is given by

$$\iint_{CS} \vec{\tau}\, dA \cdot \vec{V} = \iint_{CS} \vec{\tau} \cdot \vec{V}\, dA$$

Since the work out across the boundaries of the control volume is the negative of the work done on the control volume, then the rate of work out of the control volume due to shear stresses is given by

$$\dot{W}_{shear} = -\iint_{CS} \vec{\tau} \cdot \vec{V}\, dA$$

This integral is better expressed as three terms

$$\dot{W}_{shear} = -\iint_{CS} \vec{\tau} \cdot \vec{V}\, dA$$

$$= -\iint_{A\,(shafts)} \vec{\tau} \cdot \vec{V}\, dA - \iint_{\substack{A\,(solid \\ surface)}} \vec{\tau} \cdot \vec{V}\, dA - \iint_{A\,(ports)} \vec{\tau} \cdot \vec{V}\, dA$$

We have already accounted for the first term, since we included $\dot{W}_{shaft}$ previously. At solid surfaces, $\vec{V} = 0$, so the second term is zero (for our fixed control volume). Thus

$$\dot{W}_{shear} = - \iint_{A \text{ (ports)}} \vec{\tau} \cdot \vec{V} \, dA$$

This last term can be made zero by proper choice of control surfaces. If we choose a control surface that cuts across each port perpendicular to the flow, then $d\vec{A}$ is parallel to $\vec{V}$. Since $\vec{\tau}$ is in the plane of dA, then $\vec{\tau}$ is perpendicular to $\vec{V}$. Thus

$$\vec{\tau} \cdot \vec{V} = 0 \quad \text{and} \quad \dot{W}_{shear} = 0$$

4. *Other Work*

Electrical energy could be added to the control volume. Likewise electromagnetic energy, for example, radar or laser beams could be absorbed. In most problems, such contributions will be absent, but we should note them in our general formulation.

With all of the terms in $\dot{W}$ evaluated, we obtain

$$\dot{W} = \dot{W}_s - \iint_{CS} \sigma_{nn} \vec{V} \cdot d\vec{A} + \dot{W}_{shear} + \dot{W}_{other} \qquad (4.44)$$

4–8.2 CONTROL VOLUME EQUATION

Substituting the expression for $\dot{W}$ from Eq. 4.44 into Eq. 4.43 gives

$$\dot{Q} - \dot{W}_s + \iint_{CS} \sigma_{nn} \vec{V} \cdot d\vec{A} - \dot{W}_{shear} - \dot{W}_{other} = \frac{\partial}{\partial t} \iiint_{CV} e\rho \, d\forall + \iint_{CS} e\rho \vec{V} \cdot d\vec{A}$$

Rearranging this equation, we obtain

$$\dot{Q} - \dot{W}_s - \dot{W}_{shear} - \dot{W}_{other} = \frac{\partial}{\partial t} \iiint_{CV} e\rho \, d\forall + \iint_{CS} e\rho \vec{V} \cdot d\vec{A} - \iint_{CS} \sigma_{nn} \vec{V} \cdot d\vec{A}$$

Since $\rho = 1/v$, then

$$\iint_{CS} \sigma_{nn} \vec{V} \cdot d\vec{A} = \iint_{CS} \sigma_{nn} v \rho \vec{V} \cdot d\vec{A}$$

177   **the first law of thermodynamics**

Hence

$$\dot{Q} - \dot{W}_s - \dot{W}_{shear} - \dot{W}_{other} = \frac{\partial}{\partial t} \iiint_{CV} e\rho \, d\forall + \iint_{CS} (e - \sigma_{nn}v)\rho\vec{V} \cdot d\vec{A}$$

Viscous effects can make the normal stress, σ_{nn}, different from the negative of the thermodynamic pressure, $-p$. However, for most flows of common engineering interest, $\sigma_{nn} \simeq -p$. Then

$$\dot{Q} - \dot{W}_s - \dot{W}_{shear} - \dot{W}_{other} = \frac{\partial}{\partial t} \iiint_{CV} e\rho \, d\forall + \iint_{CS} (e + pv)\rho\vec{V} \cdot d\vec{A}$$

Finally, substituting $e = u + V^2/2 + gz$ into the last term, we obtain the familiar form of the first law formulation for a control volume

$$\dot{Q} - \dot{W}_s - \dot{W}_{shear} - \dot{W}_{other} = \frac{\partial}{\partial t} \iiint_{CV} e\rho \, d\forall + \iint_{CS} \left(u + pv + \frac{V^2}{2} + gz\right)\rho\vec{V} \cdot d\vec{A}$$

(4.45)

Each work term in Eq. 4.45 represents the rate of work out of the control volume.

Example 4.15

Air enters a machine at 14.7 psia, 70 F, with negligible velocity, and is discharged at 50 psia, 100 F through a pipe (the area is 1 ft²). The flow rate is 20 lbm/sec. The power input to the machine is 600 hp. Determine the heat transfer.

EXAMPLE PROBLEM 4.15

GIVEN:

Air enters a compressor at ① with conditions

$p_1 = 14.7$ psia $T_1 = 70$ F $V_1 \approx 0$

and is discharged at ② with conditions

$p_2 = 50$ psia $T_2 = 100$ F $A_2 = 1$ ft²

The air flow rate is 20 lbm/sec and the power input to the machine is 600 hp.

FIND:

The heat transfer.

SOLUTION:

Basic equations:
$$\dot{Q} - \dot{W_s} - \overset{= 0(5)}{\cancel{\dot{W}_{shear}}} = \overset{= 0(1)}{\cancel{\frac{\partial}{\partial t} \iiint_{CV} e\rho \, d\forall}} + \iint_{CS} \left(u + pv + \frac{V^2}{2} + gz \right) \rho \vec{V} \cdot d\vec{A}$$

$$\overset{= 0(1)}{\cancel{\frac{\partial}{\partial t} \iiint_{CV} \rho \, d\forall}} + \iint_{CS} \rho \vec{V} \cdot d\vec{A} = 0$$

Assumptions: (1) Steady flow
    (2) Properties uniform over inlet and outlet
    (3) $z_1 = z_2$, $V_1 \approx 0$
    (4) Treat air as an ideal gas, $p = \rho RT$
    (5) Area of CV at ① and ② perpendicular to velocity, thus $\dot{W}_{shear} = 0$

Under the assumptions listed, the first law becomes

$$\dot{Q} - \dot{W_s} = \iint_{CS} \left(u + pv + \frac{V^2}{2} + gz \right) \rho \vec{V} \cdot d\vec{A} = \iint_{CS} \left(h + \frac{V^2}{2} + gz \right) \rho \vec{V} \cdot d\vec{A} \quad \{h = u + pv\}$$

$$\dot{Q} = \dot{W_s} + \iint_{CS} \left(h + \frac{V^2}{2} + gz \right) \rho \vec{V} \cdot d\vec{A}.$$

For uniform properties (assumption 2) we can write

$$\dot{Q} = \dot{W_s} + \left(h_1 + \overset{\approx 0(3)}{\cancel{\frac{V_1^2}{2}}} + gz_1 \right) \{ -|\rho_1 V_1 A_1| \} + \left(h_2 + \frac{V_2^2}{2} + gz_2 \right) \{ |\rho_2 V_2 A_2| \}$$

For steady flow, from conservation of mass,

$$\iint_{CS} \rho \vec{V} \cdot d\vec{A} = 0$$

$$\therefore \quad -|\rho_1 V_1 A_1| + |\rho_2 V_2 A_2| = 0 \quad \text{that is } |\rho_1 V_1 A_1| = |\rho_2 V_2 A_2| = \dot{m}$$

Hence, we can write

$$\dot{Q} = \dot{W_s} + \dot{m} \left[(h_2 - h_1) + \frac{V_2^2}{2} + \overset{= 0(3)}{g(z_2 \cancel{- z_1})} \right]$$

Assume air behaves as an ideal gas with constant c_p. Then $h_2 - h_1 = c_p(T_2 - T_1)$, and

$$\dot{Q} = \dot{W_s} + \dot{m} \left[c_p(T_2 - T_1) + \frac{V_2^2}{2} \right]$$

From continuity $|V_2| = \dot{m}/\rho_2 A_2$. Since $p_2 = \rho_2 R T_2$ then

$$|V_2| = \frac{\dot{m}}{A_2}\frac{RT_2}{p_2} = \frac{20\,\text{lbm}}{\text{sec}} \times \frac{1}{1\,\text{ft}^2} \times \frac{53.3\,\text{ft-lbf}}{\text{lbm-R}} \times 560\,\text{R} \times \frac{\text{in.}^2}{50\,\text{lbf}} \times \frac{\text{ft}^2}{144\,\text{in.}^2} = 83\,\text{ft/sec}$$

$$\dot{Q} = \dot{W}_s + \dot{m}c_p(T_2 - T_1) + \dot{m}\frac{V_2^2}{2}$$

$$= -600\,\text{hp} \times \frac{550\,\text{ft-lbf}}{\text{sec-hp}} \times \frac{\text{Btu}}{778\,\text{ft-lbf}} + \frac{20\,\text{lbm}}{\text{sec}} \times \frac{0.24\,\text{Btu}}{\text{lbm-R}} \times 30\,\text{R}$$

$$+ \frac{20\,\text{lbm}}{\text{sec}} \times \frac{(83)^2}{2}\frac{\text{ft}^2}{\text{sec}^2} \times \frac{\text{slug}}{32.2\,\text{lbm}} \times \frac{\text{Btu}}{778\,\text{ft-lbf}} \times \frac{\text{lbf-sec}^2}{\text{slug-ft}}$$

$$\frac{d\dot{Q}}{dt} = -425\,\text{Btu/sec} + 144\,\text{Btu/sec} + 2.8\,\text{Btu/sec} = -278\,\text{Btu/sec} \quad \{\text{heat rejection}\}\ \dot{Q}$$

$$\left\{\begin{array}{l}\text{In addition to demonstrating the straight forward application of the first law, this prob-}\\ \text{lem illustrates the need for keeping units straight.}\end{array}\right\}$$

Example 4.16

A tank of volume 3 ft³ is connected to a high pressure air line; both line and tank are initially at a uniform temperature of 70 F. The initial tank pressure is 15 psig. The line pressure is 300 psia, and it is large enough so that its temperature and pressure may be assumed to remain constant. The tank temperature is monitored by a fast-response thermocouple. At the instant after the valve is opened, the tank temperature rises at the rate of 0.10 F/sec. Determine the instantaneous flowrate of air into the tank. (Heat transfer is negligible.)

EXAMPLE PROBLEM 4.16

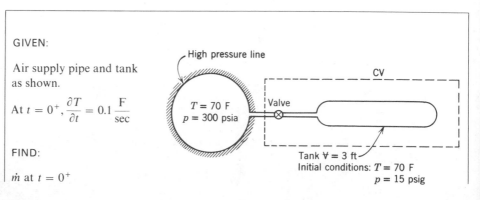

GIVEN:

Air supply pipe and tank as shown.

At $t = 0^+$, $\dfrac{\partial T}{\partial t} = 0.1\ \dfrac{\text{F}}{\text{sec}}$

FIND:

$\dot{m}$ at $t = 0^+$

High pressure line

CV

$T = 70$ F
$p = 300$ psia

Valve

Tank $\mathbb{V} = 3$ ft³
Initial conditions: $T = 70$ F
$p = 15$ psig

SOLUTION:

Choose CV shown, apply energy equation.

$$= 0(1) \quad = 0(2) \quad = 0(3) \quad = 0(4)$$

Basic equation: $\cancel{\dot{Q}} - \cancel{\dot{W}_s} - \cancel{\dot{W}_{shear}} - \cancel{\dot{W}_{other}} = \frac{\partial}{\partial t} \iiint_{CV} e\rho \, d\forall + \iint_{CS} (e + pv)\rho \vec{V} \cdot d\vec{A};$

$$\simeq 0(5) \quad \simeq 0(6)$$

$$e = u + \cancel{\frac{V^2}{2}} + \cancel{gz}$$

Assumptions: (1) $\dot{Q} = 0$ (given)
  (2) $\dot{W}_s = 0$
  (3) $\dot{W}_{shear} = 0$
  (4) $\dot{W}_{other} = 0$
  (5) Neglect kinetic energy
  (6) Neglect potential energy
  (7) Uniform flow at tank inlet
  (8) Properties uniform in tank
  (9) Ideal gas, $p = \rho RT$, $du = c_v \, dT$

Then

$$0 = \frac{\partial}{\partial t} \iiint_{CV} u_{tank}\rho \, d\forall + (u_{line} + pv)\{-|\rho VA|\}$$

But initially, T is uniform, so $u_{tank} = u_{line} = u$, and

$$0 = \frac{\partial}{\partial t} \iiint_{CV} u\rho \, d\forall + (u + pv)\{-|\rho VA|\}$$

Since properties are uniform, $\partial/\partial t$ may be replaced by d/dt, and

$$0 = \frac{d}{dt}[uM] - (u + pv)\dot{m}$$

or

$$0 = u\frac{dM}{dt} + M\frac{du}{dt} - u\dot{m} - pv\dot{m} \tag{1}$$

The term dM/dt may be evaluated from continuity:

Basic equation: $\quad 0 = \frac{\partial}{\partial t} \iiint_{CV} \rho \, d\forall + \iint_{CS} \rho \vec{V} \cdot d\vec{A}$

$$= \frac{dM}{dt} + \{-|\rho VA|\}$$

or

$$\frac{dM}{dt} = \dot{m}$$

Substituting into Eq. 1

$$0 = \cancel{u\dot{m}} + M\frac{du}{dt} - \cancel{u\dot{m}} - pv\dot{m}$$

$$0 = Mc_v\frac{dT}{dt} - pv\dot{m}$$

or

$$\dot{m} = \frac{Mc_v(dT/dt)}{pv} = \frac{\rho\forall c_v(dT/dt)}{pv} = \frac{\rho\forall c_v(dT/dt)}{RT} \tag{2}$$

But at $t = 0$, $\rho = \rho_{\text{tank}}$,

$$\rho = \rho_{\text{tank}} = \frac{p_{\text{tank}}}{RT} = \frac{(15 + 14.7)\text{lbf}}{\text{in.}^2} \times \frac{144\text{ in.}^2}{\text{ft}^2} \times \frac{\text{lbm-R}}{53.3\text{ ft-lbf}} \times \frac{1}{530\text{ R}} = 0.152\text{ lbm/ft}^3$$

Substituting in Eq. 2

$$\dot{m} = \frac{0.152\text{ lbm}}{\text{ft}^3} \times 3\text{ ft}^3 \times \frac{0.17\text{ Btu}}{\text{lbm-F}} \times \frac{778\text{ ft-lbf}}{\text{Btu}} \times \frac{0.10\text{ F}}{\text{sec}} \times \frac{\text{lbm-R}}{53.3\text{ ft-lbf}} \times \frac{1}{530\text{ R}}$$

$$\dot{m} = 0.00213\text{ lbm/sec} \tag{$\dot{m}$}$$

$\left\{\begin{array}{l}\text{This problem illustrates the application of the energy equation to an unsteady flow}\\ \text{situation.}\end{array}\right\}$

4–9 The Second Law of Thermodynamics

We also include the control volume formulation of the second law of thermodynamics for completeness.

Recall that the system formulation of the second law is

$$\left.\frac{dS}{dt}\right)_{\text{system}} \geq \frac{1}{T}\dot{Q} \tag{4.5a}$$

where the total entropy of the system is given by

$$S_{\text{system}} = \int_{\text{Mass (system)}} s\,dm = \int_{\forall\text{ (system)}} s\rho\,d\forall \tag{4.5b}$$

The recipe for relating the system and control volume formulations is

$$\left.\frac{dN}{dt}\right)_{\text{system}} = \frac{\partial}{\partial t}\iiint_{\text{CV}} \eta\rho\, d\forall + \iint_{\text{CS}} \eta\rho\vec{V}\cdot d\vec{A} \qquad (4.12)$$

where

$$N_{\text{system}} = \int_{\text{Mass (system)}} \eta\, dm = \iiint_{\forall\,(\text{system})} \eta\rho\, d\forall \qquad (4.6)$$

To derive the control volume formulation of the second law of thermodynamics, we set

$$N = S \quad \text{and} \quad \eta = s$$

From Eq. 4.12, with this substitution, we obtain

$$\left.\frac{dS}{dt}\right)_{\text{system}} = \frac{\partial}{\partial t}\iiint_{\text{CV}} s\rho\, d\forall + \iint_{\text{CS}} s\rho\vec{V}\cdot d\vec{A} \qquad (4.45)$$

From Eq. 4.5a

$$\left.\frac{dS}{dt}\right)_{\text{system}} \geq \frac{1}{T}\dot{Q}$$

In deriving the recipe, the system and the control volume coincided at time, t; thus

$$\left.\frac{1}{T}\dot{Q}\right)_{\text{system}} = \left.\frac{1}{T}\dot{Q}\right)_{\text{CV}} = \iint_{\text{CS}} \frac{1}{T}\left(\frac{\dot{Q}}{A}\right) dA$$

In light of this, Eqs. 4.5a and 4.45 yield the control volume formulation of the second law of thermodynamics

$$\frac{\partial}{\partial t}\iiint_{\text{CV}} s\rho\, d\forall + \iint_{\text{CS}} s\rho\vec{V}\cdot d\vec{A} \geq \iint_{\text{CS}} \frac{1}{T}\left(\frac{\dot{Q}}{A}\right) dA \qquad (4.46)$$

In Eq. 4.46, the term $(\dot{Q}/A)$ represents the heat flux per unit area into the control volume through the area element dA. To evaluate the term

$$\iint_{\text{CS}} \frac{1}{T}\left(\frac{\dot{Q}}{A}\right) dA$$

both the heat flux, $(\dot{Q}/A)$, and local temperature, T, must be known for each area element of the control surface.

Example 4.17

Air to be used in a climate control system is first cooled to remove water vapor and then heated to a comfortable temperature as it flows through a 1 foot square duct. The air flowrate through the duct is 7.5 lbm/sec. Heat is removed at the rate of 35 Btu/sec in a section of duct 4 ft long, where the air and duct temperature are 80 F. Heat is supplied at 25 Btu/sec in a second duct section 2 ft long, where the temperature of the duct and air are 60 F. Both processes may be considered reversible. Determine the change in specific entropy in Btu/lbm-R.

EXAMPLE PROBLEM 4.17

GIVEN:

Air flow in a duct as shown.

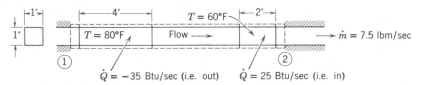

$\dot{Q} = -35$ Btu/sec (i.e. out) $\dot{Q} = 25$ Btu/sec (i.e. in)

FIND:

$s_2 - s_1$, in Btu/lbm-R.

SOLUTION:

Apply the second law of thermodynamics

Basic equation:
$$\overset{= 0(1)}{\cancel{\frac{\partial}{\partial t} \iiint_{CV} s\rho \, d\forall}} + \iint_{CS} s\rho \vec{V} \cdot d\vec{A} \geq \iint_{CS} \frac{1}{T}\left(\frac{\dot{Q}}{A}\right) dA$$

Assumptions: (1) Steady flow
 (2) Uniform flow at each section
 (3) Reversible processes, that is, equality holds in second law
 (4) T uniform at each heat transfer surface

Then

$$s_1\{-|\rho_1 V_1 A_1|\} + s_2\{|\rho_2 V_2 A_2|\} = \iint_{CS} \frac{1}{T}\left(\frac{\dot{Q}}{A}\right) dA$$

or

$$(s_2 - s_1)\dot{m} = \Sigma\left[\frac{1}{T}\left(\frac{\dot{Q}}{A}\right)A\right] = \Sigma\left[\frac{\dot{Q}}{T}\right]$$

EXAMPLE PROBLEM 4.17 (continued)

Thus

$$s_2 - s_1 = \frac{1}{\dot{m}} \Sigma \left[\frac{\dot{Q}}{T} \right]$$

$$= \frac{\text{sec}}{7.5 \text{ lbm}} \times \left[-\frac{35 \text{ Btu}}{\text{sec}} \times \frac{1}{(460 + 80) \text{ R}} + \frac{25 \text{ Btu}}{\text{sec}} \times \frac{1}{(460 + 60) \text{ R}} \right]$$

$$= \left(\frac{1}{7.5} \right) [-0.0647 + 0.0430] \frac{\text{Btu}}{\text{lbm-R}}$$

$$s_2 - s_1 = -0.00289 \text{ Btu/lbm-R} \qquad\qquad\qquad s_2 - s_1$$

This problem was included to illustrate the application of the second law of thermo-dynamics to a control volume. There are two artificial features of this problem:
  (i) reversible heat transfer—in real problems, heat transfer is almost always irreversible.
  (ii) constant temperature cooling and heating.

problems

4.1   A police investigation of tire marks shows that a car traveling along a straight level street had skidded for a total distance of 165 ft after the brakes were applied. The coefficient of friction between tires and pavement is estimated to be $\mu = 0.6$. What was the probable speed of the car when the brakes were applied?

4.2   A body of mass M is at rest on a horizontal plane. At time $t = 0$, the body is acted upon by a constant force, F_1, of 10 lbf. The force is applied for a period of 5 sec. Immediately upon removal of this force, the body is acted upon by a constant force, F_2, of 3 lbf in the opposite direction. There is no friction. Determine the length of time of action by F_2 required to bring the body to rest.

4.3   An expanding gas does 6000 ft-lbf of work while it receives 10 Btu of heat. Calculate the change in its stored energy.

4.4   During a certain process a system does 30 Btu of work, while 10 Btu of heat is removed. Then the system is restored to its initial state by means of a process in which 4 Btu of heat is added. Determine ΔE for this second process.

4.5   The average heat transfer from a person to the surroundings when not actively working is about 600 Btu/hr. Suppose that in an auditorium containing 6000 people the ventilation system fails.

   (a) How much does the internal energy of the air in the auditorium increase during the first 15 minutes after the ventilation system fails?

(b) Considering the auditorium and all the people as a system, and assuming no heat transfer to the surroundings, how much does the internal energy of the system change? How do you account for the fact that the temperature of the air increases?

4.6 A weight is dropped from a height of 10 ft above the end of a spring. When the motion of the weight is stopped by the spring, the spring has been compressed 2 ft. If the spring constant is 1 lbf/in., calculate the magnitude of the weight.

4.7 Calculate the amount of work required to accelerate a 3000 lbm automobile from rest to a speed of 50 ft/sec on a level highway if there are no frictional effects.

4.8 A cylinder fitted with a piston contains 5 lbm of H_2O at a pressure of 200 psia and a quality of 80 percent. The piston is restrained by a spring arranged so that for zero volume in the cylinder the spring is fully extended. The spring force is proportional to the spring displacement. The weight of the piston may be neglected so that the force on the spring is exactly balanced by the pressure forces on the H_2O.

Heat is transferred to the H_2O until its volume is 150 percent of the initial volume.

(a) What is the final pressure?
(b) What is the quality (if saturated) or temperature (if superheated) in the final state.
(c) Draw a p-V diagram and determine the work delivered to the spring.
(d) Determine the heat transfer.

4.9 Fluid with a density of 2 slug/ft³ is flowing steadily through a rectangular box. If

$$A_1 = 0.05 \text{ ft}^2 \qquad A_2 = 0.01 \text{ ft}^2 \qquad A_3 = 0.06 \text{ ft}^2$$
$$\vec{V}_1 = 40\hat{\imath} \text{ ft/sec} \qquad \vec{V}_2 = -80\hat{\jmath} \text{ ft/sec}$$

determine the velocity $\vec{V}_3$ (Fig. 4.7).

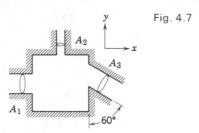

Fig. 4.7

4.10 Consider steady, incompressible flow through the device shown in Fig. 4.8. Determine the volumetric flowrate through port 3.

4/basic equations in integral form for a control volume

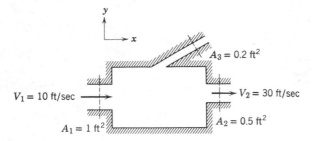

Fig. 4.8

4.11 Air at atmospheric conditions (15 psia and 40 F) enters a compressor at a rate of 500 ft^3/min. The air is discharged at 120 psia and 140 F.

If the velocity in the discharge line must be limited to 60 ft/sec, calculate the required diameter of the line.

4.12 In the incompressible flow through the device shown in Fig. 4.9, velocities may be considered uniform over the inlet and outlet sections. If the fluid flowing is

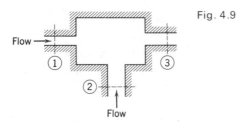

Fig. 4.9

water, determine an expression for the mass flowrate at section ③. The following conditions are known:

$$A_1 = 1.0 \text{ ft}^2 \qquad A_2 = 2.0 \text{ ft}^2 \qquad A_3 = 1.5 \text{ ft}^2$$
$$V_1 = 10 \text{ ft/sec} \qquad V_2 = 20 + 10 \cos(4\pi t) \text{ ft/sec}$$

4.13 A flow field is given by $\vec{V} = 2y\hat{i} + \hat{k}$, where $\vec{V}$ is in ft/sec. The volumetric flow rate, Q, through a surface is given by

$$Q = \iint_A \vec{V} \cdot d\vec{A}$$

Evaluate the volumetric flow through the shaded surface in Fig. 4.10. All dimensions are in feet.

187   **problems**

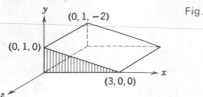

Fig. 4.10

(0, 1, -2)
(0, 1, 0)
(3, 0, 0)

4.14 Water is flowing steadily through a pipe of length L and radius $R = 3$ in. (Fig. 4.11). The velocity distribution across the outlet is given by

$$u = 10\left[1 - \frac{r^2}{R^2}\right] \text{ ft/sec}$$

Calculate the value of the uniform inlet velocity, U.

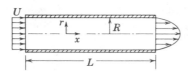

Fig. 4.11

4.15 Water flowing out of a circular pipe is assumed to have a linearly varying velocity distribution as shown in Fig. 4.12. What is the average velocity of the outward flow in terms of V_{max}?

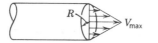

Fig. 4.12

4.16 Water enters a two-dimensional channel of height, $2h$, with a uniform velocity of 10 ft/sec. At the outlet of the channel the velocity distribution is given by

$$\frac{u}{U_\mathfrak{C}} = 1 - \left(\frac{y}{h}\right)^2$$

The coordinate y is measured from the centerline of the channel. Determine the exit centerline velocity, $U_\mathfrak{C}$.

4.17 A tank of 15 ft³ contains compressed air. A valve is opened and air escapes with a velocity of 900 ft/sec through an opening of 0.2 in.² area. Air temperature passing through the opening is 10 F and the pressure is 53 psia. Find the rate of change of density of the air in the tank at this moment.

4.18 Air enters a tank through an area of 0.2 ft^2 with a velocity of 15 ft/sec and a density of 0.03 slug/ft^3. Air leaves with a velocity of 5 ft/sec and a density equal to that in the tank. The initial density of the air in the tank is 0.02 slug/ft^3. The total tank volume is 20 ft^3. Exit area is 0.4 ft^2. Find the initial rate of change of density in the tank.

4.19 The vessel shown in Fig. 4.13 is 6 ft long in the direction normal to the diagram. There is no flow into the vessel, but a chemical reaction that occurs in the vessel

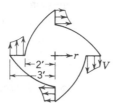

Fig. 4.13

generates gas that leaves through the four openings (each 1 ft by 6 ft in cross-section area) as shown. The velocity relative to the vessel and the density of the gas leaving vary with radius as

$$V = \frac{10}{r} \quad \text{and} \quad \rho = 0.0020 + 0.001r$$

where V is in ft/sec, ρ is in slug/ft^3, and r is in feet. Determine the rate of change of mass in the vessel in slug/sec.

4.20 A section of pipe carrying water contains an expansion chamber with a free surface whose area is 2 ft^2 (Fig. 4.14). The inlet and outlet pipes are both 1 ft^2 in area.

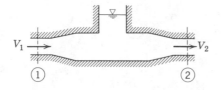

Fig. 4.14

At a given instant, the velocity at section ① is 10 ft/sec into the chamber. Water flows out at section ② at 12 ft^3/sec. Both flows are uniform.

Find the rate of change of free surface level at the given instant, in ft/sec. Indicate whether the level rises or falls.

4.21 A cylindrical tank, 1 ft in diameter, drains through a hole in its bottom. At the instant when the water depth is 2 ft, the flowrate from the tank is observed to be 0.25 slug/sec. Determine the rate of change of water level at this instant.

4.22 Evaluate the net rate of efflux of momentum through the control surface of Problem 4.9.

4.23 For the conditions of Problem 4.14, evaluate the ratio of the x direction momentum flux at the pipe outlet to that at the inlet.

4.24 For the conditions of Problem 4.16, evaluate the ratio of the x direction momentum flux at the channel outlet to that at the inlet.

4.25 Consider the steady flow of water between two parallel plates a distance h feet apart (Fig. 4.15). At station ① the velocity is uniform across the width; the velocity distribution is assumed to be linear at station ②. The flow is identical in all planes

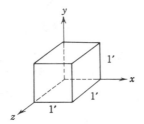

Fig. 4.15

parallel to the plane of the diagram. Calculate the ratio of the x direction momentum flux at station ② to that at station ① for the assumed velocity distributions.

4.26 Consider the flow of an incompressible fluid with vector velocity field:

$$\vec{V} = (x + 2t)\hat{i} - y\hat{j}$$

For the control volume shown (a 1 ft cube, Fig. 4.16), evaluate the rate of change of momentum within the control volume.

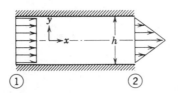

Fig. 4.16

4.27 Oil is being loaded into a barge through a 6 in. diameter pipe (Fig. 4.17). The oil,

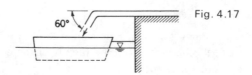

Fig. 4.17

4/basic equations in integral form for a control volume

$\rho = 2$ slug/ft^3, leaves the pipe with a uniform velocity of 20 ft/sec. Determine the force of the rope on the barge.

4.28   A jet of water issuing from a stationary nozzle with a velocity of 40 ft/sec (jet area = 0.5 ft^2) strikes a turning vane mounted on a cart as shown in Fig. 4.18. The vane turns the jet through an angle $\theta = 50°$. Determine the value of W required to hold the cart stationary.

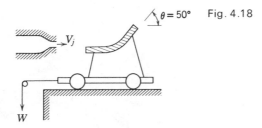

Fig. 4.18

4.29   If the vane angle, θ, of Problem 4.28 is adjustable, plot the weight, W, needed to hold the cart stationary as a function of θ for $0 \le \theta \le 180°$.

4.30   A large tank is affixed to a cart as shown in Fig. 4.19. Water issues from the tank through a 1 in.^{2} nozzle at a velocity of 20 ft/sec. The water level in the tank is maintained constant by adding water through a vertical pipe. Determine the tension in the wire holding the cart stationary.

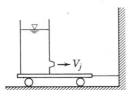

Fig. 4.19

4.31   A turning vane (Fig. 4.20), which deflects the water through an angle of 60°, is attached to the cart under the conditions of Problem 4.30. Determine

(a)   the tension in the wire holding the cart stationary, and

(b)   the force of the vane on the cart.

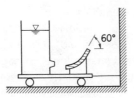

Fig. 4.20

4.32　If the turning vane of Problem 4.31 deflects the water through an angle of 90°, determine the tension in the wire holding the cart stationary.

4.33　A large irrigation sprinkler unit, mounted on a cart, discharges water with a velocity of 120 ft/sec at an angle of 30° to the horizontal (Fig. 4.21). The 2 in. diameter nozzle is 10 ft above the ground. Calculate the magnitude of the moment that tends to overturn the cart.

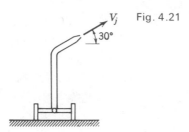

Fig. 4.21

4.34　Shown in Fig. 4.22 is an adjustable fire-fighting nozzle assembly. In the box marked M is the connection that makes it possible to turn and elevate the nozzle. For the position shown, calculate the horizontal force exerted by this assembly on the supply pipe when the supply pressure is 70 psig and the jet velocity of water leaving the nozzle, as a free jet at atmospheric pressure, is 100 fps. The nozzle diameter is 1 in.

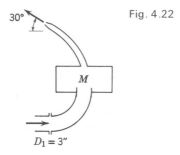

Fig. 4.22

4.35　Water flows steadily through the double nozzle assembly shown (Fig. 4.23) in a horizontal plane. The cross-section area of the supply line is 0.10 ft², and the outlet cross-section area of each of the two nozzles is 0.01 ft². The pressure of the water entering at section ① is 20 psig. The water velocity at each nozzle outlet is 50 ft/sec. Calculate the net horizontal force of the nozzle assembly on the flange of the supply pipe at section ①.

　4/basic equations in integral form for a control volume

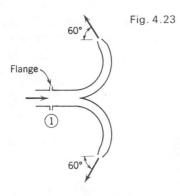

Fig. 4.23

4.36   Water is flowing steadily through the 180° elbow shown in Fig. 4.24. At the inlet
to the elbow the pressure, $p_1 = 14$ psig. The water discharges to atmospheric
pressure. Properties are assumed to be uniform over the inlet and outlet areas.
$A_1 = 4$ in.², $A_2 = 1$ in.², $V_1 = 10$ ft/sec. Find the horizontal component of the
force required to hold the elbow in place.

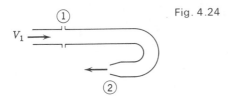

Fig. 4.24

4.37   Water flows steadily through a fire hose and nozzle. The hose is 3 in. inside diameter,
and the nozzle tip is 1 in. i.d.; water pressure in the hose is 75 psig, and the stream
leaving the nozzle is uniform. The exit velocity and pressure are 106 ft/sec, and
atmospheric, respectively.
    Find the force transmitted by the coupling between the nozzle and hose, in
pounds. Indicate whether the coupling is in tension or compression.

4.38   Consider the steady adiabatic flow of air through a long straight pipe with a
cross-section area of 0.5 ft². At the inlet, the air is at 30 psia, 140 F, and has a
velocity of 500 ft/sec. At the exit, the air is at 11.3 psia and has a velocity of
985 ft/sec. Calculate the axial force of the air on the pipe. (Be sure to make the
direction clear.)

4.39   A gas flows steadily through a heated porous pipe of constant 2 ft² cross-section
area. At the pipe inlet, the pressure is 50 psia, the density is 0.010 slug/ft³, and
the mean velocity is 500 fps. The fluid passing through the porous wall leaves in a
direction normal to the pipe axis, and the total flowrate through the porous wall

is 2 slug/sec. At the pipe outlet, the pressure is 40 psia and the density is 0.005 slug/ft³. Determine the force in the axial direction of the fluid on the pipe.

4.40   A fluid of constant density, ρ, enters a pipe of radius R with a uniform velocity, V. At a downstream section the velocity varies with radius r according to the equation

$$u = 2V\left(1 - \frac{r^2}{R^2}\right)$$

The pressure at sections ① (inlet) and ② (downstream) are p_1 and p_2, respectively. Show that the frictional force, F, of the pipe walls on the fluid between sections ① and ② is

$$F = \pi R^2[-(p_1 - p_2) + \tfrac{1}{3}\rho V^2]$$

in a direction opposing the flow.

4.41   A water jet pump has a jet area of 0.10 ft² and a jet velocity of 90 ft/sec (Fig. 4.25). The jet is within a secondary stream of water having a velocity of 10 ft/sec. The

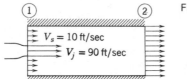

Fig. 4.25

$V_s = 10$ ft/sec
$V_j = 90$ ft/sec

total area of the pipe (the sum of the jet and secondary stream areas) is 0.75 ft². The water is thoroughly mixed leaving the jet pump. The pressures of the jet and secondary stream are the same at the pump inlet. Determine:

(a) The velocity at the pump exit.
(b) The pressure change, p_1-p_2.

4.42   A circular cylinder inserted across a stream of flowing water deflects the stream through an angle θ as shown in Fig. 4.26. (This is termed the "Coanda effect.") For $a = 0.5$ in., $b = 0.1$ in., $V_j = 10$ ft/sec, and $\theta = 20°$, determine the horizontal component of the force on the cylinder due to the flowing water.

4.43   The air intake duct of a jet engine being tested on a stationary stand takes in air at an average velocity of 600 ft/sec. Air enters the intake duct, and exhaust gases exit the tail at atmospheric pressure. Fuel enters the top of the engine at the rate of 1 slug of fuel to 50 slug of intake air. The intake duct area is 2 ft². The exist velocity of the exhaust gases is 1200 fps. The density of the entering air is 0.00238 slug/ft³. Find the force exerted on the test stand.

194   4/basic equations in integral form for a control volume

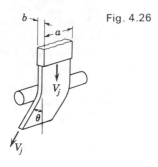

Fig. 4.26

4.44 The General Electric CF6-50A jet engine has produced a thrust force of 52,000 lbf on the test stand. A typical installation is shown in Fig. 4.27, together with some test data. The mass flowrate of fuel may be neglected compared with the air flowrate. For the given conditions, compute the air flowrate through the engine in lbm/sec.

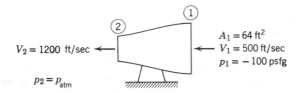

$V_2 = 1200$ ft/sec

$p_2 = p_{atm}$

$A_1 = 64$ ft^2
$V_1 = 500$ ft/sec
$p_1 = -100$ psfg

Fig. 4.27

4.45 A jet of water is directed against a vane (see Fig. 4.28), which could be a blade in a turbine or in any other piece of hydraulic machinery. The water leaves the stationary 2-in. diameter nozzle with a velocity of 60 ft/sec and enters the vane tangent to the surface at A. The inside surface of the vane at B makes an angle of 150° with the x direction. Compute the resultant net force on the vane when the vane is moving with a constant velocity of 20 ft/sec in the positive x direction.

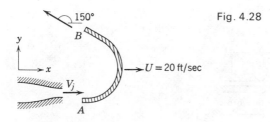

Fig. 4.28

$150°$
B
$U = 20$ ft/sec
V_j
A

195   **problems**

4.46 Water from a stationary nozzle impinges on a moving vane with a turning angle of 120°. The vane moves away from the nozzle with a constant velocity $U = 30$ ft/sec and receives a jet that leaves the nozzle with velocity $V_j = 100$ ft/sec. The nozzle has an exit area of 0.04 ft². Find the net force on the moving vane.

4.47 A water jet, issuing from a stationary nozzle, encounters a curved vane (curved through an angle of 90°) that is moving away from the nozzle at a constant velocity of 50 ft/sec. The jet has a cross-section area of 1.0 in.² and a velocity of 100 ft/sec. Determine the net force on the vane.

4.48 A jet of oil ($\gamma = 50$ lbf/ft³) strikes a curved blade that turns the fluid through an angle of 180°. The jet area is 2.0 in.² and the velocity relative to the stationary nozzle is 60 ft/sec. The blade moves toward the nozzle at 30 ft/sec. Determine the net force on the blade.

4.49 A jet of lye solution (specific gravity = 1.10) 2 in. in diameter has an absolute velocity of 40 ft/sec. It strikes a single flat plate that is moving away from the nozzle with an absolute velocity of 15 ft/sec. The plate makes an angle of 60° with the horizontal. Calculate the total force acting on the plate. Assume no friction force along the plate surface.

4.50 A sled, weighing 30 lbf, is moving along a horizontal track; the coefficient of rolling friction between sled and track is 0.02. At the instant when the sled velocity is 100 ft/sec, a small air jet affixed to the sled is activated in an effort to reduce the sled velocity. The air jet discharges to the atmosphere through a nozzle whose area is 1 in.² The air discharges with a velocity of 480 ft/sec relative to the sled and a density of 0.002 slug/ft³. Calculate the sled acceleration at the instant following activation of the jet.

4.51 A rocket sled weighing 3 tons, including 1 ton of fuel, rests on a level section of ground. At time $t = 0$, the solid fuel of the rocket is ignited and the rocket burns fuel at the rate of 150 lbm/sec. The exit velocity of the exhaust gas relative to the rocket is 3500 ft/sec. Neglecting friction and air resistance, calculate the acceleration of the sled at time $t = 10$ sec.

4.52 A rocket sled weighing 10,000 lbf, traveling 600 mph, is to be braked by lowering a scoop into a water trough (Fig. 4.29). The scoop is 6 in. wide. Determine the acceleration when the scoop is immersed to a depth of 3 in. below the water surface.

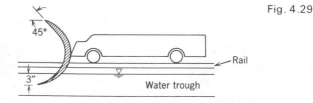

Fig. 4.29

4.53   A box car is moving to the left with a velocity of 15 ft/sec and is accelerating in the same direction at 0.5 ft/sec². A stream of grain falls into the car from a vertical stationary pipe. The velocity of the grain is 24 ft/sec in a downward direction relative to the pipe. The mass flow of grain leaving the pipe is 50 lbm/sec.

    Determine the force required to move the car as indicated at the instant when the total mass of car and contents is 15,000 lbm. Neglect friction between car and track.

4.54   A vehicle is moving at a velocity of 100 fps along level ground under the action of a constant force, $F = 1000$ lbf. At time $t = 0$, mass begins leaving the vehicle through a hole in the bottom. Assuming the mass leaves the vehicle vertically at a rate of 10 lbm/sec and the vehicle continues to move under the action of the constant force, F, determine the velocity of the vehicle after 20 sec. Initial mass of the vehicle is 2000 lbm.

4.55   For the conditions of problem 4.52 determine the time required (after lowering the scoop into the water) to bring the sled to a velocity of 20 mph.

4.56   Neglecting air resistance, what velocity would a vertically directed rocket attain in 10 sec if it starts from rest, initially weighs 200 lbm, burns 10 lbm/sec, and ejects gas at 3220 fps relative to the rocket?

4.57   A "home-made" solid propellant rocket has a mass of 20 lbm; 15 lbm of this is fuel. The rocket is directed vertically upward from rest, burns fuel at a constant rate of 0.5 lbm/sec, and ejects exhaust gas at a velocity of 1500 ft/sec relative to the rocket. Assume that the pressure at the exit is atmospheric and that air resistance may be neglected. Calculate

(a) The rocket velocity after 20 sec.
(b) The distance traveled by the rocket in the 20 sec.

4.58   A manned space vehicle with a total mass of 100 slugs is traveling in level flight (Fig. 4.30). A retro-rocket on the vehicle is fired for 10 sec. The rocket discharges

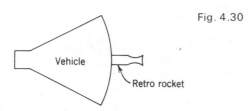

Fig. 4.30

0.5 slug/sec to ambient pressure with a velocity of 5000 ft/sec relative to the vehicle. Neglect any frictional drag forces.

(a) Calculate the deceleration of the vehicle at the end of the rocket firing, that is, at $t = 10$ sec.

(b) At $t = 10$ sec, determine the force exerted by the rocket on the vehicle if the mass of the rocket is 10 slugs at the instant of firing.

4.59   The Boeing 727 aircraft has 3 engines mounted in the tail of the plane. Each engine is equipped with a "thrust reverser" that provides for reversal of the engine exhaust as shown in Fig. 4.31. On touching down at the end of the runway the plane is

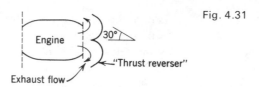

Fig. 4.31

traveling at a velocity of 186 ft/sec and the brakes fail. The total mass of the plane is 131,000 lbm.

$$|\dot{m}V| \text{ out of each engine} = 14{,}400 \text{ lbf}$$
$$|\dot{m}V| \text{ into each engine is negligible}$$

Neglecting the drag force of the air on the plane, determine:

(a) The time required to bring the plane to a stop.
(b) The distance the plane will travel in that time.

4.60   A rocket sled weighs 4 tons including 1 ton of fuel. The motion resistance in the track on which the sled rides and that of the air total KV, where K is 10 lbf/ft/sec, and V is the velocity of the sled in ft/sec. The exit velocity of the exhaust gas relative to the rocket is 4,000 ft/sec, and it burns fuel at the rate of 200 lbm/sec. Compute the velocity of the sled after 10 sec.

4.61   Consider the vehicle shown in Fig. 4.32. Starting from rest, it is propelled by a hydraulic catapult (liquid jet). The jet strikes the curved surface and makes a

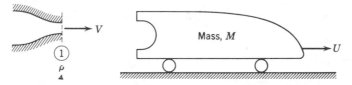

Fig. 4.32

180° turn, leaving horizontally. Air and rolling resistance may be neglected. Using the notation shown:

(a) Obtain an equation for the acceleration of the vehicle at any time, t.
(b) Determine the time required for the vehicle to reach $U = V/2$.

4.62 The mass of the cart in problem 4.61 is 200 lbm. The jet of water leaves the nozzle (area of 0.01 ft²) with a velocity of 100 ft/sec. Determine the velocity of the cart 5 sec after the jet is directed against the cart.

4.63 A jet, $\rho = 1.92$ slug/ft³, issues from a stationary nozzle; the jet is deflected by a vane through 135°. The vane is attached to the cart as shown in Fig. 4.33. Assume

Fig. 4.33

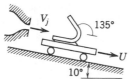

that the cart moves without friction down the 10° plane. The cart weighs 200 lbf. The jet leaves the nozzle at atmospheric pressure with velocity $V_j = 200$ ft/sec relative to the nozzle. If the area of the jet is $A_j = 1$ in.², determine the velocity of the cart 5 sec after the jet is directed against the vane.

4.64 A rocket cart that is initially at rest and weighs 1610 lbf is to be fired and is to have a constant acceleration of 20 ft/sec² (Fig. 4.34). To accomplish this, the

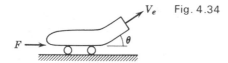

V_e Fig. 4.34

exhaust gases will be deflected through an angle θ that varies as a function of time. This will account for a frictional force that is proportional to the velocity squared, $F = 0.002V^2$ where F is in lbf and V is in ft/sec. The rocket exhausts gases at the rate of 1 slug/sec at constant velocity, $V_e = 1000$ ft/sec, relative to the vehicle.

(a) Find an expression for $\cos \theta$ as a function of time, t.
(b) Determine t at which $\cos \theta$ is a minimum.
(c) Find θ_{max}.

4.65 A turbine is supplied by 20 ft³/sec of water from a 12 in. pipe; the discharge pipe has a 16 in. diameter. Determine the pressure drop across the turbine if it delivers 75 hp.

4.66 A pump draws water from a reservoir through a 6 in. diameter suction pipe and delivers it to a 3 in. diameter discharge pipe. The end of the suction pipe is 5 ft below the free surface of the reservoir. The pressure gage on the discharge pipe (5 ft above the reservoir surface) reads 25 psig. The average velocity in the discharge pipe is 10 ft/sec. If the pump has an efficiency of 75 percent, determine the horse-power required to drive the pump.

4.67   A pump system in a dishwasher circulates 55 gallons of water per minute. The total head, $p/\rho + V^2/2$, leaving the pump is 120 in. of water. The head at the pump inlet may be neglected. If the power supplied to the pump is 0.4 hp, determine its efficiency.

4.68   Air enters a compressor at 14.7 psia, 60 F with a velocity of 200 ft/sec and leaves at 30.0 psia, 160 F, with a velocity of 400 ft/sec. The flowrate is 2.0 lbm/sec. The cooling water circulating around the compressor casing removes 8.0 Btu/lbm of air. Determine the power required by the compressor.

4.69   Air enters a compressor at 14 psia, 80 F with negligible velocity and is discharged at 70 psia, 500 F with a velocity of 500 ft/sec. If the power input is 3200 hp and the flowrate is 20 lbm/sec, determine the heat transfer in Btu/lbm of air.

4.70   Air is drawn from the atmosphere (14.7 psia, 60 F) into a turbomachine. At the exit, conditions are 75 psig and 260 F. The exit velocity is 300 ft/sec, and the mass flowrate is 0.05 slug/sec. Flow is steady and there is no heat transfer. Compute the shaft work interaction with the surroundings, in ft-lbf/sec.

4.71   Sand, at the rate of 100 lbm/sec, falls from a hopper onto a moving conveyor belt as shown in Fig. 4.35. Any friction in the drive pulleys and support rollers may be neglected, as they are supported on air bearings. End effects at the discharge end may also be neglected. If $\theta = 30°$, $L = 500$ ft, $R = 3$ ft, and $V = 10$ ft/sec, deter-mine the power required to drive the conveyor.

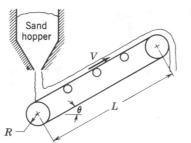

Fig. 4.35

4.72   The total weight of the helicopter-type craft shown in Fig. 4.36 is 3000 lbf. The pressure of the air is atmospheric at both inlet and outlet. Assume that the flow is steady and one dimensional. Treat the air as being incompressible at standard conditions (14.7 psia, 59 F) and calculate for a hovering position:

(a) The velocity of air leaving the craft.
(b) The minimum power that must be delivered to the air by the propeller.

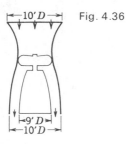

Fig. 4.36

4.73   A fire truck draws water from an open constant level reservoir through a hose of 0.2 ft² cross-section. The water is discharged from the pump through a 0.2 ft² hose to a surge tank pressurized to 120 psig with air. The pump adds 970 ft-lbf/slug of energy to the fluid; the water level in the surge tank is 3 ft above the pump center-line. From the surge tank the water flows steadily through a hose of 0.04 ft² area to a nozzle of 0.01 ft² area. The nozzle is held at the same level as the water free surface in the surge tank. Neglect all frictional losses and sketch the system. Find:

    (a)   Velocity of the water leaving the nozzle.
    (b)   Elevation of the pump above the reservoir.
    (c)   Power required to drive the pump if it has an efficiency of 91 percent.
    (d)   Maximum height above the reservoir that the stream of water can reach.

4.74   All major harbors are equipped with "fire-boats" for combating ship fires. A 3 in. diameter hose is attached to the discharge of a 10 hp pump on such a boat. The nozzle attached to the end of the hose has a diameter of 1 in. If the nozzle discharge is held 8 ft above the surface of the water determine

    (a)   The volume flowrate through the nozzle.
    (b)   The maximum height to which the water will rise.
    (c)   The force on the boat if the water jet is directed horizontally over the stern.

introduction to differential analysis of fluid motion

In Chapter 4, we developed the basic equations in integral form for a control volume. The integral equations are particularly useful when we are interested in the gross behavior of a flow field and its effect on various devices. However, the integral approach does not provide a detailed (i.e. point by point) knowledge of the flow field.

To obtain this detailed knowledge we must apply the equations of fluid motion in differential form. In this chapter we shall develop differential equations for the conservation of mass and Newton's second law of motion. Since we are interested in formulating differential equations, our analysis is in terms of infinitesimal systems and control volumes.

5–1 Review of the Field Concept

In Chapter 2, we found that the continuum assumption—that is, the assumption that a fluid could be treated as a continuous distribution of matter—led directly to a field representation of fluid properties. The property fields are then defined by continuous functions of the space coordinates and time. We may speak of a scalar field [$\rho = \rho(x, y, z, t)$], a vector field [$\vec{V} = \vec{V}(x, y, z, t)$] or a tensor field (the stress field).

We shall derive the differential equation for the conservation of mass in both rectangular and cylindrical coordinates. In both cases the derivation is carried out by applying conservation of mass to a differential control volume.

5–2.1 RECTANGULAR COORDINATE SYSTEM

In rectangular coordinates, the control volume chosen is an infinitesimal cube with sides of length dx, dy, dz as shown in Fig. 5.1. The density at the center, O, of the control volume is ρ and the velocity there is $\vec{V} = \hat{\imath}u + \hat{\jmath}v + \hat{k}w$.

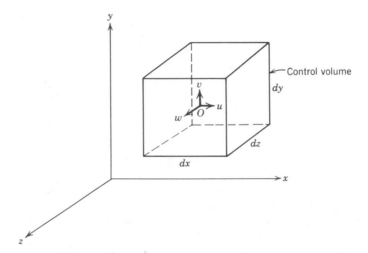

Fig. 5.1 *Differential control volume in rectangular coordinates.*

In order to determine the values of the properties at each of the six faces of the control surface, we must use a Taylor series expansion of the properties about the point O. For example, at the right face,

$$\rho\Big)_{x+dx/2} = \rho + \left(\frac{\partial\rho}{\partial x}\right)\frac{dx}{2} + \left(\frac{\partial^2\rho}{\partial x^2}\right)\frac{1}{2!}\left(\frac{dx}{2}\right)^2 + \cdots$$

Neglecting higher order terms we can write

$$\rho\Big)_{x+dx/2} = \rho + \left(\frac{\partial\rho}{\partial x}\right)\frac{dx}{2}$$

5/introduction to differential analysis of fluid motion

Similarly,

$$u\Big)_{x+dx/2} = u + \left(\frac{\partial u}{\partial x}\right)\frac{dx}{2}$$

The corresponding terms at the left face are

$$\rho\Big)_{x-dx/2} = \rho + \left(\frac{\partial \rho}{\partial x}\right)\left(-\frac{dx}{2}\right) = \rho - \left(\frac{\partial \rho}{\partial x}\right)\frac{dx}{2}$$

$$u\Big)_{x-dx/2} = u + \left(\frac{\partial u}{\partial x}\right)\left(-\frac{dx}{2}\right) = u - \left(\frac{\partial u}{\partial x}\right)\frac{dx}{2}$$

A word statement of the conservation of mass is:

$$\left[\begin{array}{c} \text{Net rate of mass efflux} \\ \text{through the control surface} \end{array}\right] + \left[\begin{array}{c} \text{Rate of change of mass} \\ \text{inside the control volume} \end{array}\right] = 0$$

In order to evaluate the first term in this equation, we must evaluate the mass flux through each of the six surfaces of the control surface, that is, we must evaluate $\iint_{cs} \rho \vec{V} \cdot d\vec{A}$. The details of this evaluation are shown in Table 5.1.

We see that the net rate of mass efflux through the control surface is given by

$$\left[\frac{\partial \rho u}{\partial x} + \frac{\partial \rho v}{\partial y} + \frac{\partial \rho w}{\partial z}\right] dx\, dy\, dz$$

The mass inside the control volume at any instant is the product of the mass per unit volume, ρ, and the volume, $dx\, dy\, dz$. Thus the rate of change of mass inside the control volume is given by

$$\frac{\partial \rho}{\partial t} dx\, dy\, dz$$

In rectangular coordinates the differential equation for the conservation of mass is then

$$\frac{\partial \rho u}{\partial x} + \frac{\partial \rho v}{\partial y} + \frac{\partial \rho w}{\partial z} + \frac{\partial \rho}{\partial t} = 0 \qquad (5.1a)$$

Equation 5.1a may be written more compactly in vector notation, since

$$\frac{\partial \rho u}{\partial x} + \frac{\partial \rho v}{\partial y} + \frac{\partial \rho w}{\partial z} = \nabla \cdot \rho \vec{V}$$

the continuity equation

table 5.1 *Mass Flux through the Control Surface of a Rectangular Differential Control Volume*

SURFACE	$\iint \rho \vec{V} \cdot d\vec{A}$

Left
$(-x)$
$$= -\left[\rho - \left(\frac{\partial \rho}{\partial x}\right)\frac{dx}{2}\right]\left[u - \left(\frac{\partial u}{\partial x}\right)\frac{dx}{2}\right] dy\,dz = -\rho\,u\,dy\,dz + \frac{1}{2}\left[u\left(\frac{\partial \rho}{\partial x}\right) + \rho\left(\frac{\partial u}{\partial x}\right)\right] dx\,dy\,dz$$

Right
$(+x)$
$$= \left[\rho + \left(\frac{\partial \rho}{\partial x}\right)\frac{dx}{2}\right]\left[u + \left(\frac{\partial u}{\partial x}\right)\frac{dx}{2}\right] dy\,dz = \rho\,u\,dy\,dz + \frac{1}{2}\left[u\left(\frac{\partial \rho}{\partial x}\right) + \rho\left(\frac{\partial u}{\partial x}\right)\right] dx\,dy\,dz$$

Bottom
$(-y)$
$$= -\left[\rho - \left(\frac{\partial \rho}{\partial y}\right)\frac{dy}{2}\right]\left[v - \left(\frac{\partial v}{\partial y}\right)\frac{dy}{2}\right] dx\,dz = -\rho\,v\,dx\,dz + \frac{1}{2}\left[v\left(\frac{\partial \rho}{\partial y}\right) + \rho\left(\frac{\partial v}{\partial y}\right)\right] dx\,dy\,dz$$

Top
$(+y)$
$$= \left[\rho + \left(\frac{\partial \rho}{\partial y}\right)\frac{dy}{2}\right]\left[v + \left(\frac{\partial v}{\partial y}\right)\frac{dy}{2}\right] dx\,dz = \rho\,v\,dx\,dz + \frac{1}{2}\left[v\left(\frac{\partial \rho}{\partial y}\right) + \rho\left(\frac{\partial v}{\partial y}\right)\right] dx\,dy\,dz$$

Back
$(-z)$
$$= -\left[\rho - \left(\frac{\partial \rho}{\partial z}\right)\frac{dz}{2}\right]\left[w - \left(\frac{\partial w}{\partial z}\right)\frac{dz}{2}\right] dx\,dy = -\rho\,w\,dx\,dy + \frac{1}{2}\left[w\left(\frac{\partial \rho}{\partial z}\right) + \rho\left(\frac{\partial w}{\partial z}\right)\right] dx\,dy\,dz$$

Front
$(+z)$
$$= \left[\rho + \left(\frac{\partial \rho}{\partial z}\right)\frac{dz}{2}\right]\left[w + \left(\frac{\partial w}{\partial z}\right)\frac{dz}{2}\right] dx\,dy = \rho\,w\,dx\,dy + \frac{1}{2}\left[w\left(\frac{\partial \rho}{\partial z}\right) + \rho\left(\frac{\partial w}{\partial z}\right)\right] dx\,dy\,dz$$

- - - - - - - - - -

Then,

$$\iint_{CS} \rho \vec{V} \cdot d\vec{A} = \left[\left\{u\left(\frac{\partial \rho}{\partial x}\right) + \rho\left(\frac{\partial u}{\partial x}\right)\right\} + \left\{v\left(\frac{\partial \rho}{\partial y}\right) + \rho\left(\frac{\partial v}{\partial y}\right)\right\} + \left\{w\left(\frac{\partial \rho}{\partial z}\right) + \rho\left(\frac{\partial w}{\partial z}\right)\right\}\right] dx\,dy\,dz$$

or

$$\iint_{CS} \rho \vec{V} \cdot d\vec{A} = \left[\frac{\partial \rho u}{\partial x} + \frac{\partial \rho v}{\partial y} + \frac{\partial \rho w}{\partial z}\right] dx\,dy\,dz$$

where ∇ in rectangular coordinates is given by

$$\nabla = \hat{\imath}\frac{\partial}{\partial x} + \hat{\jmath}\frac{\partial}{\partial y} + \hat{k}\frac{\partial}{\partial z}$$

Then the conservation of mass may be written as

$$\nabla \cdot \rho\vec{V} + \frac{\partial \rho}{\partial t} = 0 \qquad\qquad (5.1b)$$

Two flow cases for which the differential continuity equation may be simplified are worthy of note. For incompressible flow, ρ = constant, that is the density is neither a function of space coordinates, nor time. For incompressible flow, the continuity equation simplifies to

$$\frac{\partial u}{\partial x} + \frac{\partial v}{\partial y} + \frac{\partial w}{\partial z} = 0$$

or

$$\nabla \cdot \vec{V} = 0$$

Thus the velocity field, $\vec{V}(x, y, z, t)$, for incompressible flow must satisfy $\nabla \cdot \vec{V} = 0$.

For steady flow, all fluid properties are, by definition, independent of time. Thus at most $\rho = \rho(x, y, z)$, and for steady flow, the continuity equation can be written as

$$\frac{\partial \rho u}{\partial x} + \frac{\partial \rho v}{\partial y} + \frac{\partial \rho w}{\partial z} = 0$$

or

$$\nabla \cdot \rho \vec{V} = 0$$

5–2.2 CYLINDRICAL COORDINATE SYSTEM

In cylindrical coordinates, a suitable differential control volume is shown in Fig. 5.2. The density at the center, O, of the control volume is ρ and the velocity there is

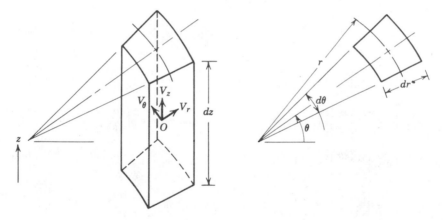

Fig 5.2 *Differential control volume in cylindrical coordinates.*

$\vec{V} = \hat{\imath}_r V_r + \hat{\imath}_\theta V_\theta + \hat{\imath}_z V_z$, where $\hat{\imath}_r$, $\hat{\imath}_\theta$, $\hat{\imath}_z$ are unit vectors in the r, θ, z directions, respectively, and V_r, V_θ, V_z are the velocity components in the r, θ, z directions, respectively.

The conservation of mass states that:

$$\left[\begin{array}{c} \text{Net rate of mass efflux} \\ \text{through the control surface} \end{array} \right] + \left[\begin{array}{c} \text{Rate of change of mass} \\ \text{inside the control volume} \end{array} \right] = 0$$

To evaluate the first term in this equation, we must evaluate the mass flux through each of the six faces of the control surface, that is, we must evaluate $\iint_{cs} \rho \vec{V} \cdot d\vec{A}$.

The properties at each of the six faces of the control surface are obtained from a Taylor series expansion of the properties about the point 0. The details of the mass flux evaluation are shown in Table 5.2.

table 5.2 *Mass Flux through the Control Surface of a Cylindrical Differential Control Volume*

SURFACE	$\iint \rho \vec{V} \cdot d\vec{A}$
Inside $(-r)$	$= -\left[\rho - \left(\frac{\partial \rho}{\partial r}\right)\frac{dr}{2}\right]\left[V_r - \left(\frac{\partial V_r}{\partial r}\right)\frac{dr}{2}\right]\left(r - \frac{dr}{2}\right) d\theta\, dz$
	$= -\rho V_r r\, d\theta\, dz + \rho V_r \frac{dr}{2} d\theta\, dz + \rho\left(\frac{\partial V_r}{\partial r}\right) r \frac{dr}{2} d\theta\, dz + V_r\left(\frac{\partial \rho}{\partial r}\right) r \frac{dr}{2} d\theta\, dz$
Outside $(+r)$	$= \left[\rho + \left(\frac{\partial \rho}{\partial r}\right)\frac{dr}{2}\right]\left[V_r + \left(\frac{\partial V_r}{\partial r}\right)\frac{dr}{2}\right]\left(r + \frac{dr}{2}\right) d\theta\, dz$
	$= \rho V_r r\, d\theta\, dz + \rho V_r \frac{dr}{2} d\theta\, dz + \rho\left(\frac{\partial V_r}{\partial r}\right) r \frac{dr}{2} d\theta\, dz + V_r\left(\frac{\partial \rho}{\partial r}\right) r \frac{dr}{2} d\theta\, dz$
Front $(-\theta)$	$= -\left[\rho - \left(\frac{\partial \rho}{\partial \theta}\right)\frac{d\theta}{2}\right]\left[V_\theta - \left(\frac{\partial V_\theta}{\partial \theta}\right)\frac{d\theta}{2}\right] dr\, dz = -\rho V_\theta\, dr\, dz + \rho\left(\frac{\partial V_\theta}{\partial \theta}\right)\frac{d\theta}{2} dr\, dz + V_\theta\left(\frac{\partial \rho}{\partial \theta}\right)\frac{d\theta}{2} dr\, dz$
Back $(+\theta)$	$= \left[\rho + \left(\frac{\partial \rho}{\partial \theta}\right)\frac{d\theta}{2}\right]\left[V_\theta + \left(\frac{\partial V_\theta}{\partial \theta}\right)\frac{d\theta}{2}\right] dr\, dz = \rho V_\theta\, dr\, dz + \rho\left(\frac{\partial V_\theta}{\partial \theta}\right)\frac{d\theta}{2} dr\, dz + V_\theta\left(\frac{\partial \rho}{\partial \theta}\right)\frac{d\theta}{2} dr\, dz$
Bottom $(-z)$	$= -\left[\rho - \left(\frac{\partial \rho}{\partial z}\right)\frac{dz}{2}\right]\left[V_z - \left(\frac{\partial V_z}{\partial z}\right)\frac{dz}{2}\right] r\, d\theta\, dr = -\rho V_z r\, d\theta\, dr + \rho\left(\frac{\partial V_z}{\partial z}\right)\frac{dz}{2} r\, d\theta\, dr + V_z\left(\frac{\partial \rho}{\partial z}\right)\frac{dz}{2} r\, d\theta\, dr$
Top $(+z)$	$= \left[\rho + \left(\frac{\partial \rho}{\partial z}\right)\frac{dz}{2}\right]\left[V_z + \left(\frac{\partial V_z}{\partial z}\right)\frac{dz}{2}\right] r\, d\theta\, dr = \rho V_z r\, d\theta\, dr + \rho\left(\frac{\partial V_z}{\partial z}\right)\frac{dz}{2} r\, d\theta\, dr + V_z\left(\frac{\partial \rho}{\partial z}\right)\frac{dz}{2} r\, d\theta\, dr$

Then,

$$\iint_{cs} \rho \vec{V} \cdot d\vec{A} = \left[\rho V_r + r\left\{\rho\left(\frac{\partial V_r}{\partial r}\right) + V_r\left(\frac{\partial \rho}{\partial r}\right)\right\} + \left\{\rho\left(\frac{\partial V_\theta}{\partial \theta}\right) + V_\theta\left(\frac{\partial \rho}{\partial \theta}\right)\right\} + r\left\{\rho\left(\frac{\partial V_z}{\partial z}\right) + V_z\left(\frac{\partial \rho}{\partial z}\right)\right\}\right] dr\, d\theta\, dz$$

or

$$\iint_{cs} \rho \vec{V} \cdot d\vec{A} = \left[\rho V_r + r\frac{\partial \rho V_r}{\partial r} + \frac{\partial \rho V_\theta}{\partial \theta} + r\frac{\partial \rho V_z}{\partial z}\right] dr\, d\theta\, dz$$

We see that the net rate of mass efflux through the control surface is given by

$$\left[\rho V_r + r\frac{\partial \rho V_r}{\partial r} + \frac{\partial \rho V_\theta}{\partial \theta} + r\frac{\partial \rho V_z}{\partial z} \right] dr \, d\theta \, dz$$

The mass inside the control volume at any instant is the product of the mass per unit volume, ρ, and the volume, $r \, d\theta \, dr \, dz$. Thus the rate of change of mass inside the control volume is given by

$$\frac{\partial \rho}{\partial t} r \, d\theta \, dr \, dz$$

In cylindrical coordinates the differential equation for the conservation of mass is then

$$\rho V_r + r\frac{\partial \rho V_r}{\partial r} + \frac{\partial \rho V_\theta}{\partial \theta} + r\frac{\partial \rho V_z}{\partial z} + r\frac{\partial \rho}{\partial t} = 0$$

Dividing through by r,

$$\frac{\rho V_r}{r} + \frac{\partial \rho V_r}{\partial r} + \frac{1}{r}\frac{\partial \rho V_\theta}{\partial \theta} + \frac{\partial \rho V_z}{\partial z} + \frac{\partial \rho}{\partial t} = 0$$

or

$$\frac{1}{r}\frac{\partial r \rho V_r}{\partial r} + \frac{1}{r}\frac{\partial \rho V_\theta}{\partial \theta} + \frac{\partial \rho V_z}{\partial z} + \frac{\partial \rho}{\partial t} = 0 \qquad (5.2)$$

In cylindrical coordinates the vector operator, ∇, is given by

$$\nabla = \hat{i}_r \frac{\partial}{\partial r} + \hat{i}_\theta \frac{1}{r}\frac{\partial}{\partial \theta} + \hat{i}_z \frac{\partial}{\partial z}$$

Thus in vector notation the conservation of mass is given by[1]

$$\nabla \cdot \rho \vec{V} + \frac{\partial \rho}{\partial t} = 0$$

For incompressible flow, $\rho =$ constant, and Eq. 5.2 reduces to

$$\frac{1}{r}\frac{\partial r V_r}{\partial r} + \frac{1}{r}\frac{\partial V_\theta}{\partial \theta} + \frac{\partial V_z}{\partial z} = 0$$

[1] In carrying out the divergence operation in cylindrical coordinates it is well to note that

$$\frac{\partial \hat{i}_r}{\partial \theta} = \hat{i}_\theta \quad \text{and} \quad \frac{\partial \hat{i}_\theta}{\partial \theta} = -\hat{i}_r$$

(This may be shown using the method of Example Problem 4.12.)

the continuity equation

For steady flow, Eq. 5.2 reduces to

$$\frac{1}{r}\frac{\partial r\rho V_r}{\partial r} + \frac{1}{r}\frac{\partial \rho V_\theta}{\partial \theta} + \frac{\partial \rho V_z}{\partial z} = 0$$

Example 5.1

For a two-dimensional flow, the x component of velocity is given by $u = x^2 - 2x + 2y$. Determine a possible y component for steady, incompressible flow. How many possible y components are there?

EXAMPLE PROBLEM 5.1

GIVEN:

Two-dimensional flow in which $u = x^2 - 2x + 2y$.

FIND:

Possible v for steady, two-dimensional, incompressible flow.

SOLUTION:

Basic equation: $\qquad \nabla \cdot \rho \vec{V} + \dfrac{\partial \rho}{\partial t} = 0$

For incompressible flow, $\rho = $ constant, and we can write

$$\nabla \cdot \vec{V} = 0$$

In rectangular coordinates

$$\frac{\partial u}{\partial x} + \frac{\partial v}{\partial y} + \frac{\partial w}{\partial z} = 0$$

Assume two-dimensional flow in the xy plane. Then partial derivatives with respect to z are zero, and

$$\frac{\partial u}{\partial x} + \frac{\partial v}{\partial y} = 0, \quad \therefore \quad \frac{\partial v}{\partial y} = -\frac{\partial u}{\partial x}$$

With $\quad u = x^2 - 2x + 2y,$

$$\frac{\partial v}{\partial y} = -2x + 2 \quad \{\text{This gives an expression for the rate of change of } v \text{ holding } x \text{ constant.}\}$$

EXAMPLE PROBLEM 5.1 (continued)

On integrating we get

$$v = -2xy + 2y + f(x) \quad \left\{ \begin{array}{l} \text{The function of } x, \text{ that is, } f(x), \text{ appears because} \\ \text{we had the partial derivative of } v \text{ with respect to } y. \end{array} \right\}$$

Since any function $f(x)$ is allowable, there are an infinite number of expressions for v that satisfy the differential continuity equation under the given conditions. The simplest expression for v would be obtained by setting $f(x) = 0$, that is,

$$v = -2xy + 2y$$

$\left\{ \begin{array}{l} \text{This "artificial" problem illustrates the use of the differential continuity equation and} \\ \text{points up the difference between a partial and a total derivative.} \end{array} \right\}$

Example 5.2

A compressible flow field is described by

$$\rho \vec{V} = [ax\hat{\imath} - bxy\hat{\jmath}]e^{-kt}$$

where x and y are coordinates in feet, t is time in seconds, and a, b, and k are constants with appropriate units so that ρ and $\vec{V}$ are in slug/ft^3 and ft/sec, respectively. Calculate the rate of change of density per unit time at the point $x = 3$ ft, $y = 2$ ft, $z = 2$ ft, for $t = 0$.

EXAMPLE PROBLEM 5.2

GIVEN:

A compressible flow field

$$\rho \vec{V} = [ax\hat{\imath} - bxy\hat{\jmath}]e^{-kt}$$

x, y —coordinates in ft
t —time in sec
a —constant with units of slug/ft^3-sec
b —constant with units of slug/ft^4-sec
k —constant with units of sec^{-1}

FIND:

$\partial \rho / \partial t$ at point (3, 2, 2) at time $t = 0$

211   the continuity equation

EXAMPLE PROBLEM 5.2 (continued)

SOLUTION:

Basic equation: $\nabla \cdot \rho \vec{V} + \dfrac{\partial \rho}{\partial t} = 0$

$$\therefore \quad \frac{\partial \rho}{\partial t} = -\nabla \cdot \rho \vec{V} = -\left[\hat{\imath}\frac{\partial}{\partial x} + \hat{\jmath}\frac{\partial}{\partial y} + \hat{k}\frac{\partial}{\partial z}\right] \cdot [ax\hat{\imath} - bxy\hat{\jmath}]e^{-kt}$$

$$= -[a - bx]e^{-kt}$$

$$\frac{\partial \rho}{\partial t} = (bx - a)e^{-kt}$$

At $t = 0$, at the point (3, 2, 2)

$$\frac{\partial \rho}{\partial t} = (3 \text{ ft} \times b \text{ slug/ft}^4\text{-sec} - a \text{ slug/ft}^3\text{-sec})e^{-k(0)}$$

$$\frac{\partial \rho}{\partial t} = (3b - a)\frac{\text{slug/ft}^3}{\text{sec}} \longleftarrow$$

$\left\{ \begin{array}{c} \text{This "artificial" problem illustrates the use of the differential} \\ \text{continuity equation in vector form.} \end{array} \right\}$

**5–3 Stream Function for Two-Dimensional Incompressible Flow

It is convenient to have a means of concisely describing the form of any particular pattern of flow. An adequate description should portray the notion of the shape of the streamlines (including the boundaries) and the scale of the velocity at representative points in the flow. A mathematical device that serves this purpose is the stream function, ψ. The stream function is formulated as a relation between the streamlines and the statement of the conservation of mass.

For a two-dimensional incompressible flow in the xy plane, the conservation of mass, Eq. 5.1a, can be written as

$$\frac{\partial u}{\partial x} + \frac{\partial v}{\partial y} = 0 \tag{5.3}$$

If a continuous function, $\psi(x, y, t)$, called the stream function, is defined such that

$$u \equiv \frac{\partial \psi}{\partial y} \quad \text{and} \quad v \equiv -\frac{\partial \psi}{\partial x} \tag{5.4}$$

** This section may be omitted without loss of continuity in the text material.

then the continuity equation, Eq. 5.3, is satisfied exactly since

$$\frac{\partial u}{\partial x} + \frac{\partial v}{\partial y} = \frac{\partial^2 \psi}{\partial x\, \partial y} - \frac{\partial^2 \psi}{\partial y\, \partial x} = 0$$

Recall that streamlines are lines drawn in the flow field, such that, at a given instant of time, they are tangent to the direction of flow at every point in the flow field. Thus if $d\vec{r}$ is an element of length along a streamline, the equation of the streamline is given by

$$\vec{V} \times d\vec{r} = 0 = (\hat{i}u + \hat{j}v) \times (\hat{i}\, dx + \hat{j}\, dy)$$
$$= \hat{k}(u\, dy - v\, dx)$$

That is, the equation of a streamline in a two-dimensional flow is

$$u\, dy - v\, dx = 0$$

Substituting for the velocity components u and v in terms of the stream function, ψ, from Eq. 5.4, then along a streamline

$$\frac{\partial \psi}{\partial x}\, dx + \frac{\partial \psi}{\partial y}\, dy = 0 \tag{5.5}$$

Since $\psi = \psi(x, y, t)$, then at an instant, t_0, $\psi = \psi(x, y, t_0)$; at this instant, a change in ψ may be evaluated as though $\psi = \psi(x, y)$. Thus at any instant

$$d\psi = \frac{\partial \psi}{\partial x}\, dx + \frac{\partial \psi}{\partial y}\, dy \tag{5.6}$$

On comparing Eqs. 5.5 and 5.6, we see that along an instantaneous streamline, $d\psi = 0$; that is, ψ is a constant along a streamline. Since the differential of ψ is exact, the integral of $d\psi$ between any two points in a flow field, that is, $\psi_2 - \psi_1$, is dependent only on the end points of integration.

From the definition of a streamline, we recognize that there can be no flow across a streamline. Thus, if the streamlines in a two-dimensional, incompressible flow field, at a given instant of time, are as shown in Fig. 5.3, the rate of flow between streamlines ψ_1 and ψ_2 across the lines AB, BC, DE, and EF must be equal.

The volumetric flow rate, Q, between the streamlines ψ_1 and ψ_2 can be evaluated by considering the flow across AB or across BC. For a unit depth, the flow rate across AB is

$$Q = \int_{y_1}^{y_2} u\, dy = \int_{y_1}^{y_2} \frac{\partial \psi}{\partial y}\, dy$$

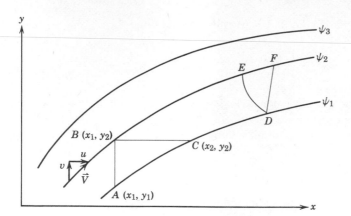

Fig. 5.3 *Instantaneous streamlines in a two-dimensional flow.*

Along AB, x = constant, and $d\psi = \partial\psi/\partial y\, dy$. Therefore

$$Q = \int_{y_1}^{y_2} \frac{\partial\psi}{\partial y}\, dy = \int_{\psi_1}^{\psi_2} d\psi = \psi_2 - \psi_1$$

For a unit depth, the flow rate across BC is

$$Q = \int_{x_1}^{x_2} v\, dx = -\int_{x_1}^{x_2} \frac{\partial\psi}{\partial x}\, dx$$

Along BC, y = constant, and $d\psi = \partial\psi/\partial x\, dx$. Therefore

$$Q = -\int_{x_1}^{x_2} \frac{\partial\psi}{\partial x}\, dx = -\int_{\psi_2}^{\psi_1} d\psi = \psi_2 - \psi_1$$

Thus the volumetric rate of flow between any two streamlines can be written as the difference between the constant values of ψ defining the two streamlines.

For a two-dimensional, incompressible flow in the $r\theta$ plane, the conservation of mass, Eq. 5.2, can be written as

$$\frac{\partial r V_r}{\partial r} + \frac{\partial V_\theta}{\partial \theta} = 0 \tag{5.7}$$

The stream function $\psi(r, \theta, t)$ is then defined so that

$$V_r \equiv \frac{1}{r}\frac{\partial\psi}{\partial\theta} \quad \text{and} \quad V_\theta \equiv -\frac{\partial\psi}{\partial r} \tag{5.8}$$

With ψ defined according to Eq. 5.8, the continuity equation, Eq. 5.7, is satisfied exactly.

Example 5.3

Given the velocity field, $\vec{V} = -3y\hat{\imath} + 3x\hat{\jmath}$, determine the family of ψ functions that will yield this velocity field.

EXAMPLE PROBLEM 5.3

GIVEN:

Velocity field, $\vec{V} = -3y\hat{\imath} + 3x\hat{\jmath}$

FIND:

Stream function, ψ

SOLUTION:

From Eqs. 5.4 defining ψ, $u = \partial\psi/\partial y$ and $v = -\partial\psi/\partial x$. For the given velocity field,

$$u = -3y = \frac{\partial\psi}{\partial y}$$

Integrating with respect to y,

$$\psi = \int \frac{\partial\psi}{\partial y} dy = \int -3y\, dy = -\tfrac{3}{2}y^2 + f(x)$$

where $f(x)$ is arbitrary. However, $f(x)$ may be evaluated using the equation for v. Thus, since $v = 3x$,

$$v = -\frac{\partial\psi}{\partial x} = -\frac{\partial f(x)}{\partial x} = -\frac{df(x)}{dx} = 3x$$

Then

$$f(x) = \int \frac{df}{dx} dx = \int -3x\, dx = -\tfrac{3}{2}x^2 + c$$

Substituting,

$$\psi = -\tfrac{3}{2}x^2 - \tfrac{3}{2}y^2 + c \qquad\qquad \psi$$

Check:

$$u = \frac{\partial \psi}{\partial y} = \frac{\partial}{\partial y}(-\tfrac{3}{2}x^2 - \tfrac{3}{2}y^2 + c) = -3y$$

$$v = -\frac{\partial \psi}{\partial x} = -\frac{\partial}{\partial x}(-\tfrac{3}{2}x^2 - \tfrac{3}{2}y^2 + c) = 3x$$

$$\left\{ \begin{array}{c} \text{This problem illustrates the relation between the} \\ \text{velocity field and the stream function.} \end{array} \right\}$$

5–4 Motion of a Fluid Element (Kinematics)

Before formulating the effects of forces on fluid motion (dynamics), let us consider, first, only the motion (kinematics) of a fluid element in a flow field. For convenience, we follow an infinitesimal element of fixed identity (mass), as shown in Fig. 5.4.

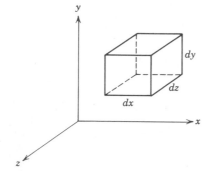

Fig. 5.4 *Infinitesimal element of fluid.*

As the infinitesimal element of mass dm moves in a flow field several things may happen to it. Perhaps the most obvious of these is that the element translates, that is, it undergoes a linear displacement from a location x, y, z to a different location x_1, y_1, z_1. The element may also undergo a rotation, that is, the orientation of the element as shown in Fig. 5.4, wherein the sides of the element are parallel to the coordinate axes x, y, z, may change as a result of pure rotation about any one (or all three) of the coordinate axes. In addition the element may deform. The deformation may be subdivided into two parts—linear and angular deformation. Linear deformation involves a change in shape of the element without a change in the orientation of the element, that is, a deformation in which planes of the element that were originally perpendicular (for example, the top and side of the element) remain perpendicular. Angular deformation involves a distortion of the element in which planes that were originally perpendicular are no longer perpendicular. In general a fluid element may undergo a combination of translation, rotation, and linear and angular deformation during the course of its motion.

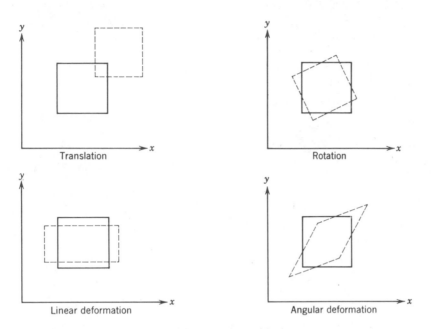

Fig. 5.5 *Pictorial representation of the components of fluid motion.*

These four components of fluid motion are illustrated pictorially in Fig. 5.5 for motion in the *xy* plane. For a general three-dimensional flow, similar motions of the particle would be depicted in the *yz* and *xz* planes. For pure translation or rotation, the fluid element retains its shape, that is, there is no deformation. Thus shear stresses do not arise as a result of pure translation or rotation (recall from Chapter 2 that in a Newtonian fluid the shear stress is directly proportional to the rate of angular deformation). We postpone further discussion of deformation and consider first the question of fluid rotation.

**5-5 Fluid Rotation

The rotation, $\bar{\omega}$, of a fluid particle is defined as the average angular velocity of any two mutually perpendicular line elements of the particle. Rotation is a vector quantity. A particle moving in a general three-dimensional flow field may rotate

** This section may be omitted without loss of continuity in the text material.

about all three coordinate axes. Thus, in general,

$$\vec{\omega} = \hat{\imath}\omega_x + \hat{\jmath}\omega_y + \hat{k}\omega_z$$

where ω_x is the rotation about the x axis, ω_y is the rotation about the y axis, and ω_z is the rotation about the z axis. The positive sense of rotation is given by the right-hand rule.

To obtain a mathematical expression for fluid rotation, consider the motion of a fluid element in the xy plane. The components of the velocity at every point in the flow field are given by $u(x, y)$ and $v(x, y)$. The rotation of a fluid element in such a flow field is illustrated in Fig. 5.6. The two mutually perpendicular lines oa and ob will rotate to the position shown in time Δt only if the velocities at the points a and b are different from the velocity at o.

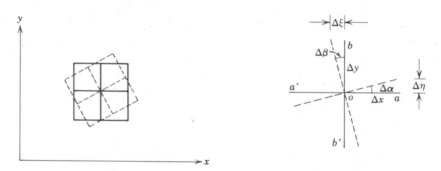

Fig. 5.6 *Rotation of a fluid element in a two-dimensional flow field.*

Consider first the rotation of the line oa of length Δx. Rotation of this line is due to variations of the y component of velocity, that is, variations in v. If the y component of velocity at the point o is taken as v_o, then the y component of velocity at point a can be written, using a Taylor series expansion, as

$$v = v_o + \frac{\partial v}{\partial x}\Delta x$$

The angular velocity of the line oa is given by

$$\omega_{oa} = \lim_{\Delta t \to 0} \frac{\Delta \alpha}{\Delta t} = \lim_{\Delta t \to 0} \frac{\Delta \eta / \Delta x}{\Delta t}$$

Since

$$\Delta \eta = \frac{\partial v}{\partial x}\Delta x\,\Delta t$$

$$\omega_{oa} = \lim_{\Delta t \to 0} \frac{(\partial v/\partial x)\,\Delta x\,\Delta t/\Delta x}{\Delta t} = \frac{\partial v}{\partial x}$$

Rotation of the line *ob*, of length Δy, results from variations of the x component of velocity, that is, variations in u. If the x component of velocity at the point o is taken as u_o, then the x component of velocity at point b can be written, using a Taylor series expansion, as

$$u = u_o + \frac{\partial u}{\partial y} \Delta y$$

The angular velocity of the line *ob* is given by

$$\omega_{ob} = \lim_{\Delta t \to 0} \frac{\Delta \beta}{\Delta t} = \lim_{\Delta t \to 0} \frac{\Delta \xi / \Delta y}{\Delta t}$$

Since

$$\Delta \xi = -\frac{\partial u}{\partial y} \Delta y \, \Delta t$$

$$\omega_{ob} = \lim_{\Delta t \to 0} \frac{-(\partial u/\partial y) \, \Delta y \, \Delta t / \Delta y}{\Delta t} = -\frac{\partial u}{\partial y}$$

(The negative sign is introduced to give a positive value of ω_{ob}. According to our sign convention, counterclockwise rotation is positive.)

The rotation of the fluid element about the z axis is the average angular velocity of the two mutually perpendicular line elements, *oa* and *ob*, in the xy plane.

$$\omega_z = \frac{1}{2}\left(\frac{\partial v}{\partial x} - \frac{\partial u}{\partial y}\right)$$

By considering the rotation of two mutually perpendicular lines in the yz and xz planes, one can show that

$$\omega_x = \frac{1}{2}\left(\frac{\partial w}{\partial y} - \frac{\partial v}{\partial z}\right)$$

and

$$\omega_y = \frac{1}{2}\left(\frac{\partial u}{\partial z} - \frac{\partial w}{\partial x}\right)$$

Then

$$\vec{\omega} = \hat{i}\omega_x + \hat{j}\omega_y + \hat{k}\omega_z = \frac{1}{2}\left[\hat{i}\left(\frac{\partial w}{\partial y} - \frac{\partial v}{\partial z}\right) + \hat{j}\left(\frac{\partial u}{\partial z} - \frac{\partial w}{\partial x}\right) + \hat{k}\left(\frac{\partial v}{\partial x} - \frac{\partial u}{\partial y}\right)\right] \quad (5.9)$$

We recognize the term in the square brackets as

$$\text{curl } \vec{V} = \nabla \times \vec{V}$$

Then, in vector notation we can write

$$\vec{\omega} = \tfrac{1}{2}\nabla \times \vec{V} \tag{5.10}$$

The factor of 1/2 can be eliminated in Eq. 5.10 by defining a quantity called the vorticity, $\vec{\zeta}$, to be twice the rotation.

$$\vec{\zeta} = 2\vec{\omega} = \nabla \times \vec{V} \tag{5.11}$$

In cylindrical coordinates,

$$\vec{V} = \hat{\imath}_r V_r + \hat{\imath}_\theta V_\theta + \hat{\imath}_z V_z$$

and

$$\nabla = \hat{\imath}_r \frac{\partial}{\partial r} + \hat{\imath}_\theta \frac{1}{r}\frac{\partial}{\partial\theta} + \hat{\imath}_z \frac{\partial}{\partial z}$$

The vorticity, in cylindrical coordinates, is then[3]

$$\nabla \times \vec{V} = \hat{\imath}_r \left(\frac{1}{r}\frac{\partial V_z}{\partial\theta} - \frac{\partial V_\theta}{\partial z} \right) + \hat{\imath}_\theta \left(\frac{\partial V_r}{\partial z} - \frac{\partial V_z}{\partial r} \right) + \hat{\imath}_z \left(\frac{1}{r}\frac{\partial r V_\theta}{\partial r} - \frac{1}{r}\frac{\partial V_r}{\partial\theta} \right) \tag{5.12}$$

**5–6 Irrotational Flow

An irrotational flow is one in which there is no rotation, that is fluid elements moving in the flow field do not undergo any rotation. For $\vec{\omega} = 0$, $\nabla \times \vec{V} = 0$ and from Eq. 5.9,

$$\frac{\partial w}{\partial y} - \frac{\partial v}{\partial z} = \frac{\partial u}{\partial z} - \frac{\partial w}{\partial x} = \frac{\partial v}{\partial x} - \frac{\partial u}{\partial y} = 0 \tag{5.13}$$

In cylindrical coordinates, the irrotationality condition requires that

$$\frac{1}{r}\frac{\partial V_z}{\partial\theta} - \frac{\partial V_\theta}{\partial z} = \frac{\partial V_r}{\partial z} - \frac{\partial V_z}{\partial r} = \frac{1}{r}\frac{\partial r V_\theta}{\partial r} - \frac{1}{r}\frac{\partial V_r}{\partial\theta} = 0 \tag{5.14}$$

5–6.1 VELOCITY POTENTIAL

In Section 5–3 we formulated the stream function, ψ, as a relation between the streamlines and the statement of the conservation of mass for two-dimensional, incompressible flow.

---

[3] In carrying out the curl operation, recall that $\hat{\imath}_r$ and $\hat{\imath}_\theta$ are functions of θ (see footnote[1] on page 209).
** This section may be omitted without loss of continuity in the text material. (Note that Sections 5-3 and 5-5 contain background material needed for study of this section.)

5/introduction to differential analysis of fluid motion

We can also formulate a relation called the potential function, ϕ, for a velocity field subject to the irrotationality condition. To do so, we must make use of the fundamental vector identity[4]

$$\text{curl (grad } \phi) = \nabla \times \nabla\phi = 0 \qquad (5.15)$$

which holds if ϕ is any scalar function (of the space coordinates and time) having continuous first and second derivatives.

Then for an irrotational flow, in which $\nabla \times \vec{V} = 0$, a scalar function, ϕ, must exist such that the gradient of ϕ is equal to the velocity vector, $\vec{V}$. In order that the positive direction of flow be in the direction of decreasing ϕ (analogous to the positive direction of heat transfer being defined in the direction of decreasing temperature), we define ϕ such that

$$\vec{V} \equiv -\nabla\phi \qquad (5.16)$$

Thus

$$u = -\frac{\partial\phi}{\partial x} \qquad v = -\frac{\partial\phi}{\partial y} \qquad w = -\frac{\partial\phi}{\partial z} \qquad (5.17)$$

With the potential function defined in this way, the irrotationality condition, Eq. 5.13, is satisfied identically.

In cylindrical coordinates,

$$\nabla = \hat{i}_r \frac{\partial}{\partial r} + \hat{i}_\theta \frac{1}{r}\frac{\partial}{\partial\theta} + \hat{i}_z \frac{\partial}{\partial z}$$

From Eq. 5.16, then in cylindrical coordinates

$$V_r = -\frac{\partial\phi}{\partial r} \qquad V_\theta = -\frac{1}{r}\frac{\partial\phi}{\partial\theta} \qquad V_z = -\frac{\partial\phi}{\partial z} \qquad (5.18)$$

5–6.2 STREAM FUNCTION AND VELOCITY POTENTIAL FOR IRROTATIONAL, TWO-DIMENSIONAL, INCOMPRESSIBLE FLOW

For a two-dimensional, incompressible, irrotational flow we have expressions for the velocity components u and v in terms of both the stream function, ψ, and the velocity potential, ϕ.

$$u = \frac{\partial\psi}{\partial y} \qquad v = -\frac{\partial\psi}{\partial x} \qquad (5.4)$$

$$u = -\frac{\partial\phi}{\partial x} \qquad v = -\frac{\partial\phi}{\partial y} \qquad (5.17)$$

[4] The proof of this identity may be found in any book treating vector analysis, or it may be proved by expanding Eq. 5.15 into components (Appendix D).

On substituting for u and v from Eq. 5.4 into the irrotationality condition,

$$\frac{\partial v}{\partial x} - \frac{\partial u}{\partial y} = 0 \qquad (5.13)$$

we obtain

$$\frac{\partial^2 \psi}{\partial x^2} + \frac{\partial^2 \psi}{\partial y^2} = 0 \qquad (5.19a)$$

On substituting for u and v from Eq. 5.17 into the continuity equation,

$$\frac{\partial u}{\partial x} + \frac{\partial v}{\partial y} = 0 \qquad (5.3)$$

we obtain

$$\frac{\partial^2 \phi}{\partial x^2} + \frac{\partial^2 \phi}{\partial y^2} = 0 \qquad (5.19b)$$

Equations 5.19a and b are forms of Laplace's equation—an equation that arises in many areas of the physical sciences and engineering. Any function ψ or ϕ that satisfies Laplace's equation represents a possible incompressible, irrotational flow field.

In Section 5–3 we showed that the stream function, ψ, is constant along a streamline. For $\psi =$ constant, $d\psi = (\partial\psi/\partial x)\, dx + (\partial\psi/\partial y)\, dy = 0$. The slope of a streamline, that is, the slope of a line of constant ψ, is given by

$$\left.\frac{dy}{dx}\right|_\psi = -\frac{\partial\psi/\partial x}{\partial\psi/\partial y} = -\frac{-v}{u} = \frac{v}{u} \qquad (5.20)$$

Along a line of constant ϕ, $d\phi = 0$, and hence

$$d\phi = \frac{\partial\phi}{\partial x}\, dx + \frac{\partial\phi}{\partial y}\, dy = 0$$

Consequently,

$$\left.\frac{dy}{dx}\right|_\phi = -\frac{\partial\phi/\partial x}{\partial\phi/\partial y} = -\frac{u}{v} \qquad (5.21)$$

The last equality of Eq. 5.21 follows from use of Eq. 5.17. Comparing Eqs. 5.20 and 5.21, we see that the slope of a constant ψ line at any point is the negative reciprocal of the slope of the constant ϕ line at that point. Consequently lines of constant ψ and constant ϕ are orthogonal.

5–6.3 IRROTATIONAL FLOW AND VISCOSITY

Before leaving the subject of irrotational flow, we should make a few additional comments. The velocity potential, ϕ, only exists for an irrotational flow. The stream function, ψ, is defined to satisfy the continuity equation; the stream function is not subject to the restriction of irrotational flow.

Under what conditions might we expect to have an irrotational flow? A fluid particle moving, without rotation, in a flow field cannot develop a rotation under the action of a body force or normal surface (i.e. pressure) forces; the development of rotation in a fluid particle, initially without rotation, requires the action of a shear stress on the surface of the particle. Since shear stress is proportional to the rate of angular deformation, then a particle that is initially without rotation will not develop a rotation without a simultaneous angular deformation. Does it then follow that a flow field in which shear stresses are present will be a rotational flow field? We can obtain a qualitative answer to this question by considering the angular deformation of a fluid element in the xy plane as shown in Fig. 5.7. The

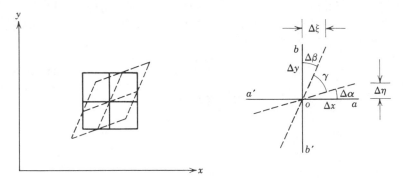

Fig. 5.7 *Angular deformation of a fluid element in a two-dimensional flow field.*

rate of angular deformation of the fluid element is the rate of change of the angle between two mutually perpendicular lines of the fluid. Referring to Fig. 5.7, the rate of angular deformation of the fluid element is the rate of change of the angle between lines *oa* and *ob*, that is, the rate of angular deformation is given by

$$\frac{d\gamma}{dt} = \frac{d\alpha}{dt} + \frac{d\beta}{dt}$$

Now,

$$\frac{d\alpha}{dt} = \lim_{\Delta t \to 0} \frac{\Delta \alpha}{\Delta t} = \lim_{\Delta t \to 0} \frac{\Delta \eta / \Delta x}{\Delta t} = \lim_{\Delta t \to 0} \frac{(\partial v / \partial x)\, \Delta x\, \Delta t / \Delta x}{\Delta t} = \frac{\partial v}{\partial x}$$

and

$$\frac{d\beta}{dt} = \lim_{\Delta t \to 0} \frac{\Delta \beta}{\Delta t} = \lim_{\Delta t \to 0} \frac{\Delta \xi / \Delta y}{\Delta t} = \lim_{\Delta t \to 0} \frac{(\partial u / \partial y)\, \Delta y\, \Delta t / \Delta y}{\Delta t} = \frac{\partial u}{\partial y}$$

irrotational flow

Consequently, the rate of angular deformation is

$$\frac{d\gamma}{dt} = \frac{\partial v}{\partial x} + \frac{\partial u}{\partial y} \tag{5.22}$$

The shear stress is related to the rate of angular deformation through the viscosity. In a viscous flow (wherein velocity gradients are present) it is highly unlikely that $\partial v/\partial x$ will be equal and opposite to $\partial u/\partial y$ throughout the flow field (consider for example, the boundary layer flow of Fig. 2.10 and the flow over a cylinder, shown in Fig. 2.11). The presence of viscous forces means the flow is rotational.[5]

The condition of irrotationality may be a valid assumption for those regions of a flow in which viscous forces are negligible.[6] (For example, such a region exists outside the boundary layer in the flow over a solid surface.) The theory for irrotational flow is developed in terms of an imaginary fluid called an ideal fluid whose viscosity is identically zero. Since, in an irrotational flow, the velocity field may be defined by the potential function, ϕ, the theory is often referred to as potential flow theory.[7]

All real fluids possess viscosity, but there are many problems for which the assumption of inviscid flow leads to a considerable simplification in the analysis and, at the same time, gives meaningful results.

Example 5.4

An incompressible flow field is characterized by the stream function

$$\psi = 3x^2 y - y^3$$

(a) Show that this flow is irrotational.
(b) Show that the magnitude of the velocity at any point in the flow field depends only on the distance of the point from the origin of coordinates.
(c) Sketch a few streamlines for the flow in the quadrant $x > 0, y > 0$.

[5] A rigorous proof employing the complete equations of motion for a fluid particle is given in *Principles of Fluid Mechanics* by W. H. Li and S. H. Lam (Reading, Mass.: Addison-Wesley, 1964), pp. 142–145.

[6] Examples of rotational and irrotational motion are shown in the film loops: S-FM014A, *Visualization of Vorticity with Vorticity Meter—Part I*; S-FM014B, *Visualization of Vorticity with Vorticity Meter— Part II.*

[7] Anyone interested in a detailed study of potential flow theory may find the following books of interest: Streeter, V. L., *Fluid Dynamics* (New York: McGraw-Hill, 1948); Valentine, H. R., *Applied Hydrodynamics* (London: Butterworths, 1959); Robertson, J. M., *Hydrodynamics in Theory and Application* (Englewood Cliffs, N.J.: Prentice-Hall, 1965).

5/introduction to differential analysis of fluid motion

GIVEN:

The stream function for an incompressible flow field is given by $\psi = 3x^2y - y^3$.

SHOW:

(a) That flow is irrotational.
(b) That magnitude of the velocity at any point depends only on the distance of the point from the origin.
(c) Sketch streamlines.

SOLUTION:

Since $\psi = \psi(x, y)$, the flow field is two dimensional.
    If the flow is to be irrotational, then

$$\omega_z = \frac{1}{2}\left[\frac{\partial v}{\partial x} - \frac{\partial u}{\partial y}\right] = 0$$

Since

$$u = \frac{\partial \psi}{\partial y}, \qquad v = -\frac{\partial \psi}{\partial x},$$

then for irrotational flow

$$\frac{\partial v}{\partial x} - \frac{\partial u}{\partial y} = 0 = -\frac{\partial^2 \psi}{\partial x^2} - \frac{\partial^2 \psi}{\partial y^2} = 0$$

Does

$$\frac{\partial^2 \psi}{\partial x^2} + \frac{\partial^2 \psi}{\partial y^2} = 0?$$

$$\frac{\partial^2}{\partial x^2}(3x^2y - y^3) + \frac{\partial^2}{\partial y^2}(3x^2y - y^3) = 6y - 6y = 0$$

∴   Flow is irrotational. ⟵_____

The magnitude of the velocity at a point is given by

$$|\vec{V}| = \sqrt{u^2 + v^2}$$

$$u = \frac{\partial \psi}{\partial y} = \frac{\partial}{\partial y}(3x^2y - y^3) = 3x^2 - 3y^2$$

$$v = -\frac{\partial \psi}{\partial x} = -\frac{\partial}{\partial x}(3x^2y - y^3) = -6xy$$

$$|\vec{V}| = \sqrt{u^2 + v^2} = \sqrt{(3x^2 - 3y^2)^2 + (-6xy)^2}$$

$$= \sqrt{9(x^4 - 2x^2y^2 + y^4) + 36x^2y^2}$$

$$= \sqrt{9(x^4 + 2x^2y^2 + y^4)} = 3\sqrt{(x^2 + y^2)^2}$$

$$|\vec{V}| = 3(x^2 + y^2) = 3r^2$$

where $r = |\vec{r}| = |x\hat{\imath} + y\hat{\jmath}|$.

A few streamlines, corresponding to lines along which ψ is constant, are sketched below.

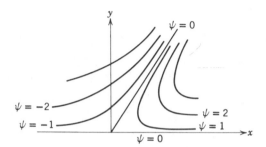

Since there is no flow across streamlines, and the flow is irrotational (i.e. without viscous effects), any streamline may be imagined to represent a solid surface. If we let the lines $\psi = 0$ represent solid surfaces, we obtain a picture of flow in a corner with either an obtuse or acute angle.

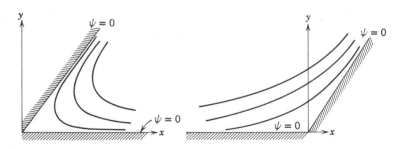

$\left\{ \begin{array}{c} \text{This problem illustrates the relations among the stream function,} \\ \text{the velocity field, and the irrotationality condition.} \end{array} \right\}$

Example 5.5

Consider the flow field given by $\psi = x^2 - y^2$.

(a) Show that the flow is irrotational.
(b) Determine the velocity potential for this flow.

EXAMPLE PROBLEM 5.5

GIVEN:

Incompressible flow field with $\psi = x^2 - y^2$.

REQUIRED:

(a) Show that the flow is irrotational.
(b) Determine the velocity potential for this flow.

SOLUTION:

We will show that the velocity field satisfies the irrotationality condition

$$\frac{\partial v}{\partial x} - \frac{\partial u}{\partial y} = 0$$

$$u = \frac{\partial \psi}{\partial y} = \frac{\partial}{\partial y}(x^2 - y^2) = -2y$$

$$v = -\frac{\partial \psi}{\partial x} = -\frac{\partial}{\partial x}(x^2 - y^2) = -2x$$

$$\frac{\partial v}{\partial x} - \frac{\partial u}{\partial y} = \frac{\partial}{\partial x}(-2x) - \frac{\partial}{\partial y}(-2y) = -2 + 2 = 0$$

∴ Flow is irrotational. ←──────────────

The velocity components can be written in terms of the velocity potential as

$$u = -\frac{\partial \phi}{\partial x} \quad \text{and} \quad v = -\frac{\partial \phi}{\partial y}$$

Consequently,

$$u = -\frac{\partial \phi}{\partial x} = -2y$$

and

$$\frac{\partial \phi}{\partial x} = 2y$$

Integrating with respect to x, $\phi = 2xy + f(y)$, where $f(y)$ is an arbitrary function of y. Then

$$v = -2x = -\frac{\partial \phi}{\partial y} = -\frac{\partial}{\partial y}[2xy + f(y)]$$

$$\therefore \quad -2x = -2x - \frac{\partial f(y)}{\partial y} = -2x - \frac{df}{dy}$$

$$\frac{df}{dy} = 0 \quad \text{and} \quad f = \text{constant}$$

$$\therefore \quad \phi = 2xy + \text{constant} \quad\quad\quad\quad\quad \phi$$

We can also show that lines of constant ψ and ϕ are orthogonal.

$$\psi = x^2 - y^2$$

For $\psi = \text{constant}$, $d\psi = 0 = 2x\,dx - 2y\,dy$

$$\therefore \quad \left.\frac{dy}{dx}\right|_{\psi=c} = \frac{x}{y}$$

$$\phi = 2xy + c$$

For $\phi = \text{constant}$, $d\phi = 0 = 2y\,dx + 2x\,dy$

$$\therefore \quad \left.\frac{dy}{dx}\right|_{\phi=c} = -\frac{y}{x}$$

$\left\{\begin{array}{l}\text{This problem illustrates the relation among the stream function,}\\ \text{the velocity field, and the velocity potential.}\end{array}\right\}$

5-7 The Momentum Equation

A dynamic equation describing fluid motion may be obtained by applying Newton's second law to a particle. To derive the differential form of the momentum equation, we will apply Newton's second law to an infinitesimal system of mass, *dm*.

Recall that Newton's second law for a finite system is given by

$$\vec{F} = \left.\frac{d\vec{P}}{dt}\right]_{\text{system}} \tag{4.2a}$$

where the linear momentum, $\vec{P}$, of the system is given by

$$\vec{P}_{\text{system}} = \int_{\text{Mass (system)}} \vec{V}\,dm \tag{4.2b}$$

Then for an infinitesimal system of mass, dm, Newton's second law can be written as

$$d\vec{F} = dm \frac{d\vec{V}}{dt}\bigg]_{\text{system}} \qquad (5.23)$$

We recognize that the force term, $d\vec{F}$, can be written in terms of the surface forces and body forces acting on the element of fixed mass, dm. Indeed we will eventually have to formulate $d\vec{F}$. For the moment we will delay that and concentrate our efforts on determining a suitable expression for $d\vec{V}/dt]_{\text{system}}$.

5–7.1 ACCELERATION OF A PARTICLE

Let us remember first that we are dealing with an element of fixed mass, dm. As discussed in Section 1–4.3, one may obtain the equation of motion for a particle by applying Newton's second law to that particle. The disadvantage of this approach is that a separate equation is required for each particle. Thus the bookkeeping for many particles becomes a problem.

5–7.2 ACCELERATION OF A PARTICLE IN A VELOCITY FIELD

A more general description of acceleration can be obtained by considering a particle moving in a velocity field. The basic hypothesis of continuum fluid mechanics has led us to a field description of fluid flow in which the properties of a flow field are defined by continuous functions of the space coordinates and time. In particular, the velocity field is given by $\vec{V} = \vec{V}(x, y, z, t)$. The field description is very powerful, since information for the entire flow is given by one equation.

The problem then is to retain the field description for fluid properties and obtain an expression for the acceleration of a fluid particle as it translates in a flow field. Stated simply, the problem is:

Given the velocity field, $\vec{V} = \vec{V}(x, y, z, t)$, find the acceleration of a fluid particle, $\vec{a}_p$.

Consider a particle moving in a velocity field. At time, t, the particle is at the position x, y, z and has a velocity corresponding to the velocity at that point in space at time t, that is,

$$\vec{V}_p]_t = \vec{V}(x, y, z, t)$$

At time, $t + dt$, the particle has moved to a new position, with coordinates $x + dx, y + dy, z + dz$, and has a velocity given by

$$\vec{V}_p]_{t + dt} = \vec{V}(x + dx, y + dy, z + dz, t + dt)$$

This is shown pictorially in Fig. 5.8.

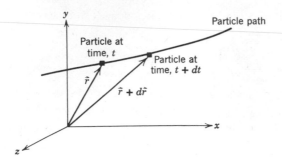

Fig. 5.8 *Motion of a particle in a flow field.*

The particle velocity at time t (position $\vec{r}$) is given by $\vec{V}_p = \vec{V}(x, y, z, t)$. Then the change in the velocity of the particle, that is, $d\vec{V}_p$, in moving from location $\vec{r}$ to $\vec{r} + d\vec{r}$ is given by

$$d\vec{V}_p = \frac{\partial \vec{V}}{\partial x} dx_p + \frac{\partial \vec{V}}{\partial y} dy_p + \frac{\partial \vec{V}}{\partial z} dz_p + \frac{\partial \vec{V}}{\partial t} dt$$

The total acceleration of the particle is given by

$$\vec{a}_p = \frac{d\vec{V}_p}{dt} = \frac{\partial \vec{V}}{\partial x} \frac{dx_p}{dt} + \frac{\partial \vec{V}}{\partial y} \frac{dy_p}{dt} + \frac{\partial \vec{V}}{\partial z} \frac{dz_p}{dt} + \frac{\partial \vec{V}}{\partial t}$$

Since

$$\frac{dx_p}{dt} = u \qquad \frac{dy_p}{dt} = v \quad \text{and} \quad \frac{dz_p}{dt} = w$$

then

$$\vec{a}_p = \frac{d\vec{V}_p}{dt} = u\frac{\partial \vec{V}}{\partial x} + v\frac{\partial \vec{V}}{\partial y} + w\frac{\partial \vec{V}}{\partial z} + \frac{\partial \vec{V}}{\partial t}$$

To remind us that calculation of the acceleration of a fluid particle in a velocity field requires a special derivative, it is given the symbol $D\vec{V}/Dt$. Thus

$$\frac{D\vec{V}}{Dt} \equiv \vec{a}_p = u\frac{\partial \vec{V}}{\partial x} + v\frac{\partial \vec{V}}{\partial y} + w\frac{\partial \vec{V}}{\partial z} + \frac{\partial \vec{V}}{\partial t} \qquad (5.24)$$

The derivative, $D\vec{V}/Dt$, defined by Eq. 5.24, is commonly called the substantial derivative to remind us that it is computed following a particle of "substance." It often is called the material or particle derivative.

From Eq. 5.24 we recognize that a fluid particle moving in a flow field may undergo an acceleration for either of two reasons. It may be accelerated because it is convected into a region of higher (or lower) velocity. For example, in the steady flow through a nozzle, in which, by definition, the velocity field is not a function of time, a fluid particle will accelerate as it moves through the nozzle. The particle is convected into a region of higher velocity. If a flow field is unsteady, a fluid particle will undergo an acceleration, a "local" acceleration, because the velocity field is a function of time.

The physical significance of the terms in Eq. 5.24 is indicated below:

$$\vec{a}_p = \underbrace{\frac{D\vec{V}}{Dt}}_{\substack{\text{Total} \\ \text{acceleration} \\ \text{of a particle}}} = \underbrace{u\frac{\partial \vec{V}}{\partial x} + v\frac{\partial \vec{V}}{\partial y} + w\frac{\partial \vec{V}}{\partial z}}_{\substack{\text{Convective} \\ \text{acceleration}}} + \underbrace{\frac{\partial \vec{V}}{\partial t}}_{\substack{\text{Local} \\ \text{acceleration}}}$$

For a two-dimensional flow, say $\vec{V} = \vec{V}(x, y, t)$, Eq. 5.24 reduces to

$$\frac{D\vec{V}}{Dt} = u\frac{\partial \vec{V}}{\partial x} + v\frac{\partial \vec{V}}{\partial y} + \frac{\partial \vec{V}}{\partial t}$$

For a one-dimensional flow, say $\vec{V} = \vec{V}(x, t)$, Eq. 5.24 becomes

$$\frac{D\vec{V}}{Dt} = u\frac{\partial \vec{V}}{\partial x} + \frac{\partial \vec{V}}{\partial t}$$

Finally, for a steady flow in three dimensions, Eq. 5.24 becomes

$$\frac{D\vec{V}}{Dt} = u\frac{\partial \vec{V}}{\partial x} + v\frac{\partial \vec{V}}{\partial y} + w\frac{\partial \vec{V}}{\partial z}$$

which is not necessarily zero. Thus a fluid particle can undergo a convective acceleration due to its motion, even in a steady velocity field.

Equation 5.24 is a vector equation. As with all vector equations, it may be written in scalar component equations. Relative to an xyz coordinate system, the scalar components of Eq. 5.24 are written:

$$a_{x_p} = \frac{Du}{Dt} = u\frac{\partial u}{\partial x} + v\frac{\partial u}{\partial y} + w\frac{\partial u}{\partial z} + \frac{\partial u}{\partial t} \qquad (5.25a)$$

$$a_{y_p} = \frac{Dv}{Dt} = u\frac{\partial v}{\partial x} + v\frac{\partial v}{\partial y} + w\frac{\partial v}{\partial z} + \frac{\partial v}{\partial t} \qquad (5.25b)$$

$$a_{z_p} = \frac{Dw}{Dt} = u\frac{\partial w}{\partial x} + v\frac{\partial w}{\partial y} + w\frac{\partial w}{\partial z} + \frac{\partial w}{\partial t} \qquad (5.25c)$$

Example 5.6

Consider the one-dimensional steady, incompressible flow through the converging channel shown. The velocity field is given by $\vec{V} = [10 + 10(x/L)]\hat{i}$. Find the x component of acceleration, that is Du/Dt, for a particle moving in the flow field. If we use the method of description of particle mechanics, the position of the particle, located at $x = 0$ at time $t = 0$, will be a function of time, $x_p = f(t)$. Determine the expression for $f(t)$ and then by taking the second derivative of the function with respect to time, obtain an expression for the x component of the particle acceleration.

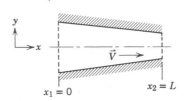

$x_1 = 0$ $x_2 = L$

EXAMPLE PROBLEM 5.6

GIVEN:

Steady, one-dimensional, incompressible flow through the converging channel shown.

$$\vec{V} = \left(10 + 10\frac{x}{L}\right)\hat{i} \text{ ft/sec}$$

$x_1 = 0$ $x_2 = L$

FIND:

(a) The x component of the acceleration of a particle moving in the flow field.
(b) For the particle located at $x = 0$ at time $t = 0$,
(1) Determine an expression for its position, x_p, as a function of time.
(2) Determine an expression for its x component of acceleration, that is, a_{x_p}, as a function of time.

SOLUTION:

The acceleration of a particle moving in a velocity field is given by

$$\frac{D\vec{V}}{Dt} = u\frac{\partial \vec{V}}{\partial x} + v\frac{\partial \vec{V}}{\partial y} + w\frac{\partial \vec{V}}{\partial z} + \frac{\partial \vec{V}}{\partial t}$$

The x component of the acceleration is given by

$$\frac{Du}{Dt} = u\frac{\partial u}{\partial x} + v\frac{\partial u}{\partial y} + w\frac{\partial u}{\partial z} + \frac{\partial u}{\partial t}$$

For the flow field given,

$$v = w = 0 \qquad u = \left(10 + 10\frac{x}{L}\right)$$

$$\therefore \quad \frac{Du}{Dt} = u\frac{\partial u}{\partial x} = \left(10 + 10\frac{x}{L}\right)\frac{10}{L} = \frac{100}{L}\left(1 + \frac{x}{L}\right) \quad \longleftarrow \qquad Du/Dt$$

$$\left\{\begin{array}{l}\text{To determine the acceleration of a particle at any point in the flow field, one} \\ \text{merely substitutes the present location of the particle into the above result.}\end{array}\right\}$$

In the second part of this problem we are interested in following a particular particle—namely, the one located at $x = 0$ at time $t = 0$, as it flows through the channel.

The x coordinate, that is, the location, of this particle will be a function of time, $x_p = f(t)$. Furthermore, $u_p = df/dt$ will be a function of time. The particle will have the velocity corresponding to its location in the velocity field. At time, $t = 0$, the particle is at $x = 0$, and its velocity $u_p = 10$ ft/sec. At some time later, time $t = t$, the particle will reach the exit, $x = L$; at that time it will have a velocity $u_p = 20$ ft/sec. To find the expression for $x_p = f(t)$, we write

$$u_p = \frac{dx_p}{dt} = \frac{df}{dt} = 10\left(1 + \frac{x}{L}\right) = 10\left(1 + \frac{f}{L}\right)$$

Separating variables

$$\frac{df}{(1 + f/L)} = 10\,dt$$

Since at time $t = 0$, the particle in question was located at $x = 0$, and at time t, this particle is located at $x_p = f$, then

$$\int_0^f \frac{df}{(1 + f/L)} = \int_0^t 10\,dt$$

$$L \ln(1 + f/L) = 10t$$

$$\ln(1 + f/L) = 10t/L$$

$$1 + f/L = e^{10t/L}$$

and

$$f = L[e^{10t/L} - 1]$$

Then the position of the particle, located at $x = 0$ at time $t = 0$, as a function of time is given by

$$x_p = f(t) = L[e^{10t/L} - 1] \quad \longleftarrow \qquad x_p$$

The x component of acceleration of this particle is given by

$$a_{x_p} = \frac{d^2 x_p}{dt^2} = \frac{d^2 f}{dt^2} = \frac{100}{L} e^{10t/L} \qquad \longleftarrow \qquad a_{x_p}$$

We now have two different ways of expressing the acceleration of the particular particle, which was located at $x = 0$ at time, $t = 0$. Note that although the flow field is steady, when we follow a particular particle, its position and acceleration (and velocity for that matter) are functions of time.

We check to see that both expressions for the acceleration give identical results.

$$a_{x_p} = \frac{100}{L} e^{10t/L} \qquad\qquad a_{x_p} = \frac{Du}{Dt} = \frac{100}{L}\left(1 + \frac{x}{L}\right)$$

(a) At time $t = 0$, $x_p = 0$ At time $t = 0$, the particle is at $x = 0$

$$a_{x_p} = \frac{100}{L} e^0 = \frac{100}{L} \qquad (a) \qquad\qquad \frac{Du}{Dt} = \frac{100}{L}(1 + 0) = \frac{100}{L} \qquad (a)$$
$$\text{Check.}$$

(b) When $x_p = \dfrac{L}{2}$, time $t = t_1$,   At $x = 0.5L$

$$x_p = \frac{L}{2} = L[e^{10t_1/L} - 1] \qquad\qquad \frac{Du}{Dt} = \frac{100}{L}(1 + 0.5)$$

$$\therefore \; e^{10t_1/L} = 1.5 \qquad\qquad\qquad \frac{Du}{Dt} = \frac{150}{L} \qquad (b)$$
$$\text{Check.}$$

and

$$a_{x_p} = \frac{100}{L} e^{10t_1/L}$$

$$a_{x_p} = \frac{100}{L}(1.5) = \frac{150}{L} \qquad\qquad (b)$$

(c) When $x_p = L$, time $t = t_2$,   At $x = L$

$$x_p = L = L[e^{10t_2/L} - 1] \qquad\qquad \frac{Du}{Dt} = \frac{100}{L}(1 + 1)$$

$$\therefore \; e^{10t_2/L} = 2$$

and $\dfrac{Du}{Dt} = \dfrac{200}{L} \qquad (c)$

$$a_{x_p} = \frac{100}{L} e^{10t_2/L} \qquad\qquad\qquad\qquad \text{Check.}$$

$$a_{x_p} = \frac{100}{L}(2) = \frac{200}{L} \qquad\qquad (c)$$

{This problem illustrates the two different methods of describing the motion of a particle.}

5–7.3 FORMULATION OF FORCES ACTING ON A FLUID PARTICLE

Having obtained an expression for the acceleration of a fluid element of mass, *dm*, moving in a velocity field, we can now write Newton's second law as the vector equation

$$dF = dm\frac{D\vec{V}}{Dt} = dm\left[u\frac{\partial\vec{V}}{\partial x} + v\frac{\partial\vec{V}}{\partial y} + w\frac{\partial\vec{V}}{\partial z} + \frac{\partial\vec{V}}{\partial t}\right] \qquad (5.26)$$

In terms of scalar component equations we would write

$$dF_x = dm\frac{Du}{Dt} = dm\left[u\frac{\partial u}{\partial x} + v\frac{\partial u}{\partial y} + w\frac{\partial u}{\partial z} + \frac{\partial u}{\partial t}\right] \qquad (5.27a)$$

$$dF_y = dm\frac{Dv}{Dt} = dm\left[u\frac{\partial v}{\partial x} + v\frac{\partial v}{\partial y} + w\frac{\partial v}{\partial z} + \frac{\partial v}{\partial t}\right] \qquad (5.27b)$$

$$dF_z = dm\frac{Dw}{Dt} = dm\left[u\frac{\partial w}{\partial x} + v\frac{\partial w}{\partial y} + w\frac{\partial w}{\partial z} + \frac{\partial w}{\partial t}\right] \qquad (5.27c)$$

We need now to obtain a suitable formulation for the force, $d\vec{F}$, or its components dF_x, dF_y, dF_z, acting on the element. Recall that the forces acting on a fluid element may be classified as body forces and surface forces; surface forces include both normal forces and tangential (shear) forces.

We shall consider the *x* component of the force acting on a differential element of mass, *dm*, and volume, $d\forall = dx\,dy\,dz$. Only those stresses that act in the *x* direction will give rise to surface forces in the *x* direction. If the stresses at the

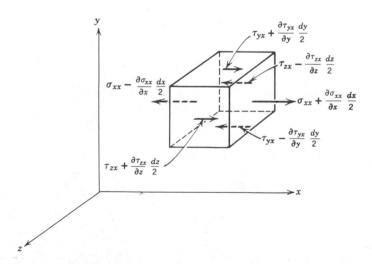

Fig. 5.9 *Stresses in the x direction on an element of fluid.*

the momentum equation

center of the differential element are taken to be σ_{xx}, τ_{yx}, and τ_{zx}, then the stresses acting in the x direction on each face of the element (obtained by a Taylor series expansion about the center of the element) are as shown in Fig. 5.9.

To obtain the net surface force in the x direction, dF_{S_x}, we must sum the forces in the x direction. Thus

$$dF_{s_x} = \left(\sigma_{xx} + \frac{\partial \sigma_{xx}}{\partial x}\frac{dx}{2}\right) dy\, dz - \left(\sigma_{xx} - \frac{\partial \sigma_{xx}}{\partial x}\frac{dx}{2}\right) dy\, dz$$

$$+ \left(\tau_{yx} + \frac{\partial \tau_{yx}}{\partial y}\frac{dy}{2}\right) dx\, dz - \left(\tau_{yx} - \frac{\partial \tau_{yx}}{\partial y}\frac{dy}{2}\right) dx\, dz$$

$$+ \left(\tau_{zx} + \frac{\partial \tau_{zx}}{\partial z}\frac{dz}{2}\right) dx\, dy - \left(\tau_{zx} - \frac{\partial \tau_{zx}}{\partial z}\frac{dz}{2}\right) dx\, dy$$

On simplifying we obtain

$$dF_{s_x} = \left(\frac{\partial \sigma_{xx}}{\partial x} + \frac{\partial \tau_{yx}}{\partial y} + \frac{\partial \tau_{zx}}{\partial z}\right) dx\, dy\, dz$$

If the body force per unit mass is designated as $\vec{B} = \hat{\imath}B_x + \hat{\jmath}B_y + \hat{k}B_z$, then the component of the body force in the x direction, dF_{B_x}, is given by $dF_{B_x} = B_x\, dm = B_x \rho\, d\forall$.

Then the net force in the x direction, dF_x, is given by

$$dF_x = dF_{s_x} + dF_{B_x} = \left(\rho B_x + \frac{\partial \sigma_{xx}}{\partial x} + \frac{\partial \tau_{yx}}{\partial y} + \frac{\partial \tau_{zx}}{\partial z}\right) dx\, dy\, dz \qquad (5.28a)$$

One can derive similar expressions for the force components in the y and z directions:

$$dF_y = dF_{s_y} + dF_{B_y} = \left(\rho B_y + \frac{\partial \tau_{xy}}{\partial x} + \frac{\partial \sigma_{yy}}{\partial y} + \frac{\partial \tau_{zy}}{\partial z}\right) dx\, dy\, dz \qquad (5.28b)$$

$$dF_z = dF_{s_z} + dF_{B_z} = \left(\rho B_z + \frac{\partial \tau_{xz}}{\partial x} + \frac{\partial \tau_{yz}}{\partial y} + \frac{\partial \sigma_{zz}}{\partial z}\right) dx\, dy\, dz \qquad (5.28c)$$

5–7.4 DIFFERENTIAL MOMENTUM EQUATION

We have now formulated expressions for the components dF_x, dF_y, and dF_z of the force, $d\vec{F}$, acting on the element of mass, dm. If we substitute these expressions (Eqs. 5.28) for the force components into Eqs. 5.27, we obtain the differential equations of motion.

$$\rho B_x + \frac{\partial \sigma_{xx}}{\partial x} + \frac{\partial \tau_{yx}}{\partial y} + \frac{\partial \tau_{zx}}{\partial z} = \rho\left(\frac{\partial u}{\partial t} + u\frac{\partial u}{\partial x} + v\frac{\partial u}{\partial y} + w\frac{\partial u}{\partial z}\right) \qquad (5.29a)$$

$$\rho B_y + \frac{\partial \tau_{xy}}{\partial x} + \frac{\partial \sigma_{yy}}{\partial y} + \frac{\partial \tau_{zy}}{\partial z} = \rho \left(\frac{\partial v}{\partial t} + u \frac{\partial v}{\partial x} + v \frac{\partial v}{\partial y} + w \frac{\partial v}{\partial z} \right) \qquad (5.29b)$$

$$\rho B_z + \frac{\partial \tau_{xz}}{\partial x} + \frac{\partial \tau_{yz}}{\partial y} + \frac{\partial \sigma_{zz}}{\partial z} = \rho \left(\frac{\partial w}{\partial t} + u \frac{\partial w}{\partial x} + v \frac{\partial w}{\partial y} + w \frac{\partial w}{\partial z} \right) \qquad (5.29c)$$

Equations 5.29 are the differential equations of motion for any fluid satisfying the continuum assumption. Before the equations can be employed in the solution of problems, suitable expressions for the stresses must be obtained. One sees that Eqs. 5.29 are simplified considerably under the assumption of frictionless flow, because all shear stresses vanish. We shall consider the case of frictionless flow in the next chapter.

problems

5.1 A velocity field is given by the expression

$$\vec{V} = Ax\hat{i} - Ay\hat{j}$$

where x and y are measured in feet, and A equals 10 ft/sec/ft.
Locate all points in the xy plane where $|\vec{V}|$ is equal to 20 ft/sec.

5.2 A velocity field is given by the expression

$$\vec{V} = U \cos \theta \left[1 - \left(\frac{a}{r} \right)^2 \right] \hat{i}_r - U \sin \theta \left[1 + \left(\frac{a}{r} \right)^2 \right] \hat{i}_\theta$$

Find all points in the $r\theta$ plane where:

(a) $V_r = 0$
(b) $V_\theta = 0$
(c) $V_r = V_\theta = 0$

5.3 A viscous liquid is sheared between two parallel discs of radius, R, one of which rotates while the other is fixed. The velocity field is purely tangential, and the velocity varies linear with z from $V_\theta = 0$ at $z = 0$ (the fixed disc) to the velocity of the rotating disc at its surface ($z = h$).
Write an expression for the velocity field between the discs.

5.4 Which of the following sets of equations represent possible two-dimensional incompressible flow cases?

(a) $u = x + y; v = x - y$
(b) $u = x + 2y; v = x^2 - y^2$
(c) $u = 4x + y; v = x - y^2$
(d) $u = xt + 2y; v = x^2 - yt^2$
(e) $u = xt^2; v = xyt + y^2$

5.5 Which of the following sets of equations represent possible two-dimensional incompressible flow cases?

(a) $u = 2x^2 + y^2 ; v = x^3 - x(y^2 - 2y)$
(b) $u = 2xy - x^2 + y ; v = 2xy - y^2 + x^2$
(c) $u = xt + 2y ; v = xt^2 - yt$
(d) $u = (x + 2y)xt ; v = (2x - y)yt$

5.6 Which of the following equations represent possible three-dimensional incompressible flow cases?

(a) $u = x + y + z^2 ; v = x - y + z ; w = 2xy + y^2 + 4$
(b) $u = xyzt ; v = -xyzt^2 ; w = (z^2/2)(xt^2 - yt)$
(c) $u = y^2 + 2xz ; v = -2yz + x^2yz ; w = (1/2)x^2z^2 + x^3y^4$

5.7 Which of the following sets of equations represent possible incompressible flow cases?

(a) $V_r = U \cos \theta ; V_\theta = -U \sin \theta$
(b) $V_r = -Q/2\pi r ; V_\theta = k/2\pi r$
(c) $V_r = U \cos \theta[1 - (a/r)^2] ; V_\theta = -U \sin \theta[1 + (a/r)^2]$

5.8 For a flow in the xy plane, the y component of velocity is given by

$$v = y^2 - 2x + 2y$$

Determine a possible x component for steady, incompressible flow. Is it also valid for unsteady, incompressible flow? Why? How many possible x components are there?

5.9 For an incompressible flow in the $r\theta$ plane, the r component of velocity is given as

$$V_r = -\frac{\Lambda \cos \theta}{r^2}$$

Determine a possible θ component of velocity. How many possible θ components are there?

5.10 The three components of velocity in a velocity field are given by

$$u = Ax + By + Cz$$
$$v = Dx + Ey + Fz$$
$$w = Gx + Hy + Jz$$

Determine the relationship among the coefficients A through J, which is necessary if this is to be a possible incompressible flow field.

5.11 The stream function for a certain incompressible flow is given as

$$\psi = Axy$$

Plot several streamlines, including $\psi = 0$. Obtain an expression for the velocity field.

5.12 The stream function for a certain incompressible flow field is given by the expression

$$\psi = -Ur \sin \theta + \frac{Q}{2\pi} \theta$$

Find the point(s) where $|\vec{V}| = 0$, and show that $\psi = 0$ there. Obtain an expression for the velocity field.

5.13 Incompressible flow around a circular cylinder of radius, a, is represented by the stream function

$$\psi = -Ur \sin \theta + \frac{Ua^2 \sin \theta}{r}$$

where U represents the freestream velocity.

Show that $V_r = 0$ along the circle, $r = a$. Locate the points along $r = a$ where $|\vec{V}| = U$.

5.14 Consider a flow with velocity components:

$$u = 0, \qquad v = -y^3 - 4z, \qquad w = 3y^2 z$$

(a) Is this one-, two-, or three-dimensional flow?
(b) Demonstrate whether this is an incompressible or compressible flow.
(c) If possible, derive a stream function for this flow.

5.15 Determine the family of ψ functions that will yield the velocity field

$$\vec{V} = (x^2 - y^2)\hat{\imath} - 2xy\hat{\jmath}$$

5.16 An incompressible frictionless flow field is specified by the stream function

$$\psi = -6x - 8y$$

(a) Sketch the streamlines $\psi = 0$ and $\psi = 8$.
(b) Indicate the direction of the resultant velocity vector at the point $(0, 0)$ on the sketch of part (a).
(c) Determine the magnitude of the flowrate between the streamlines passing through the points $(2, 2)$ and $(4, 1)$.

5.17 In a parallel one-dimensional flow in the positive x direction the velocity varies linearly from zero at $y = 0$ to 100 ft/sec at $y = 4$.

Determine an expression for the stream function, ψ. Also determine the y coordinate above which the volume flowrate is half the total between $y = 0$ and $y = 4$.

5.18 A flow is represented by the velocity field

$$\vec{V} = 10x\hat{\imath} - 10y\hat{\jmath} + 30\hat{k}$$

239 **problems**

Determine if the field is:

(a) A possible incompressible flow.
(b) Irrotational.

5.19 A flow is represented by the velocity field

$$\vec{V} = (4x^2 + 3y)\hat{\imath} + (3x - 2y)\hat{\jmath}$$

Determine if the field is:

(a) A possible incompressible flow.
(b) Irrotational.

5.20 Consider the flow field represented by the stream function

$$\psi = 10xy + 17$$

(a) Is this a possible two-dimensional incompressible flow?
(b) Is the flow irrotational?

5.21 Consider a laminar flow between closely spaced parallel plates. The velocity profile is given by

$$u = V\left(\frac{y}{h}\right)$$

Show that the rate of rotation of a fluid particle is given by

$$\omega_z = -\frac{V}{2h}$$

5.22 For a fluid in solid body rotation about the origin, the velocity field is given by

$$V_r = 0, \qquad V_\theta = r\omega, \qquad V_z = 0$$

Determine the rotation and vorticity for a particle in this flow field.

5.23 The velocity field near the core of a tornado can be represented by

$$\vec{V} = -\frac{Q}{2\pi r}\hat{\imath}_r + \frac{k}{2\pi r}\hat{\imath}_\theta$$

Is this an irrotational flow field?

5.24 A free vortex is a flow field whose streamlines are concentric circles about the origin. Thus

$$V_r = 0 \quad \text{and} \quad V_\theta = g(r)$$

for this flow field.
    Determine $g(r)$ such that the rotation of any fluid particle located at $r > 0$ is zero.

5.25   A flow field is represented by the stream function

$$\psi = x^2 - y^2$$

Show that this flow field is irrotational, and determine the velocity potential, ϕ.

5.26   Consider the flow field represented by the potential function, $\phi = -ax + by$.

(a) Is this a possible incompressible flow field?
(b) Is the flow irrotational?
(c) If a = 10 ft/sec and b = 10 ft/sec, what is the velocity of a particle at the origin?
(d) What are the coordinates of a particle, initially at the origin, after 10 sec?
(e) Sketch the lines of constant ϕ, labeling $\phi = 0$ and showing direction of increasing ϕ.

5.27   Consider the flow field represented by the potential function,

$$\phi = x^2 - y^2$$

Obtain the stream function that corresponds to this flow field.

5.28   Consider the flow field given by

$$\vec{V} = xy^2\hat{i} - \tfrac{1}{3}y^3\hat{j} + xy\hat{k}$$

Determine:

(a) The number of dimensions of the flow.
(b) If it is a possible incompressible flow.
(c) The acceleration of a particle at the point $(x, y, z) = (1, 2, 3)$.

5.29   Consider the flow field given by

$$\vec{V} = x^2y\hat{i} - 3y\hat{j} + 2z^2\hat{k}$$

Determine:

(a) The number of dimensions of the flow.
(b) If it is a possible incompressible flow.
(c) The acceleration of a particle at the point $(x, y, z) = (3, 1, 2)$.

5.30   Consider the flow field given by

$$\vec{V} = (4x^3 + 2y + xy)\hat{i} + (3x - y^3 + z)\hat{j}$$

Determine:

(a) The number of dimensions of the flow.
(b) If it is a possible incompressible flow.
(c) The acceleration of a particle at the point $(x, y, z) = (2, 2, 3)$.

5.31   Consider the flow field represented by the potential function,

$$\phi = ax^2 + bxy - ay^2$$

241   **problems**

Determine:

(a) The corresponding stream function.

(b) The acceleration of a particle at the point $(x, y) = (1, 0)$.

5.32 Consider the incompressible flow of a fluid through a nozzle, as shown. The area of the nozzle is given by

$$A = A_0(1 - bx)$$

and the inlet velocity varies according to

$$V_0 = U(1 + at)$$

where $A_0 = 1$ ft^2, $L = 4$ ft, $b = 0.1$ ft^{-1}, $a = 2$ sec^{-1}, and $U = 10$ ft/sec. The flow may be assumed one-dimensional.

Find the acceleration of a fluid particle at $x = L/2$ for $t = 0$ and 0.5 sec.

5.33 The temperature, T, in a long tunnel is known to vary approximately as

$$T = T_0 - \alpha e^{-x/L} \sin \frac{2\pi t}{\tau}$$

where T_0, α, L, and τ are constants, and x is measured from the entrance. A particle moves into the tunnel with a constant speed, U.

Find the rate of change of temperature experienced by the particle.

dynamics of incompressible inviscid flow.

6–1 *Stress Field in an Inviscid Flow*

In an inviscid flow, the viscosity, μ, is identically zero and, hence, there can be no shear stresses present. Consequently we need only concern ourselves with the normal stresses. By applying Newton's second law to the element of mass shown in Fig. 6.1, we can show that the normal stress at a point is the same in all directions, that is, the normal stress at a point is a scalar. Writing Newton's second law in the z direction we include surface and body forces,

$$dF_z = dF_{S_z} + dF_{B_z} = dm \, a_z$$

Substituting according to the stresses shown in Fig. 6.1,

$$-\sigma_{zz} \, dx \, dy + \sigma_{nn} \, ds \, dx \sin \alpha - \gamma \frac{dx \, dy \, dz}{2} = \rho \frac{dx \, dy \, dz}{2} a_z$$

Since

$$\sin \alpha = dy/ds,$$

then

$$-\sigma_{zz} \, dx \, dy + \sigma_{nn} \, dx \, dy - \gamma \frac{dx \, dy \, dz}{2} = \rho \frac{dx \, dy \, dz}{2} a_z$$

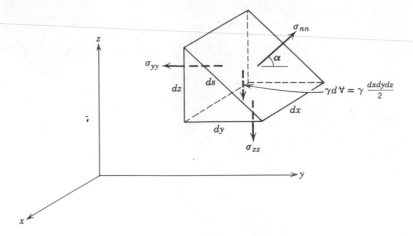

Fig. 6.1 *Differential element of mass in an inviscid flow field.*

and on dividing through by $dx\,dy$, we obtain

$$-\sigma_{zz} + \sigma_{nn} - \gamma\frac{dz}{2} = \rho\frac{dz}{2}a_z$$

Recognizing that dz is vanishingly small, we obtain

$$\sigma_{zz} = \sigma_{nn}$$

Writing Newton's second law in the y direction leads to the relation that $\sigma_{yy} = \sigma_{nn}$. Consequently we have

$$\sigma_{zz} = \sigma_{yy} = \sigma_{nn}$$

Since the shape of the element (i.e. the surface ds) was arbitrarily chosen, we conclude that for a nonviscous fluid in motion the normal stress at a point is the same in all directions (i.e. a scalar quantity). The normal stress in an inviscid flow is the negative of the thermodynamic pressure, or $\sigma_{nn} = -p$.

6–2 *Momentum Equation for Frictionless Flow: Euler's Equations*

The equations of motion for frictionless flow, called Euler's equations, are obtained from the general equations of motion (Eqs. 5.29). Since, in a frictionless flow, there can be no shear stresses present and the normal stress is the negative of the thermodynamic pressure, then the equations of motion for a frictionless flow are:

$$\rho B_x - \frac{\partial p}{\partial x} = \rho\left(\frac{\partial u}{\partial t} + u\frac{\partial u}{\partial x} + v\frac{\partial u}{\partial y} + w\frac{\partial u}{\partial z}\right) \qquad (6.1a)$$

$$\rho B_y - \frac{\partial p}{\partial y} = \rho\left(\frac{\partial v}{\partial t} + u\frac{\partial v}{\partial x} + v\frac{\partial v}{\partial y} + w\frac{\partial v}{\partial z}\right) \qquad (6.1b)$$

$$\rho B_z - \frac{\partial p}{\partial z} = \rho\left(\frac{\partial w}{\partial t} + u\frac{\partial w}{\partial x} + v\frac{\partial w}{\partial y} + w\frac{\partial w}{\partial z}\right) \qquad (6.1c)$$

We can also write the above equations as a single vector equation

$$\rho\vec{B} - \nabla p = \rho\left(\frac{\partial \vec{V}}{\partial t} + u\frac{\partial \vec{V}}{\partial x} + v\frac{\partial \vec{V}}{\partial y} + w\frac{\partial \vec{V}}{\partial z}\right)$$

or

$$\rho\vec{B} - \nabla p = \rho\frac{D\vec{V}}{Dt} \qquad (6.2)$$

For the case where the only body force is due to gravity, the body force per unit mass, $\vec{B}$, is equal to the gravity vector, $\vec{g}$, and Eq. 6.2 becomes

$$\rho\vec{g} - \nabla p = \rho\frac{D\vec{V}}{Dt} \qquad (6.3)$$

If the z coordinate is directed vertically then

$$\rho\vec{g} = -\rho g\hat{k} = -\rho g\nabla z \qquad (\text{Note } \nabla z = \hat{k})$$

and Euler's equation can be written as

$$-\frac{1}{\rho}\nabla p - g\nabla z = \frac{D\vec{V}}{Dt} \qquad (6.4)$$

The momentum equation for frictionless flow can also be written in cylindrical coordinates. The equations in component form, with gravity the only body force, are

$$g_r - \frac{1}{\rho}\frac{\partial p}{\partial r} = a_r = \frac{\partial V_r}{\partial t} + V_r\frac{\partial V_r}{\partial r} + \frac{V_\theta}{r}\frac{\partial V_r}{\partial \theta} + V_z\frac{\partial V_r}{\partial z} - \frac{V_\theta^2}{r} \qquad (6.5a)$$

$$g_\theta - \frac{1}{\rho r}\frac{\partial p}{\partial \theta} = a_\theta = \frac{\partial V_\theta}{\partial t} + V_r\frac{\partial V_\theta}{\partial r} + \frac{V_\theta}{r}\frac{\partial V_\theta}{\partial \theta} + V_z\frac{\partial V_\theta}{\partial z} + \frac{V_r V_\theta}{r} \qquad (6.5b)$$

$$g_z - \frac{1}{\rho}\frac{\partial p}{\partial z} = a_z = \frac{\partial V_z}{\partial t} + V_r\frac{\partial V_z}{\partial r} + \frac{V_\theta}{r}\frac{\partial V_z}{\partial \theta} + V_z\frac{\partial V_z}{\partial z} \qquad (6.5c)$$

If the z axis is directed vertically upward, then $g_r = g_\theta = 0$ and $g_z = -g$.

Example 6.1

Consider a flow in which the velocity field is given by $\vec{V} = \hat{i}Ax + \hat{j}Ay - \hat{k}2Az$ where A is a constant. Show that this is a possible incompressible flow. Assuming the z axis is vertical, $\rho = 2.0$ slug/ft^3, $A = 1.0$ sec^{-1}, and all dimensions are in feet, calculate the pressure gradient at the point $(1, 1, 5)$.

EXAMPLE PROBLEM 6.1

GIVEN:

Velocity field, $\vec{V} = \hat{i}Ax + \hat{j}Ay - \hat{k}2Az$, with $A = 1.0$ sec^{-1}.
The z axis is vertical and $\rho = 2.0$ slug/ft^3.

FIND:

(a) Show that velocity field represents a possible incompressible flow.
(b) Pressure gradient at $(1, 1, 5)$.

SOLUTION:

Basic equations:
$$\nabla \cdot \rho\vec{V} + \frac{\partial \rho}{\partial t} = 0$$

$$\rho\vec{B} - \nabla p = \rho\frac{D\vec{V}}{Dt} \quad \text{(neglecting viscous forces)}$$

For incompressible flow, the conservation of mass becomes $\nabla \cdot \vec{V} = 0$, or

$$\frac{\partial u}{\partial x} + \frac{\partial v}{\partial y} + \frac{\partial w}{\partial z} = 0$$

With $\vec{V} = \hat{i}Ax + \hat{j}Ay - \hat{k}2Az$, then

$$\frac{\partial(Ax)}{\partial x} + \frac{\partial(Ay)}{\partial y} + \frac{\partial(-2Az)}{\partial z} = A + A - 2A = 0$$

Therefore, the velocity field represents a possible incompressible flow.

Euler's equation states that $\rho\vec{B} - \nabla p = \rho\frac{D\vec{V}}{Dt}$.

If gravity is the only body force and the z axis is directed vertically upward, then

$$\rho\vec{B} = \rho\vec{g} = -\rho g\hat{k}$$

and Euler's equation gives

$$-\rho g\hat{k} - \nabla p = \rho\frac{D\vec{V}}{Dt}$$

and

$$\nabla p = -\rho g\hat{k} - \rho\frac{D\vec{V}}{Dt}$$

$$= -\rho g\hat{k} - \rho\left[\overset{0}{\cancel{\frac{\partial\vec{V}}{\partial t}}} + u\frac{\partial\vec{V}}{\partial x} + v\frac{\partial\vec{V}}{\partial y} + w\frac{\partial\vec{V}}{\partial z}\right]$$

$$= -\rho g\hat{k} - \rho[Ax(\hat{i}A) + Ay(\hat{j}A) + (-2Az)(-\hat{k}2A)]$$

$$= -\rho g\hat{k} - \rho[A^2x\hat{i} + A^2y\hat{j} + 4A^2z\hat{k}]$$

$$\nabla p = -\rho(g + 4A^2z)\hat{k} - \rho A^2x\hat{i} - \rho A^2y\hat{j} = -\rho[(g + 4A^2z)\hat{k} + A^2x\hat{i} + A^2y\hat{j}]$$

At the point $(1, 1, 5)$, with $\rho = 2.0$ slug/ft^3, and $A = 1.0$ sec^{-1}

$$\nabla p = -\frac{2.0\text{ slug}}{\text{ft}^3} \times [(32.2 + 20)\hat{k} + \hat{i} + \hat{j}]\frac{\text{ft}}{\text{sec}^2} \times \frac{\text{lbf-sec}^2}{\text{slug-ft}}$$

$$\nabla p = -2.0\hat{i} - 2.0\hat{j} - 104.4\hat{k}\frac{\text{lbf/ft}^2}{\text{ft}} \longleftarrow$$

$\left.\begin{array}{l}\text{This problem illustrates the calculation of the local pressure gradient from Euler's}\\ \text{equation.}\end{array}\right\}$

6–3 Euler's Equations for Fluids in Rigid Body Motion

In Chapter 3 we found that if a fluid is accelerated so that there is no relative motion between adjacent layers of the fluid, that is, when the fluid moves without deformation, no shear stresses occur. We were able to determine the pressure variation within the fluid by applying the equations of motion to an appropriate free body. We considered two specific cases. For the case of uniform linear accelera- tion, we derived the differential equation of motion, Eq. 3.3; in Example Problem 3.1 we applied the equation to a tank of water moving as a solid body. In Example Problem 3.2 we considered the case of a fluid undergoing steady rotation about a vertical axis; we derived the equation of motion for a differential element of fluid undergoing rotation.

Since Euler's equations are the equations of motion for a frictionless flow, that is, a flow in which the shear stresses are zero, they can be employed as the basic governing equations (written in the appropriate coordinate system) to solve frictionless flow problems. It is left to you as an exercise to show that the use of Euler's equations in the solution of Example Problems 3.1 and 3.2 leads to results identical to those previously obtained.

6–4 Euler's Equations in Streamline Coordinates

In Chapter 2 we pointed out that the notion of a streamline—that is, a line drawn tangent to the velocity vector at every point in the flow field—provides a convenient graphical representation. In steady flow a fluid particle will move along a streamline because, for steady flow, pathlines and streamlines coincide. Thus, in describing the motion of a fluid particle in a steady flow, the distance along a streamline is a logical coordinate to use in writing the equations of motion. "Streamline coordinates" may also be used to describe unsteady flow. Streamlines in unsteady flow give a graphical representation of the instantaneous velocity field.

For simplicity, consider the flow in the yz plane shown in Fig. 6.2. The equations of motion are to be written in terms of the coordinate, s, distance along a streamline

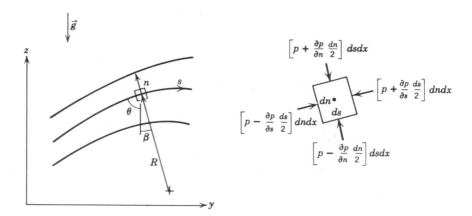

Fig. 6.2 *Fluid particle moving along a streamline.*

and the coordinate, n, distance normal to the streamline. Since the velocity vector must be tangent to the streamline, then the velocity field is given by $\vec{V} = \vec{V}(s, t)$. The pressure at the center of the fluid element is p. If we apply Newton's second law in the streamwise direction, that is, in the s direction, to the fluid element of volume $ds\, dn\, dx$, then we have, on neglecting the viscous forces

$$\left(p - \frac{\partial p}{\partial s}\frac{ds}{2}\right) dn\, dx - \left(p + \frac{\partial p}{\partial s}\frac{ds}{2}\right) dn\, dx - \rho g \cos\theta\, dn\, dx\, ds = \rho\, ds\, dn\, dx\, a_s$$

where θ is the angle between the tangent to the streamline and the vertical, and a_s is the acceleration of the fluid particle along the streamline. Then on simplifying the equation we obtain

$$-\frac{\partial p}{\partial s} - \rho g \cos \theta = \rho a_s$$

Since $\cos \theta = \partial z/\partial s$ we can write

$$-\frac{1}{\rho}\frac{\partial p}{\partial s} - g\frac{\partial z}{\partial s} = a_s$$

Along any streamline $V_s = V_s(s, t)$, and the total acceleration of a fluid particle in the streamwise direction is given by

$$a_s = \frac{DV_s}{Dt} = \frac{\partial V_s}{\partial t} + V_s\frac{\partial V_s}{\partial s}$$

The velocity is tangent to the streamline and, hence, the subscript, s, on V_s is redundant and can be dropped. Euler's equation in the streamwise direction with the z axis directed vertically is then

$$-\frac{1}{\rho}\frac{\partial p}{\partial s} - g\frac{\partial z}{\partial s} = \frac{\partial V}{\partial t} + V\frac{\partial V}{\partial s} \qquad (6.6a)$$

For steady flow, and neglecting body forces, Euler's equation in the streamwise direction reduces to

$$\frac{1}{\rho}\frac{\partial p}{\partial s} = -V\frac{\partial V}{\partial s} \qquad (6.6b)$$

which indicates that a decrease in velocity is accompanied by an increase in pressure and conversely.[1]

To obtain Euler's equation in a direction normal to the streamlines, we apply Newton's second law in the n direction to the fluid element. Again, neglecting viscous forces, we obtain

$$\left(p - \frac{\partial p}{\partial n}\frac{dn}{2}\right)ds\,dx - \left(p + \frac{\partial p}{\partial n}\frac{dn}{2}\right)ds\,dx - \rho g \cos \beta\,dn\,dx\,ds = \rho a_n\,dn\,dx\,ds$$

where β is the angle between the n direction and the vertical, and a_n is the acceleration of the fluid particle in the n direction. Then, on simplifying the equation, we obtain

$$-\frac{\partial p}{\partial n} - \rho g \cos \beta = \rho a_n$$

[1] The relationship between variations in pressure and velocity in the streamwise direction for steady incompressible inviscid flow is illustrated in the NCFMF film loop: S-FM038, *Streamwise Pressure Gradient in Inviscid Flow.*

249   **Euler's equations in streamline coordinates**

Since $\cos \beta = \partial z / \partial n$, we write

$$-\frac{1}{\rho}\frac{\partial p}{\partial n} - g\frac{\partial z}{\partial n} = a_n$$

The normal acceleration of the fluid element is toward the center of curvature of the streamline, that is, in the minus n direction; thus in the coordinate system of Fig. 6.2, the familiar centripetal acceleration is written as

$$a_n = \frac{-V^2}{R}$$

for steady flow,[2] where R is the radius of curvature of the streamline. Then, Euler's equation normal to the streamline is written for steady flow as

$$\frac{1}{\rho}\frac{\partial p}{\partial n} + g\frac{\partial z}{\partial n} = \frac{V^2}{R} \tag{6.7a}$$

For steady flow in a horizontal plane, Euler's equation normal to a streamline becomes:

$$\frac{1}{\rho}\frac{\partial p}{\partial n} = \frac{V^2}{R} \tag{6.7b}$$

which indicates that pressure increases in a direction outward from the center of curvature of the streamlines.[3] In regions where the streamlines are straight, the radius of curvature, R, of the streamlines is infinite and, hence, there is no pressure variation normal to the streamlines.

6–5 Bernoulli Equation—Integration of Euler's Equations Along a Streamline for Steady Flow

We have written the momentum equations and the continuity equation in differential form. Theoretically, for an incompressible inviscid flow these equations can be solved to give the complete velocity and pressure field. (If the density is not constant, an additional thermodynamic relation for the density is required.) While, in theory, the equations can be solved, the solution for a particular flow field may be very involved. However, we can integrate Euler's equation readily for steady flow along a streamline. This is most easily accomplished by using the form of Euler's equation along a streamline given by Eq. 6.6a. We shall derive the result in two ways in order to give added physical insight about restrictions on the results.

[2] If the flow were not steady, the streamline pattern could change with time. In that case,

$$a_n = -\frac{V_s^2}{R} + \frac{\partial V_n}{\partial t}$$

[3] The effect of streamline curvature on the pressure gradient normal to a streamline is illustrated in the NCFMF film loop: S-FM037, *Streamline Curvature and Normal Pressure Gradient.*

6-5.1 USING STREAMLINE COORDINATES

Euler's equation for steady flow along a streamline is given by

$$-\frac{1}{\rho}\frac{\partial p}{\partial s} - g\frac{\partial z}{\partial s} = V\frac{\partial V}{\partial s} \tag{6.8}$$

If a fluid particle moves a distance ds along a streamline, then

$$\frac{\partial p}{\partial s}ds = dp, \text{ the change in pressure along } s$$

$$\frac{\partial z}{\partial s}ds = dz, \text{ the change in elevation along } s, \text{ and}$$

$$\frac{\partial V}{\partial s}ds = dV, \text{ the change in velocity along } s$$

Then on multiplying Eq. 6.8 by ds, we can write

$$-\frac{dp}{\rho} - g\,dz = V\,dV \quad \text{(along } s)$$

or

$$\frac{dp}{\rho} + g\,dz + V\,dV = 0 \quad \text{(along } s)$$

Integration of this equation gives

$$\int \frac{dp}{\rho} + gz + \frac{V^2}{2} = \text{constant} \quad \text{(along } s) \tag{6.9}$$

Before Eq. 6.9 can be used, we must specify the relation between the pressure, p, and the density, ρ. For the special case of incompressible flow, $\rho = \text{constant}$, and Eq. 6.9 becomes

$$\frac{p}{\rho} + gz + \frac{V^2}{2} = \text{constant} \quad \text{(along a streamline)} \tag{6.10}$$

It is well to remember the assumptions made in arriving at the Bernoulli equation, Eq. 6.10. These were:

1. Steady flow
2. Incompressible flow
3. Frictionless flow
4. Flow along a streamline

Equation 6.10 is a powerful and useful equation, because it relates pressure changes to velocity and elevation changes along a streamline. However, it gives

Bernoulli equation —integration of Euler's equations

correct results only when applied to a flow situation where all four of the restrictions are reasonable. Therefore, keep the restrictions firmly in mind when you consider using the Bernoulli equation. (In general, the Bernoulli constant in Eq. 6.10 has different values along different streamlines.)

6–5.2 USING RECTILINEAR COORDINATES

Beginning with the vector form of Euler's equation, Eq. 6.4, one can integrate the equation along a streamline. Since the velocity field, $\vec{V}$, is specified in terms of the rectangular coordinates, x, y, z, it is convenient to employ vector notation.[4] We shall restrict the derivation to steady flow; thus, the end result of our effort should be identical to Eq. 6.9.

For steady flow, Euler's equation may be written as

$$-\frac{1}{\rho}\nabla p - g\nabla z = \frac{D\vec{V}}{Dt} = u\frac{\partial \vec{V}}{\partial x} + v\frac{\partial \vec{V}}{\partial y} + w\frac{\partial \vec{V}}{\partial z} \qquad (6.4)$$

The right-hand side of Eq. 6.4 is rewritten using vector notation as

$$u\frac{\partial \vec{V}}{\partial x} + v\frac{\partial \vec{V}}{\partial y} + w\frac{\partial \vec{V}}{\partial z} = (\vec{V}\cdot\nabla)\vec{V} \qquad (6.11)$$

It is suggested that you check this equality by expanding the right-hand side of Eq. 6.11 using the familiar dot product operation. Using Eq. 6.11, Euler's equation can be expressed as

$$-\frac{1}{\rho}\nabla p - g\nabla z = (\vec{V}\cdot\nabla)\vec{V} \qquad (6.12)$$

For steady flow the velocity field is given by $\vec{V} = \vec{V}(x, y, z)$. The streamlines are lines drawn in the flow field tangent to the velocity vector at every point. Recall again that for steady flow, streamlines, pathlines, and streaklines coincide. The motion of a particle along a streamline is governed by Eq. 6.12. In the time increment dt, the particle moves a distance $d\vec{s}$ along the streamline.

If we take the dot product of the terms in Eq. 6.12 with the distance $d\vec{s}$ along the streamline, we will obtain a scalar equation relating the pressure, p, velocity, V, and elevation, z, along the streamline. Taking the dot product of $d\vec{s}$ with Eq. 6.12 gives

$$-\frac{1}{\rho}\nabla p\cdot d\vec{s} - g\nabla z\cdot d\vec{s} = (\vec{V}\cdot\nabla)\vec{V}\cdot d\vec{s} \qquad (6.13)$$

[4] We will also have to employ the vector identity

$$(\vec{V}\cdot\nabla)\vec{V} = \tfrac{1}{2}\nabla(\vec{V}\cdot\vec{V}) - \vec{V}\times(\nabla\times\vec{V})$$

which may be verified by expanding each side into components (see Appendix D).

252   **6/dynamics of incompressible inviscid flow**

where

$$d\vec{s} = \hat{\imath}\,dx + \hat{\jmath}\,dy + \hat{k}\,dz \quad \text{(along } s\text{)}$$

Now we shall evaluate each of the three terms in Eq. 6.13.

$$-\frac{1}{\rho}\nabla p \cdot d\vec{s} = -\frac{1}{\rho}\left[\hat{\imath}\frac{\partial p}{\partial x} + \hat{\jmath}\frac{\partial p}{\partial y} + \hat{k}\frac{\partial p}{\partial z}\right] \cdot [\hat{\imath}\,dx + \hat{\jmath}\,dy + \hat{k}\,dz]$$

$$= -\frac{1}{\rho}\left[\frac{\partial p}{\partial x}\,dx + \frac{\partial p}{\partial y}\,dy + \frac{\partial p}{\partial z}\,dz\right] \quad \text{(along } s\text{)}$$

$$= -\frac{1}{\rho}\,dp \quad \text{(along } s\text{)}$$

$$-g\nabla z \cdot d\vec{s} = -g\hat{k} \cdot [\hat{\imath}\,dx + \hat{\jmath}\,dy + \hat{k}\,dz]$$

$$= -g\,dz \quad \text{(along } s\text{)}$$

Using a vector identity,[5] we can write the third term as

$$(\vec{V}\cdot\nabla)\vec{V}\cdot d\vec{s} = [\tfrac{1}{2}\nabla(\vec{V}\cdot\vec{V}) - \vec{V}\times(\nabla\times\vec{V})]\cdot d\vec{s}$$

$$= \{\tfrac{1}{2}\nabla(\vec{V}\cdot\vec{V})\}\cdot d\vec{s} - \{\vec{V}\times(\nabla\times\vec{V})\}\cdot d\vec{s}$$

The last term on the right-hand side of this equation is zero since $\vec{V}$ is parallel to $d\vec{s}$. Consequently,

$$(\vec{V}\cdot\nabla)\vec{V}\cdot d\vec{s} = \tfrac{1}{2}\nabla(\vec{V}\cdot\vec{V})\cdot d\vec{s} \quad \text{(along } s\text{)}$$

$$= \frac{1}{2}\left[\hat{\imath}\frac{\partial V^2}{\partial x} + \hat{\jmath}\frac{\partial V^2}{\partial y} + \hat{k}\frac{\partial V^2}{\partial z}\right] \cdot [\hat{\imath}\,dx + \hat{\jmath}\,dy + \hat{k}\,dz]$$

$$= \frac{1}{2}\left[\frac{\partial V^2}{\partial x}\,dx + \frac{\partial V^2}{\partial y}\,dy + \frac{\partial V^2}{\partial z}\,dz\right] \quad \text{(along } s\text{)}$$

$$= \tfrac{1}{2}d(V^2) \quad \text{(along } s\text{)}$$

Substituting for these three terms back into Eq. 6.13 yields

$$-\frac{1}{\rho}\,dp - g\,dz = \frac{1}{2}d(V^2) \quad \text{(along } s\text{)}$$

or

$$\frac{dp}{\rho} + g\,dz + \frac{1}{2}d(V^2) = 0 \quad \text{(along } s\text{)}$$

[5] This identity is proved in Appendix D.

253   **Bernoulli equation—integration of Euler's equations**

Integrating this equation, we obtain

$$\int \frac{dp}{\rho} + gz + \frac{V^2}{2} = \text{constant} \quad \text{(along } s\text{)} \tag{6.9}$$

If the density is constant

$$\frac{p}{\rho} + \frac{V^2}{2} + gz = \text{constant} \quad \text{(along a streamline)} \tag{6.10}$$

As expected, we see that the last two equations are identical to Eqs. 6.9 and 6.10 derived previously using streamline coordinates. Furthermore, Eq. 6.10 is still limited by the restrictions:

1. Steady flow
2. Incompressible flow
3. Frictionless flow
4. Flow along a streamline

Equation 6.10 can be applied between any two points on a streamline, provided the other three restrictions are satisfied. The result is

$$\frac{p_1}{\rho} + \frac{V_1^2}{2} + gz_1 = \frac{p_2}{\rho} + \frac{V_2^2}{2} + gz_2 \tag{6.11}$$

where subscripts 1 and 2 represent any two points on a streamline. Application of Eqs. 6.10 and 6.11 to typical flow situations is illustrated in Example Problems 6.2 through 6.4.

In some problems, the flow appears unsteady from one reference frame, but steady from another, which translates in the flow. Since the Bernoulli equation was derived by integrating Newton's second law for a fluid particle, it can be applied in any inertial reference frame (see the discussion of translating frames in Section 4–4.2). The procedure is illustrated in Example Problem 6.5.

Example 6.2

Air flows steadily and at low speed through a horizontal nozzle. At the nozzle inlet, the flow velocity, pressure, and area are 30 ft/sec, 14.7 psia, and 1 ft^2, respectively. At the nozzle exit, the area is 0.2 ft^2. The flow is essentially incompressible, and frictional effects are negligible. Determine the velocity and pressure at the nozzle outlet.

EXAMPLE PROBLEM 6.2

GIVEN:

Flow through a nozzle, as shown. The air flow is steady, incompressible and frictionless.

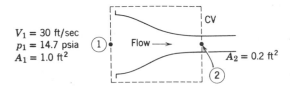

$V_1 = 30$ ft/sec
$p_1 = 14.7$ psia ①
$A_1 = 1.0$ ft²

FIND:

(a) V_2
(b) p_2

SOLUTION:

Apply the continuity equation to determine V_2, using CV as shown.

Basic equation:

$$0 = \overset{= \ 0(1)}{\cancel{\frac{\partial}{\partial t} \iiint_{CV} \rho \, d\forall}} + \iint_{CS} \rho \vec{V} \cdot d\vec{A}$$

Assumptions: (1) Steady flow
(2) Incompressible flow
(3) Uniform flow at sections ① and ②

Then

$$0 = \{-|\rho V_1 A_1|\} + \{|\rho V_2 A_2|\}$$

or

$$V_1 A_1 = V_2 A_2$$

Thus

$$V_2 = V_1 \frac{A_1}{A_2} = 30 \, \frac{\text{ft}}{\text{sec}} \times \frac{1.0 \, \text{ft}^2}{0.2 \, \text{ft}^2} = 150 \, \frac{\text{ft}}{\text{sec}} \longleftarrow \underline{ V_2}$$

Apply the Bernoulli equation along a streamline between points ① and ② to evaluate p_2.

Basic equation:

$$\frac{p_1}{\rho} + \frac{V_1^2}{2} + g z_1 = \frac{p_2}{\rho} + \frac{V_2^2}{2} + g z_2$$

Bernoulli equation — integration of Euler's equations

Assumptions: (4) Frictionless flow
 (5) Flow along a streamline
 (6) $z_1 = z_2$

Then

$$p_2 = p_1 + \frac{\rho}{2}(V_1^2 - V_2^2)$$

For air, $\rho = 0.00238$ slug/ft^3 at standard conditions.

$$p_2 = 14.7 \frac{\text{lbf}}{\text{in.}^2} + 0.00238 \frac{\text{slug}}{\text{ft}^3} \times \frac{1}{2} \times \frac{\text{lbf-sec}^2}{\text{slug-ft}} \left[(30)^2 \frac{\text{ft}^2}{\text{sec}^2} - (150)^2 \frac{\text{ft}^2}{\text{sec}^2} \right] \frac{\text{ft}^2}{144 \text{ in.}^2}$$

$$= 14.7 - 0.112 \frac{\text{lbf}}{\text{in.}^2}$$

$$p_2 = 14.6 \text{ psia} \qquad\qquad\qquad\qquad\qquad\qquad p_2$$

{
This problem illustrates a typical application of the Bernoulli equation. Note that if the flow streamlines are straight at the nozzle inlet and exit, the pressure will be uniform at those sections.
}

Example 6.3

A U tube acting as a siphon is shown below. In addition to the data given on the figure, the atmospheric pressure is 14.4 psia, and the fluid is water with a density of 1.94 slug/ft^3. If the flow is frictionless as a first approximation, and the fluid issues from the bottom of the siphon as a free jet at atmospheric pressure, determine (after listing the necessary assumptions):

(a) the velocity of the free jet, in fps, and
(b) the absolute pressure of the fluid at point A in the flow, in psia.

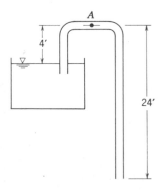

EXAMPLE PROBLEM 6.3

GIVEN:

Water flowing through a siphon as shown

$$p_{atm} = 14.4 \text{ psia} \qquad \rho = 1.94 \text{ slug/ft}^3$$

FIND:

(a) Velocity of water leaving as a free jet.
(b) Pressure at point (A) in the flow.

SOLUTION:

Basic equation:

$$\frac{p}{\rho} + \frac{V^2}{2} + gz = \text{constant}$$

Assumptions: (1) Neglect friction
(2) Steady flow
(3) Incompressible flow
(4) Flow along a streamline
(5) Reservoir is large compared to pipe

Apply Bernoulli equation between points ① and ②

$$\frac{p_1}{\rho} + \frac{V_1^2}{2} + gz_1 = \frac{p_2}{\rho} + \frac{V_2^2}{2} + gz_2$$

Since Area$_{reservoir} \gg$ Area$_{pipe}$, $V_1 \approx 0$. Also $p_1 = p_2 = p_{atm}$, so

$$gz_1 = \frac{V_2^2}{2} + gz_2 \quad \text{and} \quad V_2^2 = 2g(z_1 - z_2)$$

$$V_2 = \sqrt{2g(z_1 - z_2)} = \sqrt{2 \times 32.2 \text{ ft/sec}^2 \times 20 \text{ ft}} = 35.8 \text{ ft/sec} \qquad \underleftarrow{\quad V_2 \quad}$$

To determine the pressure at location (A), we write Bernoulli equation between ① and (A)

$$\frac{p_1}{\rho} + \frac{V_1^2}{2} + gz_1 = \frac{p_A}{\rho} + \frac{V_A^2}{2} + gz_A$$

Again $V_1 \approx 0$ and from conservation of mass $V_A = V_2$. Hence

$$\frac{p_A}{\rho} = \frac{p_1}{\rho} + gz_1 - \frac{V_2^2}{2} - gz_A = \frac{p_1}{\rho} + g(z_1 - z_A) - \frac{V_2^2}{2}$$

$$p_A = p_1 + \rho g(z_1 - z_A) - \rho \frac{V_1^2}{2}$$

$$= 14.4 \frac{lbf}{in.^2} - 1.94 \frac{slug}{ft^3} \times 32.2 \frac{ft}{sec^2} \times 4\,ft - 1.94 \frac{slug}{ft^3} \times \frac{(35.8)^2}{2} \frac{ft^2}{sec^2}$$

$$= 14.4 \frac{lbf}{in.^2} - 250 \frac{slug}{ft\text{-}sec^2} - 1250 \frac{slug}{ft\text{-}sec^2}$$

$$= 14.4 \frac{lbf}{in.^2} - 1500 \frac{slug}{ft\text{-}sec^2} \times \frac{ft^2}{144\,in.^2} \times \frac{lbf\text{-}sec^2}{slug\text{-}ft}$$

$$p_A = 14.4 \frac{lbf}{in.^2} - 10.4 \frac{lbf}{in.^2} = 4.0\,psia \qquad\qquad \overleftarrow{\hspace{3cm}} \qquad p_A$$

$$\left\{ \begin{array}{c} \text{This problem illustrates a straightforward application of the} \\ \text{Bernoulli equation with elevation changes included.} \end{array} \right\}$$

Example 6.4

Water flows under a sluice gate on a horizontal bed at the inlet to a flume. Above the gate, the water level is 1.5 ft, and the velocity is negligible. At the vena contracta below the gate, the flow streamlines are straight, and the depth is 2 in. Hydrostatic pressure distributions and uniform flow may be assumed at each section, and friction is negligible. Determine the flow velocity downstream from the gate, and the discharge in cubic feet per second per foot of width.

EXAMPLE PROBLEM 6.4

GIVEN:

Flow of water under a sluice gate. Flow is uniform and frictionless, and pressure distribution is hydrostatic at sections ① and ②.

FIND:

(a) V_2.
(b) Q in ft³/sec/ft of width.

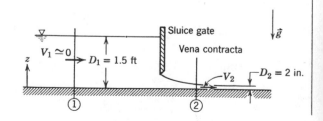

SOLUTION:

The flow satisfies all conditions necessary to apply the Bernoulli equation. The question is, what streamline do we use?

Basic equation:
$$\frac{p_1}{\rho} + \frac{V_1^2}{2} + gz_1 = \frac{p_2}{\rho} + \frac{V_2^2}{2} + gz_2 \quad.$$

Assumptions: (1) Steady flow
(2) Incompressible flow
(3) Frictionless flow
(4) Flow along a streamline
(5) Hydrostatic pressure distribution:

$$\frac{dp}{dz} = -\gamma, \text{ so} \qquad p = p_{atm} + \gamma(D - z), \text{ or} \qquad \frac{p}{\rho} = \frac{p_{atm}}{\rho} + g(D - z)$$

Substituting this relation into the Bernoulli equation,

$$\frac{p_{atm}}{\rho} + g(D_1 - z_1) + \frac{V_1^2}{2} + gz_1 = \frac{p_{atm}}{\rho} + g(D_2 - z_2) + \frac{V_2^2}{2} + gz_2$$

or

$$\frac{V_1^2}{2} + gD_1 = \frac{V_2^2}{2} + gD_2$$

This result implies that $V^2/2 + gD = $ constant, and the constant has the same value along *any* streamline for this flow. Solving for V_2,

$$V_2 = \sqrt{2g(D_1 - D_2) + V_1^2}$$

But $V_1^2 \simeq 0$, so

$$V_2 = \sqrt{2g(D_1 - D_2)} = \sqrt{2 \times 32.2 \frac{\text{ft}}{\text{sec}^2} \left(1.5 \text{ ft} - 2 \text{ in.} \times \frac{\text{ft}}{12 \text{ in.}} \right)}$$

$$V_2 = \sqrt{(2)(32.2)(1.33)} = 9.23 \frac{\text{ft}}{\text{sec}} \longleftarrow \hspace{3cm} V_2$$

Assumption: (6) Uniform flow at a section

For uniform flow, $Q = VA = VDw$, or

$$\frac{Q}{w} = VD = V_2 D_2 = 9.23 \frac{\text{ft}}{\text{sec}} \times 2 \text{ in.} \times \frac{\text{ft}}{12 \text{ in.}} = 1.58 \frac{\text{ft}^2}{\text{sec}}$$

$$\frac{Q}{w} = 1.58 \frac{\text{ft}^3}{\text{sec}} \text{ (per foot of width)} \longleftarrow \hspace{3cm} Q/w$$

Example 6.5

A Piper Cub flies at 90 mph in standard air at an altitude of 5000 ft. At a certain point close to the wing, the air speed *relative* to the wing is 200 ft/sec. Compute the pressure at this point.

EXAMPLE PROBLEM 6.5

GIVEN:

Aircraft in flight at 90 mph at 5000 ft altitude in standard air.

$$V_{air} = 0$$

FIND:

Pressure, p_A, at point A,

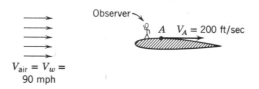

SOLUTION:

Flow is unsteady when observed from a fixed frame, that is, by an observer on the ground. However, an observer *on* the wing sees the steady flow sketched below:

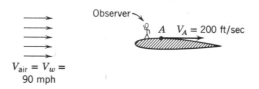

Thus the Bernoulli equation can be applied along a streamline in the moving observer's inertial reference frame.

Basic equation:
$$\frac{p_{air}}{\rho} + \frac{V_{air}^2}{2} + gz_{air} = \frac{p_A}{\rho} + \frac{V_A^2}{2} + gz_A$$

Assumptions: (1) Steady flow
 (2) Incompressible flow ($V < 300$ ft/sec)
 (3) Frictionless flow
 (4) Flow along a streamline
 (5) Neglect Δz

Values for pressure and density may be found from Table A.4. Thus at 5000 ft,

$$\frac{p}{p_0} = 0.832 \quad \text{and} \quad \frac{\rho}{\rho_0} = 0.862$$

Consequently

$$p = 0.832p_0 = (0.832)(2116 \text{ lbf/ft}^2) = 1761 \text{ lbf/ft}^2$$

and

$$\rho = 0.862\rho_0 = (0.862)(0.00238 \text{ slug/ft}^3) = 0.00215 \text{ slug/ft}^3$$

Solving for p_A,

$$p_A = p_{air} + \frac{\rho}{2}(V_{air}^2 - V_A^2)$$

$$= 1761\frac{\text{lbf}}{\text{ft}^2} + 0.00215\frac{\text{slug}}{\text{ft}^3} \times \frac{1}{2}\left[\left(\frac{90 \text{ mi}}{\text{hr}} \times \frac{5280 \text{ ft}}{\text{mi}} \times \frac{\text{hr}}{3600 \text{ sec}}\right)^2 - (200)^2\frac{\text{ft}^2}{\text{sec}^2}\right] \times \frac{\text{lbf-sec}^2}{\text{slug-ft}}$$

$$p_A = 1761 - 24.3 = 1737 \text{ lbf/ft}^2$$

or

$$p_A = 12.0 \text{ psia} \qquad\qquad\qquad\qquad\qquad\qquad\qquad\qquad\qquad\qquad\qquad\qquad p_A$$

6–6 Static, Stagnation, and Dynamic Pressures

The pressure, p, which we have used in deriving the Bernoulli equation, Eq. 6.10, is the thermodynamic pressure, which is commonly called the static pressure. The static pressure is that pressure which would be measured by an instrument moving with the flow. However, such a measurement is rather difficult to make in a practical situation! How do we measure static pressure experimentally?

In Section 6–5.1, we showed that there was no pressure variation normal to flow streamlines when those streamlines were straight. This fact makes it possible

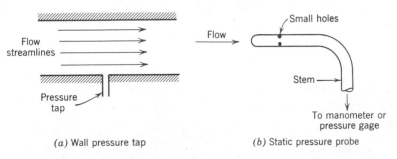

(a) Wall pressure tap   (b) Static pressure probe

Fig. 6.3 *Measurement of static pressure.*

static, stagnation, and dynamic pressures

to measure the static pressure in a flowing fluid using a wall pressure "tap," placed in a region where the flow streamlines are straight, as shown in Fig. 6.3a. The pressure tap is a small hole, drilled carefully in the wall, with its axis perpendicular to the surface. If the hole is perpendicular to the duct wall and free from burrs, accurate measurements of static pressure can then be made by connecting the tap to a suitable measuring instrument.

In a fluid stream far from a wall, or where streamlines are curved, accurate static pressure measurements can be made by careful use of a static pressure probe, shown in Fig. 6.3b. Such probes must be designed so that the measuring holes are placed correctly with respect to the probe tip and stem to avoid erroneous results. In use, the measuring section must be aligned with the flow direction.

Static pressure probes, such as that shown in Fig. 6.3b, and in a variety of other forms, are available commercially in sizes as small as $\frac{1}{16}$ in. in diameter.

The stagnation pressure is that value obtained when a flowing fluid is decelerated to zero velocity by a frictionless process. Stagnation pressure is measured in the laboratory using a probe with a hole that faces directly upstream, as shown in Fig. 6.4. Such a probe is called a stagnation pressure probe, or pitot tube. Again, the measuring section should be aligned with the flow direction.

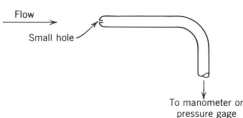

Fig. 6.4 *Measurement of stagnation pressure.*

In incompressible flow, the Bernoulli equation can be used to relate changes in velocity and pressure along a streamline when the flow is not affected by friction. Neglecting elevation differences, Eq. 6.10 becomes

$$\frac{p}{\rho} + \frac{V^2}{2} = \text{constant}$$

If the static pressure is p at a point in the flow where the velocity is V, then the stagnation pressure, p_0, may be computed from

$$\frac{p_0}{\rho} + \frac{V_0^2}{2} = \frac{p}{\rho} + \frac{V^2}{2}$$

or

$$19.1 = 17.2^{-1} \tfrac{1}{2}$$

$$p_0 = p + \tfrac{1}{2}\rho V^2 \qquad\qquad (6.14)$$

since the stagnation pressure is obtained where the stagnation velocity, V_0, is zero.

Equation 6.14 is a mathematical statement of the definition of stagnation pressure, valid for incompressible flow. The term $\tfrac{1}{2}\rho V^2$ is generally called the dynamic pressure. Solving for the dynamic pressure,

$$\tfrac{1}{2}\rho V^2 = p_0 - p$$

and for the velocity

$$V = \sqrt{\frac{2(p_0 - p)}{\rho}} \qquad\qquad (6.15)$$

Thus if the stagnation pressure and the static pressure could be measured at a point, Eq. 6.15 would give the local flow velocity.

We have seen that static pressure at a point can be measured with a static pressure tap or probe (Fig. 6.3). If we knew the stagnation pressure at the same point, then the flow velocity could be computed from Eq. 6.15. Two possible experimental setups are shown in Fig. 6.5.

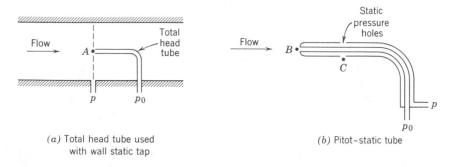

(a) Total head tube used with wall static tap

(b) Pitot–static tube

Fig. 6.5 *Simultaneous measurement of stagnation and static pressure.*

In Fig. 6.5a, the static pressure corresponding to point A is read from the wall static pressure tap. The stagnation pressure is measured directly at A by the total head tube, as shown. (The stem of the total head tube is placed downstream from the static pressure tap so that there will be no effect on the reading.)

Two probes are often combined, as in the pitot-static tube shown in Fig. 6.5b. The inner tube is used to measure the stagnation pressure at point B, while the static pressure at C is sensed by the small holes in the outer tube. In flow fields

263   **static, stagnation, and dynamic pressures**

$$8.27 \qquad 7.4476$$

where the static pressure variation in the streamwise direction is small, the pitot-static tube may be used to infer the velocity at point B in the flow, by assuming $p_B = p_C$, and using Eq. 6.15. (Note that when $p_B \neq p_C$, this procedure will give erroneous results.)

The definition and calculation of the stagnation pressure for compressible flow will be discussed in Section 9–3.1.

Example 6.6

A pitot probe is inserted in a water flow to measure the flow velocity. The tube is inserted so that it points upstream into the flow and the pressure sensed by the probe is the stagnation pressure. The static pressure is measured at the same point in the flow, using a wall pressure tap. If the fluid in the manometer tube is mercury, determine the flow velocity.

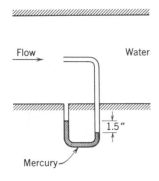

EXAMPLE PROBLEM 6.6

GIVEN:

A pitot tube inserted in a flow as shown. The flowing fluid is water and the manometer fluid is mercury.

FIND:

The flow velocity.

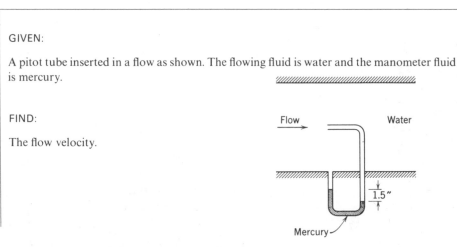

6/dynamics of incompressible inviscid flow

SOLUTION:

Basic equation: $\dfrac{p}{\rho} + \dfrac{V^2}{2} + gz = \text{constant}$

Assumptions: (1) Steady flow
(2) Frictionless flow
(3) Flow along a streamline
(4) Incompressible flow

Writing Bernoulli's equation along the stagnation streamline, taking $z = 0$, the constant can be evaluated. Designating the constant as the stagnation pressure, p_0 (the pressure at the tube opening where the velocity has been reduced, without friction, to zero), then

$$\frac{p_0}{\rho} = \frac{p}{\rho} + \frac{V^2}{2}$$

Solving for V,

$$V = \sqrt{\frac{2(p_0 - p)}{\rho}}$$

from the diagram, $p_0 - p = \gamma_{Hg}h - \gamma_{H_2O}h$, where $h = 1.5$ in., and $\gamma_{Hg} = SG_{Hg}\gamma_{H_2O}$.

$p_0 - p = SG_{Hg}\gamma_{H_2O}h - \gamma_{H_2O}h = \gamma_{H_2O}h(SG_{Hg} - 1)$

$$= 62.4\,\frac{lbf}{ft^3} \times 1.5\,\text{in.} \times \frac{ft}{12\,\text{in.}} \times (13.6 - 1)$$

$p_0 - p = 98\ \text{lbf/ft}^2$

$$V = \sqrt{\frac{2(p_0 - p)}{\rho}} = \sqrt{2 \times 98\,\frac{lbf}{ft^2} \times \frac{ft^3}{1.94\,\text{slug}} \times \frac{\text{slug-ft}}{lbf\text{-sec}^2}}$$

$V = 10.1\ \text{ft/sec}$ $\underleftarrow{\hspace{3cm}}$ V

{This problem illustrates the use of a pitot tube in determining the velocity at a point.}

6–7 Relation between the First Law of Thermodynamics and the Bernoulli Equation

The Bernoulli equation, Eq. 6.10, was obtained by integrating Newton's second law for a fluid particle, along a streamline, for steady, incompressible, frictionless flow. Thus Eq. 6.10 was derived from conservation of momentum.

An equation identical in form to Eq. 6.10 (although requiring very different restrictions) may be obtained from the first law of thermodynamics. Our objective in this section is to reduce the energy equation to the form of the Bernoulli equation given by Eq. 6.10. Having arrived at this form, we shall compare the restrictions on the two equations. This procedure will help us to understand more clearly the restrictions on the use of Eq. 6.10.

Consider steady, incompressible flow in the absence of shear forces. We consider a control volume of cross-section, A, bounded by streamlines along its periphery. Such a control volume, shown in Fig. 6.6, is often referred to as a stream tube.

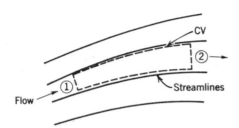

Fig. 6.6 *Flow through a stream tube.*

Basic equation:

$$\dot{Q} - \overset{= 0(1)}{\cancel{\dot{W}_s}} - \overset{= 0(2)}{\cancel{\dot{W}_{shear}}} - \overset{= 0(3)}{\cancel{\dot{W}_{other}}} = \overset{= 0(4)}{\cancel{\frac{\partial}{\partial t} \iiint_{CV} e\rho \, dV}} + \iint_{CS} (e + pv)\rho \vec{V} \cdot d\vec{A} \quad (4.45)$$

$$e = u + \frac{V^2}{2} + gz$$

Restrictions: (1) $\dot{W}_s = 0$
(2) $\dot{W}_{shear} = 0$
(3) $\dot{W}_{other} = 0$
(4) Steady flow
(5) Uniform flow and properties at each section

Under these restrictions, Eq. 4.45 becomes

$$0 = \left(u_1 + p_1 v_1 + \frac{V_1^2}{2} + gz_1 \right)\{-|\rho_1 V_1 A_1|\}$$

$$+ \left(u_2 + p_2 v_2 + \frac{V_2^2}{2} + gz_2 \right)\{|\rho_2 V_2 A_2|\} - \dot{Q}$$

6/dynamics of incompressible inviscid flow

But from continuity under these restrictions,

$$0 = \overbrace{\frac{\partial}{\partial t} \iiint_{CV} \rho \, dV}^{= 0(4)} + \iint_{CS} \rho \vec{V} \cdot d\vec{A}$$

or

$$0 = \{-|\rho_1 V_1 A_1|\} + \{|\rho_2 V_2 A_2|\}$$

that is,

$$\dot{m} = \rho_1 V_1 A_1 = \rho_2 V_2 A_2$$

Also

$$\dot{Q} = \frac{\delta Q}{dt} = \frac{\delta Q}{dm} \frac{dm}{dt} = \frac{\delta Q}{dm} \dot{m}$$

Thus, from the energy equation

$$0 = \left[\left(p_1 v_1 + \frac{V_1^2}{2} + gz_1 \right) - \left(p_2 v_2 + \frac{V_2^2}{2} + gz_2 \right) \right] \dot{m} + \left(u_2 - u_1 - \frac{\delta Q}{dm} \right) \dot{m}$$

or

$$p_1 v_1 + \frac{V_1^2}{2} + gz_1 = p_2 v_2 + \frac{V_2^2}{2} + gz_2 + \left(u_2 - u_1 - \frac{\delta Q}{dm} \right) \qquad (6.16)$$

If we could neglect the last term in Eq. 6.16, we would have the equation we seek. Thus we apply the additional restrictions:

(6) $(u_2 - u_1 - \delta Q/dm) = 0$
(7) Incompressible flow, that is, $v_1 = v_2 = 1/\rho = $ constant

Then the energy equation, Eq. 6.16, reduces to

$$\frac{p_1}{\rho} + \frac{V_1^2}{2} + gz_1 = \frac{p_2}{\rho} + \frac{V_2^2}{2} + gz_2$$

or

$$\frac{p}{\rho} + \frac{V^2}{2} + gz = \text{constant} \qquad (6.17)$$

Equation 6.17 is identical in form to the Bernoulli equation, Eq. 6.10. The Bernoulli equation was derived from conservation of momentum (Newton's

267   **relation between the first law of thermodynamics and the Bernoulli equation**

second law), and is valid for steady, incompressible, frictionless flow along a streamline. Equation 6.17 was obtained by applying the first law of thermodynamics to a stream tube control volume, subject to restrictions 1 through 7 above. Thus, the Bernoulli equation (Eq. 6.10) and the identical form of the energy equation (Eq. 6.17) were developed from entirely different models, coming from entirely different basic concepts, and involving entirely different restrictions.

After deriving Eq. 6.17, there might still be some confusion remaining about restriction 6, that is,

$$u_2 - u_1 - \frac{\delta Q}{dm} = 0$$

Obviously, one possibility is that $u_2 = u_1$ and $\delta Q/dm = 0$. Another possibility is that the terms $u_2 - u_1$ and $\delta Q/dm$ are equal. That this is true for incompressible frictionless flow is shown in Example Problem 6.7.

For the special case considered in this section it is true that the first law of thermodynamics reduces to the Bernoulli equation. Since the Bernoulli equation is obtained by integrating Euler's equation (differential form of Newton's second law) for steady, incompressible, frictionless flow along a streamline, for this special case the first law of thermodynamics and Newton's second law do not yield separate information. However, in general the first law of thermodynamics and Newton's second law are independent equations and must be satisfied separately.

Example 6.7

Consider frictionless, incompressible flow with heat transfer. Show that

$$u_2 - u_1 = \frac{\delta Q}{dm}$$

EXAMPLE PROBLEM 6.7

GIVEN:

Frictionless, incompressible flow with heat transfer.

SHOW:

$$u_2 - u_1 = \frac{\delta Q}{dm}$$

EXAMPLE PROBLEM 6.7 (continued)

SOLUTION:

In general, the internal energy, u, can be expressed as $u = u(T, v)$. For incompressible flow, $v = $ constant, and $u = u(T)$. Thus the thermodynamic state of the fluid is determined by the single thermodynamic property, T. The internal energy change for any process, $u_2 - u_1$, depends only on the temperatures at the end states.

From the Gibbs equation, $T\,ds = du + p\,dv$, valid for a pure substance undergoing any process, we obtain

$$T\,ds = du$$

Since the internal energy change, du, is independent of the process, we take a reversible process for which $T\,ds = d(\delta Q/dm) = du$. Therefore

$$\frac{\delta Q}{dm} = u_2 - u_1$$

or

$$u_2 - u_1 - \frac{\delta Q}{dm} = 0$$

Example 6.8

Water flows steadily from a large open reservoir through the system shown. The discharge at point ③ is to atmospheric pressure. A heater around the pipe adds 100 Btu/lbm to the flow. The flow is assumed to be steady, frictionless, and incompressible. Find the temperature rise of the fluid between points ① and ②.

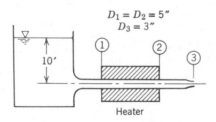

$D_1 = D_2 = 5''$
$D_3 = 3''$

10′

Heater

relation between the first law of thermodynamics and the Bernoulli equation

GIVEN:

Water flows from a large reservoir through the system shown and discharges to atmospheric pressure. The heater adds 100 Btu/lbm to the flow.

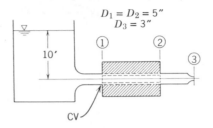

$$D_1 = D_2 = 5''$$
$$D_3 = 3''$$

FIND:

The temperature rise of the fluid between points ① and ②.

SOLUTION:

Basic equations:

$$\frac{p}{\rho} + \frac{V^2}{2} + gz = \text{constant}$$

$$0 = \underbrace{\frac{\partial}{\partial t} \iiint_{CV} \rho \, d\forall}_{= \, 0(1)} + \iint_{CS} \rho \vec{V} \cdot d\vec{A}$$

$$\dot{Q} - \underbrace{\dot{W}_s}_{= \, 0(4)} - \underbrace{\dot{W}_{shear}}_{= \, 0(4)} = \underbrace{\frac{\partial}{\partial t} \iiint_{CV} e\rho \, d\forall}_{= \, 0(1)} + \iint_{CS} \left(u + pv + \frac{V^2}{2} + gz \right) \rho \vec{V} \cdot d\vec{A}$$

Assumptions: (1) Steady flow
                (2) Frictionless flow
                (3) Incompressible flow
                (4) No shaft work, no shear work
                (5) Flow along a streamline

Under the assumptions listed, the first law of thermodynamics for the CV shown becomes

$$\dot{Q} = \iint_{CS} \left(u + pv + \frac{V^2}{2} + gz \right) \rho \vec{V} \cdot d\vec{A}$$

$$= \iint_{A_1} \left(u + pv + \frac{V^2}{2} + gz \right) \rho \vec{V} \cdot d\vec{A} + \iint_{A_2} \left(u + pv + \frac{V^2}{2} + gz \right) \rho \vec{V} \cdot d\vec{A}$$

EXAMPLE PROBLEM 6.8 (*continued*)

For uniform properties at ① and ②

$$\dot{Q} = -|\rho A_1 V_1| \left(u_1 + p_1 v + \frac{V_1^2}{2} + gz_1 \right) + |\rho V_2 A_2| \left(u_2 + p_2 v + \frac{V_2^2}{2} + gz_2 \right)$$

From conservation of mass $|\rho A_1 V_1| = |\rho A_2 V_2| = \dot{m}$

$$\dot{Q} = \dot{m} \left[u_2 - u_1 + \left(\frac{p_2}{\rho} + \frac{V_2^2}{2} + gz_2 \right) - \left(\frac{p_1}{\rho} + \frac{V_1^2}{2} + gz_1 \right) \right]$$

For frictionless, incompressible, steady flow, along a streamline,

$$\frac{p}{\rho} + \frac{V^2}{2} + gz = \text{constant}$$

$$\therefore \quad \dot{Q} = \dot{m}(u_2 - u_1)$$

Dividing through by $\dot{m}$,

$$\frac{\delta Q}{dm} = u_2 - u_1$$

$$\therefore \quad u_2 - u_1 = 100 \text{ Btu/lbm}$$

Since $u_2 - u_1 = c(T_2 - T_1)$, then

$$T_2 - T_1 = \frac{u_2 - u_1}{c} = 100 \frac{\text{Btu}}{\text{lbm}} \times \frac{\text{lbm-R}}{1 \text{ Btu}}$$

$$T_2 - T_1 = 100 \text{ R} \qquad\qquad\qquad T_2 - T_1$$

$\left\{ \begin{array}{l} \text{This problem illustrates that in general the first law of thermodynamics and the} \\ \text{Bernoulli equation are independent equations.} \end{array} \right\}$

**6–8 Bernoulli Equation Applied to Irrotational Flow

In Section 6–5.1, we integrated Euler's equation along a streamline for steady, incompressible, inviscid flow to obtain the Bernoulli equation

$$\frac{p}{\rho} + \frac{V^2}{2} + gz = \text{constant} \quad \text{(along a streamline)} \qquad (6.10)$$

Equation 6.10 can then be applied between any two points on the same stream-line. The value of the constant will vary, in general, from streamline to streamline.

If, in addition to being inviscid, steady and incompressible, the flow field is also irrotational (i.e. the velocity field is such that $2\vec{\omega} = \nabla \times \vec{V} = 0$) we can show

** This section may be omitted without loss of continuity in the text material.

that Bernoulli's equation can be applied between any two points in the flow, that is, the value of the constant in Eq. 6.10 is the same for all streamlines. To illustrate this we start with Euler's equation in vector form.

$$-\frac{1}{\rho}\nabla p - g\nabla z = (\vec{V} \cdot \nabla)\vec{V} \tag{6.12}$$

Using the vector identity

$$(\vec{V} \cdot \nabla)\vec{V} = \tfrac{1}{2}\nabla(\vec{V} \cdot \vec{V}) - \vec{V} \times (\nabla \times \vec{V})$$

we see that for irrotational flow, since $\nabla \times \vec{V} = 0$, then

$$(\vec{V} \cdot \nabla)\vec{V} = \tfrac{1}{2}\nabla(\vec{V} \cdot \vec{V})$$

and Euler's equation for irrotational flow can be written as

$$-\frac{1}{\rho}\nabla p - g\nabla z = \frac{1}{2}\nabla(\vec{V} \cdot \vec{V}) = \frac{1}{2}\nabla(V^2) \tag{6.18}$$

In the time increment dt, a fluid particle moves from the vector position $\vec{r}$ to the position $\vec{r} + d\vec{r}$; the displacement, $d\vec{r}$, is an arbitrary infinitesimal displacement in any direction. Taking the dot product of $d\vec{r} = \hat{i}\,dx + \hat{j}\,dy + \hat{k}\,dz$ with each of the terms in Eq. 6.18, we have

$$-\frac{1}{\rho}\nabla p \cdot d\vec{r} - g\nabla z \cdot d\vec{r} = \frac{1}{2}\nabla(V^2) \cdot d\vec{r}$$

and hence

$$-\frac{dp}{\rho} - g\,dz = \frac{1}{2}d(V^2)$$

or

$$\frac{dp}{\rho} + g\,dz + \frac{1}{2}d(V^2) = 0$$

Integrating this equation,

$$\int \frac{dp}{\rho} + gz + \frac{V^2}{2} = \text{constant} \tag{6.19}$$

For incompressible flow, $\rho = \text{constant}$, and

$$\frac{p}{\rho} + gz + \frac{V^2}{2} = \text{constant} \tag{6.20}$$

Since $d\vec{r}$ was an arbitrary displacement, then for a steady, incompressible, inviscid flow that is also irrotational, Eq. 6.20 is valid between any two points in the flow field.

**6–9 Unsteady Bernoulli Equation—Integration of Euler's Equation along a Streamline

It is not necessary to restrict the development of the Bernoulli equation to steady flows. The purpose of this section is to develop the corresponding equation for unsteady flow along a streamline, and to illustrate its use.

The momentum equation for frictionless flow was found in Section 6–2 to be

$$-\frac{1}{\rho}\nabla p - g\nabla z = \frac{D\vec{V}}{Dt} \tag{6.4}$$

Equation 6.4 is a vector equation. It can be converted to a scalar equation by taking the dot product with $d\vec{s}$, where $d\vec{s}$ is an element of distance along a streamline. Thus

$$-\frac{1}{\rho}\nabla p \cdot d\vec{s} - g\nabla z \cdot d\vec{s} = \frac{D\vec{V}}{Dt}\cdot d\vec{s} = \frac{DV_s}{Dt}ds = V_s\frac{\partial V_s}{\partial s}ds + \frac{\partial V_s}{\partial t}ds \tag{6.21}$$

The terms become

$$\nabla p \cdot d\vec{s} = dp, \text{ the change in pressure along } s$$

$$\nabla z \cdot d\vec{s} = dz, \text{ the change in } z \text{ along } s, \text{ and}$$

$$\frac{\partial V_s}{\partial s}ds = dV_s, \text{ the change in } V_s \text{ along } s$$

Substituting into Eq. 6.21, we obtain

$$-\frac{dp}{\rho} - g\,dz = V_s\,dV_s + \frac{\partial V_s}{\partial t}ds \tag{6.22}$$

Integrating along a streamline from point 1 to point 2,

$$\int_1^2 \frac{dp}{\rho} + \frac{V_2^2 - V_1^2}{2} + g(z_2 - z_1) + \int_1^2 \frac{\partial V_s}{\partial t}ds = 0 \tag{6.23}$$

For incompressible flow, the density is constant. For this special case, Eq. 6.23 becomes

$$\frac{p_1}{\rho} + \frac{V_1^2}{2} + gz_1 = \frac{p_2}{\rho} + \frac{V_2^2}{2} + gz_2 + \int_1^2 \frac{\partial V_s}{\partial t}ds \tag{6.24}$$

(along a streamline)

** This section may be omitted without loss of continuity in the text material.

To evaluate the integral term in Eq. 6.24, the variation in the quantity $\partial V_s/\partial t$ must be known as a function of s, the distance along the streamline measured from point 1. (For steady flow, $\partial V_s/\partial t = 0$, and Eq. 6.24 reduces to Eq. 6.10.)

Let us review the restrictions on Eq. 6.24. They were:

1. Incompressible flow
2. Frictionless flow
3. Flow along a streamline

Consequently, Eq. 6.24 may be applied to any flow where these restrictions are compatible with the physical situation.

Application of Eq. 6.24 is illustrated in Example Problem 6.9.

Example 6.9

A long pipe is connected to a large reservoir that is initially filled with water to a depth of 10 ft. The pipe is 6 in. in diameter and 20 ft long. As a first approximation, friction may be neglected. Determine the flow velocity leaving the pipe as a function of time after a cap is removed from its free end. The reservoir is large enough that the change in its level may be neglected.

EXAMPLE PROBLEM 6.9

GIVEN:

Pipe and large reservoir as shown.

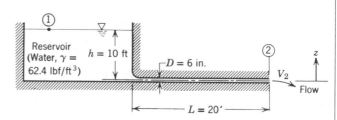

FIND:

$V_2(t)$

SOLUTION:

Apply the Bernoulli equation to the unsteady flow along a streamline from point ① to point ②.

Basic equation:
$$\cancel{\frac{p_1}{\rho}} + \overset{\simeq 0(5)}{\cancel{\frac{V_1^2}{2}}} + gz_1 = \cancel{\frac{p_2}{\rho}} + \frac{V_2^2}{2} + \cancel{gz_2} + \overset{= 0(6)}{\int_1^2 \frac{\partial V_s}{\partial t}\, ds}$$

Assumptions: (1) Incompressible flow
(2) Frictionless flow
(3) Flow along a streamline from ① to ②
(4) $p_1 = p_2 = p_{atm}$
(5) $V_1^2 \simeq 0$
(6) $z_2 = 0$
(7) $z_1 = h = $ constant
(8) Neglect velocity in reservoir, except for a small region near the inlet to the tube

Then

$$gz_1 = gh = \frac{V_2^2}{2} + \int_1^2 \frac{\partial V_s}{\partial t} ds$$

In view of assumption (8), the integral becomes

$$\int_1^2 \frac{\partial V_s}{\partial t} ds \simeq \int_0^L \frac{\partial V_s}{\partial t} ds$$

In the tube, $V_s = V_2$ everywhere, so

$$\int_0^L \frac{\partial V_s}{\partial t} ds = \int_0^L \frac{dV_2}{dt} ds = L\frac{dV_2}{dt}$$

Substituting

$$gh = \frac{V_2^2}{2} + L\frac{dV_2}{dt}$$

Separating variables

$$\frac{dV_2}{2gh - V_2^2} = \frac{dt}{2L}$$

Integrating between limits $V = 0$ at $t = 0$ and $V = V_2$ at $t = t$,

$$\int_0^{V_2} \frac{dV}{2gh - V^2} = \left[\frac{1}{\sqrt{2gh}} \tanh^{-1}\left(\frac{V}{\sqrt{2gh}}\right) \right]_0^{V_2} = \frac{t}{2L}$$

Since $\tanh^{-1}(0) = 0$, we obtain

$$\frac{1}{\sqrt{2gh}} \tanh^{-1}\left(\frac{V_2}{\sqrt{2gh}}\right) = \frac{t}{2L}$$

or

$$\frac{V_2}{\sqrt{2gh}} = \tanh\left(\frac{t}{2L}\sqrt{2gh}\right) \quad \longleftarrow \quad V_2(t)$$

For the given conditions,

$$\sqrt{2gh} = \sqrt{2 \times 32.2 \text{ ft/sec}^2 \times 10 \text{ ft}} = 25.4 \text{ ft/sec}$$

and

$$\frac{t}{2L}\sqrt{2gh} = \frac{1}{2} \times \frac{1}{20 \text{ ft}} \times 25.4 \frac{\text{ft}}{\text{sec}} \times t = 0.635t$$

The results are then

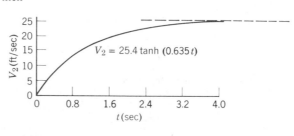

$$V_2 = 25.4 \tanh (0.635t)$$

problems

6.1 By expanding the expression $(\vec{V} \cdot \nabla)\vec{V}$ into components, show that

$$\frac{D\vec{V}}{Dt} = (\vec{V} \cdot \nabla)\vec{V} + \frac{\partial \vec{V}}{\partial t}$$

6.2 A velocity field is given as

$$\vec{V} = Ay\hat{i} + Ax\hat{j}$$

where $A = 10$ ft/sec/ft, and the coordinates x and y are given in feet. The fluid density is 1.5 slug/ft³.

(a) Calculate the acceleration of a fluid particle at the point $(x, y) = (1, 0)$.
(b) Determine the pressure gradient at the same point if $\vec{B} = -g\hat{j}$.

6.3 Consider the flow field with velocity in ft/sec given by

$$\vec{V} = (2xy - x^2)\hat{i} + (2xy - y^2)\hat{j}$$

The density is 2 slug/ft³, and the body force per unit mass is $\vec{B} = -g\hat{j}$.
     Determine the acceleration of a fluid particle and also the pressure gradient at the point $(x, y) = (1, 1)$.

6.4 The x component of velocity in an incompressible flow field is given in ft/sec by

$$u = 6x$$

At the point $(x, y) = (2, 0)$, the y component of velocity is $v = 0$. $w = 0$ everywhere.

(a) Obtain an equation for the y component of velocity.
(b) Find the acceleration of a fluid particle at the point $(2, 0)$.

6.5 An incompressible flow field is given by

$$\vec{V} = (x + 2y)\hat{i} - y\hat{j}$$

Find the magnitude and direction of the acceleration of a fluid particle at the point $(x, y) = (1, 2)$.

6.6 A flow of water is described by the following velocity field in ft/sec:

$$\vec{V} = (10x + 5t)\hat{i} + (-10y + 5t)\hat{j}$$

where x and y are in ft, and t is in sec.

(a) Compute the acceleration of a fluid particle at the point $(x, y) = (1, 5)$ at $t = 10$ sec.
(b) Evaluate $\partial p/\partial x$ under the same conditions. If $B_x = 0$.

6.7 A velocity field in a fluid with density equal to 3 slug/ft³, is given in ft/sec by

$$\vec{V} = (x - 2y)t\hat{i} + (y - 2x)t\hat{j}$$

where x and y are in ft, and t is in sec. Body forces are negligible.
Evaluate ∇p at the point $(x, y) = (1, 2)$ at $t = 1$ sec.

6.8 In a frictionless, incompressible flow, the velocity field in ft/sec, and the body force are given by

$$\vec{V} = x\hat{i} - y\hat{j}$$
$$\vec{B} = -g\hat{k}$$

The pressure is p_0 at the point $(x, y, z) = (0, 0, 0)$.
Obtain an expression for the pressure field, $p(x, y, z)$.

6.9 Example Problem 3.1 read: "As a result of a promotion, you are transferred from your present location. Your wife insists that you transport her fish tank in the back of the station wagon. The tank is $12 \times 24 \times 12$ in. How much water should you leave in the tank to be reasonably sure that it will not spill over during the trip?"
Solve this problem using the Euler equations.

6.10 Example Problem 3.2 read: "A cylindrical container, partially filled with liquid, is rotated at constant angular velocity, ω, about its axis. After a short period of time there is no relative motion, that is, the liquid rotates with the cylinder as if the system were a rigid body. Determine the shape of the free surface."
Solve this problem using the Euler equations. (The cylinder radius is R.)

6.11   The cylindrical container of Problem 6.10 is initially half full of liquid.

Determine the maximum angular speed at which the container can be spun without spilling liquid over the top. Use the Euler equations.

6.12   Consider a steady fluid motion with rigid body swirl about the z axis. Assume $\vec{g} = -g\hat{k}$.

Show that Eqs. 6.5 reduce to

$$\frac{\partial p}{\partial r} = \frac{\rho V_\theta^2}{r}$$

$$\frac{\partial p}{\partial \theta} = 0$$

$$\frac{\partial p}{\partial z} = -\rho g$$

6.13   The flow field in a forced, or rigid body, vortex is given by the expression

$$\vec{V} = \omega r \hat{i}_\theta$$

where $\omega = 10 \text{ sec}^{-1}$, and r is measured in feet. Assume a frictionless fluid with $\rho = 2 \text{ slug/ft}^3$.

(a)   Express the radial pressure gradient, $\partial p/\partial r$, as a function of r.
(b)   Evaluate the pressure change between $r_1 = 1$ ft and $r_2 = 2$ ft.

6.14   Air at 20 psia and 100 F flows around a smooth corner at the inlet to a diffuser. The air velocity is 150 ft/sec, and the radius of curvature of the streamlines is 3 in.

Determine the magnitude of the centripetal acceleration in "G's" experienced by a fluid particle rounding the corner. Evaluate the pressure gradient, $\partial p/\partial r$.

6.15   The flow field in a free, or irrotational, vortex is given by the expression

$$\vec{V} = \frac{k}{2\pi r}\hat{i}_\theta; \qquad r > 0$$

where r is measured in feet. Assume a frictionless fluid with $\rho = 2 \text{ slug/ft}^3$, and $k = 20\pi \text{ ft}^2/\text{sec}$.

(a)   Express the radial pressure gradient, $\partial p/\partial r$, as a function of r.
(b)   Evaluate the pressure change between $r_1 = 1$ ft and $r_2 = 2$ ft.

6.16   Steady, frictionless, and incompressible flow from right to left over a stationary circular cylinder of radius, a, is given by the field

$$\vec{V} = U\left[\left(\frac{a}{r}\right)^2 - 1\right]\cos\theta\hat{i}_r + U\left[\left(\frac{a}{r}\right)^2 + 1\right]\sin\theta\hat{i}_\theta$$

Consider flow along the streamline forming the cylinder surface, that is, $r = a$.
Express the acceleration components, a_s and a_n in terms of angle, θ.

6.17   The flow area of a horizontal air duct is reduced smoothly from 0.75 to 0.25 ft^2. The flowrate is steady at 1.5 lbm/sec of air at 40 psia, 70 F. Frictionless flow may be assumed.

Determine the pressure change over the length of duct in which the area is reduced.

6.18   Water flows in a circular pipe. At one station the diameter is 1 ft, the static pressure is 40 psia, the velocity is 10 ft/sec, and the elevation is 30 ft above ground level. The elevation at a station downstream is 0 ft, and the pipe diameter is 0.5 ft.

Find the pressure (psia) at the downstream station. Frictional effects may be neglected.

6.19   Water flows steadily up the vertical 4 in. diameter pipe (Fig. 6.7) and out the nozzle, which is 2 in. in diameter, discharging to atmospheric pressure. The stream velocity at the nozzle exit must be 60 ft/sec.

Calculate the gage pressure required at section ① assuming frictionless flow.

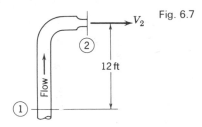

Fig. 6.7

6.20   A siphon is shown in Fig. 6.8. The water flowing may be considered without friction. The water flowrate is 1.0 ft^3/sec, its temperature is 68 F and the pipe diameter is 3 in.

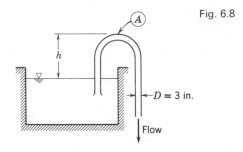

Fig. 6.8

Compute the maximum allowable height, h, so that the pressure at point A is above the vapor pressure of the water.

279   **problems**

6.21   Water flows from a very large tank through a 2 in. diameter tube (Fig. 6.9). The dark fluid in the manometer is mercury.

Determine the velocity in the pipe and the rate of discharge from the tank.

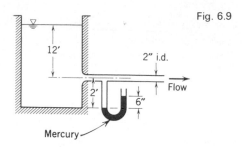

Fig. 6.9

6.22   The flow system of parallel discs shown in Fig. 6.10 contains water. As a first approximation, friction may be neglected. Determine: (a) the volume flowrate, and (b) the pressure at point $\textcircled{C}$.

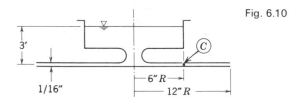

Fig. 6.10

6.23   The tank shown in Fig. 6.11 has a well-rounded orifice with area A_j. At time $t = 0$, the water level is at height h_0.

Develop an expression for the water height, h, at any later time, t.

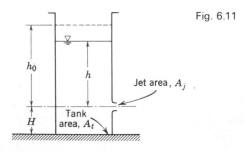

Fig. 6.11

6.24 Consider the steady, frictionless, incompressible flow of air over the wing of an airplane. The air approaching the wing is at 10 psia, 40 F, and has a velocity of 200 ft/sec relative to the wing. At a certain point in the flow, the gage pressure is −0.40 psi. Calculate the velocity of the air at this point relative to the wing.

6.25 A sea water inlet for reactor cooling is to be located on the outer hull of a nuclear submarine. The maximum submerged speed of the sub is 35 knots. At the location of the inlet, the water speed parallel to the hull is 20 knots when the sub moves at maximum speed.

Determine the maximum static pressure that might be expected at the inlet.

6.26 A pitot-static tube is used to measure the velocity of moving air at a point in a flow system. To ensure that flow may be assumed incompressible for calculations of engineering accuracy, the velocity is to be maintained at 330 ft/sec or less. Properties of standard air may be assumed.

Determine the manometer deflection, in inches of water, that corresponds to the maximum desirable velocity.

6.27 An open-circuit wind tunnel draws in air from the atmosphere through a well-contoured nozzle. In the test section, where the flow is straight and nearly uniform, a static pressure tap is drilled into the tunnel wall. A manometer connected to the tap shows that the static pressure within the tunnel is 1.8 in. of water below atmospheric. Assume the air is incompressible, and at 85 F, 14.6 psia.

Calculate the velocity in the wind tunnel test section.

6.28 A jet of air from a nozzle is blown at right angles against a wall in which a pressure tap is located. A manometer connected to the tap reads a pressure of 0.14 in. of mercury above atmospheric pressure.

Determine the approximate velocity of the air leaving the nozzle if it is at 40 F and 14.7 psia.

6.29 A smoothly contoured nozzle is connected to the end of a garden hose. At the nozzle inlet, where the velocity is negligible, the water pressure is 25 psig. Pressure at the nozzle exit is atmospheric.

Assuming that the water remains in a single stream that has negligible aerodynamic drag, estimate the maximum height above the nozzle outlet that the stream could reach.

6.30 A stream of liquid moving at low speed leaves a nozzle pointed directly downward. The velocity may be considered uniform across the nozzle section, and the effects of friction may be ignored. At the nozzle exit, located at elevation z_0, the jet velocity and area are V_0 and A_0, respectively.

Determine the variation of jet area with elevation for $z < z_0$.

6.31 Water is discharged from the system shown in Fig. 6.12 through the pipes ① and ②. The area of ① is 1.0 in.² and the area of ② is 2.0 in.² There is 1.8 Btu of head

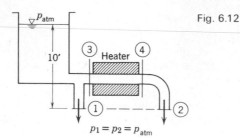

Fig. 6.12

added at the heater per slug of fluid flowing through the heater. Neglect frictional effects and any pressure drop across the heater.

(a) Calculate the volume flowrate at section ①.
(b) If area ③ equals area ④ and both are at the same elevation, compute the change in specific internal energy across the heater.
(c) Calculate the volume flowrate at section ②.

6.32 Consider an incompressible flow of standard air, with velocity field given by

$$\vec{V} = 10x\hat{\imath} - 10y\hat{\jmath} + 30\hat{k}$$

where $\vec{V}$ is in ft/sec, and the coordinates are in feet. Neglect gravity.
Determine the pressure change between the points $(0, 0, 0)$ and $(3, 1, 0)$.

6.33 Consider the flow of water represented by the velocity field

$$\vec{V} = 3y\hat{\imath} + 3x\hat{\jmath}$$

Is it possible to calculate the pressute change between the points $(0, 0, 0)$ and $(1, 1, 1)$? If possible, do so.

6.34 An incompressible flow field is given as

$$\vec{V} = 6x^2y\hat{\imath} - 6xy^2\hat{\jmath}$$

Evaluate the rotation of the flow. What can be said about the use of Bernoulli's equation for this flow?

6.35 Consider the flow field of water given by

$$\vec{V} = 3x^2y^2\hat{\imath} - 2xy^3\hat{\jmath}$$

(a) Determine the stream function for this flow.
(b) Determine the fluid rotation.
(c) Neglecting gravity, is it possible to calculate the pressure difference between the points $(0, 0, 0)$ and $(1, 1, 1)$. If so, do it; if not, why not?

6.36   Determine whether the Bernoulli equation can be applied between different radii
        for the following vortex flows:

        (a) $\vec{V} = \omega r \hat{i}_\theta$

        (b) $\vec{V} = \dfrac{k}{2\pi r}\hat{i}_\theta$

6.37   Two circular discs of radius, R, are separated by a distance, b. The upper disc
        moves toward the lower one at velocity, V. The space between the discs is filled
        with a frictionless, incompressible fluid, which is squeezed out as the discs come
        together. Assume that at any radial station, the velocity is uniform across the gap
        width, b. However, note that b is a function of time. The pressure surrounding the
        discs is atmospheric.
            Determine the gage pressure at $r = 0$.

6.38   Apply the Bernoulli equation to the U-tube manometer of constant area shown
        in Fig. 6.13. Assume that the manometer is initially deflected, and then released.
            Obtain a differential equation for l as a function of time.

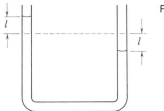

Fig. 6.13

dimensional analysis and similitude

7–1 Introduction

Up to now we have concentrated our efforts on the analytic aspects of fluid flow. In effecting solutions to some of the problems, it has been necessary for us to make some assumptions which we recognize are questionable. For example, in applying the control volume equations to the solution of the rocket motion in Example Problem 4.11, we neglected the effect of air resistance (i.e. the drag force) acting on the rocket. In Chapter 6 we made the assumption of frictionless flow in order to obtain approximate solutions to Example Problems 6.2 and 6.3. What do we do when these approximate solutions are not adequate? We could use the complete equations of motion, Eqs. 5.29. Although the normal and shear stresses in Eqs. 5.29 can be formulated in terms of the velocity field, the pressure field, and the fluid viscosity, solution of the equations is at best extremely difficult and, in many cases, virtually impossible even by the most sophisticated methods. Consequently, it often is necessary to rely on experimental results.

The history of the development of fluid mechanics has depended heavily on experimental results because so few real flows can be solved exactly by analytical methods alone. Solutions of real problems involve a combination of analysis and experimental information. First, the real physical flow situation is approximated with a mathematical model that is simple enough to yield a solution. Then experimental measurements are made to check the analytical results. Based on the measurements, refinements in the analysis can be made, and so on. The experimental results are an essential link in this iterative design process. Empirical designs developed without analysis or careful review of available experimental data are often high in cost and poor or inadequate in performance.

However, experimental work in the laboratory is both time-consuming and expensive. One obvious goal is to obtain the most information from the fewest experiments. Dimensional analysis is an important tool that often helps us to achieve this goal. The dimensionless parameters that we obtain can also be used to correlate data for succinct presentation using the minimum possible number of plots.

When experimental testing of a full-size prototype is either impossible or prohibitively expensive (as happens so often), the only feasible way of attacking the problem is through model testing in the laboratory. If we are to predict the prototype behavior from measurements on the model, it is obvious that we cannot run just any test on any model. The model flow and the prototype flow must be similar, i.e. related by known scaling laws. We shall investigate the conditions necessary to obtain this similarity of model and prototype flows following the discussion of dimensional analysis.

7–2 Nature of Dimensional Analysis

Most phenomena in fluid mechanics depend in a complex way on geometric and flow parameters. For example, consider the drag force on a stationary smooth sphere immersed in a uniform stream of flowing fluid. What experiments need to be conducted to determine the drag force on the sphere? In order to answer this question, we have to determine what parameters are important in determining the drag force. Clearly we would expect the drag force to depend on the size of the sphere (characterized by the diameter, D), the fluid velocity, V, and the fluid viscosity, μ. In addition, the mass of the fluid as characterized by the density, ρ, might also be important. Representing the drag force by F, we can write the symbolic equation

$$F = f(D, V, \rho, \mu)$$

While we may have neglected parameters on which the drag force depends, such as spin or acceleration (or have included parameters on which it does not depend), we have formulated the problem of determining the drag force for a stationary sphere in terms of quantities that are both controllable and measurable in the laboratory.

Let us imagine a series of experiments to determine the form of dependence of F on the variables D, V, ρ, and μ. After building a suitable experimental facility, the work could begin. To obtain a curve of F versus V for a fixed ρ, μ, and D, we might need tests at 10 values of V. To explore the diameter effect, each test would be repeated for spheres of 10 different diameters. Then the procedure would be repeated 10 times for ρ and μ in turn. Simple arithmetic shows that 10^4 separate experiments would be needed. If each test takes $\frac{1}{2}$ hour and we work 8 hours per day, the testing will require $2\frac{1}{2}$ years to complete. Needless to say, there also would be some difficulty in presenting the data. By plotting F versus V with D as a parameter for each combination of density and viscosity values, all of the data could be presented on a total of 100 sheets of graph paper. The utility of such results would be limited at best.

Fortunately we can obtain meaningful results with significantly less effort through the use of dimensional analysis. As we shall see in Section 7–4, all the data for the drag force on a smooth sphere can be plotted as a functional relation between only two nondimensional parameters in the form

$$\frac{F}{\rho V^2 D^2} = f_1\left(\frac{\rho V D}{\mu}\right)$$

The form of the function still must be determined experimentally. However, rather than conducting 10^4 experiments, we could establish the nature of the function as accurately with only 10 different experiments. The time saved in performing only 10 rather than 10^4 experiments is obvious. Even more important is the greater experimental convenience. No longer must we find fluids with 10 different values of density and viscosity. Nor must we make 10 spheres of different diameters. Instead, only the *ratio* $\rho V D/\mu$ must be varied. This can be accomplished by changing the velocity, for example.

Next, we shall formally state the theorem on which these results are based. Then we shall present a detailed procedure for obtaining appropriate parameters for any given physical phenomenon. (Note that this technique is not restricted to use for solving problems in fluid mechanics.)

7–3 Buckingham Pi Theorem

Given a physical problem in which the dependent parameter is a function of $n - 1$ independent parameters, we may express the relationship among the variables in functional form as

$$q_1 = f(q_2, q_3, \ldots, q_n)$$

where q_1 is the dependent parameter, and $q_2, q_3, \ldots, q_n$ are the $n - 1$ independent parameters. Mathematically we can express the functional relationship in the equivalent form

$$g(q_1, q_2, \ldots, q_n) = 0$$

287   **Buckingham Pi theorem**

where g is an unspecified function, different from f. For the drag on a sphere we wrote the symbolic equation

$$F = f(D, V, \rho, \mu)$$

We could just as well write

$$g(F, D, V, \rho, \mu) = 0$$

The Buckingham Pi theorem states that: Given a relation among n parameters of the form

$$g(q_1, q_2, \ldots, q_n) = 0$$

then the n parameters may be grouped into $n - m$ independent dimensionless ratios, or Π parameters, expressible in functional form by

$$G(\Pi_1, \Pi_2, \ldots, \Pi_{n-m}) = 0$$

or

$$\Pi_1 = G_1(\Pi_2, \Pi_3, \ldots, \Pi_{n-m})$$

The number m is usually,[1] but not always, equal to the minimum number of independent dimensions required to specify the dimensions of all the parameters q_1, q_2, $\ldots$, q_n.

The theorem does not predict the functional form of G or G_1. The functional relation among the independent, dimensionless Π parameters must be determined experimentally.

A Π parameter is not independent if it can be formed from a product or quotient of the other parameters of the problem. For example, if

$$\Pi_5 = \frac{2\Pi_1}{\Pi_2 \Pi_3}$$

or

$$\Pi_6 = \frac{\Pi_1^{3/4}}{\Pi_3^2}$$

then neither Π_5 nor Π_6 is independent of the other dimensionless parameters.

Dimensional analysis of a problem is performed in three steps:

1. An appropriate list of parameters is selected.
2. The dimensionless Π parameters are obtained by using the Pi theorem.
3. The functional relation among the Π parameters is determined experimentally.

In the next section, a detailed procedure for applying the Pi theorem is developed.

[1] See Example Problem 7.3.

7–4.1 SELECTION OF PARAMETERS

Some experience is necessary to select a list that includes all parameters that affect a given flow phenomenon. Students, who do not have this experience, are often troubled by the need to apply engineering judgment in an apparent massive dose. However, it is really difficult to go wrong if a generous selection is made.

If you suspect that a phenomenon depends on a given parameter, include it. If your suspicion is correct, experiments will show that the parameter must be included to get consistent results. If the parameter is an extraneous factor, an extra Π parameter may result, but experiments will show that it may be eliminated from consideration.

Therefore, *don't be afraid to include **all** the parameters you **feel** are important.*

7–4.2 PROCEDURE FOR DETERMINING THE Π GROUPS

Six steps are given below for determining the Π parameters. They are almost foolproof when followed.

Step 1. *List all the parameters involved.* (Let n be the number of parameters.) If all the pertinent parameters are not included, some relation may finally be obtained, but it will not give the complete story. If parameters that actually have no effect on the physical phenomenon are included, either the process of dimensional analysis will show that these do not enter the relation sought or one or more dimensionless groups will be obtained that experiments will show to be extraneous.

Step 2. *Select a set of fundamental (primary) dimensions,* for example, MLt or FLt. (Note that for heat transfer problems you may also need T for temperature, and in electrical systems, q for charge.)

Step 3. *List the dimensions of all parameters in terms of primary dimensions.* (Let r be the number of primary dimensions.) Either force *or* mass may be selected as a primary dimension.

Step 4. *Select from the list of parameters a number of repeating parameters equal to the number of primary dimensions, r, and including all the primary dimensions.* No two repeating parameters should have the same net dimensions differing by only a single exponent; for example, do not include a length (L) and a moment of inertia of an area (L^4). The repeating parameters chosen may appear in all the dimensionless groups obtained; consequently do *not* include the dependent parameter among those selected in this step.

Step 5. *Set up dimensional equations combining the parameters selected in Step 4 with each of the other parameters in turn to form dimensionless groups.* (There will be $(n - r)$ equations.) Solve the dimensional equations to obtain the $(n - r)$ dimensionless groups.

Check to see that each group obtained is dimensionless. If mass was initially selected as a primary dimension, it is wise to check the groups using force as a primary dimension, and vice versa.

The functional relationship among the Π parameters must be determined experimentally.

The detailed procedure for determining the dimensionless Π parameters is illustrated in Example Problems 7.1 and 7.2.

Example 7.1

As noted in Section 7-2, the drag force, F, on a sphere depends on the relative velocity, V, the sphere diameter, D, the fluid density, ρ, and the fluid viscosity, μ. Obtain a set of dimensionless groups that can be used to correlate experimental data.

EXAMPLE PROBLEM 7.1

GIVEN:

$F = f(\rho, V, D, \mu)$ for a smooth sphere.

FIND:

An appropriate set of dimensionless groups.

SOLUTION:

(Circled numbers refer to steps in the procedure for determining dimensionless Π parameters.)

① $F \quad V \quad D \quad \rho \quad \mu$ $n = 5$ parameters

② Select primary dimensions M, L, t

③ $F \quad V \quad D \quad \rho \quad \mu$

$$\frac{ML}{t^2} \quad \frac{L}{t} \quad L \quad \frac{M}{L^3} \quad \frac{M}{Lt} \qquad r = 3 \text{ primary dimensions}$$

④ ρ, V, D $m = r = 3$ repeating parameters

⑤ Then $n - m = 2$ dimensionless groups will result. Setting up dimensional equations.

$$\Pi_1 = \rho^a V^b D^c F$$

$$= \left(\frac{M}{L^3}\right)^a \left(\frac{L}{t}\right)^b L^c \left(\frac{ML}{t^2}\right) = M^0 L^0 t^0$$

EXAMPLE PROBLEM 7.1 (continued)

Summing exponents,

$$M: a + 1 = 0 \qquad a = -1$$
$$L: -3a + b + c + 1 = 0 \qquad c = -2 \left.\begin{matrix} \\ \\ \\ \end{matrix}\right\} \text{Therefore } \Pi_1 = \frac{F}{\rho V^2 D^2}$$
$$t: -b - 2 = 0 \qquad b = -2$$

Similarly

$$\Pi_2 = \rho^d V^e D^f \mu$$

$$= \left(\frac{M}{L^3}\right)^d \left(\frac{L}{t}\right)^e (L)^f \left(\frac{M}{Lt}\right) = M^0 L^0 t^0$$

$$M: d + 1 = 0 \qquad d = -1$$
$$L: -3d + e + f - 1 = 0 \qquad f = -1 \left.\begin{matrix} \\ \\ \\ \end{matrix}\right\} \text{Therefore } \Pi_2 = \frac{\mu}{\rho V D}$$
$$t: -e - 1 = 0 \qquad e = -1$$

⑥ Check using F, L, t dimensions

$$\Pi_1 = \frac{F}{\rho V^2 D^2} = F \frac{L^4}{Ft^2} \left(\frac{t}{L}\right)^2 \frac{1}{L^2} = [1]$$

where [ ] means "dimensions of", and

$$\Pi_2 = \frac{\mu}{\rho V D} = \frac{Ft}{L^2} \frac{L^4}{Ft^2} \frac{t}{L} \frac{1}{L} = [1]$$

The functional relationship is $\Pi_1 = f(\Pi_2)$, or

$$\frac{F}{\rho V^2 D^2} = f\left(\frac{\mu}{\rho V D}\right)$$

as noted before. The form of the function, f, must be determined experimentally.

Example 7.2

The pressure drop, Δp, for steady, incompressible viscous flow through a straight horizontal pipe depends on the pipe length, l, the average velocity, $\bar{V}$, the viscosity, μ, the pipe diameter, D, the density, ρ, and the average variation, e, of the inside radius (the average "roughness" height). Determine a set of dimensionless groups that can be used in the correlation of data.

detailed procedure for use of Buckingham Pi theorem

EXAMPLE PROBLEM 7.2

GIVEN:

$\Delta p = f(\rho, \bar{V}, D, l, \mu, e)$ for flow in a circular pipe.

FIND:

A suitable set of dimensionless groups.

SOLUTION:

(Circled numbers refer to steps in the procedure for determining dimensionless Π parameters.)

① Δp ρ μ $\bar{V}$ l D e $\qquad n = 7$ parameters

② Choose primary dimensions M, L, t

③ Δp ρ μ $\bar{V}$ l D e

$\dfrac{M}{Lt^2}$ $\dfrac{M}{L^3}$ $\dfrac{M}{Lt}$ $\dfrac{L}{t}$ L L L $\qquad r = 3$ primary dimensions

④ $\rho, \bar{V}, D \qquad m = r = 3$ repeating parameters

⑤ Then $n - m = 4$ dimensionless groups will result. Setting up dimensional equations,

$$\Pi_1 = \rho^a \bar{V}^b D^c \Delta p$$

$$= \left(\frac{M}{L^3}\right)^a \left(\frac{L}{t}\right)^b (L)^c \left(\frac{M}{Lt^2}\right) = M^0 L^0 t^0$$

$$\Pi_2 = \rho^d \bar{V}^e D^f \mu$$

$$= \left(\frac{M}{L^3}\right)^d \left(\frac{L}{t}\right)^e (L)^f \frac{M}{Lt} = M^0 L^0 t^0$$

$M: 0 = a + 1 \qquad\qquad a = -1$
$L: 0 = -3a + b + c - 1 \qquad b = -2$
$t: 0 = -b - 2 \qquad\qquad c = 0$

$M: 0 = d + 1 \qquad\qquad d = -1$
$L: 0 = -3d + e + f - 1 \qquad e = -1$
$t: 0 = -e - 1 \qquad\qquad f = -1$

Therefore $\Pi_1 = \rho^{-1} \bar{V}^{-2} D^0 \Delta p = \dfrac{\Delta p}{\rho \bar{V}^2}$

Therefore $\Pi_2 = \dfrac{\mu}{\rho \bar{V} D}$

$$\Pi_3 = \rho^g \bar{V}^h D^i l$$

$$= \left(\frac{M}{L^3}\right)^g \left(\frac{L}{t}\right)^h (L)^i L = M^0 L^0 t^0$$

$$\Pi_4 = \rho^j \bar{V}^k D^l e$$

$$= \left(\frac{M}{L^3}\right)^j \left(\frac{L}{t}\right)^k (L)^l L = M^0 L^0 t^0$$

$M: 0 = g \qquad\qquad g = 0$
$L: 0 = -3g + h + i + 1 \qquad h = 0$
$t: 0 = -h \qquad\qquad i = -1$

$M: 0 = j \qquad\qquad j = 0$
$L: 0 = -3j + k + l + 1 \qquad k = 0$
$t: 0 = -k \qquad\qquad l = -1$

Therefore $\Pi_3 = \dfrac{l}{D}$

Therefore $\Pi_4 = \dfrac{e}{D}$

EXAMPLE PROBLEM 7.2 (*continued*)

⑥ Check, using F, L, t dimensions

$$\Pi_1 = \frac{\Delta p}{\rho \overline{V}^2} = \frac{F}{L^2} \frac{L^4}{Ft^2} \frac{t^2}{L^2} = [1]$$

$$\Pi_2 = \frac{\mu}{\rho \overline{V} D} = \frac{Ft}{L^2} \frac{L^4}{Ft^2} \frac{t}{L} \frac{1}{L} = [1]$$

$$\Pi_3 = \frac{l}{D} = \frac{L}{L} = [1]$$

$$\Pi_4 = \frac{e}{D} = \frac{L}{L} = [1]$$

Finally, the functional relationship is

$$\Pi_1 = f(\Pi_2, \Pi_3, \Pi_4)$$

$$\frac{\Delta p}{\rho \overline{V}^2} = f\left(\frac{\mu}{\rho \overline{V} D}, \frac{l}{D}, \frac{e}{D}\right)$$

$\{$ Experiments in many laboratories have shown that this relationship correlates the $\}$
data well. We shall discuss this result in greater detail in Section 8–8.1.

7–4.3 COMMENTS ON THE PROCEDURE

The procedure outlined above, where m is taken equal to r (the fewest independent dimensions required to specify the dimensions of all parameters involved) almost always produces the correct number of dimensionless Π parameters. In a few cases, trouble arises because the number of primary dimensions differs when variables are expressed in terms of different systems of dimensions. The value of m can be established with certainty by determining the rank of the dimensional matrix; m is equal to the rank of the dimensional matrix. The procedure is illustrated in Example Problem 7.3.

The $n - m$ dimensionless groups obtained from the Buckingham procedure are independent, but not unique. If a different set of repeating parameters is chosen, different groups will result. Only the tradition of common usage has dictated which groups are preferred.

If $n - m = 1$, then a single dimensionless Π parameter is obtained. In this case, the Buckingham Pi theorem indicates the single Π parameter must be a constant.

Example 7.3

When a small tube is dipped into a pool of liquid, surface tension causes a *meniscus* to form at the free surface which is elevated or depressed depending on the contact angle at the liquid-solid-gas interface. Experiments indicate that the magnitude of this *capillary effect*, Δh, is a function of the tube diameter, D, liquid specific weight, γ, and surface tension, σ. Determine the number of independent Π parameters that can be formed, and obtain a set.

EXAMPLE PROBLEM 7.3

GIVEN:

$$\Delta h = f(D, \gamma, \sigma)$$

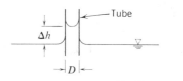

Tube

Δh

Liquid
(Specific weight $= \gamma$
Surface tension $= \sigma$)

FIND:

(a) Number of independent Π parameters.
(b) Evaluate one set.

SOLUTION:

(Circled numbers refer to steps in the procedure for determining dimensionless Π parameters)

① Δh D γ σ $n = 4$ parameters

② Choose primary dimensions (use both M, L, t and F, L, t dimensions to illustrate the problem of determining m):

③ (a) M, L, t

Δh D γ σ

L L $\dfrac{M}{L^2 t^2}$ $\dfrac{M}{t^2}$

$r = 3$ primary dimensions

(b) F, L, t

Δh D γ σ

L L $\dfrac{F}{L^3}$ $\dfrac{F}{L}$

$r = 2$ primary dimensions

Thus we ask, "Is m equal to r?" Let us check the dimensional matrix to find out.

	Δh	D	γ	σ
M	0	0	1	1
L	1	1	-2	0
t	0	0	-2	-2

	Δh	D	γ	σ
F	0	0	1	1
L	1	1	-3	-1

The rank of a matrix is determined by the order of its smallest nonzero determinant.

$$\begin{vmatrix} 0 & 1 & 1 \\ 1 & -2 & 0 \\ 0 & -2 & -2 \end{vmatrix} = 0 + (-1)(-2) \\ + (1)(-2) \equiv 0$$

$$\begin{vmatrix} -2 & 0 \\ -2 & -2 \end{vmatrix} = 4 \neq 0 \quad \therefore \quad m = 2 \\ m \neq r$$

④ $m = 2$. Choose D, γ as repeating parameters.

⑤ $n - m = 2$ dimensionless groups will result.

$$\Pi_1 = D^a \gamma^b \Delta h$$

$$= (L)^a \left(\frac{M}{L^2 t^2}\right)^b (L) = M^0 L^0 t^0$$

$M: b + 0 = 0 \qquad\qquad b = 0$
$L: a - 2b + 1 = 0 \qquad a = -1$
$t: -2b + 0 = 0$

$$\therefore \quad \Pi_1 = \frac{\Delta h}{D}$$

$$\Pi_2 = D^c \gamma^d \sigma$$

$$= (L)^c \left(\frac{M}{L^2 t^2}\right)^d \frac{M}{t^2} = M^0 L^0 t^0$$

$M: d + 1 = 0 \qquad\qquad d = -1$
$L: c - 2d = 0 \qquad\qquad c = -2$
$t: -2d - 2 = 0$

$$\therefore \quad \Pi_2 = \frac{\sigma}{D^2 \gamma}$$

⑥ Check, using F, L, t dimensions

$$\Pi_1 = \frac{L}{L} = [1]$$

$$\Pi_2 = \frac{F}{L} \frac{1}{L^2} \frac{L^3}{F} = [1]$$

$$\begin{vmatrix} 1 & 1 \\ -3 & -1 \end{vmatrix} = -1 + 3 = 2 \neq 0$$

$$\therefore \quad m = 2 \\ m = r$$

$m = 2$. Choose D, γ as repeating parameters.

$n - m = 2$ dimensionless groups will result.

$$\Pi_1 = D^e \gamma^f \Delta h$$

$$= (L)^e \left(\frac{F}{L^3}\right)^f L = F^0 L^0 t^0$$

$F: f = 0$
$L: e - 3f + 1 = 0 \qquad e = -1$

$$\therefore \quad \Pi_1 = \frac{\Delta h}{D}$$

$$\Pi_2 = D^g \gamma^h \sigma$$

$$= (L)^g \left(\frac{F}{L^3}\right)^h \frac{F}{L} = F^0 L^0 t^0$$

$F: h + 1 = 0 \qquad\qquad h = -1$
$L: g - 3h - 1 = 0 \qquad g = -2$

$$\Pi_2 = \frac{\sigma}{D^2 \gamma}$$

Check, using M, L, t dimensions

$$\Pi_1 = \frac{L}{L} = [1]$$

$$\Pi_2 = \frac{M}{t^2} \frac{1}{L^2} \frac{L^2 t^2}{M} = [1]$$

Therefore both systems of dimensions yield the same dimensionless Π parameters. The predicted functional relationship is

$$\Pi_1 = f(\Pi_2)$$

$$\frac{\Delta h}{D} = f\left(\frac{\sigma}{D^2 \gamma}\right)$$

Over the years, several hundred different important dimensionless groups have been identified. Following tradition, each such group has been given the name of a prominent scientist or engineer, usually the one who pioneered its use. In spite of the plethora of names we encounter, several are so important and occur so frequently that we should take time to learn their definitions. Understanding their physical meaning also gives insight into the phenomena we study. The student should learn the material the following short list contains.

7–5.1 THE REYNOLDS NUMBER

In the 1880s, Osborne Reynolds, the British engineer, studied the transition between laminar and turbulent flow in a tube. He discovered that the parameter bearing his name

$$Re = \frac{\rho \overline{V} D}{\mu} = \frac{\overline{V} D}{\nu}$$

is a criterion by which the state of a flow may be determined. Later experiments have shown that the Reynolds number is a key parameter for other flow cases as well. Thus, in general,

$$Re = \frac{\rho V L}{\mu} = \frac{V L}{\nu}$$

where L is a characteristic length descriptive of the flow field.

The physical significance of the Reynolds number may be seen more readily by rewriting it in the form

$$Re = \frac{\rho V L}{\mu} = \frac{\rho V L}{\mu} \frac{V}{V} \frac{L}{L} \frac{1}{L/L} = \frac{\rho V^2 L^2}{(\mu V/L) L^2}$$

In its final form, this expression may be interpreted as follows:

$$\rho V^2 L^2 \sim (\text{dynamic pressure}) \times (\text{area}) \sim \text{inertia force}$$

$$\frac{\mu V}{L} L^2 \sim (\text{viscous stress}) \times (\text{area}) \sim \text{viscous force}$$

and

$$Re \sim \frac{\text{inertia forces}}{\text{viscous forces}}$$

Thus the Reynolds number may be considered as a ratio of inertia forces to viscous forces.

7-5.2 THE MACH NUMBER

In the 1870s, the Austrian physicist Ernst Mach introduced the parameter

$$M \equiv \frac{V}{c}$$

where V is the flow speed and c is the local sonic speed. Analysis and experiments have shown that the Mach number is a key parameter that characterizes compressibility effects in a fluid flow.

Rearranging slightly, the Mach number may be written

$$M = \frac{V}{c} = \sqrt{\frac{\rho V^2}{\rho c^2}}$$

which may be interpreted as a ratio of inertia forces to forces due to compressibility. For truly incompressible flow (under some conditions even liquids are quite compressible), $c = \infty$ so $M = 0$.

7-5.3 THE FROUDE NUMBER

William Froude was a naval architect. Together with his son, Robert Edmund Froude, he discovered that the parameter

$$Fr = \frac{V}{\sqrt{gL}}$$

was significant for flows with free surface effects.

Squaring the Froude number gives

$$Fr^2 = \frac{V^2}{gL} = \frac{\rho V^2 L^2}{\rho g L^3}$$

which may be interpreted as the ratio of inertia forces to gravity forces.

7-5.4 THE EULER NUMBER (Pressure Coefficient)

In aerodynamic and other model testing, it is convenient to present pressure data in dimensionless form. The ratio

$$Eu(= C_p) \equiv \frac{\Delta p}{\frac{1}{2}\rho V^2}$$

is formed, where Δp is the local pressure minus freestream pressure, and ρ and V are properties of the freestream flow. This ratio has been named after Leonhard Euler, the French mathematician who did much of the early analytical work in fluid mechanics.

physical meaning of common dimensionless groups

The Saturn V booster used for the recent Apollo moon shots is a very large vehicle (about 360 feet long!), and enormous forces are applied to it by even a gentle breeze. Failure of the vehicle or its support structure due to wind loading would be catastrophic, so data on expected wind loads were required before the first booster and launch tower were designed and erected. Existing theories for wind loads were not adequate, so test data were required. The impossibility of testing a full-scale mockup of the booster is obvious. Therefore model tests were needed.

To be useful, a model test must yield data that can be scaled to obtain the forces, moments, and dynamic loads that would exist on the full scale prototype. What are the conditions that must be met to insure the similarity of model and prototype flows?

Perhaps the most obvious requirement is that the model and the prototype must be geometrically similar. Geometric similarity requires that the model and the prototype be the same shape, and that all linear dimensions of the model be related to the corresponding dimensions of the prototype by a constant scale factor.

A second requirement is that the model and prototype flows must be kinematically similar. If two flows are kinematically similar, then the velocities at corresponding points in the two flows are in the same direction and are related in magnitude by a constant scale factor. Thus two flows that are kinematically similar also have streamline patterns related by a constant scale factor. Since the boundaries form the bounding streamlines, flows that are kinematically similar must be geometrically similar.

In principle, kinematic similarity would require that a wind tunnel of infinite cross section be used to measure data for drag on an object, in order to model correctly the performance in an infinite flow field. In practice, this restriction may be relaxed considerably, permitting the use of finite size equipment.

Kinematic similarity requires that the regimes of flow be the same for model and prototype. If compressibility or cavitation effects, which may change even the qualitative patterns of flow, are not present in the prototype flow, they must be avoided in the model flow.

When two flows have a force distribution such that at corresponding points in the flows, identical types of forces are parallel and are related in magnitude by a constant scale factor at all corresponding points, the flows are dynamically similar.

The requirements for dynamic similarity are most restrictive; two flows must possess both geometric and kinematic similarity in order to be dynamically similar.

To establish the conditions required for complete dynamic similarity, all forces that are important in the flow situation must be considered. Thus the effects of viscous forces, of pressure forces, of surface tension forces, and so on, must be considered. Then test conditions must be established so that all important forces are related by the same scale factor between model and prototype flows. When dynamic similarity exists, data measured in a model flow may be related quantitatively to conditions in the prototype flow. What then are the conditions for insuring dynamic similarity between model and prototype flows?

The Buckingham Pi theorem is used to obtain the governing dimensionless groups for a flow phenomenon; to achieve dynamic similarity between geometrically similar flows we must duplicate all but one of these dimensionless groups.

For example, in considering the drag force on a sphere in Example Problem 7.1, we began with

$$F = f(D, V, \rho, \mu)$$

The Buckingham Pi theorem predicted the functional relation

$$\frac{F}{\rho V^2 D^2} = f_1\left(\frac{\rho V D}{\mu}\right)$$

In Section 7-5 we have shown that the dimensionless parameters can be viewed as ratios of forces. Thus in considering a model flow and a prototype flow about a sphere (the flows are geometrically similar), the flows also will be dynamically similar if

$$\left(\frac{\rho V D}{\mu}\right)_{model} = \left(\frac{\rho V D}{\mu}\right)_{prototype}$$

Furthermore if

$$Re_{model} = Re_{prototype}$$

then

$$\left(\frac{F}{\rho V^2 D^2}\right)_{model} = \left(\frac{F}{\rho V^2 D^2}\right)_{prototype}$$

and the results determined from the model study can be used to predict the drag on the full-scale prototype.

Note that the actual force due to the fluid on the object is not the same in both cases, but its dimensionless value is. Note also that the two tests can be run using different fluids, if desired, as long as the Reynolds numbers are matched. For experimental convenience, test data can be measured in a wind tunnel in air, and the results used to predict drag in water, as illustrated in Example Problem 7.4.

Example 7.4

The drag of a sonar transducer is to be predicted, based on wind tunnel test data. The prototype, a 1 ft diameter sphere, is to be towed at 5 knots (nautical miles per hour) in sea water. The model is 6 in. in diameter.

Determine the required test speed in air. If the drag of the model at test conditions in 5.58 lbf, estimate the drag of the prototype.

EXAMPLE PROBLEM 7.4

GIVEN:

Sonar transducer to be tested in a wind tunnel.

FIND:

$-D_p = 1$ ft $\quad\quad\quad\quad\quad -D_m = 6$ in.

(a) V_m $\quad\quad V_p = 5$ knots $\quad\quad\to F_p$ $\quad\quad\quad\to V_m$ $\quad\to F_m = 5.58$ lbf
(b) F_p

SOLUTION:

If no cavitation or compressibility effects are present in the prototype or model flows, kinematic similarity may be obtained, provided the wind tunnel cross section is large enough. (Experience shows that $A_{tunnel} > 15\, A_{model}$ is satisfactory.) Then

$$\frac{F}{\rho V^2 D^2} = f\left(\frac{\rho D V}{\mu}\right)$$

and the test should be run at

$$Re_{model} = Re_{prototype}$$

to ensure dynamic similarity. For sea water, $\rho = 1.98$ slug/ft^3 and $v = 1.4 \times 10^{-5}$ ft^2/sec. At prototype conditions,

$$V_p = \frac{5\text{ nmi}}{\text{hr}} \times \frac{6080\text{ ft}}{\text{nmi}} \times \frac{\text{hr}}{3600\text{ sec}} = 8.4\,\frac{\text{ft}}{\text{sec}}$$

$$Re_p = \frac{V_p D_p}{v_p} = \frac{8.4\text{ ft}}{\text{sec}} \times 1\text{ ft} \times \frac{\text{sec}}{1.4 \times 10^{-5}\text{ ft}^2} = 6 \times 10^5$$

The model test conditions must duplicate this Reynolds number. Thus

$$Re_m = \frac{V_m D_m}{v_m} = 6 \times 10^5$$

For air at STP, $\rho = 0.00238$ slug/ft^3 and $v = 1.8 \times 10^{-4}$ ft^2/sec. The wind tunnel must be operated at

$$V_m = Re_m \frac{v_m}{D_m} = 6 \times 10^5 \times \frac{1.8 \times 10^{-4}\text{ ft}^2}{\text{sec}} \times \frac{1}{0.5\text{ ft}}$$

$$V_m = 216\,\frac{\text{ft}}{\text{sec}} \quad\longleftarrow \quad\quad\quad\quad\quad\quad\quad\quad\quad\quad\quad\quad\quad V_m$$

This value is low enough for compressibility effects to be ignored.

At these test conditions, the model and prototype flows are dynamically similar. Hence

$$\left.\frac{F}{\rho V^2 D^2}\right|_m = \left.\frac{F}{\rho V^2 D^2}\right|_p$$

and

$$F_p = F_m \frac{\rho_p}{\rho_m} \frac{V_p^2}{V_m^2} \frac{D_p^2}{D_m^2} = 5.58 \text{ lbf} \times \frac{1.98}{0.00238} \times \frac{(8.4)^2}{(216)^2} \times \frac{1}{(0.5)^2}$$

$$F_p = 28 \text{ lbf} \qquad\qquad\qquad\qquad\qquad\qquad\qquad\qquad\qquad F_p$$

If cavitation were expected—that is, if the sonar probe were operated near the free surface of the sea water—then additional parameters would have to be included.

{This problem demonstrates the calculation of prototype values from model test data.}

7–7 Similitude Established from the Differential Equations

A more rigorous and broader approach in determining the conditions under which two different flows are similar is through the use of the governing differential equations.[2]

Similitude or dynamic similarity may be present when two physical phenomena are governed by identical differential equations and boundary conditions. Similitude is obtained when the governing equations and boundary conditions have the same dimensionless form. Dynamic similarity is obtained by duplicating the dimensionless coefficients of the equations between prototype and model.

Establishment of similitude from the differential equations that describe the flow is a rigorous procedure. If one begins from the correct equations and performs each step correctly, one can be sure that all appropriate variables have been included. However, in this text we shall not derive the complete differential equations describing viscous fluid motion (the Navier–Stokes equations). Therefore we shall not treat the procedure in detail.

In contrast to use of the differential equations, success in the use of the Buckingham Pi theorem is determined by the insight used to select the parameters affecting the problem. If a complete set is chosen, the results will be complete. If one or more important variables is omitted, the results will be without meaning. Additional variables can be included if there is any uncertainty. As more experience is gained with fluid flow phenomena, the selection process becomes

[2] For a detailed discussion of similitude and the use of the governing equations, refer to *Similitude and Approximation Theory* by S. J. Kline (New York: McGraw-Hill, 1965).

easier. Experience also gives more insight into the physical significance of each dimensionless group.

problems

7.1   The Reynolds number for flow in a pipe is defined as $Re = \rho \bar{V} D / \mu$, where $\bar{V}$ is the average velocity, and D is the pipe diameter. Using both M, L, t and F, L, t systems of dimensions, show that the Reynolds number is dimensionless.

7.2   On a standard day, sonic speed at sea level is approximately 1120 ft/sec; at 28,000 ft, it is approximately 1000 ft/sec. A jet aircraft is capable of level flight at a Mach number of 1.8 at sea level, and 2.3 at 28,000 ft.
        Determine the corresponding air speeds in mph.

7.3   For free surface flow in a wide channel, the Froude number is defined as $Fr = V / \sqrt{gD}$. Flow in a laboratory flume occurs at a depth of 4 in. with a velocity of 1.5 ft/sec or at a depth of 0.5 in. with a velocity of 8 ft/sec.
        Calculate the Froude numbers corresponding to these flow conditions.

7.4   The *Weber number* can be considered the ratio of inertia forces to surface tension forces. It is defined as $We = \rho V^2 L / \sigma$. For wave motion, the characteristic length used is the wavelength, λ. On a water table where the depth is 0.25 in., and the velocity is 0.75 ft/sec, waves are observed with $\lambda = 6$ in. The water temperature is 68 F and 14.7 psia.
        Show that the Weber number is dimensionless. Determine its magnitude for the water table flow.

7.5   Cavitation phenomena are known to depend on the *cavitation number*, defined as

$$Ca = \frac{p - p_v}{\frac{1}{2}\rho V^2}$$

where p, ρ, and V are conditions in the liquid stream far ahead of a model, and p_v is the liquid vapor pressure at the test temperature. Consider a model test performed in a water tunnel at 68 F.
        Determine the flow velocity at which the cavitation number equals unity.

7.6   At very low velocities, the drag on an object is independent of fluid density. Thus, the drag, F, on a small sphere is a function only of velocity, fluid viscosity, and sphere diameter, D.
        Use dimensional analysis to express the drag as a function of these variables. Use μ, V, and D as repeating variables.

7.7   Experiments show that the pressure drop due to flow through a sudden contraction in a circular duct may be expressed as

$$\Delta p = p_1 - p_2 = f(\rho, \mu, \bar{V}, d, D)$$

where the geometric variables are defined in Fig. 7.1.

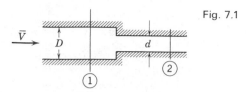

Fig. 7.1

You are asked to organize some experimental data.

Evaluate the resulting dimensionless parameters, using ρ, V, and D as repeating variables.

7.8 The power, $\dot{P}$, required to drive a fan is believed to depend on the fluid density, the volumetric flowrate, Q, the impeller diameter, D, and angular velocity, ω.

Use dimensional analysis to evaluate the dependence of $\dot{P}$ on other variables. Choose ρ, D, and ω as repeating variables.

7.9 The radiator fan in an automobile is a source of considerable noise. The sound power, $\mathcal{P}$ (energy per unit time), radiated by a fan depends on its diameter and angular velocity, and on the fluid density, ρ, and sonic speed, c.

Determine the dependence of $\mathcal{P}$ on ω, using dimensional analysis. Choose ρ, ω, and D as repeating variables.

7.10 Small droplets of liquid are formed when a liquid jet breaks up in spray and fuel injection processes. The resulting droplet diameter, d, is thought to depend on the liquid density, viscosity and surface tension, as well as the jet velocity, V, and diameter, D.

How many dimensionless ratios are required to characterize this process? Determine these ratios, using ρ, D, and V as repeating variables.

7.11 Assume that the resistance, R, of a flat plate immersed in a fluid depends on the fluid density and viscosity, the velocity, and the width, b, and height, h, of the plate.

Find a convenient set of coordinates for organizing data.

7.12 The velocity, V, of a free surface gravity wave in deep water is a function of wavelength, λ, depth, D, water density, and the acceleration of gravity.

Use dimensional analysis to find the functional dependence of V on the other variables. Choose ρ, D, and g as repeating variables. Express V in the simplest form possible.

7.13 The power, $\dot{P}$, required to drive a propeller is known to depend on the following variables:

$$V = \text{freestream velocity}$$
$$D = \text{propeller diameter}$$
$$\omega = \text{angular velocity of propeller}$$
$$\mu = \text{fluid viscosity}$$
$$\rho = \text{fluid density}$$
$$c = \text{speed of sound in fluid}$$

(a) How many dimensionless groups are required to characterize this situation?

(b) Obtain these dimensionless groups (do *not* use μ as a repeating variable).

7.14 Approximately 90 percent of the resistance of surface ships is due to wave motion, and so in model tests the Froude number is matched between model and prototype. The remaining 10 percent of the resistance is due to viscous forces, which depend on the Reynolds number. Consequently, in ship model testing, Froude numbers are matched, but no attempt is made to match Reynolds numbers.

Determine the kinematic viscosity required in a model fluid to match Reynolds numbers for length scales of 1:10, 1:50, and 1:100. Does any liquid exist with the required properties?

7.15 Capillary waves are formed on a liquid free surface as a result of surface tension. They have short wavelengths. The velocity of a capillary wave depends on the surface tension, wavelength, and liquid density.

Use dimensional analysis to find the wave velocity as a function of these variables.

7.16 An automobile is to travel through standard air at a speed of 80 mph. To determine the pressure distribution, a scale model 1/5 the length of the full-size car is to be tested in water.

Determine the water velocity that should be used. What factors must be considered to assure kinematic similarity in the tests?

7.17 The capillary rise, Δh, of a liquid in a circular tube of diameter, D, depends on the surface tension, σ, and the specific weight, γ, of the fluid. The significant variables found from dimensional analysis are

$$\Pi_1 = \Delta h \sqrt{\frac{\gamma}{\sigma}}, \qquad \Pi_2 = \frac{\sigma}{\gamma D^2}$$

The capillary rise for liquid A is 1.0 in. in a tube of 0.010 in. diameter.

What will be the rise for liquid B (having the same surface tension but four times the density of A) in a tube of diameter 0.005 in.?

7.18 The drag force F on a boat can be expressed through dimensional analysis as follows:

$$\frac{F}{\rho V^2 L^2} = f\left(\frac{\rho V L}{\mu}, \frac{V^2}{Lg}\right)$$

(a) If we wish to run experiments on a model to determine the drag on the prototype, what relations between the model and the prototype must be satisfied?

(b) If the model is constructed to 1/20 scale and the prototype is to operate at 20 ft/sec, at what velocity should the model test be run?

(c) If the prototype is to operate in water at 70 F, what should be the kinematic viscosity of the fluid used in the model test?

7.19   An airship is to operate at 50 ft/sec in air at standard conditions. A model is constructed to 1/20 scale and tested in a wind tunnel at the same air temperature to determine drag.

(a) What criterion should be considered to obtain dynamic similarity?
(b) If the model is tested at 50 ft/sec, what pressure should be used in the wind tunnel?
(c) If the model drag is 50 lbf, what will be the drag of the prototype?

7.20   The power input, $\dot{P}$, to an impeller depends on the viscosity, μ, density, ρ, and velocity, V, of the fluid and on the diameter, D, and rotational speed, ω, of the impeller. The following nondimensional parameters were found:

$$\Pi_1 = \frac{\dot{P}}{\rho\omega^3 D^5}, \qquad \Pi_2 = \frac{\rho V D}{\mu}, \qquad \Pi_3 = \frac{D\omega}{V}$$

Tests of a 1/5 scale model (operating with the same incompressible fluid at the same temperature as the prototype) required 100 watts of power at maximum speed.

(a) Have all the Π parameters been included? If not, how many would you expect?
(b) Determine the maximum power requirement of the prototype.

7.21   The power input to an axial flow pump is thought to depend on the volume flowrate, head, speed and diameter of the pump, and density of the fluid, that is

$$\dot{P} = f(Q, H, N, D, \rho)$$

where

$$
\begin{aligned}
Q &= \text{volume flowrate}\\
H &= \text{head (energy per unit mass)}\\
N &= \text{angular speed}\\
D &= \text{diameter}\\
\rho &= \text{density}
\end{aligned}
$$

An axial flow pump is required to deliver 25 ft³/sec of water at a head of 150 ft-lbf/slug. The diameter of the rotor is 1 ft, and it is to be driven at 500 rpm. The prototype is to be modeled on a small test apparatus having a 3 hp, 1000 rpm power supply.

For similar performance between the prototype and the model, calculate the head, flowrate, and the diameter of the model.

7.22   The drag force, F, on a submarine moving far below the free surface in an incompressible flow is a function of density, ρ, viscosity, μ, velocity, V, and maximum crosssection area, A. It has been suggested that dimensional analysis will yield the following results:

$$\frac{F}{\rho V^2 A} = g\left(\frac{\rho V A}{\mu}\right)$$

(a) Is this the correct number of Π parameters? Why or why not?
(b) Are these Π parameters dimensionally correct? If not, correct them.
(c) A submarine model is scaled down so that all lengths are 1/10 those of the prototype. The model is to be tested in sea water; the model test is to be used in predicting the prototype drag at a velocity of 15 knots.
    1. At what velocity should the model test be run?
    2. If this velocity could be achieved, what would be the relationship between the drag force of the model and that of the prototype?

7.23   The drag of an airfoil at zero angle of attack is a function of the flow density, viscosity, and velocity, in addition to a length parameter. A 1/10 scale model of an airfoil was tested in a wind tunnel at a Reynolds number of 5.5×10^6, based on chord length. Test conditions in the wind tunnel air stream were 60 F and 140 psia. The prototype airfoil has a chord length of 6 ft, and it is to be flown in air at 60 F, 14 psia.
    Determine:
(a) The velocity at which the wind tunnel model was tested.
(b) The corresponding prototype velocity.

7.24   In some low-speed ranges, vortices are shed from the rear of bluff cylinders placed across the flow. The vortices alternately leave the top and bottom of the cylinder, as shown in Fig. 7.2, causing an alternating force normal to the velocity. The vortex shedding frequency, f, is thought to depend on ρ, V, d, and μ.

Fig. 7.2

(a) Use dimensional analysis to develop a functional relationship for f.
(b) Vortex shedding occurs in standard air on two cylinders with a diameter ratio of 2. Determine the velocity ratio for dynamic similarity, and the ratio of vortex shedding frequencies.

7.25   A model propeller 2 ft in diameter is tested in a wind tunnel. Air approaches the propeller at 150 ft/sec when it rotates at 2000 rpm. The thrust and torque measured under these conditions are 25 lbf and 15 ft-lbf, respectively. A prototype ten times as large as the model is to be built. At a dynamically similar operating point, the approach air speed is to be 400 ft/sec.

306   **7/dimensional analysis and similitude**

Calculate the speed, thrust, and torque of the prototype propeller under these conditions, neglecting the effect of viscosity, but including density.

7.26 A continuous belt moving vertically through a bath of viscous liquid drags a layer of liquid of thickness, h, along with it. The volumetric rate of liquid loss, Q, is assumed to depend on μ, ρ, g, h, and V, where V is the belt speed.

Apply dimensional analysis to predict the form of dependence of Q on other variables. Use ρ, V, and h as repeating variables.

7.27 The power loss, $\dot{W}_L$, in a journal bearing depends on the diameter, D, and clearance, c, of the bearing, in addition to its angular speed, ω. The lubricant viscosity and mean pressure are also important.

Determine the functional form of dependence of $\dot{W}_L$ on these parameters.

7.28 The boundary layer thickness, δ, on a smooth flat plate in an incompressible flow without pressure gradients depends on the freestream velocity, U, the fluid density, ρ, the fluid viscosity, μ, and the distance from the leading edge of the plate, x.

Express these variables in dimensionless form.

7.29 A disc rotates near a fixed surface. The radius of the disc is R, and the space between the disc and the surface is filled with a fluid of viscosity μ. The spacing between disc and surface is h feet, and the disc rotates at angular velocity ω.

Find the dependence between torque on the disc, T, and the other variables.

7.30 Independent variables in a turbomachine are the impeller diameter, D, angular speed, ω, and fluid viscosity and density. Dependent properties are volume flow-rate, Q, head, H (energy per unit mass), and power input, $\dot{P}$. Use ρ, μ, and D as repeating variables in a dimensional analysis.

(a) Determine the dimensionless ratios that characterize this problem.
(b) Under what conditions will flows in two different machines be similar?
(c) Assume for a moment that viscous effects are unimportant. Determine the speed of operation for machine 2 for the same flow as machine 1 if $D_2/D_1 = 2$. What will be the head ratio?

incompressible viscous flow

Part A. Introduction

Viscous effects can have a profound influence on the behavior of fluid flows. Viscous flows are divided into the general categories of internal and external flows. In addition, laminar or turbulent flow regimes are possible, depending on the flow Reynolds number and external conditions. Thus this chapter begins with a brief discussion of internal and external, laminar and turbulent flows.

The rest of the chapter is devoted to the solution of practical flow cases. The objectives of the analysis are to determine the flowrate, pressure loss, and power requirements for flows in pipes and ducts, and to evaluate the lift and drag on bodies moving through fluids. We shall employ the same basic tools we have used before: the basic equations in control volume form that we developed in Chapter 4.

After deriving the differential form of the momentum equation (Eqs. 5.29) we studied the dynamics of inviscid flow in Chapter 6. The resulting differential equations of motion, the Euler equations, were valid for flows in which the shear stresses could be neglected. In this chapter we are interested in flows for which viscous forces are not negligible. Since all fluids possess viscosity, the presence of velocity gradients necessitates the presence of viscous stresses. We will thus consider flows with both velocity gradients and shear stresses. Such flows were discussed qualitatively in Chapter 2 (Section 2–5.1). It would be helpful for you to reread that section at this point.

8–1 Internal and External Flows

Flows completely bounded by solid surfaces (i.e. duct flows) are called internal flows. Figure 8.1 illustrates laminar flow in the entrance region of a circular pipe. The flow is uniform at the pipe entrance with velocity U_0. Because of the no-slip condition at the wall, we know that the velocity at the wall must be zero along the entire length of the pipe. A boundary layer develops along the walls of the channel.

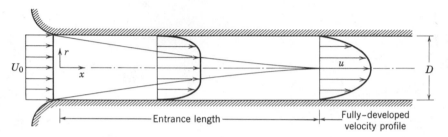

Fig. 8.1 *Flow in the entrance region of a pipe.*

The solid surface exerts a retarding shear force on the flow; thus the speed of the fluid in the neighborhood of the surface is reduced. At successive stations along the pipe, the effect of the solid surface is felt farther out into the flow.

For incompressible flow, the velocity at the pipe centerline must increase with distance from the inlet in order to satisfy the continuity equation. However the average velocity at any cross section

$$\overline{V} = \frac{1}{A} \int_{\text{Area}} u \, dA$$

must equal U_0, so

$$\overline{V} = U_0 = \text{constant} \tag{8.1}$$

Sufficiently far from the pipe entrance, the boundary layer growing around the circumference reaches the pipe centerline. The distance downstream from the entrance to the location at which the boundary layer reaches the centerline is called the entrance length. Beyond the entrance length the velocity profile no longer changes with increasing distance, x, and the flow is fully-developed. The actual shape of the fully-developed velocity profile depends on whether the flow is laminar or turbulent. In Fig. 8.1 the profile is shown qualitatively for a laminar flow.

For laminar flow, the entrance length, L, is a function of the Reynolds number,

$$\frac{L}{D} \simeq 0.06 \frac{\rho D \overline{V}}{\mu}$$

where D is the pipe diameter, $\overline{V}$ is the average velocity, ρ is the fluid density, and μ is the fluid viscosity. As pointed out in Chapter 2, laminar flow in a pipe may be expected only for Reynolds numbers less than about 2300. Thus the entrance length for laminar pipe flow may be as long as

$$L \simeq 0.06 \, Re \, D \leq (0.06)(2300)D = 138D$$

that is, more than 100 pipe diameters. If the flow is turbulent, enhanced mixing among fluid layers[1] causes more rapid growth of the boundary layer. Experiments show that the mean velocity profile becomes fully-developed within 25 to 40 pipe diameters from the entrance. However, the details of the turbulent motion may not be fully-developed for 80 or more pipe diameters. Fully-developed internal flows will be treated in Parts B and C of this chapter.

U_∞—Uniform velocity field upstream

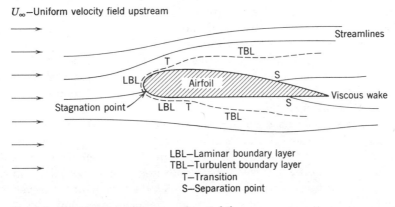

LBL—Laminar boundary layer
TBL—Turbulent boundary layer
T—Transition
S—Separation point

Fig. 8.2 *Details of viscous flow around an airfoil.*

External flows are flows over bodies immersed in an unbounded flow. The flow over a semi-infinite flat plate (Fig. 2.10) and the flow over a cylinder (Fig. 2.11*a*) are examples of external flows. These were discussed qualitatively in Chapter 2. A more detailed sketch of viscous flow over an airfoil is shown in Fig. 8.2. Although viscous effects are confined to a thin boundary layer (the boundary layer thickness

[1] This mixing is illustrated extremely well in the introductory portion of the film *Turbulence*, R. W. Stewart, principal.

in Fig. 8.2 is exaggerated greatly), they are responsible for flow separation and wake formation, which alters the entire pressure distribution around the airfoil.

Forces and moments on objects are due to both pressure and viscous shear forces on the object, caused by the flow. In Parts D and E of this chapter, we develop approximate analyses and correlations of experimental data so that forces and moments can be determined for some simple viscous flow situations.

8–2 Laminar and Turbulent Flows

As discussed previously in Section 2–5.2, the nature of pipe flow (laminar or turbulent) is determined by the value of the Reynolds number,

$$Re = \frac{\rho D \bar{V}}{\mu}$$

One can demonstrate by the classic Reynolds experiment the qualitative difference between the nature of laminar and turbulent flow. In this experiment water flows from a large reservoir through a clear tube. A thin filament of dye injected at the entrance to the tube allows visual observation of the flow. At low flow rates (low Reynolds numbers) the dye injected into the flow remains in a single filament; there is little dispersion of dye throughout the flow field because the flow is laminar. A laminar flow is one in which the fluid flows in laminae or layers; there is no macroscopic mixing of adjacent fluid layers.

As the flowrate through the tube is increased, the dye filament becomes unstable and breaks up into a totally random motion; the line of dye is stretched and twisted into myriad entangled threads, and it quickly disperses throughout the entire flow field. This behavior of turbulent flow is due to small, high frequency, velocity fluctuations superimposed on the mean motion of a turbulent flow, mixing of fluid particles from adjacent layers of fluid results in rapid dispersion of the dye.

One can obtain a more quantitative picture of the difference between laminar and turbulent flow by examining the output from a sensitive velocity measuring device immersed in the flow. If one measures the x component of velocity at a fixed location in a pipe for both laminar and turbulent flow, the traces of velocity versus time would be as shown in Fig. 8.3. For steady laminar flow, the velocity at a point remains constant with time. In turbulent flow the velocity trace indicates a random fluctuation of the instantaneous velocity, u, about the time mean velocity, $\bar{u}$. Thus we consider the instantaneous velocity, u, as the sum of the time mean velocity, $\bar{u}$, and the fluctuating component u', that is

$$u = \bar{u} + u'$$

Because the mean velocity, $\bar{u}$, does not vary with time, we would designate the flow as steady.

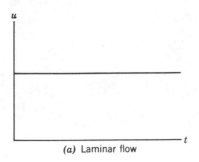

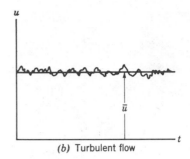

(a) Laminar flow   (b) Turbulent flow

Fig. 8.3 *Velocity at a point.*

Although many turbulent flows of interest are steady in the mean (i.e. $\bar{u}$ is not a function of time) the presence of the random, high frequency velocity fluctuations makes the analysis of turbulent flows extremely difficult. In a one-dimensional laminar flow, the shear stress is related to the velocity gradient by the simple relation

$$\tau_{yx} = \mu \frac{du}{dy} \tag{2.8}$$

For a turbulent flow in which the mean velocity field is one-dimensional, no such simple relation is valid. In turbulent flow there is no universal relationship between the stress field and the mean velocity field. Thus in turbulent flows we must rely very heavily on semi-empirical theories and on experimental data.

Part B. Fully-Developed Laminar Flow

There are relatively few viscous flow problems for which we can obtain closed form analytical solutions, but the method of solution is important. In this section we consider a few classic examples of fully-developed laminar flows. Our interest is in obtaining detailed information about the velocity field. Knowledge of the velocity field permits calculation of shear stress, pressure drop, and flowrate.

Rather than develop complete differential equations of motion for the flow of a viscous fluid, we shall derive the governing equations from first principles for each flow field of interest. Since we are interested in the details of the flow field, our aim will be to derive differential equations that describe the flow. In every case we shall begin by applying the familiar control volume formulation of Newton's second law (we assume you are very familiar with it by now) to a suitably chosen differential control volume.

8–3.1 BOTH PLATES STATIONARY

Fluid in high pressure hydraulic systems often leaks through the annular gap between a piston and cylinder. For very small gaps (typically about 0.0002 in.), this flow field may be imagined as flow between infinite parallel plates. To calculate the leakage flowrate, we must first determine the velocity field for such a flow.

Let us consider the fully-developed laminar flow between infinite parallel plates. The plates are separated by a distance, *a*, as shown in Fig. 8.4. The plates are considered infinite in the *z* direction, with no variation of any fluid property in this

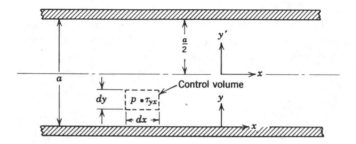

Fig 8.4 *Laminar flow between stationary infinite parallel plates.*

direction. The flow is further assumed to be steady and incompressible. Before starting our analysis, what do we know about the flow field? For one thing we know that the *x* component of velocity must be zero at both the upper and lower plates as a result of the no-slip condition at the wall, that is the boundary conditions are:

$$\text{at } y = 0 \qquad u = 0$$

$$\text{at } y = a \qquad u = 0$$

Since the flow is fully-developed, the velocity cannot vary with *x* and, hence, depends on *y* only, that is, $u = u(y)$. Furthermore, there is no component of velocity in either the *y* or *z* directions (i.e. $v = w = 0$).

For our analysis, we select a differential control volume of size $d\forall = dx\,dy\,dz$, and apply the *x* component of the momentum equation.

Basic equation:

$$F_{S_x} + \overset{= \,0(3)}{\cancel{F_{B_x}}} = \overset{= \,0(1)}{\cancel{\frac{\partial}{\partial t}}} \iiint_{CV} u\rho\,d\forall + \iint_{CS} u\rho\vec{V}\cdot d\vec{A} \tag{4.19a}$$

Assumptions: (1) Steady flow
(2) Fully-developed flow
(3) $F_{B_x} = 0$

For fully-developed flow, the net momentum flux through the control surface is zero. (The momentum flux through the right face of the control surface is equal in magnitude but opposite in sign to the momentum flux through the left face; there is no momentum flux through any of the remaining faces of the control volume.) Since there are no body forces in the x direction the momentum equation reduces to

$$F_{S_x} = 0 \tag{8.2}$$

The next step is to sum the forces acting on the control volume in the x direction. We recognize that there are normal forces (pressure forces) acting on the left and right faces; there are tangential forces (shear forces) acting on the top and bottom faces.

If the pressure at the center of the element is p, then the force on the left face is

$$\left(p - \frac{\partial p}{\partial x} \frac{dx}{2} \right) dy \, dz,$$

and the force on the right face is

$$-\left(p + \frac{\partial p}{\partial x} \frac{dx}{2} \right) dy \, dz$$

If the shear stress at the center of the element is τ_{yx}, then the shear force on the bottom face is

$$-\left(\tau_{yx} - \frac{d\tau_{yx}}{dy} \frac{dy}{2} \right) dx \, dz$$

and the shear force on the top face is

$$\left(\tau_{yx} + \frac{d\tau_{yx}}{dy} \frac{dy}{2} \right) dx \, dz$$

Note that in expanding the shear stress, τ_{yx}, in a Taylor series about the center of the element, we have used the total derivative rather than a partial derivative. We did this because we recognized that τ_{yx} is only a function of y, since $u = u(y)$.

Having evaluated the forces acting on each face of the control volume, we can now substitute them back into Eq. 8.2

$$\left(p - \frac{\partial p}{\partial x} \frac{dx}{2} \right) dy \, dz - \left(p + \frac{\partial p}{\partial x} \frac{dx}{2} \right) dy \, dz - \left(\tau_{yx} - \frac{d\tau_{yx}}{dy} \frac{dy}{2} \right) dx \, dz$$

$$+ \left(\tau_{yx} + \frac{d\tau_{yx}}{dy} \frac{dy}{2} \right) dx \, dz = 0$$

315 **fully-developed laminar flow between infinite parallel plates**

This equation simplifies to

$$-\frac{\partial p}{\partial x} + \frac{d\tau_{yx}}{dy} = 0$$

or

$$\frac{d\tau_{yx}}{dy} = \frac{\partial p}{\partial x} \qquad (8.3)$$

Equation 8.3 must hold true for all values of x and y. This requires that

$$\frac{d\tau_{yx}}{dy} = \frac{\partial p}{\partial x} = \text{constant}$$

Integrating this equation, we obtain

$$\tau_{yx} = \left(\frac{\partial p}{\partial x}\right) y + c_1$$

which states that the shear stress is linear with y. Since

$$\tau_{yx} = \mu \frac{du}{dy} \qquad (2.8)$$

then

$$\mu \frac{du}{dy} = \left(\frac{\partial p}{\partial x}\right) y + c_1$$

and

$$u = \frac{1}{2\mu}\left(\frac{\partial p}{\partial x}\right) y^2 + \frac{c_1}{\mu} y + c_2 \qquad (8.4)$$

To evaluate the constants c_1 and c_2 we must use the boundary conditions. At $y = 0$, $u = 0$. Consequently, $c_2 = 0$. At $y = a$, $u = 0$. Consequently,

$$0 = \frac{1}{2\mu}\left(\frac{\partial p}{\partial x}\right) a^2 + \frac{c_1}{\mu} a$$

This gives

$$c_1 = -\frac{1}{2}\left(\frac{\partial p}{\partial x}\right) a$$

and hence

$$u = \frac{1}{2\mu}\left(\frac{\partial p}{\partial x}\right) y^2 - \frac{1}{2\mu}\left(\frac{\partial p}{\partial x}\right) ay$$

or

$$u = \frac{a^2}{2\mu}\left(\frac{\partial p}{\partial x}\right)\left[\left(\frac{y}{a}\right)^2 - \left(\frac{y}{a}\right)\right] \tag{8.5}$$

At this point we have the velocity profile. What else can we learn about the flow?

Shear Stress Distribution

The shear stress distribution is given by

$$\tau_{yx} = \left(\frac{\partial p}{\partial x}\right)y + c_1 = \left(\frac{\partial p}{\partial x}\right)y - \frac{1}{2}\left(\frac{\partial p}{\partial x}\right)a$$

$$\tau_{yx} = a\left(\frac{\partial p}{\partial x}\right)\left[\frac{y}{a} - \frac{1}{2}\right] \tag{8.6a}$$

Volumetric Flowrate

The volumetric flowrate is given by

$$Q = \int \vec{V} \cdot d\vec{A}$$

For a depth *l* in the *z* direction

$$Q = \int_0^a ul\, dy$$

or

$$\frac{Q}{l} = \int_0^a \frac{1}{2\mu}\left(\frac{\partial p}{\partial x}\right)(y^2 - ay)\, dy$$

Thus the volumetric flowrate per depth *l* is given by

$$\frac{Q}{l} = -\frac{1}{12\mu}\left(\frac{\partial p}{\partial x}\right)a^3 \tag{8.6b}$$

Flowrate as a Function of Pressure Drop

Since $\partial p/\partial x$ is a constant, then

$$\frac{p_2 - p_1}{L} = \frac{-\Delta p}{L} = \frac{\partial p}{\partial x}$$

fully-developed laminar flow between infinite parallel plates

Substituting into the expression for volumetric flowrate,

$$\frac{Q}{l} = -\frac{1}{12\mu}\left[\frac{-\Delta p}{L}\right]a^3$$

or

$$\frac{Q}{l} = \frac{\Delta p a^3}{12\mu L} \tag{8.6c}$$

Average Velocity

The average velocity, $\bar{V}$, is given by

$$\bar{V} = \frac{Q}{A} = -\frac{1}{12\mu}\left(\frac{\partial p}{\partial x}\right)a^3 l/la$$

$$\bar{V} = -\frac{1}{12\mu}\left(\frac{\partial p}{\partial x}\right)a^2 \tag{8.6d}$$

Point of Maximum Velocity

To find the point of maximum velocity, we set du/dy equal to zero and solve for the corresponding value of y. From Eq. 8.5

$$\frac{du}{dy} = \frac{a^2}{2\mu}\left(\frac{\partial p}{\partial x}\right)\left[\frac{2y}{a^2} - \frac{1}{a}\right]$$

Thus

$$\frac{du}{dy} = 0 \quad \text{at} \quad y = \frac{a}{2}$$

At

$$y = \frac{a}{2}, \quad u = u_{max} = -\frac{1}{8\mu}\left(\frac{\partial p}{\partial x}\right)a^2 = \frac{3}{2}\bar{V} \tag{8.6e}$$

Transformation of Coordinates

In deriving the above relations, the origin of coordinates, $y = 0$, was taken at the bottom plate. We could just as easily have taken the origin at the centerline of the channel. If we denote the coordinates with origin at the channel centerline as x, y', the boundary conditions are

$$u = 0 \quad \text{at} \quad y' = \frac{a}{2}$$

$$u = 0 \quad \text{at} \quad y' = -\frac{a}{2}$$

8/incompressible viscous flow

To obtain the velocity profile in terms of x, y' we substitute $y = y' + a/2$, into Eq. 8.5. The result is

$$u = \frac{a^2}{2\mu}\left(\frac{\partial p}{\partial x}\right)\left[\left(\frac{y'}{a}\right)^2 - \frac{1}{4}\right] \tag{8.7}$$

This equation shows that the velocity profile we have determined is parabolic, as shown in Fig. 8.5.

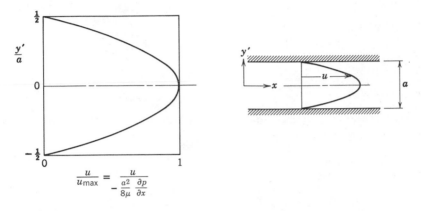

Fig. 8.5 *Dimensionless velocity profile for fully developed laminar flow between infinite parallel plates.*

All of the results in this section are valid for laminar flow only, since Newton's law of viscosity was used to relate shear stress to velocity gradient. Experiments show that transition to turbulent flow occurs in this flow case for Reynolds number values (defined as $Re = \rho \bar{V} a/\mu$) greater than approximately 1400. Consequently, the Reynolds number should be checked after using Eqs. 8.6 to assure a valid solution.

Example 8.1

A hydraulic system operates at 3000 psig and 130 F. The hydraulic fluid is SAE 10 engine oil (SG = 0.88). A control valve consists of a piston 1 in. in diameter, fitted to a cylinder with a mean radial clearance of 0.0002 in. Determine the leakage flow-rate in in.^{3}/min if the pressure on the low pressure side of the piston is 200 psig. (The piston is 0.5 in. long.)

319   **fully-developed laminar flow between infinite parallel plates**

GIVEN:

Flow of hydraulic oil between piston and cylinder, as shown.
Fluid is SAE 10 oil at 130 F (SG = 0.88).

FIND:

Q in in.^{3}/min

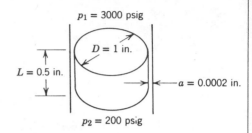

$p_1 = 3000$ psig

$D = 1$ in.

$L = 0.5$ in.

$a = 0.0002$ in.

$p_2 = 200$ psig

SOLUTION:

The gap width is very small, so the flow may be imagined as flow between parallel plates.
Equation 8.6c is applied.

Computing equation:

$$\frac{Q}{l} = \frac{\Delta p a^3}{12\mu L} \tag{8.6c}$$

Assumptions: (1) Laminar flow
(2) Steady flow
(3) Incompressible flow
(4) Fully developed flow (note $L/a = 0.5/0.0002 = 2500!$)

The width, l, is approximately $l = \pi D$. Thus

$$Q = \frac{\pi D \Delta p a^3}{12\mu L}$$

For SAE 10 oil at 130 F, $\mu = 1.9 \times 10^{-4}$ lbf-sec/ft^2, from Fig. A2, Appendix A. Thus

$$Q = \frac{\pi}{12} \times 1 \text{ in.} \times \frac{(3000 - 200) \text{ lbf}}{\text{in.}^2} \times (0.0002)^3 \text{ in.}^3 \times \frac{\text{ft}^2}{1.9 \times 10^{-4} \text{ lbf-sec}}$$

$$\times \frac{1}{0.5 \text{ in.}} \times \frac{60 \text{ sec}}{\text{min}} \times \frac{144 \text{ in.}^2}{\text{ft}^2}$$

$$Q = 0.534 \frac{\text{in.}^3}{\text{min}} \longleftarrow \hspace{5cm} Q$$

To assure that flow is laminar, we should also check the flow Reynolds number.

$$\bar{V} = \frac{Q}{A} = \frac{Q}{\pi D a}$$

$$= \frac{0.534 \text{ in.}^3}{\text{min}} \times \frac{1}{\pi} \times \frac{1}{1 \text{ in.}} \times \frac{1}{0.0002 \text{ in.}} \times \frac{\text{min}}{60 \text{ sec}} \times \frac{\text{ft}}{12 \text{ in.}}$$

$$\bar{V} = 1.18 \frac{\text{ft}}{\text{sec}}$$

and

$$Re = \frac{\rho \bar{V} a}{\mu} = \frac{SG \, \rho_{H_2O} \bar{V} a}{\mu}$$

$$= 0.88 \times \frac{1.94 \text{ slug}}{\text{ft}^3} \times \frac{1.18 \text{ ft}}{\text{sec}} \times 0.0002 \text{ in.} \times \frac{\text{ft}}{12 \text{ in.}} \times \frac{\text{ft}^2}{1.9 \times 10^{-4} \text{ lbf-sec}} \times \frac{\text{lbf-sec}^2}{\text{slug-ft}}$$

$$Re = 0.178$$

Thus, flow is surely laminar, since $Re \ll 1200$.

8–3.2 UPPER PLATE MOVING WITH VELOCITY U

A second laminar flow case of practical importance is flow in a journal bearing. In such a bearing, an inner cylinder, or journal, rotates inside a stationary member. At light loads, the centers of the two members essentially coincide, and the small clearance gap is symmetric. Since the gap is normally small, it is reasonable to "unfold" the bearing and consider the flow field between infinite parallel plates.

Let us now consider a case where the upper plate is moving to the right with a constant velocity, U, as shown in Fig. 8.6. All we have done in going from a

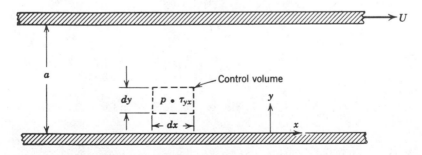

Fig. 8.6 *Laminar flow between infinite parallel plates: upper plate moving with velocity, U.*

fully-developed laminar flow between infinite parallel plates

stationary upper plate to a moving upper plate is to change one of the boundary conditions. The boundary conditions for the moving plate case are:

$$u = 0 \quad \text{at} \quad y = 0$$

$$u = U \quad \text{at} \quad y = a$$

Since it is only the boundary conditions that have been changed, there is no need to repeat the entire analysis of Section 8–3.1. The analysis leading to Eq. 8.4 is equally valid for the moving plate case. Thus the velocity distribution is given by

$$u = \frac{1}{2\mu}\left(\frac{\partial p}{\partial x}\right)y^2 + \frac{c_1}{\mu}y + c_2 \tag{8.4}$$

and our only task is to evaluate the constants c_1 and c_2 by employing the appropriate boundary conditions.

At $y = 0$, $u = 0$. Consequently, $c_2 = 0$.

At $y = a$, $u = U$. Consequently,

$$U = \frac{1}{2\mu}\left(\frac{\partial p}{\partial x}\right)a^2 + \frac{c_1}{\mu}a$$

Thus

$$c_1 = \frac{U\mu}{a} - \frac{1}{2}\left(\frac{\partial p}{\partial x}\right)a$$

and,

$$u = \frac{1}{2\mu}\left(\frac{\partial p}{\partial x}\right)y^2 + \frac{Uy}{a} - \frac{1}{2\mu}\left(\frac{\partial p}{\partial x}\right)ay$$

$$= \frac{Uy}{a} + \frac{1}{2\mu}\left(\frac{\partial p}{\partial x}\right)(y^2 - ay)$$

$$u = \frac{Uy}{a} + \frac{a^2}{2\mu}\left(\frac{\partial p}{\partial x}\right)\left[\left(\frac{y}{a}\right)^2 - \left(\frac{y}{a}\right)\right] \tag{8.8}$$

It is reassuring to note that Eq. 8.8 reduces to Eq. 8.5 for a stationary upper plate. From Eq. 8.8, for zero pressure gradient (i.e. for $\partial p/\partial x = 0$) the velocity distribution is linear with y. This was the case treated earlier in Chapter 2.

From the velocity distribution of Eq. 8.8 we can obtain additional information about the flow.

Shear Stress Distribution

The shear stress distribution is given by $\tau_{yx} = \mu(du/dy)$

$$\tau_{yx} = \mu\frac{U}{a} + \frac{a^2}{2}\left(\frac{\partial p}{\partial x}\right)\left[\frac{2y}{a^2} - \frac{1}{a}\right]$$

$$\tau_{yx} = \mu\frac{U}{a} + a\left(\frac{\partial p}{\partial x}\right)\left[\frac{y}{a} - \frac{1}{2}\right] \tag{8.9a}$$

Volumetric Flowrate

The volumetric flow is given by $Q = \int \vec{V} \cdot d\vec{A}$. For a depth l in the z direction

$$Q = \int_0^a ul\, dy$$

or

$$\frac{Q}{l} = \int_0^a \left[\frac{Uy}{a} + \frac{1}{2\mu}\left(\frac{\partial p}{\partial x}\right)(y^2 - ay)\right] dy$$

Thus, the volumetric flow per depth l is given by

$$\frac{Q}{l} = \frac{Ua}{2} - \frac{1}{12\mu}\left(\frac{\partial p}{\partial x}\right)a^3 \tag{8.9b}$$

Average Velocity

The average velocity, $\bar{V}$, is given by

$$\bar{V} = \frac{Q}{A} = l\left[\frac{Ua}{2} - \frac{1}{12\mu}\left(\frac{\partial p}{\partial x}\right)a^3\right]\Big/ la$$

$$\bar{V} = \frac{U}{2} - \frac{1}{12\mu}\left(\frac{\partial p}{\partial x}\right)a^2 \tag{8.9c}$$

Point of Maximum Velocity

To find the point of maximum velocity, we set du/dy equal to zero and solve for the corresponding value of y. From Eq. 8.8

$$\frac{du}{dy} = \frac{U}{a} + \frac{a^2}{2\mu}\left(\frac{\partial p}{\partial x}\right)\left[\frac{2y}{a^2} - \frac{1}{a}\right]$$

fully-developed laminar flow between infinite parallel plates

Thus

$$\frac{du}{dy} = 0 \quad \text{at} \quad y = \frac{a}{2} - \frac{U/a}{(1/\mu)(\partial p/\partial x)}$$

There is no simple relation between the maximum velocity, u_{max}, and the mean velocity, $\bar{V}$, for this flow case.

Equation 8.8 suggests that the velocity profile may be treated as a combination of a linear and a parabolic velocity profile; the last term in Eq. 8.8 is identical with Eq. 8.5. The result is a family of velocity profiles, depending on the values of U and $(1/\mu)(\partial p/\partial x)$, a few of which are sketched in Fig. 8.7. (As shown in Fig. 8.7, some reverse flow, i.e. flow in the negative x direction, can occur when $\partial p/\partial x > 0$.)

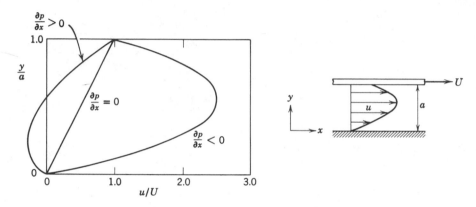

Fig. 8.7 *Dimensionless velocity profile for fully-developed laminar flow between infinite parallel plates: upper plate moving with velocity, U.*

Again, all of the results developed in this section are valid for laminar flow only. Experiments show that transition to turbulent flow occurs (for $\partial p/\partial x = 0$) at a Reynolds number value of approximately 1500, where $Re = \rho U a/\mu$ for this flow case. Not much information is available for the case where the pressure gradient is not zero.

Example 8.2

A crankshaft journal bearing in an automobile engine is lubricated by SAE 40 engine oil at 210 F. The bearing is 3 in. in diameter, has a diametral clearance of 0.0025 in., and rotates at 3600 rpm. It is 1.25 in. wide. The bearing is under no load, so the clearance is symmetric. Determine the torque required to turn the journal, and the power dissipated.

GIVEN:

Journal bearing, as shown. Note that the gap width, a, is *half* the diametral clearance.

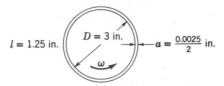

$l = 1.25$ in. $D = 3$ in. $a = \dfrac{0.0025}{2}$ in.

Lubricant is SAE 40 oil at 210 F. Speed: 3600 rpm.

FIND:

(a) Torque, T, in in.-lbf.
(b) Power dissipated, in hp.

SOLUTION:

Torque on the journal is due to viscous shear in the oil film. The gap width is small, so the flow may be represented by flow between infinite parallel planes:

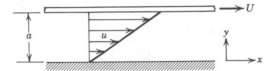

Computing equation:

$$\tau_{yx} = \mu \frac{U}{a} + a \left[\overset{= \ 0(6)}{\frac{\partial p}{\partial x}} \right] \left[\frac{y}{a} - \frac{1}{2} \right] \tag{8.9a}$$

Assumptions: (1) Laminar flow
(2) Steady flow
(3) Incompressible flow
(4) Fully developed flow
(5) Infinite width ($l/a = 1.25/0.00125 = 1000$, so this is a reasonable assumption)
(6) $\partial p / \partial x = 0$ (i.e. flow is symmetric in the actual bearing at no load)

Then

$$\tau_{yx} = \mu \frac{U}{a} = \mu \frac{\omega R}{a} = \mu \frac{\omega D}{2a}$$

fully-developed laminar flow between infinite parallel plates

For SAE 40 oil at 210 F, $\mu = 1.71 \times 10^{-4}$ lbf-sec/ft², from Fig. A2, Appendix A. Thus

$$\tau_{yx} = \frac{1.71 \times 10^{-4} \text{ lbf-sec}}{\text{ft}^2} \times \frac{3600 \text{ rev}}{\text{min}} \times \frac{2\pi \text{ rad}}{\text{rev}} \times \frac{\text{min}}{60 \text{ sec}} \times 3 \text{ in.} \times \frac{1}{2} \times \frac{1}{0.00125 \text{ in.}}$$

$$\tau_{yx} = 77.5 \frac{\text{lbf}}{\text{ft}^2}$$

Since $\tau_{yx} > 0$, it acts to the *left* on the upper plate; it is a minus y surface. The total shear force is given by the shear stress times the area. It is applied to the journal surface. Therefore

$$T = FR = \tau_{yx}\pi DlR = \frac{\pi}{2}\tau_{yx}D^2 l$$

$$= \frac{\pi}{2} \times \frac{77.5 \text{ lbf}}{\text{ft}^2} \times (3)^2 \text{ in.}^2 \times \frac{\text{ft}^2}{144 \text{ in.}^2} \times 1.25 \text{ in.}$$

$$T = 9.5 \text{ in.-lbf} \qquad\qquad\qquad\qquad\qquad\qquad\qquad\qquad\qquad\qquad T$$

The power dissipated in the bearing is

$$\dot{W} = FU = FR\left(\frac{U}{R}\right) = FR\left(\frac{\omega R}{R}\right) = T\omega$$

$$= 9.5 \text{ in.-lbf} \times \frac{3600 \text{ rev}}{\text{min}} \times \frac{\text{min}}{60 \text{ sec}} \times \frac{2\pi \text{ rad}}{\text{rev}} \times \frac{\text{ft}}{12 \text{ in.}} \times \frac{\text{hp-sec}}{550 \text{ ft-lbf}}$$

$$\dot{W} = 0.54 \text{ hp} \qquad\qquad\qquad\qquad\qquad\qquad\qquad\qquad\qquad\qquad\qquad \dot{W}$$

To assure laminar flow, check the Reynolds number value.

$$Re = \frac{\rho U a}{\mu} = \frac{SG \, \rho_{H_2O} U a}{\mu} = \frac{SG \, \rho_{H_2O} \omega R a}{\mu}$$

$$= 0.88 \times \frac{1.94 \text{ slug}}{\text{ft}^3} \times \frac{(3600)2\pi \text{ rad}}{60 \text{ sec}} \times 1.5 \text{ in.} \times 0.00125 \text{ in.} \times \frac{\text{ft}^2}{1.71 \times 10^{-4} \text{ lbf-sec}}$$

$$\times \frac{\text{ft}^2}{144 \text{ in.}^2} \times \frac{\text{lbf-sec}^2}{\text{slug-ft}}$$

$$Re = 48.9$$

Therefore the flow is laminar, since $Re \ll 1500$.

8–4 Fully-Developed Laminar Flow in a Pipe

As a final example of fully-developed laminar flow cases, let us consider fully-developed laminar flow in a pipe. Here the flow is axisymmetric and it is therefore most convenient to work in cylindrical coordinates. We shall again employ a differential control volume, but this time since the flow is axisymmetric, the control volume will be a differential annular ring as shown in Fig. 8.8. The annular differential control volume is of length dx and has thickness dr.

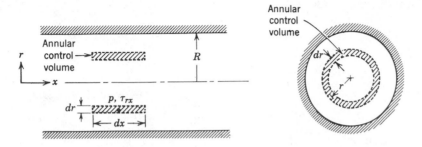

Fig. 8.8 *Control volume for analysis of fully-developed laminar flow in a pipe.*

For a fully-developed steady flow, the momentum equation in the x direction (Eq. 4.19a) reduces to

$$F_{s_x} = 0$$

The next step is to sum the forces acting on the control volume in the x direction. We know that there are normal forces (pressure forces) acting on the left and right ends of the control volume, and that there are tangential forces (shear forces) acting on the inner and outer cylindrical surfaces.

If the pressure at the center of the annular control volume is p, then the force on the left end is

$$\left(p - \frac{\partial p}{\partial x}\frac{dx}{2}\right)2\pi r\,dr$$

The force on the right end is

$$-\left(p + \frac{\partial p}{\partial x}\frac{dx}{2}\right)2\pi r\,dr$$

If the shear stress at the center of the annular control volume is τ_{rx}, then the shear force on the inner cylindrical surface is

$$-\left(\tau_{rx} - \frac{d\tau_{rx}}{dr}\frac{dr}{2}\right)2\pi\left(r - \frac{dr}{2}\right)dx$$

327   **fully-developed laminar flow in a pipe**

The shear force on the outer cylindrical surface is

$$\left(\tau_{rx} + \frac{d\tau_{rx}}{dr}\frac{dr}{2}\right)2\pi\left(r + \frac{dr}{2}\right)dx$$

The summation of the x components of the forces acting on the control volume must be zero. This leads to the condition:

$$-\frac{\partial p}{\partial x}2\pi r\, dr\, dx + \tau_{rx}\, dr2\pi\, dx + \frac{d\tau_{rx}}{dr}r\, dr2\pi\, dx = 0$$

Dividing this equation through by $2\pi r\, dr\, dx$ gives

$$\frac{\partial p}{\partial x} = \frac{\tau_{rx}}{r} + \frac{d\tau_{rx}}{dr} = \frac{1}{r}\frac{d(r\tau_{rx})}{dr} \tag{8.10}$$

Since τ_{rx} is only a function of r (this is the reason for using the total rather than the partial derivative of τ_{rx} in the force components above) we recognize that Eq. 8.10 holds true for all values of r and x only if each side of the equation is equal to a constant. Equation 8.10 can be written as

$$\frac{1}{r}\frac{d(r\tau_{rx})}{dr} = \frac{\partial p}{\partial x} = \text{constant}$$

or

$$\frac{d(r\tau_{rx})}{dr} = r\frac{\partial p}{\partial x}$$

On integrating this equation, we obtain

$$r\tau_{rx} = \frac{r^2}{2}\left(\frac{\partial p}{\partial x}\right) + c_1$$

or

$$\tau_{rx} = \frac{r}{2}\left(\frac{\partial p}{\partial x}\right) + \frac{c_1}{r}$$

Since

$$\tau_{rx} = \mu\frac{du}{dr}$$

then

$$\mu\frac{du}{dr} = \frac{r}{2}\left(\frac{\partial p}{\partial x}\right) + \frac{c_1}{r}$$

and

$$u = \frac{r^2}{4\mu}\left(\frac{\partial p}{\partial x}\right) + \frac{c_1}{\mu}\ln r + c_2 \qquad (8.11)$$

At this point we need to evaluate the constants c_1 and c_2. However we have only the one boundary condition that $u = 0$ at $r = R$. What do we do? Before throwing in the towel, let us look at the solution for the velocity profile given by Eq. 8.11. Although we do not know the value of the velocity at the pipe centerline, we do know from physical considerations that the velocity must be finite at $r = 0$. The only way that this can be true is for c_1 to be zero. Thus from physical considerations we conclude that $c_1 = 0$, and hence

$$u = \frac{r^2}{4\mu}\left(\frac{\partial p}{\partial x}\right) + c_2$$

The constant c_2 is evaluated by using the available boundary condition at the pipe wall: at $r = R$, $u = 0$. Consequently

$$0 = \frac{R^2}{4\mu}\left(\frac{\partial p}{\partial x}\right) + c_2$$

This gives

$$c_2 = -\frac{R^2}{4\mu}\left(\frac{\partial p}{\partial x}\right)$$

and hence

$$u = \frac{r^2}{4\mu}\left(\frac{\partial p}{\partial x}\right) - \frac{R^2}{4\mu}\left(\frac{\partial p}{\partial x}\right) = \frac{1}{4\mu}\left(\frac{\partial p}{\partial x}\right)(r^2 - R^2)$$

or

$$u = -\frac{R^2}{4\mu}\left(\frac{\partial p}{\partial x}\right)\left[1 - \left(\frac{r}{R}\right)^2\right] \qquad (8.12)$$

Because we have the velocity profile there are a number of additional features of the flow that we can obtain.

Shear Stress Distribution

The shear stress is given by

$$\tau_{rx} = \mu\frac{du}{dr} = \frac{r}{2}\left(\frac{\partial p}{\partial x}\right) \qquad (8.13a)$$

329   **fully-developed laminar flow in a pipe**

Volumetric Flowrate

$$Q = \int \vec{V} \cdot d\vec{A}$$

$$= \int_0^R u 2\pi r \, dr$$

$$= \int_0^R \frac{1}{4\mu}\left(\frac{\partial p}{\partial x}\right)(r^2 - R^2) 2\pi r \, dr$$

$$Q = -\frac{\pi R^4}{8\mu}\left(\frac{\partial p}{\partial x}\right) \tag{8.13b}$$

Flowrate as a Function of Pressure Drop

In fully-developed flow, the pressure gradient, $\partial p/\partial x$, is constant. Therefore $(p_2 - p_1)/L = -\Delta p/L = \partial p/\partial x$. Substituting into Eq. 8.13b for the volumetric flowrate,

$$Q = -\frac{\pi R^4}{8\mu}\left[\frac{-\Delta p}{L}\right]$$

or

$$Q = \frac{\pi \Delta p R^4}{8\mu L} = \frac{\pi \Delta p D^4}{128\mu L} \tag{8.13c}$$

for laminar flow in a horizontal pipe.

Average Velocity

The average velocity, $\bar{V}$, is given by

$$\bar{V} = \frac{Q}{A} = \frac{Q}{\pi R^2}$$

$$\bar{V} = -\frac{R^2}{8\mu}\left(\frac{\partial p}{\partial x}\right) \tag{8.13d}$$

Point of Maximum Velocity

To find the point of maximum velocity, we set du/dr equal to zero and solve for the corresponding value of r. From Eq. 8.12

$$\frac{du}{dr} = \frac{1}{2\mu}\left(\frac{\partial p}{\partial x}\right) r$$

Thus

$$\frac{du}{dr} = 0 \quad \text{at} \quad r = 0$$

At $r = 0$,

$$u = u_{max} = -\frac{R^2}{4\mu}\left(\frac{\partial p}{\partial x}\right) = 2\overline{V} \qquad (8.13e)$$

The parabolic velocity profile, given by Eq. 8.12 for fully-developed laminar pipe flow, was sketched in Fig. 8.1.

Example 8.3

A simple and accurate viscometer can be made from a length of capillary tubing. If the flowrate and pressure drop are measured, and the tube geometry is known, the viscosity can be computed from Eq. 8.13c. A test of a certain liquid in a capillary viscometer gave the following data:

> Flowrate   52.8 cm³/min
> Tube diameter 0.020 in.
> Tube length   3 ft
> Pressure drop   150 psi

Determine the viscosity of the liquid in centipoise.

EXAMPLE PROBLEM 8.3

GIVEN:

Flow in a capillary viscometer. The flowrate is $Q = 52.8 \ \text{cm}^3/\text{min}$.

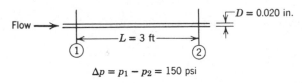

$$\Delta p = p_1 - p_2 = 150 \text{ psi}$$

FIND:

The fluid viscosity in centipoise (cp).

SOLUTION:

Equation 8.13c may be applied.

Computing equation: $\quad Q = \dfrac{\pi \Delta p R^4}{8\mu L}$ $\qquad\qquad\qquad\qquad\qquad$ (8.13c)

Assumptions: (1) Laminar flow
$\qquad\qquad\quad$ (2) Steady flow
$\qquad\qquad\quad$ (3) Incompressible flow

Then

$$\mu = \frac{\pi \Delta p R^4}{8QL} = \frac{\pi}{8} \times \frac{150\ \text{lbf}}{\text{in.}^2} \times (0.01)^4\ \text{in.}^4 \times \frac{\text{min}}{52.8\ \text{cm}^3} \times \frac{1}{3\ \text{ft}} \times \frac{60\ \text{sec}}{\text{min}} \times \frac{\text{slug-ft}}{\text{lbf-sec}^2}$$

$$\times \frac{32.2\ \text{lbm}}{\text{slug}} \times \frac{453\ \text{gm}}{\text{lbm}} \times \frac{(2.54)^2\ \text{cm}^2}{\text{in.}^2}$$

$$\mu = 2.1 \times 10^{-2}\ \frac{\text{gm}}{\text{cm-sec}}$$

or

$$\mu = 2.1\ \text{cp.} \hskip 6cm \mu$$

Check the Reynolds number. Assume the fluid density is similar to that of water, that is, 1 gm/cm^3

$$\bar{V} = \frac{Q}{A} = \frac{4Q}{\pi D^2} = \frac{4}{\pi} \times \frac{52.8\ \text{cm}^3}{\text{min}} \times \frac{\text{min}}{60\ \text{sec}} \times \frac{1}{(0.020)^2\ \text{in.}^2} \times \frac{\text{in.}^2}{(2.54)^2\ \text{cm}^2}$$

$$\bar{V} = 43.5\ \frac{\text{cm}}{\text{sec}}$$

Then

$$Re = \frac{\rho \bar{V} D}{\mu} = \frac{1\ \text{gm}}{\text{cm}^3} \times \frac{43.5\ \text{cm}}{\text{sec}} \times 0.020\ \text{in.} \times \frac{2.54\ \text{cm}}{\text{in.}} \times \frac{\text{cm-sec}}{0.021\ \text{gm}} = 105$$

Consequently, flow is laminar, since $Re \ll 2300$.

Example 8.4

A viscous, incompressible, Newtonian liquid flows in steady, laminar flow down a vertical wall as shown in the sketch. The thickness, δ, of the liquid film is constant. Since the free liquid surface is exposed to atmospheric pressure, there is no pressure gradient. Apply the momentum equation to the differential control volume of volume *dx dy dz* to derive the velocity distribution through the liquid film.

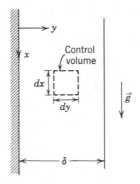

EXAMPLE PROBLEM 8.4

GIVEN:

Fully-developed laminar flow of an incompressible, Newtonian fluid down a vertical wall; thickness, δ, of the liquid film is constant. $dp/dx = 0$.

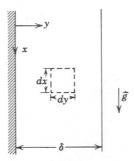

FIND:

Expression for the velocity distribution through the film.

SOLUTION:

The x component of the momentum equation for a control volume is

$$F_{S_x} + F_{B_x} = \frac{\partial}{\partial t} \iiint_{CV} u\rho \, d\forall + \iint_{CS} u\rho \vec{V} \cdot d\vec{A}$$

For steady flow,

$$\frac{\partial}{\partial t} \iiint_{CV} u\rho \, d\forall = 0$$

For fully developed flow,

$$\iint_{CS} u\rho \vec{V} \cdot d\vec{A} = 0$$

Thus, the momentum equation for the present case reduces to

$$F_{S_x} + F_{B_x} = 0$$

The body force, F_{B_x}, is given by $F_{B_x} = \rho g \, d\forall = \rho g \, dx \, dy \, dz$. The surface forces acting on the differential control volume are shear forces on the vertical surfaces. Since $dp/dx = 0$, there is no net pressure force acting on the element.

If the shear stress at the center of the differential control volume is τ_{yx}, then,

$$\text{Shear stress on left face is } \tau_{yx_L} = \left(\tau_{yx} - \frac{d\tau_{yx}}{dy}\frac{dy}{2}\right),$$

and

$$\text{Shear stress on right face is } \tau_{yx_R} = \left(\tau_{yx} + \frac{d\tau_{yx}}{dy}\frac{dy}{2}\right)$$

The direction of the shear stress vectors is taken consistent with the sign convention of Section 2–4.2. Thus on the left face, τ_{yx_L} acts upward, and on the right face, τ_{yx_R} acts downward.

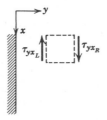

The surface forces are obtained by multiplying each shear stress by the area over which it acts.

Substituting into $F_{S_x} + F_{B_x} = 0$, we obtain

$$-\tau_{yx_L} \, dx \, dz + \tau_{yx_R} \, dx \, dz + \rho g \, dx \, dy \, dz = 0$$

or

$$-\left(\tau_{yx} - \frac{d\tau_{yx}}{dy}\frac{dy}{2}\right) dx \, dz + \left(\tau_{yx} + \frac{d\tau_{yx}}{dy}\frac{dy}{2}\right) dx \, dz + \rho g \, dx \, dy \, dz = 0$$

and

$$\frac{d\tau_{yx}}{dy} + \rho g = 0$$

or

$$\frac{d\tau_{yx}}{dy} = -\rho g$$

Since

$$\tau_{yx} = \mu \frac{du}{dy}, \quad \text{then} \quad \mu \frac{d^2 u}{dy^2} = -\rho g$$

and

$$\frac{d^2 u}{dy^2} = -\frac{\rho g}{\mu}$$

Integrating with respect to y

$$\frac{du}{dy} = -\frac{\rho g}{\mu} y + c_1$$

Integrating again,

$$u = -\frac{\rho g}{\mu} \frac{y^2}{2} + c_1 y + c_2$$

To evaluate the constants c_1 and c_2, we must apply appropriate boundary conditions:

$$\text{(i)} \quad y = 0 \qquad u = 0$$

$$\text{(ii)} \quad y = \delta \qquad \frac{du}{dy} = 0$$

From boundary condition (i) $c_2 \doteq 0$

From boundary condition (ii) $0 = -\frac{\rho g}{\mu} \delta + c_1$

$$c_1 = \frac{\rho g}{\mu} \delta$$

Hence

$$u = -\frac{\rho g}{\mu} \frac{y^2}{2} + \frac{\rho g}{\mu} \delta y$$

$$u = \frac{\rho g}{\mu} \delta^2 \left[\left(\frac{y}{\delta} \right) - \frac{1}{2} \left(\frac{y}{\delta} \right)^2 \right] \longleftarrow \qquad u(y)$$

$\left\{ \begin{array}{l} \text{The purpose of this problem is to illustrate the application of the momentum equation to} \\ \text{a differential control volume for a fully-developed laminar flow.} \end{array} \right\}$

Part C. Flow in Pipes and Ducts

Our main purpose in this section is to determine the pressure changes in pipes, ducts, and piping systems. All of the pressure loss in a pipe system results from friction. However, to simplify analysis, loss will be divided into major losses (due to friction in constant area portions of the system) and minor losses (due to flow through valves, tees, elbows and area changes). Since ducts of circular cross section are most common in engineering applications, the basic analysis will be performed for circular geometries. (The results can be extended to other geometries by introducing the hydraulic diameter, which is treated in Section 8–9.3.)

In developing the relations for the major losses due to friction in a constant area duct, we shall deal with fully-developed flows, that is, flows in which the velocity profile is unvarying in the direction of flow. Thus we will not account for the pressure change that occurs in the entrance region of a pipe as a result of a change in the velocity profile (Fig. 8.1); instead, the entrance loss will be treated as a minor loss.

8–5 Velocity Profiles in Pipe Flow

Pipe flow may be either laminar or turbulent, depending on the Reynolds number. For fully-developed laminar pipe flow, the velocity profile is parabolic, as shown in Section 8-4. Thus

$$u = -\frac{R^2}{4\mu}\left(\frac{\partial p}{\partial x}\right)\left[1 - \left(\frac{r}{R}\right)^2\right]$$ (8.12)

But $u = u_{max} = U$ at $r = 0$, that is,

$$U = -\frac{R^2}{4\mu}\left(\frac{\partial p}{\partial x}\right)$$

so that

$$\frac{u}{U} = 1 - \left(\frac{r}{R}\right)^2$$ (8.14)

for laminar pipe flow. Furthermore, according to Eq. 8.13e, the average velocity is equal to half the centerline velocity, that is,

$$\frac{\overline{V}}{U} = \frac{1}{2}$$ (8.15)

Except for flow of very viscous fluids in small diameter ducts, internal flows in general are turbulent. As noted in the discussion of laminar and turbulent flows (Section 8–2), in turbulent flow there is no universal relationship between the

8/incompressible viscous flow

stress field and the mean velocity field. Thus, in turbulent flows, we are forced to rely on experimental data.

We can represent the velocity profile for turbulent flow through a smooth pipe by the empirical equation

$$\frac{u}{U} = \left(1 - \frac{r}{R}\right)^{1/n} \tag{8.16}$$

where the exponent, n, varies with the Reynolds number. The value of the exponent, n, is 6 at a Reynolds number of 4.0×10^3. For a Reynolds number of 1.1×10^5, n is equal to 7; the value of n increases to 10 for a Reynolds number of 3.2×10^6. Since the average velocity $\overline{V} = Q/A$ and

$$Q = \int \vec{V} \cdot d\vec{A}$$

the ratio of the average velocity to the centerline velocity may be calculated from Eq. 8.16. The result is

$$\frac{\overline{V}}{U} = \frac{2n^2}{(n+1)(2n+1)} \tag{8.17}$$

Equation 8.17 predicts that as n increases (due to increasing Reynolds number) the ratio of the average velocity to the centerline velocity increases, that is, the velocity profile becomes more blunt, or "fuller." (For $n = 6$, $\overline{V}/U = 0.79$, while for $n = 10$, $\overline{V}/U = 0.87$.)

In Fig. 8.9 the velocity profiles for both laminar flow (Eq. 8.14) and turbulent flow (Eq. 8.16) are plotted for the same value of the Reynolds number ($Re = 4.0 \times 10^3$). Both profiles have the same average velocity and, hence, the same flow rate; they do not have the same centerline velocity. From Fig. 8.9 we also see that the turbulent profile has a much steeper slope at the wall.

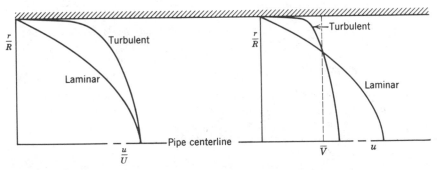

Fig. 8.9 *Velocity profiles for laminar and turbulent flow in a smooth circular pipe at Reynolds number, $Re = 4 \times 10^3$.*

velocity profiles in pipe flow

8–6 Shear Stress Distribution in Pipe Flow

In fully-developed steady flow in a horizontal pipe, be it laminar or turbulent, the pressure drop is balanced only by shear forces at the pipe wall. This can be seen by applying the momentum equation to a cylindrical control volume in the flow, Fig. 8.10. The x component of the momentum equation is

Basic equation:

$$= 0(1) \quad = 0(2) \qquad\qquad = 0(3,4)$$

$$F_{S_x} + F_{B_x} = \frac{\partial}{\partial t}\iiint_{CV} u\rho \, d\forall + \iint_{CS} u\rho \vec{V} \cdot d\vec{A} \qquad (4.19a)$$

Assumptions: (1) Horizontal pipe, $F_{B_x} = 0$
(2) Steady flow
(3) Incompressible flow
(4) Fully-developed flow

Then

$$F_{S_x} = 0$$

The surface forces acting on the control volume are shown in Fig. 8.10. The pressure at the center of the element has been taken as p; the pressure at each

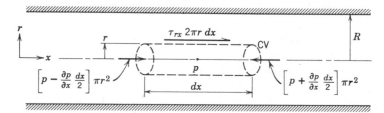

Fig. 8.10 *Control volume for analysis of shear stress in fully-developed flow in a circular pipe.*

end of the element has been obtained from a Taylor series expansion of p about the center of the element. The shear force acts on the circumferential surface of the element. The direction of the stress has been assumed such that the stress is positive. Thus

$$F_{S_x} = \left(p - \frac{\partial p}{\partial x}\frac{dx}{2}\right)\pi r^2 - \left(p + \frac{\partial p}{\partial x}\frac{dx}{2}\right)\pi r^2 + \tau_{rx}2\pi r \, dx = 0$$

8/incompressible viscous flow

or

$$-\frac{\partial p}{\partial x}\,dx\pi r^2 + \tau_{rx}2\pi r\,dx = 0$$

Therefore

$$\tau_{rx} = \frac{r}{2}\frac{\partial p}{\partial x} \tag{8.18}$$

Thus we see that the shear stress on the fluid varies linearly across the pipe, from zero at the centerline to a maximum at the pipe wall. If we denote the wall value of shear stress as τ_w, Eq. 8.18 shows that at the surface of the pipe

$$\tau_w = -[\tau_{rx}]_{r=R} = -\frac{R}{2}\frac{\partial p}{\partial x} \tag{8.19}$$

Equation 8.19 relates the wall shear stress to the axial pressure gradient. The momentum equation was used to derive it, but no assumption was made about a relation between stress and the velocity field. Consequently, Eq. 8.19 is applicable to both laminar and turbulent pipe flow.

   If we now could relate the stress field to the mean velocity field we could determine analytically the pressure drop over a length of pipe for fully-developed flow. Such a relation between the stress field and the mean velocity field does exist for laminar flow and was used in Section 8-4. The resulting equation, Eq. 8.13c, was first discovered experimentally by Jean Louis Poiseuille, a French physician, and independently by Gotthif H. L. Hagen, a German engineer, in the 1850s.

8–7 *Energy Considerations in Pipe Flow*

   Thus far in our discussion of viscous flow, all derived results have been obtained from an application of the momentum equation for a control volume. We have,

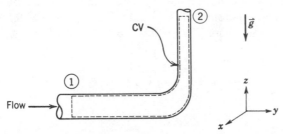

Fig. 8.11 *Control volume and coordinates for energy analysis of flow through a 90-degree reducing elbow.*

of course, also used the control volume formulation of the conservation of mass. Nothing has been said about the conservation of energy—the first law of thermodynamics. Additional insight into the nature of the pressure losses in internal viscous flows can be obtained from the energy equation. Consider, for example, the steady flow through the section of a piping system including a reducing elbow shown in Fig. 8.11. The control volume boundaries are shown as dashed lines. They are normal to the flow at sections ① and ② and coincide with the inside pipe wall elsewhere.

Basic equation:

$$\overset{=\ 0(1)}{\dot{Q}} - \overset{=\ 0(2)}{\dot{W}_s} - \overset{=\ 0(3)}{\dot{W}_{shear}} = \frac{\partial}{\partial t} \iiint_{CV} e\rho \, d\forall + \iint_{CS} (e + pv)\rho \vec{V} \cdot d\vec{A} \qquad (4.45)$$

$$e = u + \frac{V^2}{2} + gz$$

Assumptions: (1) $\dot{W}_s = 0$

(2) $\dot{W}_{shear} = 0$ (although shear stresses are present at the walls of the elbow, the velocities are zero at the walls)

(3) Steady flow

(4) Incompressible flow

(5) Uniform internal energy and pressure across sections ① and ②

Then the energy equation reduces to

$$\dot{Q} = \dot{m}(u_2 - u_1) + \dot{m}\left(\frac{p_2}{\rho} - \frac{p_1}{\rho}\right) + \dot{m}g(z_2 - z_1) + \int_{A_2} \frac{V_2^2}{2}\rho V_2 \, dA_2$$

$$- \int_{A_1} \frac{V_1^2}{2}\rho V_1 \, dA_1 \qquad (8.20)$$

Note that we have not assumed the velocity to be uniform at sections ① and ② since we know for viscous flows the velocity at a section cannot be uniform. However, it is convenient to introduce the average velocity into Eq. 8.20 so that we can eliminate the integral signs. To do this we define a kinetic energy flux coefficient, α, so that

$$\int_A \frac{V^2}{2}\rho V \, dA = \alpha \int_A \frac{\overline{V}^2}{2}\rho V \, dA = \alpha \dot{m}\frac{\overline{V}^2}{2} \qquad (8.21)$$

With this definition of α, Eq. 8.20 can be written as

$$\dot{Q} = \dot{m}(u_2 - u_1) + \dot{m}\left(\frac{p_2}{\rho} - \frac{p_1}{\rho}\right) + \dot{m}g(z_2 - z_1) + \dot{m}\left(\frac{\alpha_2 \overline{V}_2^2}{2} - \frac{\alpha_1 \overline{V}_1^2}{2}\right)$$

Dividing through by the mass rate of flow, then

$$\frac{\delta Q}{dm} = u_2 - u_1 + \frac{p_2}{\rho} - \frac{p_1}{\rho} + gz_2 - gz_1 + \frac{\alpha_2 \bar{V}_2^2}{2} - \frac{\alpha_1 \bar{V}_1^2}{2}$$

On rearranging this equation we have

$$\left(\frac{p_1}{\rho} + \alpha_1 \frac{\bar{V}_1^2}{2} + gz_1\right) = \left(\frac{p_2}{\rho} + \alpha_2 \frac{\bar{V}_2^2}{2} + gz_2\right) + (u_2 - u_1) - \frac{\delta Q}{dm} \quad (8.22a)$$

8–7.1 HEAD LOSS

In Eq. 8.22a, the term

$$\left(\frac{p}{\rho} + \alpha \frac{\bar{V}^2}{2} + gz\right)$$

represents the flux of mechanical energy per unit mass at a flow cross section. The term $u_2 - u_1 - \delta Q/dm$ is equal to the difference in mechanical energy flux between sections ① and ②. Thus, it represents the (irreversible) conversion of mechanical energy at section ① to unwanted thermal energy $(u_2 - u_1)$ and loss of energy via heat transfer $(-\delta Q/dm)$. Thus we identify this group of terms as the total head loss, h_{l_T}. Then

$$\left(\frac{p_1}{\rho} + \alpha_1 \frac{\bar{V}_1^2}{2} + gz_1\right) - \left(\frac{p_2}{\rho} + \alpha_2 \frac{\bar{V}_2^2}{2} + gz_2\right) = h_{l_T} \quad (8.22b)$$

If the flow were frictionless, the velocity at a section would be uniform $(\alpha_1 = \alpha_2 = 1)$ and Bernoulli's equation would tell us that the headloss was zero. In incompressible frictionless flow the only change in internal energy of the flow is brought about through the effects of heat transfer; there is no conversion of mechanical energy $(p/\rho + V^2/2 + gz)$ to internal energy. For viscous flow in a pipe, one effect of friction may be to increase the internal energy of the flow, as shown by Eq. 8.22a.

For laminar flow (velocity profile given by Eq. 8.12) $\alpha = 2.0$. However, there is no need to use Eq. 8.22b since we can calculate the pressure drop analytically as was done in Section 8–4.

In turbulent flow the velocity profile is quite flat as shown in Fig. 8.9. We can use Eq. 8.21 together with Eqs. 8.16 and 8.17 to determine the value of α. Substituting the power law velocity profile of Eq. 8.16 into Eq. 8.21, we obtain (after a little algebra) the expression

$$\alpha = \left[\frac{U}{\bar{V}}\right]^3 \frac{2n^2}{(3+n)(3+2n)} \quad (8.23)$$

The value of $\bar{V}/U$ is determined from Eq. 8.17. Since the exponent, n, in the power law profile is a function of the Reynolds number, α also varies with the Reynolds number. For $n = 6$ ($Re = 4.0 \times 10^3$), $\alpha = 1.08$; for $n = 10$ ($Re = 3.2 \times 10^6$), $\alpha = 1.03$. Because α is reasonably close to unity over a large range of Reynolds number and the change of kinetic energy is usually small compared to the total head loss, α is normally set equal to unity and Eq. 8.22b is written as

$$\left(\frac{p_1}{\rho} + \frac{\bar{V}_1^2}{2} + gz_1 \right) - \left(\frac{p_2}{\rho} + \frac{\bar{V}_2^2}{2} + gz_2 \right) = h_{l_T} \tag{8.24}$$

Recall that we have defined the total head loss, h_{l_T}, as the sum of the major head loss, h_l, and minor head loss, h_{l_m}. Head loss has dimensions of energy per unit mass $[FL/M]$. In the $FLtT$ system of dimensions, this is equivalent to dimensions of $[L^2/t^2]$.

You might wonder why the energy loss, h_{l_T}, is called a "head" loss. As the empirical science of hydraulics developed in the 19th century, it was common practice to express the energy balance in terms of energy per unit *weight* of flowing fluid (i.e. water) rather than energy per unit *mass* as in Eq. 8.24. To obtain dimensions of energy per unit weight, we divide each term in Eq. 8.24 by the acceleration of gravity, g. Then the net dimensions of h_{l_T} are $[(L^2/t^2)(t^2/L)] = [L]$, or feet of flowing fluid. Since the term head loss is in common use, we shall also use it here. Remember that its physical interpretation is a loss in mechanical energy, expressed per unit mass of flowing fluid.

Equation 8.24 can be used to calculate the pressure difference between any two points in a piping system, provided the head loss, h_{l_T}, is known. We shall turn our attention to this in the next section.

8–8 Calculation of Head Loss

Recall that total head loss, h_{l_T}, is regarded as the sum of major losses, h_l, due to frictional effects in fully developed flow in constant area tubes, and minor losses, h_{l_m}, due to transitions, entrances, and so on. Therefore let us consider these two cases separately.

8–8.1 MAJOR LOSSES: FRICTION FACTOR

The energy balance, expressed by Eq. 8.22b, can be used to evaluate the major head loss. For fully developed flow through a constant area pipe, $h_{l_m} = 0$, and $\alpha_1(\bar{V}_1^2/2) = \alpha_2(\bar{V}_2^2/2)$. Then Eq. 8.22b becomes

$$\frac{p_1 - p_2}{\rho} = g(z_2 - z_1) + h_l \tag{8.25}$$

If the pipe is horizontal, then $z_2 = z_1$ and

$$\frac{p_1 - p_2}{\rho} = \frac{\Delta p}{\rho} = h_l \qquad (8.26)$$

Thus the major head loss can be expressed as the pressure loss for flow through a horizontal pipe of constant area.

Head loss for flow in a constant area duct depends only on the details of the flow through the duct, since it represents the amount of energy converted by frictional effects from mechanical to thermal energy. Therefore, the head loss obtained from Eq. 8.26 for flow in a horizontal pipe is also valid for flow at the same flowrate in an inclined pipe.

a. Laminar Flow

In laminar flow the pressure drop may be computed analytically for fully-developed flow in a horizontal pipe. Thus from Eq. 8.13c

$$\Delta p = \frac{128\mu L Q}{\pi D^4} = \frac{128\mu L \overline{V}(\pi D^2/4)}{\pi D^4} = 32\frac{L}{D}\frac{\mu\overline{V}}{D}$$

Substituting in Eq. 8.26,

$$h_l = 32\frac{L}{D}\frac{\mu\overline{V}}{\rho D} = \frac{L}{D}\frac{\overline{V}^2}{2}\left(64\frac{\mu}{\rho\overline{V}D}\right) = \frac{L}{D}\frac{\overline{V}^2}{2}\left(\frac{64}{Re}\right) \qquad (8.27)$$

(We shall see the reason for writing h_l in this form shortly.)

b. Turbulent Flow

In turbulent flow we cannot evaluate the pressure drop analytically and, hence, we must resort to experimental results and utilize dimensional analysis as an aid in correlating the experimental data. In turbulent flow, the pressure drop, Δp, due to friction in a horizontal constant area pipe is known to depend on the pipe diameter, D, the pipe length, L, the pipe roughness, e, the average flow velocity, $\overline{V}$, the fluid density, ρ, and the fluid viscosity, μ. In functional form

$$\Delta p = \Delta p(D, L, e, \overline{V}, \rho, \mu)$$

We applied dimensional analysis to this problem in Example Problem 7.2. The results were a correlation of the form

$$\frac{\Delta p}{\rho\overline{V}^2} = f\left(\frac{\mu}{\rho\overline{V}D}, \frac{L}{D}, \frac{e}{D}\right)$$

343   **calculation of head loss**

We recognize that $\mu/\rho\bar{V}D = 1/Re$, so we could just as well write

$$\frac{\Delta p}{\rho\bar{V}^2} = \phi\left(Re, \frac{L}{D}, \frac{e}{D}\right)$$

Substituting in Eq. 8.26, we see that

$$h_l = \bar{V}^2\phi\left(Re, \frac{L}{D}, \frac{e}{D}\right)$$

Thus dimensional analysis gives the form of the functional relationship, but we must resort to experiment to obtain actual values.

Experiments show that the nondimensional head loss is directly proportional to L/D. Hence we can write

$$\frac{h_l}{\bar{V}^2} = \frac{L}{D}\phi_1\left(Re, \frac{e}{D}\right)$$

Since the function ϕ_1 is as yet undetermined, it is permissible to introduce a constant into the left side of the above equation. The number $\frac{1}{2}$ is introduced into the denominator such that the head loss is nondimensionalized on the kinetic energy per unit mass of flow. Then

$$\frac{h_l}{\frac{1}{2}\bar{V}^2} = \frac{L}{D}\phi_2\left(Re, \frac{e}{D}\right)$$

The unknown function $\phi_2(Re, e/D)$ is defined as the friction factor, f, that is,

$$f \equiv \phi_2\left(Re, \frac{e}{D}\right)$$

and

$$h_l = f\frac{L}{D}\frac{\bar{V}^2}{2} \tag{8.28}$$

The friction factor, f, is determined experimentally. The results, published by L. F. Moody, are shown in Fig. 8.12.

To determine head loss for flow with known conditions, the flow Reynolds number is evaluated first. The value of relative roughness, e/D, for the flow is obtained from Fig. 8.13. Then the friction factor, f, is read from the appropriate curve in Fig. 8.12, at the known Re and e/D. Finally the head loss is found using Eq. 8.28.

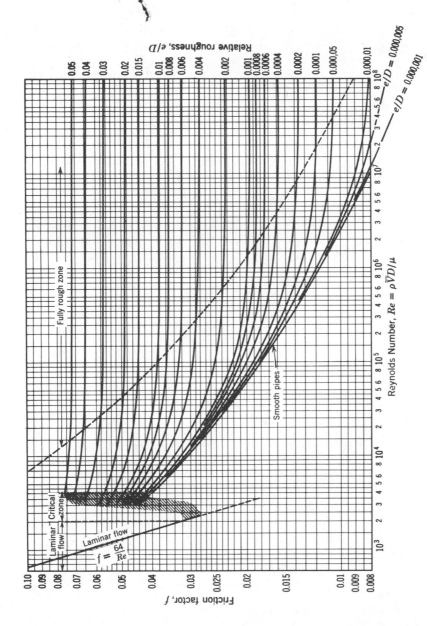

Fig. 8.12 *Friction factor for fully-developed flow in circular pipes (data from Ref. 1, used by permission).*

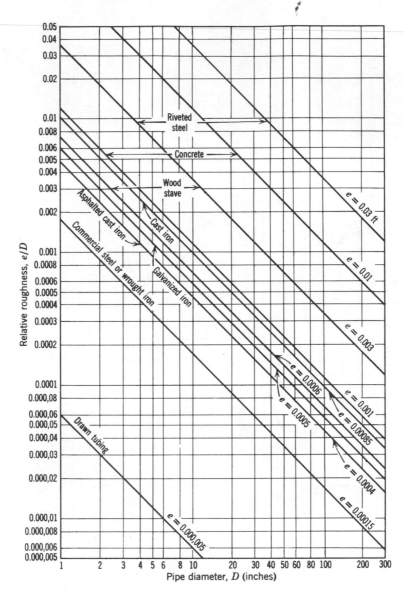

Fig. 8.13 *Relative roughness values for pipes of common engineering materials (data from Ref. 1, used by permission).*

Several features of Fig. 8.12 require some discussion. The friction factor for laminar flow may be obtained by comparing Eqs. 8.27 and 8.28. Thus

$$h_l = \frac{L}{D} \frac{\overline{V}^2}{2} \left(\frac{64}{Re} \right) = f \frac{L}{D} \frac{\overline{V}^2}{2}$$

Consequently, for laminar flow

$$f_{\text{laminar}} = \frac{64}{Re} \tag{8.29}$$

Thus in laminar flow, the friction factor is a function of Reynolds number only; it is independent of roughness. Although we took no notice of roughness in deriving Eq. 8.27, experimental results verify that friction factor is a function only of Reynolds number in laminar flow.

The Reynolds number in a pipe may be changed most easily by varying the average flow velocity. If the flow in a pipe is originally laminar, increasing the velocity until the critical Reynolds number is reached causes transition to occur; the laminar flow gives way to turbulent flow. The effect of transition on the velocity profile was discussed in Section 8–5. Figure 8.9 shows that the velocity gradient at the tube wall is much larger for turbulent flow than for laminar flow.[2] This change in velocity profile causes the wall shear stress to increase sharply, with the same effect on the friction factor.

As Reynolds number is increased above the transition value, the velocity profile continues to become fuller, as noted in Section 8–5; the friction factor at first tends to follow the smooth pipe curve, along which friction factor is a function of Reynolds number only. However, as Reynolds number increases, the velocity profile becomes still fuller. The size of the thin "viscous sublayer" near the tube wall decreases. As roughness elements begin to poke through this layer, the effect of roughness becomes important, and the friction factor becomes a function of Reynolds number *and* relative roughness.

At very large Reynolds number, most of the roughness elements on the tube wall protrude through the viscous sublayer; the drag and, hence, the pressure loss, depend only on the size of the roughness elements. This flow regime is termed the "fully rough" flow regime; friction factor depends on e/D only.

To summarize the preceding discussion, we see that as Reynolds number is increased, the friction factor will decrease as long as the flow remains laminar. At transition, f increases sharply. In the turbulent flow regime, the friction factor decreases gradually along the smooth pipe curve, and finally levels out at a constant value for extremely large Reynolds number. (The only exception to these trends is that friction factors for pipes with $e/D \gtrsim 0.001$ in turbulent flow fall above the smooth pipe curve.)

[2] The power law profile (Eq. 8.16) predicts an infinite velocity gradient at the tube wall, which is obviously not correct. However, it fits the central portion of the profile well.

Figure 8.13 also needs some explanation. All of the e/D values given are those for new pipes, in relatively good condition. Over long periods of service corrosion takes place and, particularly in hard water areas, lime deposits and rust scale form on pipe walls. Corrosion can weaken pipes, eventually leading to failure. Deposit formation appreciably increases wall roughness, and also decreases the effective diameter. These factors combine to cause e/D to increase by a factor of 2 or 3 for old pipes. An example is shown in Fig. 8.14.

Fig. 8.14 *Pipe section removed after 40 years of service as a water line, showing formation of scale.*

8–8.2 MINOR LOSSES

The flow in a piping system may be required to pass through a variety of fittings, bends, or abrupt changes in area. Additional head losses are encountered primarily as a result of flow separation. (Energy is eventually dissipated by violent mixing in the separated zones.) These losses will be minor (hence the term minor losses) if the particular piping system includes long lengths of constant area pipe. The minor head loss may be expressed as

$$h_{l_m} = K \frac{\bar{V}^2}{2} \qquad (8.30a)$$

where the *loss coefficient*, K, must be determined experimentally for each situation. Minor head loss may also be expressed as

$$h_{l_m} = f \frac{L_e}{D} \frac{\bar{V}^2}{2} \qquad (8.30b)$$

where L_e is an *equivalent length* of straight pipe.

Experimental data for minor loss coefficients are plentiful, but they are scattered among a variety of sources. Consequently, data for some commonly encountered situations are summarized below.[3]

a. Inlets and Entrance Length

Appreciable head loss can occur when a poorly designed entrance is used for a pipe flow system. If the inlet has sharp corners, flow separation occurs at the corners, and a *vena contracta* is formed.[4] Losses result from the unconfined mixing

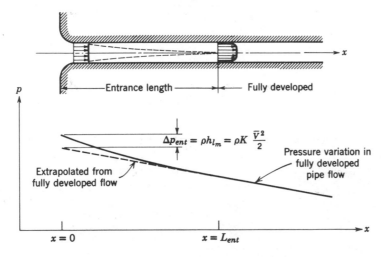

Fig. 8.15 *Variation of static pressure in a pipe inlet section, illustrating the definition of* h_{l_m}.

---

[3] It is worth noting here that many data found in textbooks have been taken from very old sources, and that sometimes even original sources give different values for the same flow configuration. Some of the data given here reflect the authors' "best guess" values; where uncertainties exist, a range of values is given. In each case, the source of the data we have included is identified.

[4] This behavior is well illustrated in film loop S-FM016, *Flow from a Reservoir to a Duct.*

as the flow stream decelerates to fill the pipe again. The head loss can be reduced appreciably by rounding the inlet corners, as shown by the loss coefficients given in Table 8.1.

As the data in Table 8.1 show, even a well-rounded inlet causes additional head loss, compared to fully-developed pipe flow. Part of the additional loss is due to the increased momentum of the fully-developed velocity profile, as compared to the uniform inlet profile. The rest of the additional loss arises from the larger wall shear stress in the part of the inlet where the wall boundary layer is thin. The variation of static pressure in an inlet section is shown qualitatively in Fig. 8.15. (Note that h_{l_m} is a head loss in addition to that due to an assumed fully-developed flow in the length L_{ent}.)

table 8.1 *Loss Coefficients for Pipe Entrances (Data from Ref. 2, except $^+$, from Ref. 3)*

ENTRANCE TYPE	DIAGRAM	LOSS COEFFICIENT, K*
Reentrant		0.78
Square-edged		0.34$^+$
Slightly rounded		0.2–0.25
Well rounded†		0.04

* Based on $h_{l_m} = K(\bar{V}^2/2)$, where $\bar{V}$ is the mean velocity in the pipe.
† $r/R \simeq 0.35$.

b. Enlargements and Contractions

Loss coefficient data for sudden expansions and contractions in circular ducts are given in Fig. 8.16. Note that both loss coefficients are based on the larger value of $\bar{V}^2/2$. Thus losses for a sudden expansion are based on $\bar{V}_1^2/2$, and those for a contraction are based on $\bar{V}_2^2/2$.

Losses due to area change can be reduced somewhat by installing a nozzle or diffuser between the two sections of straight pipe. Data for nozzles are sparse; a

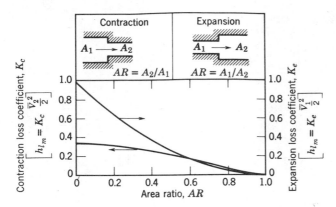

Fig. 8.16 *Loss coefficients for flow through sudden area changes (Ref. 3).*

few representative values are given in Table 8.2. These data suggest that losses in gradual contractions are independent of area ratio; this is likely to be true only for moderate area ratios. Losses in diffusers depend on a number of geometric and flow variables. They are most commonly presented in terms of a pressure

table 8.2 *Loss Coefficients for Gradual Contractions (Data from Ref. 2)*

DIAGRAM	INCLUDED ANGLE, θ, DEGREES	LOSS COEFFICIENT, $K*$
Flow θ A_1 A_2	30	0.02
	45	0.04
	60	0.07

* Based on $h_{l_m} = K(\bar{V}_2^2/2)$.

recovery coefficient, C_p, defined as the ratio of static pressure rise to inlet dynamic pressure, that is,

$$C_p \equiv \frac{p_2 - p_1}{\frac{1}{2}\rho \bar{V}_1^2} \tag{8.31}$$

The definition of C_p may be related to head loss easily. If gravity is neglected,

351   **calculation of head loss**

Eq. 8.24 reduces to

$$\left[\frac{p_1}{\rho} + \frac{\overline{V}_1^2}{2}\right] - \left[\frac{p_2}{\rho} + \frac{\overline{V}_2^2}{2}\right] = h_{l_T} = h_{l_m}$$

Thus

$$h_{l_m} = \frac{\overline{V}_1^2}{2} - \frac{\overline{V}_2^2}{2} - \frac{p_2 - p_1}{\rho}$$

$$= \frac{\overline{V}_1^2}{2}\left[\left(1 - \frac{\overline{V}_2^2}{\overline{V}_1^2}\right) - \frac{p_2 - p_1}{\frac{1}{2}\rho\overline{V}_1^2}\right] = \frac{\overline{V}_1^2}{2}\left[\left(1 - \frac{\overline{V}_2^2}{\overline{V}_1^2}\right) - C_p\right]$$

From continuity, $A_1\overline{V}_1 = A_2\overline{V}_2$, so

$$h_{l_m} = \frac{\overline{V}_1^2}{2}\left[1 - \left(\frac{A_1}{A_2}\right)^2 - C_p\right]$$

or

$$h_{l_m} = \frac{\overline{V}_1^2}{2}\left[\left(1 - \frac{1}{(AR)^2}\right) - C_p\right] \tag{8.32}$$

Data for conical diffusers with fully-developed turbulent pipe flow at the diffuser inlet are presented in Fig. 8.17 as a function of geometry. Similar data correlations for plane wall and annular (Ref. 5) and for radial diffusers (Ref. 6) are available in the literature. Diffuser pressure recovery with uniform inlet flow is somewhat better than for fully-developed inlet flow.

Since the static pressure rises in the direction of flow in a diffuser, the flow may separate from the walls. For some geometries, the outlet flow is distorted, and sometimes pulsations occur. The flow regime behavior of plane wall diffusers is illustrated well in the film *Flow Visualization* (S. J. Kline, principal).[5] For wide angle diffusers, vanes or splitters can be used to suppress stall and improve pressure recovery (Ref. 7).

c. Exits

The kinetic energy per unit mass of flow, $\overline{V}^2/2$, is completely dissipated by mixing when flow discharges from a duct into a large plenum chamber. The situation corresponds to flow through an abrupt expansion with $AR = 0$, Fig. 8.16. Several examples are sketched and loss values given in Table 8.3. These results suggest that no improvement in exit loss coefficient is possible; addition of a diffuser can reduce the value of $\overline{V}^2/2$ considerably, see Example Problem 8.7.

[5] Film loop S-FM049, *Flow Regimes in Subsonic Diffusers* was edited from this film. See also film loop S-FM065, *Wide Angle Diffuser with Suction.*

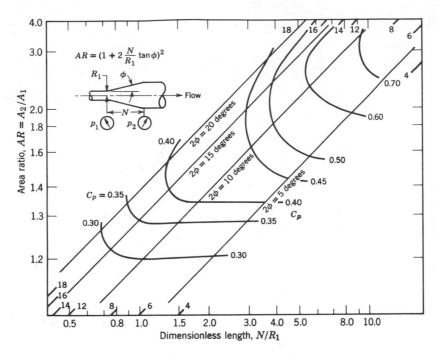

Fig. 8.17 *Pressure recovery data for conical diffuser with fully-developed pipe flow at inlet (data from Ref. 4).*

table 8.3 *Loss Coefficients for Pipe Exits (Data from Ref. 2)*

EXIT TYPE	DIAGRAM	LOSS COEFFICIENT, K^*
Projecting pipe		1.0
Square-edged		1.0
Rounded		1.0

* Based on $h_{l_m} = K(\bar{V}^2/2)$.

353   **calculation of head loss**

d. Pipe Bends

Head loss is increased in fully-developed pipe flow through bends as a result of secondary flows.[6] The additional head loss is expressed most conveniently by an equivalent length of straight pipe. The equivalent length depends on the relative radius of curvature of the bend, as shown in Fig. 8.18, which is an averaged

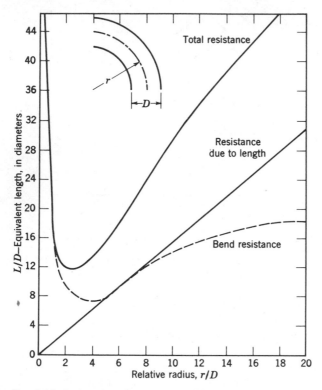

Fig. 8.18 *Design chart for resistance of 90-degree bends in circular pipe with fully-developed turbulent flow (Ref. 2).*

design curve for 90-degree bends. Reference 2 also gives an approximate procedure for computing the resistance of bends with other turning angles.

Because they are simple and inexpensive to erect in the field, miter bends are often used, especially in large pipe systems. Design data for miter bends are given in Fig. 8.19.

[6] Secondary flows are shown in film loop S-FM019, *Secondary Flow in a Bend.*

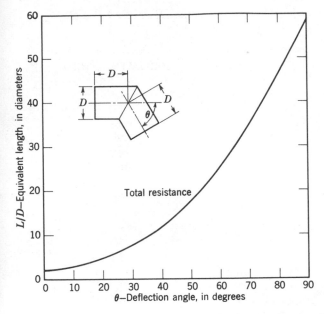

Fig. 8.19 *Design chart for resistance of miter bends in circular pipe with fully-developed turbulent flow (Ref. 2).*

e. Valves and Fittings

Losses for flow through valves and fittings are conveniently expressed in terms of an equivalent length of straight pipe. Some representative data are given in Table 8.4.

In practice, insertion losses for fittings and valves vary considerably, depending on the care used in fabricating the pipe system. If burrs from cutting pipe sections are allowed to remain, they cause local flow obstructions, which increase losses appreciably.

Although the losses discussed in this section were termed "minor losses," they can be a large fraction of the overall system loss. Thus a system for which calculations are to be made should be checked carefully to make sure all losses have been identified and their magnitudes estimated. If calculations are made carefully, the results will be of satisfactory engineering accuracy, that is, variations in loss of plus or minus 10 percent from computed values may be expected.

8–9 Solution of Pipe Flow Problems

Once the head loss has been obtained using the methods of Section 8–8, pipe flow problems can be solved using the energy equation, Eq. 8.24. The same basic

table 8.4 *Representative Equivalent Lengths in Pipe Diameters (L_e/D) for Valves and Fittings (Data from Ref. 2)*

FITTING TYPE	DESCRIPTION	EQUIVALENT LENGTH, $L_e/D*$
Globe Valve	Fully Open	350
Gate Valve	Fully Open	13
	$\frac{3}{4}$ Open	35
	$\frac{1}{2}$ Open	160
	$\frac{1}{4}$ Open	900
Check Valve		50–100
90° Std. Elbow		30
45° Std. Elbow		16
90° Elbow	Long Radius	20
90° Street Elbow		50
45° Street Elbow		26
Tee	Flow through run	20
	Flow through branch	60
Return bend	Close pattern	50

* Based on $h_{l_m} = f\frac{L_e}{D}\frac{\bar{V}^2}{2}$

techniques are used, even for complex piping systems, but let us first consider single path pipe flow problems.

8–9.1 SINGLE PATH SYSTEMS

Equation 8.28 shows that head loss depends on friction factor, pipe length and diameter, and velocity (or flowrate). Since the friction factor in turn depends on Reynolds number (i.e. velocity and diameter for given properties), diameter and roughness (constant for a given pipe material), we may write

$$h_l = \phi_3(L, \bar{V}, D)$$

or

$$h_l = L\phi_4(Q, D) \qquad (8.33)$$

Equation 8.33 contains four parameters, h_l, L, Q, and D. Any one of these can be the unknown quantity in practical flow situations. Thus, four cases are possible:

(a) L, Q, and D known. h_l unknown.
(b) h_l, Q, and D known. L unknown.
(c) h_l, L, and D known. Q unknown.
(d) h_l, L, and Q known. D unknown.

Cases **a** and **b** may be solved directly by applying the continuity and energy equations, and using loss data from Section 8–8. Solutions for cases **c** and **d** make use of the same equations and data, but require iterative solutions. Each case is discussed and illustrated by an example below.

a. L, Q, and D Known. h_l Unknown

A friction factor is obtained from the Moody chart using Re and e/D values computed from the given data. The head loss is computed from Eq. 8.28. The procedure is illustrated in Example Problems 8.5 through 8.7.

b. h_l, Q, and D Known. L Unknown

The procedure of case **a** is followed, except Eq. 8.28 is solved for L. The procedure is illustrated in Example Problem 8.8.

c. h_l, L, and D Known. Q Unknown

Most pipe flows of engineering interest have relatively large values of Reynolds number. Thus, even though the Reynolds number cannot be calculated because Q is not known, a good first guess for friction factor is the value in the fully rough region of Fig. 8.12. Using the assumed value for f, the energy equation, Eq. 8.24, is then solved for $\bar{V}$. The Reynolds number is computed for this value of $\bar{V}$, and the first guess for f is checked. Since f is a rather weak function of Reynolds number, more than two iterations are seldom required. A flow of this type is evaluated in Example Problem 8.9.

d. h_l, L, and Q Known. D Unknown

When a fluid handling device is available and the geometry of the piping system is known, the problem is to determine the smallest (and hence least costly) pipe size that can deliver the desired flowrate. Since the pipe diameter is unknown, neither the Reynolds number nor relative roughness can be computed, and an iterative solution is again required. Calculations begin by evaluating the head loss (approximately, if necessary) from Eq. 8.24. Since $\bar{V} = Q/A$, Eq. 8.28 is written in terms of Q, that is,

$$h_l = f\frac{L}{D}\frac{\bar{V}^2}{2} = f\frac{L}{D}\frac{1}{2}\left(\frac{Q}{A}\right)^2 = f\frac{L}{D}\frac{8}{\pi}\frac{Q^2}{D^4} = \frac{8fLQ^2}{\pi D^5}$$

Solving for D,

$$D = \left[\frac{8f\,LQ^2}{\pi h_l}\right]^{1/5} \tag{8.34}$$

Often a good first approximation to the required diameter can be obtained by guessing an approximate value for friction factor and using Eq. 8.34. The guess for f is then checked by computing Re and e/D, and using the Moody diagram, Fig. 8.12.

When calculations converge, the resulting value of D may not correspond to a standard pipe size. The next larger standard size should be selected. See Example Problem 8.10.

Some data for standard pipe sizes are given in Table 8.5. For data on extra heavy (250 psig) or double extra heavy (350 psig) pipes, consult a handbook (e.g. Ref. 2). Note that for standard pipe in nominal sizes larger than 4 in., the actual diameter and nominal diameter are essentially the same. Pipe larger than 12 in. nominal diameter is produced in 14, 16, 18, 20, 22, and 24 in. sizes, and in multiples of 6 in. for still larger sizes.

table 8.5 *Data for Commercial Wrought Steel Pipe in Standard Wall Thickness (Schedule 40, for cold water service to 125 psig, Ref. 2)*

NOMINAL PIPE SIZE (in.)	INSIDE DIAMETER (in.)
$\frac{1}{8}$	0.269
$\frac{1}{4}$	0.364
$\frac{3}{8}$	0.493
$\frac{1}{2}$	0.622
$\frac{3}{4}$	0.824
1	1.049
$1\frac{1}{2}$	1.610
2	2.067
$2\frac{1}{2}$	2.469
3	3.068
$3\frac{1}{2}$	3.548
4	4.026
5	5.047
6	6.065
8	8.071
10	10.020
12	12.090

Example 8.5

Water flows through a 6 in. diameter horizontal pipe ($e/D = 0.0002$) at a volumetric flowrate of 3.24 ft³/sec. Determine the pressure drop over a 30 ft length of the pipe. The flow is fully developed.

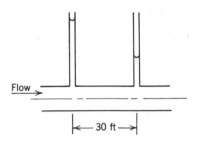

Flow

|← 30 ft →|

EXAMPLE PROBLEM 8.5

GIVEN:

Water flowing through a 6 in. diameter horizontal pipe ($e/D = 0.0002$) at a rate of 3.24 ft³/sec.

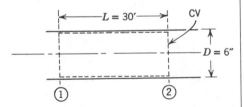

FIND:

Pressure drop over 30 ft length.

SOLUTION:

For steady incompressible flow, in the absence of work, the first law of thermodynamics may be written

$$\left(\frac{p_1}{\rho} + \frac{\bar{V}_1^2}{2} + gz_1\right) - \left(\frac{p_2}{\rho} + \frac{\bar{V}_2^2}{2} + gz_2\right) = h_{l_T} = h_l + h_{l_m}$$

where

$$h_l = f\frac{L}{D}\frac{\bar{V}^2}{2} \quad \text{and} \quad h_{l_m} = K\frac{\bar{V}^2}{2}$$

For incompressible flow, $\rho = c$, and since $A_1 = A_2$, then $\overline{V}_1 = \overline{V}_2$. Since the pipe is horizontal, $z_1 = z_2$. For a straight section of pipe, $h_{l_m} = 0$. Thus

$$\frac{p_1}{\rho} - \frac{p_2}{\rho} = h_l = f \frac{L}{D} \frac{\overline{V}^2}{2}$$

or

$$p_1 - p_2 = f \frac{L}{D} \frac{\rho \overline{V}^2}{2}$$

where

$$f = f\left(Re, \frac{e}{D}\right)$$

$$Re = \frac{\rho D \overline{V}}{\mu}$$

For water at 70 F, $\rho = 1.94 \text{ slug/ft}^3$, $\mu = 2.04 \times 10^{-5} \text{ lbf-sec/ft}^2$

$$\overline{V} = \frac{Q}{A} = \frac{Q}{\pi D^2/4} = \frac{4Q}{\pi D^2} = \frac{4}{\pi} \times \frac{3.24 \text{ ft}^3}{\text{sec}} \times \frac{1}{(0.5)^2 \text{ ft}^2} = 16.4 \frac{\text{ft}}{\text{sec}}$$

$$Re = \frac{\rho 4Q}{\mu \pi D} = \frac{4}{\pi} \times \frac{1.94 \text{ slug}}{\text{ft}^3} \times \frac{\text{ft}^2}{2.04 \times 10^{-5} \text{ lbf-sec}} \times \frac{3.24 \text{ ft}^3}{\text{sec}} \times \frac{1}{0.5 \text{ ft}} \times \frac{\text{lbf-sec}^2}{\text{ft-slug}}$$

$$Re = 7.7 \times 10^5$$

$\frac{e}{D} = 0.0002$, so from Fig. 8.12, $f = 0.015$.

Then

$$p_1 - p_2 = f \frac{L}{D} \rho \frac{\overline{V}^2}{2}$$

$$= 0.015 \times \frac{30 \text{ ft}}{0.5 \text{ ft}} \times \frac{1.94 \text{ slug}}{\text{ft}^3} \times \frac{1}{2} \times \frac{(16.4)^2 \text{ ft}^2}{\text{sec}^2} \times \frac{\text{lbf-sec}^2}{\text{slug-ft}} \times \frac{\text{ft}^2}{144 \text{ in.}^2}$$

$$p_1 - p_2 = 1.63 \text{ lbf/in.}^2 \qquad\qquad\qquad p_1 - p_2$$

$$\left\{ \begin{array}{c} \text{This problem illustrates the method for calculating the} \\ \text{head loss and pressure drop in a pipe.} \end{array} \right\}$$

Example 8.6

What level, h, must be maintained in the reservoir to produce a volumetric flowrate of 1.0 ft³/sec of water? The inside diameter of the smooth pipe is 3 in. and the pipe length is 300 ft. The loss coefficient, K, for the inlet is $K = 0.5$. The water discharges to the atmosphere.

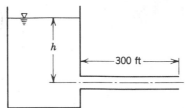

EXAMPLE PROBLEM 8.6

GIVEN:

Volumetric water flow of 1.0 ft³/sec through a 3 in. diameter pipe, with $L = 300$ ft, attached to a constant level reservoir. Inlet loss coefficient $K = 0.5$.

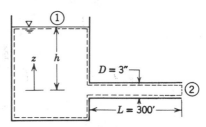

FIND:

Reservoir level, h, to maintain the flow.

SOLUTION:

For steady incompressible flow, in the absence of work, the first law of thermodynamics may be written

$$\left(\frac{p_1}{\rho} + \frac{\bar{V}_1^2}{2} + gz_1\right) - \left(\frac{p_2}{\rho} + \frac{\bar{V}_2^2}{2} + gz_2\right) = h_{l_T} = h_l + h_{l_m} \tag{1}$$

where

$$h_l = f\frac{L}{D}\frac{\bar{V}^2}{2} \qquad h_{l_m} = K\frac{\bar{V}^2}{2}$$

For the given problem, $p_1 = p_2 = p_{atm}$, $\bar{V}_1 \simeq 0$, and $\bar{V}_2 = \bar{V}$. Taking $z_2 = 0$, then $z_1 = h$.

Simplifying Eq. (1)

$$gh - \frac{\bar{V}^2}{2} = f\frac{L}{D}\frac{\bar{V}^2}{2} + K\frac{\bar{V}^2}{2}$$

EXAMPLE PROBLEM 8.6 (continued)

Then

$$h = \frac{1}{g}\left[f\frac{L}{D}\frac{\bar{V}^2}{2} + K\frac{\bar{V}^2}{2} + \frac{\bar{V}^2}{2}\right] = \frac{\bar{V}^2}{2g}\left[f\frac{L}{D} + K + 1\right]$$

Since

$$\bar{V} = \frac{Q}{A} = \frac{4Q}{\pi D^2}$$

$$h = \frac{8Q^2}{\pi^2 D^4 g}\left[f\frac{L}{D} + K + 1\right] \quad \text{where} \quad f = f\left(Re, \frac{e}{D}\right)$$

For water at 70 F, $\rho = 1.94\ \text{slug/ft}^3$, $\mu = 2.04 \times 10^{-5}\ \text{lbf-sec/ft}^2$. Thus

$$Re = \frac{\rho D \bar{V}}{\mu} = \frac{4\rho Q}{\mu \pi D}$$

$$= \frac{4}{\pi} \times \frac{1.94\ \text{slug}}{\text{ft}^3} \times \frac{1.0\ \text{ft}^3}{\text{sec}} \times \frac{\text{ft}^2}{2.04 \times 10^{-5}\ \text{lbf-sec}} \times \frac{1}{0.25\ \text{ft}} \times \frac{\text{lbf-sec}^2}{\text{slug-ft}}.$$

$$Re = 4.84 \times 10^5$$

Assume smooth pipe, then from Fig. 8.12, $f = 0.0133$

$$h = \frac{8Q^2}{\pi^2 D^4 g}\left[f\frac{L}{D} + K + 1\right]$$

$$= \frac{8}{\pi^2} \times \frac{(1)^2\ \text{ft}^6}{\text{sec}^2} \times \frac{1}{(0.25)^4\ \text{ft}^4} \times \frac{\text{sec}^2}{32.2\ \text{ft}}\left[0.0133 \times \frac{300}{0.25} + 0.5 + 1.0\right]$$

$$h = 6.42[16 + 0.5 + 1.0]\ \text{ft} = 112\ \text{ft} \qquad\qquad\qquad\qquad h$$

{This problem illustrates the method for calculating the total head loss.}

Example 8.7

Air is supplied to a steel-making process through a 6 in. duct. The duct, as installed, terminates abruptly into a large plenum chamber. A new young engineer suggests that appreciable power can be saved by replacing the existing duct, which includes two 90° bends (center line radius = 2 ft), with a combination duct-diffuser. There is room for a diffuser with area ratio, $AR = 1.6$. The proposed duct would eliminate the elbows, 8 ft of duct, and an additional piece of duct as long as the diffuser. The air speed required in the smooth duct is 150 ft/sec, and the pressure is essentially atmospheric. How much pumping power would be saved by making the changes suggested by the young engineer?

GIVEN:

Flow systems shown. $L_{new} + N + 8 \text{ ft} = L_{old}$.

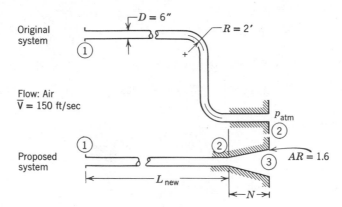

FIND:

Decrease in pumping power for new system.

SOLUTION:

Determine the head loss for both systems, then compute pumping power using the energy equation.

Basic equation:

$$h_{l_T} = h_l + h_{l_m} = f\frac{L}{D}\frac{\overline{V}^2}{2} + h_{l_m}$$

Assumptions: (1) Entrance losses the same for both systems
(2) Incompressible flow

For the proposed system,

$$h_{l_T} = f\frac{L_{new}}{D}\frac{\overline{V}_1^2}{2} + h_{l_m, \text{diffuser}} + K_{\text{exit}}\frac{\overline{V}_3^2}{2}$$

$$Re = \frac{\overline{V}D}{\nu} = \frac{150 \text{ ft}}{\text{sec}} \times \frac{1}{2}\text{ft} \times \frac{\text{sec}}{2 \times 10^{-4} \text{ ft}^2} = 375{,}000$$

For smooth pipe, $f = 0.014$, from Fig. 8.12.

EXAMPLE PROBLEM 8.7 (continued)

The best diffuser with $AR = 1.6$, from Fig. 8.15, has a pressure recovery coefficient, $C_p = 0.52$ at $N/R_1 = 7.0$. Thus $N = (7.0)R_1 = (7)(\frac{1}{4}) = 1.75$ ft. From Eq. 8.32,

$$h_{l_m,\,\text{diffuser}} = \frac{\bar{V}_2^2}{2}\left[1 - \frac{1}{(AR)^2} - C_p\right] = \frac{\bar{V}_2^2}{2}[1 - 0.39 - 0.52] = 0.09\frac{\bar{V}_2^2}{2}$$

The exit loss is

$$h_{l_m,\,\text{exit}} = K\frac{\bar{V}_3^2}{2} = K\left(\frac{A_2}{A_3}\right)^2\frac{\bar{V}_2^2}{2} = (1)\left(\frac{1}{1.6}\right)^2\frac{\bar{V}_2^2}{2} = 0.39\frac{\bar{V}_2^2}{2}$$

Combining and substituting, with $\bar{V}_1 = \bar{V}_2$,

$$h_{l_T}(\text{new}) = \left[f\frac{L_{\text{new}}}{D} + 0.09 + 0.39\right]\frac{\bar{V}_2^2}{2} = \left[f\frac{L_{\text{new}}}{D} + 0.48\right]\frac{\bar{V}_1^2}{2}$$

For the present system,

$$h_{l_T} = f\frac{L_{\text{old}}}{D}\frac{\bar{V}_1^2}{2} + 2f\frac{L_{e,\,\text{bend}}}{D}\frac{\bar{V}_1^2}{2} + K_{\text{exit}}\frac{\bar{V}_1^2}{2}$$

From Fig. 8.17, with $\dfrac{r}{D} = \dfrac{2}{(\frac{1}{2})} = 4$, $\dfrac{L_{e,\,\text{bend}}}{D} = 13.5$

Thus

$$h_{l_T}(\text{old}) = \left[f\frac{L_{\text{old}}}{D} + 27f + 1.0\right]\frac{\bar{V}_1^2}{2}$$

Combining these results,

$$h_{l_T}(\text{old}) - h_{l_T}(\text{new}) = \left[f\frac{(L_{\text{old}} - L_{\text{new}})}{D} + 27f + 0.52\right]\frac{\bar{V}_1^2}{2}$$

or

$$\Delta h_{l_T} = \left[f\frac{(N + 8)}{D} + 27f + 0.52\right]\frac{\bar{V}_1^2}{2}$$

$$= \left[0.014 \times \frac{9.75\text{ ft}}{0.5\text{ ft}} + (27)(0.014) + 0.52\right]\frac{(150)^2}{2}\text{ ft}^2/\text{sec}^2$$

$$\Delta h_{l_T} = 1.32 \times 10^4\text{ ft}^2/\text{sec}^2$$

From the energy equation, the change in power required is given by

$$\Delta\dot{W}_{\text{in}} = \dot{m}\Delta h_{l_T} = \rho A\bar{V}\Delta h_{l_T}$$

$$= \frac{0.00238\text{ slug}}{\text{ft}^3} \times \frac{\pi}{4}\left(\frac{1}{2}\right)^2\text{ ft}^2 \times \frac{150\text{ ft}}{\text{sec}} \times \frac{1.32 \times 10^4\text{ ft}^2}{\text{sec}^2} \times \frac{\text{lbf-sec}^2}{\text{slug-ft}} \times \frac{\text{hp-sec}}{550\text{ ft-lbf}}$$

$$\Delta\dot{W}_{\text{in}} = 1.68\text{ hp} \longleftarrow \qquad\qquad\qquad\qquad\qquad\qquad\qquad \Delta\dot{W}_{\text{in}}$$

{This problem illustrates use of the empirically determined minor loss coefficients.}

Example 8.8

A compressed air drill requires an air supply of 0.50 lbm/sec at a pressure of 95 psig at the drill. The hose from the air compressor to the drill is $1\frac{1}{2}$ in. inside diameter. The maximum compressor discharge pressure is 100 psig. Neglect changes in density and any effects due to hose curvature. Air leaves the compressor at 100 F. Calculate the longest hose that may be used.

EXAMPLE PROBLEM 8.8

GIVEN:

Air flow through a line of length, L, and diameter $D = 1.5$ in.

$p_1 = 100$ psig $p_2 = 95$ psig

$T_1 = 100$ F $\dot{m} = 0.50$ lbm/sec $\rho \approx$ constant

FIND:

Allowable length of hose.

SOLUTION:

For steady incompressible flow, in the absence of work, the first law of thermodynamics may be written

$$\left(\frac{p_1}{\rho} + \frac{\bar{V}_1^2}{2} + gz_1\right) - \left(\frac{p_2}{\rho} + \frac{\bar{V}_2^2}{2} + gz_2\right) = h_{l_T} = h_l + h_{l_m} \tag{1}$$

where

$$h_l = f\frac{L}{D}\frac{\bar{V}^2}{2} \qquad h_{l_m} = K\frac{\bar{V}^2}{2}$$

For $\rho = c$, then $\bar{V}_1 = \bar{V}_2$, since $A_1 = A_2$. Since p_1 and p_2 are given, neglect minor losses. Neglect changes in elevation, that is, $z_1 = z_2$. Then Eq. (1) can be written as

$$\frac{p_1 - p_2}{\rho} = f\frac{L}{D}\frac{\bar{V}^2}{2}$$

and

$$L = \frac{(p_1 - p_2)}{\rho}\frac{2D}{f\bar{V}^2}$$

$$\rho = \rho_1 = \text{constant} = \frac{p_1}{RT_1} = \frac{114.7\text{ lbf}}{\text{in.}^2} \times \frac{\text{lbm-R}}{53.3\text{ ft-lbf}} \times \frac{1}{560\text{ R}} \times \frac{144\text{ in.}^2}{\text{ft}^2} \quad \begin{cases} p_\text{atm} \text{ assumed} \\ \text{at } 14.7 \text{ psia.} \end{cases}$$

$\rho = 0.550\text{ lbm/ft}^3$

365   **solution of pipe flow problems**

EXAMPLE PROBLEM 8.8 (continued)

$\dot{m} = \rho \overline{V} A = \text{constant, so}$

$$\overline{V} = \frac{\dot{m}}{\rho A} = \frac{4\dot{m}}{\rho \pi D^2} = \frac{4}{\pi} \times \frac{0.5\,\text{lbm}}{\text{sec}} \times \frac{\text{ft}^3}{0.550\,\text{lbm}} \times \frac{1}{(1.5)^2\,\text{in.}^2} \times \frac{144\,\text{in.}^2}{\text{ft}^2}$$

$\overline{V} = 74\,\text{ft/sec}$

For air at 100 F, $\mu = 3.96 \times 10^{-7}\,\text{lbf-sec/ft}^2$.

$$Re = \frac{\rho D \overline{V}}{\mu} = \frac{0.550\,\text{lbm}}{\text{ft}^3} \times \frac{1.5\,\text{ft}}{12} \times \frac{74\,\text{ft}}{\text{sec}} \times \frac{1\,\text{ft}^2}{3.96 \times 10^{-7}\,\text{lbf-sec}} \times \frac{1\,\text{slug}}{32.2\,\text{lbm}} \times \frac{\text{lbf-sec}^2}{\text{ft-slug}}$$

$Re = 4.0 \times 10^5$

Assume smooth pipe, then from Fig. 8.12, $f = 0.0138$

$$L = \frac{(p_1 - p_2)}{\rho} \frac{2D}{f \overline{V}^2}$$

$$= \frac{5.0\,\text{lbf}}{\text{in.}^2} \times \frac{\text{ft}^3}{0.550\,\text{lbm}} \times \frac{2}{0.0138} \times \frac{1.5\,\text{ft}}{12} \times \frac{\text{sec}^2}{(74)^2\,\text{ft}^2} \times \frac{144\,\text{in.}^2}{\text{ft}^2} \times \frac{32.2\,\text{lbm}}{\text{slug}} \times \frac{\text{slug-ft}}{\text{lbf-sec}^2}$$

$L = 140\,\text{ft}$ L

{ This problem illustrates the method for calculating the head loss over a section of straight pipe, when length is unknown. Note that the relative change in density for this problem, $\Delta\rho/\rho \simeq \Delta p/p$, is only 5 percent. Thus the assumption of incompressible flow is reasonable. }

Example 8.9

A fire protection system is supplied from a water tower 80 ft tall. The longest pipe in the system is 600 ft long, and is made of cast iron about 20 years old. The pipe contains one gate valve; other minor losses may be neglected. The pipe diameter is 4 in. Determine the maximum rate of flow through this pipe, in gallons per minute.

EXAMPLE PROBLEM 8.9

GIVEN:

Fire protection system, as shown.

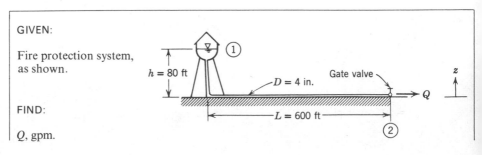

FIND:

Q, gpm.

SOLUTION:

Apply the energy equation for pipe flow (Eq. 8.24) between sections ① and ②, using the datum shown.

Basic equations:
$$\left(\frac{p_1}{\rho} + \frac{\overset{\approx\ 0(5)}{\cancel{\overline{V_1^2}}}}{2} + gz_1\right) - \left(\frac{p_2}{\rho} + \frac{\overline{V_2^2}}{2} + gz_2\right) = h_{l_T}$$

$$h_{l_T} = f\frac{L}{D}\frac{\overline{V_2^2}}{2} + h_{l_m}$$

Assumptions: (1) No shaft or other work
(2) Steady flow
(3) Incompressible flow
(4) $p_1 = p_2 = p_{atm}$
(5) $\overline{V_1} \simeq 0$
(6) $h_{l_m} = f\frac{L_e}{D}\frac{\overline{V_2^2}}{2}$; from Table 8.4, $\frac{L_e}{D} = 13$ for a fully open gate valve
(7) Neglect length of standpipe, h

Combining equations and substituting,

$$h_{l_T} = f\frac{L}{D}\frac{\overline{V_2^2}}{2} + 13f\frac{\overline{V_2^2}}{2} = g(z_1 - z_2) - \frac{\overline{V_2^2}}{2}$$

Thus

$$\frac{\overline{V_2^2}}{2}\left[f\left(\frac{L}{D} + 13\right) + 1\right] = g(z_1 - z_2)$$

or

$$\overline{V_2} = \left[\frac{2g(z_1 - z_2)}{f(L/D + 13) + 1}\right]^{1/2}; \qquad \frac{L}{D} = \frac{600}{1/3} = 1800, \qquad z_1 - z_2 = h$$

Since $\overline{V_2}$ is not known, we cannot compute Re. But we can assume a value of friction factor in the fully rough flow region. From Fig. 8.13, $e/D \simeq 0.0025$ for cast iron pipe. Since the pipe is quite old, choose $e/D = 0.004$. Then from Fig. 8.12, guess $f \simeq 0.028$. Then

$$\overline{V_2} = \left[\frac{2 \times 32.2\ \text{ft/sec}^2 \times 80\ \text{ft}}{0.028(1800 + 13) + 1}\right]^{1/2} = \left[\frac{5150}{50.8 + 1}\right]^{1/2} = [99.5]^{1/2} = 9.98\ \text{ft/sec}$$

Now check the assumed value for f.

$$Re = \frac{\rho\overline{V}D}{\mu} = \frac{\overline{V}D}{\nu} = \frac{9.98\ \text{ft}}{\text{sec}} \times \frac{1}{3}\text{ft} \times \frac{\text{sec}}{1.2 \times 10^{-5}\ \text{ft}^2} = 277,000$$

For $e/D = 0.004$, $f = 0.029$ from Fig. 8.12. Using this value,

$$\overline{V} = \left[\frac{5150}{0.029(1813) + 1}\right]^{1/2} = \left[\frac{5150}{52.6 + 1}\right]^{1/2} = [95.9]^{1/2} = 9.78\ \text{ft/sec}$$

Thus, convergence is satisfactory. The volumetric flowrate is

$$Q = \bar{V}_2 A = \bar{V}_2 \frac{\pi D^2}{4} = \frac{9.78 \text{ ft}}{\text{sec}} \times \frac{\pi}{4} \left(\frac{1}{3}\right)^2 \text{ft}^2 \times \frac{7.48 \text{ gal}}{\text{ft}^3} \times \frac{60 \text{ sec}}{\text{min}}$$

$$Q = 383 \text{ gpm} \qquad\qquad\qquad\qquad\qquad\qquad\qquad\qquad\qquad\qquad Q$$

{ This problem illustrates the procedure for solving pipe flow problems in which the
flowrate is unknown. Note that the velocity and, hence, the flowrate, is essentially
proportional to $1/\sqrt{f}$, so our choice of a larger value of e/D to account for aging was
not too critical. }

Example 8.10

Spray heads in an agricultural spraying system are to be supplied with water
from an engine-driven pump. At its most efficient operating point, the pump
output is 1500 gpm at a discharge pressure of 65 psig. The water is to be supplied
through 500 ft of pipe. For satisfactory operation, the sprinklers must operate at
30 psig or higher pressure. Minor losses and elevation changes may be neglected.
Determine the smallest standard pipe size that can be used.

EXAMPLE PROBLEM 8.10

GIVEN:

Water supply system, as shown.

FIND:

Smallest standard D.

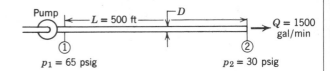

$$p_1 = 65 \text{ psig} \qquad\qquad\qquad\qquad\qquad\qquad p_2 = 30 \text{ psig}$$

SOLUTION:

Δp, L, and Q are known. D is unknown, so iteration will be required. Determine h_{l_T} from
Eq. 8.24, apply Eq. 8.34:

Computing equations: $\left(\dfrac{p_1}{\rho} + \dfrac{\bar{V}_1^2}{2} + gz_1\right) - \left(\dfrac{p_2}{\rho} + \dfrac{\bar{V}_2^2}{2} + gz_2\right) = h_{l_T}$

$$D = \left[\frac{8fLQ^2}{\pi h_l}\right]^{1/5}$$

Assumptions: (1) Steady flow
(2) Incompressible flow
(3) $h_{l_T} = h_l$, that is, $h_{l_m} = 0$
(4) $z_1 = z_2$
(5) $\dfrac{\overline{V}_1^2}{2} = \dfrac{\overline{V}_2^2}{2}$

Then

$$h_l = \frac{p_1 - p_2}{\rho} = \frac{(65 - 30)\,\text{lbf}}{\text{in.}^2} \times \frac{144\,\text{in.}^2}{\text{ft}^2} \times \frac{\text{ft}^3}{1.94\,\text{slug}} \times \frac{\text{slug-ft}}{\text{lbf-sec}^2} = 2600\,\text{ft}^2/\text{sec}^2$$

$$Q = \frac{1500\,\text{gal}}{\text{min}} \times \frac{\text{min}}{60\,\text{sec}} \times \frac{\text{ft}^3}{7.48\,\text{gal}} = 3.34\,\text{ft}^3/\text{sec}$$

Guess a trial value for f. If $D = 5$ in., then $e/D = 0.00035$ for commercial steel, from Fig. 8.13. Then from Fig. 8.12, $f \simeq 0.015$, from the fully rough flow region. Then

$$D = \left[\frac{8fLQ^2}{\pi h_l}\right]^{1/5} = \left[\frac{8}{\pi}(0.015)500\,\text{ft} \times \frac{(3.34)^2\,\text{ft}^6}{\text{sec}^2} \times \frac{\text{sec}^2}{2600\,\text{ft}^2}\right]^{1/5}$$

$$D = [0.0822]^{1/5} = 0.606\,\text{ft}$$

or

$$D = 7.28\,\text{in.}$$

Check 6 in. pipe. From Fig. 8.13, $e/D = 0.0003$, and $f \simeq 0.015$ from Fig. 8.12. These data will produce the same result. Therefore, choose $D = 8$ in. (nominal) pipe. Now check:

$$Re = \frac{\rho \overline{V} D}{\mu} = \frac{\overline{V} D}{\nu} = \frac{QD}{\nu A} = \frac{4Q}{\pi \nu D}$$

$$Re = \frac{4}{\pi} \times \frac{3.34\,\text{ft}^3}{\text{sec}} \times \frac{\text{sec}}{1.2 \times 10^{-5}\,\text{ft}^2} \times \frac{1}{8.07\,\text{in.}} \times \frac{12\,\text{in.}}{\text{ft}} = 529,000$$

$$\frac{e}{D} = 0.00022, \text{ from Fig. 8.13.}$$

Using these data, $f = 0.0154$, from Fig. 8.12. Thus

$$h_l = f\frac{L}{D}\frac{\overline{V}^2}{2} = \frac{8fLQ^2}{\pi D^5}$$

$$h_l = \frac{8}{\pi}(0.0154)500\,\text{ft} \times \frac{(3.34)^2\,\text{ft}^6}{\text{sec}^2} \times \frac{1}{(8.07/12)^5\,\text{ft}^5} = 1610\,\text{ft}^2/\text{sec}^2$$

and

$$\Delta p = \rho h_l = \frac{1.94 \text{ slug}}{\text{ft}^3} \times \frac{1610 \text{ ft}^2}{\text{sec}^2} \times \frac{\text{ft}^2}{144 \text{ in.}^2} \times \frac{\text{lbf-sec}}{\text{slug-ft}} = 21.6 \text{ lbf/in.}^2$$

Thus, the criterion on pressure drop is satisfied for

$$D = 8 \text{ in.} \quad \text{(nominal pipe size)}$$ D

**8–9.2 MULTIPLE PATH SYSTEMS

In many practical situations, such as water supply or fire protection systems, complex pipe networks must be analyzed. The basic techniques developed in Section 8–9.1 are used to analyze flow in multiple path pipe systems also. The procedure is analogous to that used in solving direct current electric circuits. A representative pipe system is shown in Fig. 8.20.

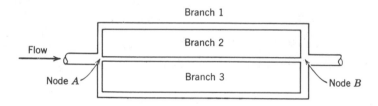

Fig. 8.20 *Simple multiple path pipe flow system.*

The pipe system shown in Fig. 8.20 has two nodes, labeled A and B, and three branches. The total flowrate into the system must be distributed among the branches. Consequently, the flowrate through each branch is unknown. However, the pressure drop for each branch is the same, that is, equal to $p_A - p_B$. This information is sufficient to permit an iterative solution for the flowrate in each branch, as shown in Example Problem 8.11.

The quantities, flowrate and pressure drop, are respectively analogous to the current and voltage in an electric circuit. However, the simple linear relation between voltage and current given by Ohm's law does not apply to the fluid flow system. Instead, the flow pressure drop is approximately proportional to the

** This section may be omitted without loss of continuity in the text material.

square of the flowrate. This nonlinearity makes iterative solutions necessary, and the resulting calculations are quite lengthy and tedious. A number of schemes have been developed for use with a digital computer. For an example, see Reference 8.

Example 8.11

The irrigation system of Example 8.10 is to be extended by adding on three branches, each made from 6 in. pipe, as shown in the sketch below. As before, the minimum pressure at the sprinklers must be at least 30 psig (it may be assumed to be the same at the end of each branch). Minor losses at elbows should be included, but elevation and kinetic energy terms may be neglected. Determine the actual pressure at B, and the flowrate in each branch, in gpm.

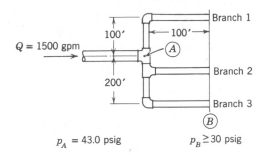

$$p_A = 43.0 \text{ psig} \qquad\qquad p_B \geq 30 \text{ psig}$$

EXAMPLE PROBLEM 8.11

GIVEN:

Pipe network shown schematically below.

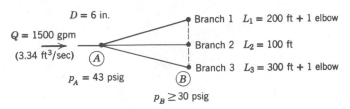

FIND:

(a) Q_1, Q_2, Q_3
(b) p_B

SOLUTION:

Apply Eq. 8.24 to each branch.

Computing equation: $\left(\dfrac{p_A}{\rho} + \dfrac{\bar{V}_A^2}{2} + gz_A\right) - \left(\dfrac{p_B}{\rho} + \dfrac{\bar{V}_B^2}{2} + gz_B\right) = h_{l_T}$

Assumptions: (1) Steady flow
(2) Incompressible flow
(3) $z_A = z_B$
(4) Neglect changes in $\bar{V}^2/2$

Therefore

$$h_{l_T} = \frac{p_A - p_B}{\rho}$$

or the head loss is the same for each branch. We know from continuity that

$$Q = Q_1 + Q_2 + Q_3$$

Formulating the head loss for each branch,

$$h_{l_{T_1}} = f\frac{L_1}{D}\frac{\bar{V}_1^2}{2} + f\frac{L_e}{D}\frac{\bar{V}_1^2}{2} = f\left(\frac{L_1}{D} + \frac{L_e}{D}\right)\frac{\bar{V}_1^2}{2} = f(400 + 30)\frac{\bar{V}_1^2}{2} = h_{l_T}$$

$$h_{l_{T_2}} = f\frac{L_2}{D}\frac{\bar{V}_2^2}{2} = f(200)\frac{\bar{V}_2^2}{2} = h_{l_T}$$

$$h_{l_{T_3}} = f\frac{L_3}{D}\frac{\bar{V}_3^2}{2} + f\frac{L_e}{D}\frac{\bar{V}_3^2}{2} = f\left(\frac{L_3}{D} + \frac{L_e}{D}\right)\frac{\bar{V}_3^2}{2} = f(600 + 30)\frac{\bar{V}_3^2}{2} = h_{l_T}$$

since the equivalent length for an elbow is $L_e/D = 30$, from Table 8.4. To obtain some idea of the flow division among branches, we could either guess, or solve for $\bar{V}_1$, $\bar{V}_2$, and $\bar{V}_3$, using an approximate value for f. For 6 in. pipe, $e/D = 0.0003$, from Fig. 8.13. Then $f \simeq 0.015$ from Fig. 8.12. Then

$$\bar{V}_1 \simeq \sqrt{\frac{2h_{l_T}}{430f}} = \sqrt{\frac{2}{(430)(0.015)}}\sqrt{h_{l_T}} = 0.56\sqrt{h_{l_T}}$$

$$\bar{V}_2 \simeq \sqrt{\frac{2h_{l_T}}{200f}} = \sqrt{\frac{2}{(200)(0.015)}}\sqrt{h_{l_T}} = 0.82\sqrt{h_{l_T}}$$

and

$$\bar{V}_3 \simeq \sqrt{\frac{2h_{l_T}}{630f}} = \sqrt{\frac{2}{(630)(0.015)}}\sqrt{h_{l_T}} = 0.46\sqrt{h_{l_T}}$$

Then

$$\bar{V}_1 + \bar{V}_2 + \bar{V}_3 = \frac{Q}{A} = 1.84\sqrt{h_{l_T}}$$

where A is the area of a 6 in. pipe. Thus

$$Q_1 = A\bar{V}_1 \simeq \left(\frac{0.56}{1.84}\right)Q = 456 \text{ gpm}$$

$$Q_2 = A\bar{V}_2 \simeq \left(\frac{0.82}{1.84}\right)Q = 669 \text{ gpm}$$

$$Q_3 = A\bar{V}_3 \simeq \left(\frac{0.46}{1.84}\right)Q = 375 \text{ gpm}$$

and $Q_1 + Q_2 + Q_3 = 1500$ gpm.

Now let us calculate the actual corresponding values of head loss.

$$A = \frac{\pi}{4}D^2 = \frac{\pi}{4}\left(\frac{1}{2}\right)^2 \text{ ft}^2 = 0.197 \text{ ft}^2$$

$$\bar{V} = \frac{Q}{A} = \frac{(Q)\text{gal}}{\text{min}} \times \frac{\text{min}}{60 \text{ sec}} \times \frac{\text{ft}^3}{7.48 \text{ gal}} \times \frac{1}{0.197 \text{ ft}^2} = \frac{Q}{88} \text{ ft/sec}$$

Then

$$\bar{V}_1 = \frac{456}{88} = 5.19 \text{ ft/sec}; \qquad Re_1 = \frac{\bar{V}_1 D}{\nu} = 2.2 \times 10^5; \qquad f_1 = 0.0178$$

$$h_{l_{T_1}} = f_1(430)\frac{\bar{V}_1^2}{2} = (0.0178)(430)\frac{(5.2)^2}{2} = 103 \text{ ft}^2/\text{sec}^2$$

$$\bar{V}_2 = \frac{669}{88} = 7.58 \text{ ft/sec}; \qquad Re_2 = \frac{\bar{V}_2 D}{\nu} = 3.2 \times 10^5; \qquad f_2 = 0.017$$

$$h_{l_{T_2}} = f_2(200)\frac{\bar{V}_2^2}{2} = (0.017)(200)\frac{(7.58)^2}{2} = 97.6 \text{ ft}^2/\text{sec}^2$$

and

$$\bar{V}_3 = \frac{375}{88} = 4.27 \text{ ft/sec}; \qquad Re_3 = \frac{\bar{V}_3 D}{\nu} = 1.8 \times 10^5; \qquad f_3 = 0.018$$

$$h_{l_{T_3}} = f_3(630)\frac{\bar{V}_3^2}{2} = (0.018)(630)\frac{(4.27)^2}{2} = 103 \text{ ft}^2/\text{sec}^2$$

Comparison of $h_{l_{T_1}}$, $h_{l_{T_2}}$ and $h_{l_{T_3}}$ shows that a slight adjustment should be made. However, the head loss values are within plus or minus 3 percent of 100 ft^2/sec^2, which is probably within the accuracy that can be expected. Thus

$$Q_1 \simeq 460 \text{gpm} \qquad\qquad\qquad\qquad Q_1$$
$$Q_2 \simeq 670 \text{ gpm} \qquad\qquad\qquad\qquad Q_2$$
$$Q_3 \simeq 370 \text{ gpm} \qquad\qquad\qquad\qquad Q_3$$

Taking $100 \text{ ft}^2/\text{sec}^2$ as the mean head loss,

$$\Delta p = \rho h_{l_T} = \frac{1.94 \text{ slug}}{\text{ft}^3} \times \frac{100 \text{ ft}^2}{\text{sec}^2} \times \frac{\text{ft}^2}{144 \text{ in.}^2} \times \frac{\text{lbf-sec}^2}{\text{slug-ft}} = 1.35 \text{ lbf/in.}^2$$

Hence $p_B = p_A - \Delta p \simeq 42 \text{ psig.}$ $\longleftarrow$ $\qquad\qquad\qquad\qquad\qquad\qquad p_B$

$\left\{\begin{array}{l}\text{This problem illustrates the general method used to solve multiple path pipe flow} \\ \text{problems.}\end{array}\right\}$

**8–9.3 NONCIRCULAR DUCTS

The empirical correlations for pipe flow may also be used for computations involving noncircular ducts, provided their cross sections are not too exaggerated. Thus ducts of square or rectangular cross section may be treated, but not if the ratio of height to width is greater than about 3 or 4.

The correlations for turbulent pipe flow developed in Section 8–8 are extended for use with noncircular geometries by introducing the hydraulic diameter,

$$D_h \equiv \frac{4A}{P_w} \tag{8.35}$$

in place of the diameter, D. In Eq. 8.35, A is the cross-section area, and P_w is the "wetted perimeter," that is, the length of wall in contact with the flowing fluid at any cross section. The factor 4 is introduced so that the hydraulic diameter will equal the duct diameter for a circular geometry. For a circular duct, $A = \pi D^2/4$ and $P_w = \pi D$, so that

$$D_h = \frac{4A}{P_w} = \frac{4\left(\frac{\pi}{4}\right)D^2}{\pi D} = D$$

For a rectangular duct of width, b, and height, h, then $A = bh$ and $P_w = 2(b + h)$, so

$$D_h = \frac{4bh}{2(b + h)}$$

Defining the *aspect ratio, ar,* as $ar = h/b$, then

$$D_h = \frac{2h}{1 + ar}$$

for rectangular ducts. For a square duct, $ar = 1$, and $D_h = h$.

** This section may be omitted without loss of continuity in the text material.

As noted, the hydraulic diameter concept can be applied in the approximate range $1/3 < ar < 3$. Under these conditions, the correlations of Section 8–8 give acceptably accurate results for rectangular ducts. Since such ducts are easy and cheap to fabricate from sheet metal, they are commonly used in air conditioning, heating, and ventilating applications. Extensive data on losses for air flow are available (e.g. see Refs. 9 and 10).

Losses due to secondary flows increase rapidly for more extreme geometries, so the correlations are not applicable to wide, flat ducts, or to ducts of triangular or other irregular shapes. One must resort to use of experimental data for specific situations when precise design information is required.

Part D. Boundary Layers

8–10 The Boundary Layer Concept

The concept of a boundary layer was first introduced by Ludwig Prandtl, a German aerodynamicist, in 1904.[7]

Prior to Prandtl's historic breakthrough, the science of fluid mechanics had been developing in two rather different directions. Theoretical hydrodynamics evolved from Euler's equations (Leonhard Euler, 1755) of motion for a frictionless, nonviscous fluid. Since the results of hydrodynamics contradicted many experimental observations, practicing engineers developed their own empirical art of hydraulics. This was based on experimental data and differed significantly from the purely mathematical approach of theoretical hydrodynamics.

Although the complete equations describing the motion of a viscous fluid (the Navier–Stokes equations developed by Navier, 1827, and independently by Stokes, 1845) were known prior to Prandtl, the mathematical difficulties in solving these equations (except for a few simple cases) prohibited a theoretical treatment of viscous flows. Prandtl showed that many viscous flows can be analyzed by dividing the flow into two regions, one close to solid boundaries, the other covering the rest of the flow. Only in the thin region adjacent to a solid boundary (the boundary layer), is the effect of viscosity important. In the region outside of the boundary layer, the effect of viscosity is negligible and the fluid may be treated as inviscid.

The boundary layer concept provided the link that had been missing between theory and practice. Furthermore the boundary layer concept permitted the solution of viscous flow problems that would have been impossible through application of the Navier–Stokes equations to the complete flow field.[8] Thus the introduction

---

[7] Prandtl presented the boundary layer concept in a paper, "Fluid Motion with Very Small Friction (in German)," before the Mathematical Congress in Heidelberg.

[8] Today, computer solutions of the Navier–Stokes equations are common.

of the boundary layer concept marked the beginning of the modern era of fluid mechanics.

The development of a boundary layer on a solid surface was discussed in Section 2–5.1. The development of a laminar boundary layer on a flat plate was illustrated in Fig. 2.10. In the boundary layer both viscous and inertia forces are important. Consequently it is not surprising that the Reynolds number (which represents the ratio of inertia to viscous forces) is significant in characterizing boundary layer flows. The characteristic length used in the Reynolds number is taken either as the length in the flow direction over which the boundary layer has grown, or as some measure of the boundary layer thickness.

As for flow in a duct, boundary layer flows may be laminar or turbulent. Furthermore, there is no unique value of Reynolds number at which transition from laminar to turbulent flow in a boundary layer occurs. Among the factors that affect boundary layer transition are: pressure gradient, surface roughness, heat transfer, body forces, and free stream disturbances. However, consideration of these effects is beyond the scope of this text.

In many real flow situations, a boundary layer develops over a long, essentially flat surface. Examples include flow over ship and submarine hulls, aircraft wings, and atmospheric motions over flat terrain. Since the basic features of all these flows are illustrated in the simpler case of flow over a flat plate, let us consider this first.

For incompressible flow over a smooth flat plate (zero pressure gradient) in the absence of heat transfer, transition from laminar to turbulent flow in the boundary layer occurs at a Reynolds number, $Re_x = \rho U x / \mu$, in the range $3 \times 10^6 < Re_x < 4 \times 10^6$. For air at standard conditions, with a free stream velocity, $U = 250$ ft/sec, this corresponds to a length, x, along the plate of $1.7 < x$ (ft) < 2.6. A qualitative picture of the boundary layer growth over a flat plate is shown in Fig. 8.21. The boundary layer is laminar for a short distance downstream from the leading edge; transition occurs over a region of the plate rather than at a single line across the plate. The transition region extends downstream to the location where the boundary

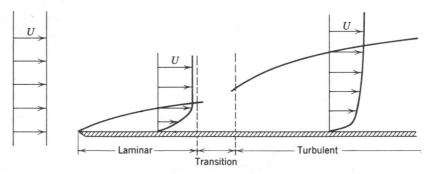

Fig. 8.21 *Boundary layer on a flat plate (vertical thickness exaggerated greatly)*

layer flow becomes completely turbulent. In the qualitative picture of Fig. 8.21, we have shown the turbulent boundary layer growing at a faster rate than the laminar layer. In later sections of this chapter we shall show that this is indeed true. The velocity profiles in the laminar and turbulent boundary layers are qualitatively the same as the respective profiles found in pipe flow (Fig. 8.9).

8–11 Displacement Thickness

The boundary layer is the region adjacent to a solid surface in which viscous forces are important. Since the velocity profile merges smoothly and asymptotically into the free stream, the boundary-layer thickness, δ, is difficult to measure. The thickness, δ, is usually defined as the distance from the surface to the point where the velocity is within 1 percent of the freestream velocity.

The effect of the boundary layer is to displace the streamlines in the flow outside the boundary layer away from the wall. A boundary-layer displacement thickness, δ^*, can be defined as the distance by which the solid surface would have to be displaced to maintain the same mass flow rate in a hypothetical frictionless flow. The displacement thickness concept is illustrated in Fig. 8.22. The actual, viscous,

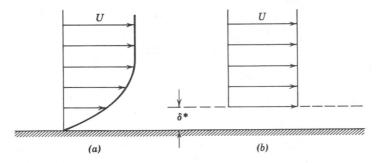

Fig. 8.22 *Definition of displacement thickness,* δ^*.

velocity profile, (a), has been replaced by the hypothetical, frictionless profile, (b), displaced by a distance, δ^*, from the wall. For a width of plate, b:

(i) The mass flow rate for the actual flow (with viscous boundary layer) is given by

$$\dot{m}_a = \int_0^\infty \rho u b \, dy$$

(ii) the mass flow rate for the hypothetical frictionless flow is given by

$$\dot{m}_b = \int_{\delta^*}^\infty \rho U b \, dy$$

The displacement thickness, δ^*, is defined such that $\dot{m}_a = \dot{m}_b$, that is,

$$\int_0^\infty \rho u b \, dy = \int_{\delta^*}^\infty \rho U b \, dy$$

For incompressible flow we can write

$$\int_0^\infty u \, dy = \int_{\delta^*}^\infty U \, dy$$

Then,

$$\int_0^\infty u \, dy = \int_0^\infty U \, dy - \int_0^{\delta^*} U \, dy$$

and

$$\int_0^{\delta^*} U \, dy = \int_0^\infty (U - u) \, dy$$

$$U\delta^* = \int_0^\infty (U - u) \, dy$$

Finally,

$$\delta^* = \int_0^\infty \left(1 - \frac{u}{U}\right) dy \qquad (8.36)$$

or

$$\delta^* \simeq \int_0^\delta \left(1 - \frac{u}{U}\right) dy$$

since $u \simeq U$ and, hence, the integrand is essentially zero for $y \geq \delta$. Application of the displacement thickness concept is illustrated in Example Problem 8.12.

Example 8.12

Air is flowing between the parallel flat plates shown below. The velocity is uniform at the entrance and has a magnitude of 16.4 ft/sec. The boundary-layer thickness has been found to be approximated by: $\delta = 0.0376 x^{0.8}$. The width of the plates, w, is much larger than the distance between the plates, h, and therefore end effects may be neglected. The boundary-layer velocity profile is given by:

$$\frac{u}{U} = \left(\frac{y}{\delta}\right)^{1/7}$$

where U is the core velocity. Note that U is a function of x.

Assume that the flow is incompressible, with $\rho = 0.00238 \text{ slug/ft}^3$.

Determine the pressure drop between the inlet and a point 16 ft downstream of the inlet, in units of lbf/ft^2.

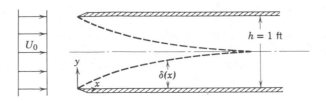

EXAMPLE PROBLEM 8.12

GIVEN:

Air flow between parallel plates
Boundary layer thickness on either wall is given by $\delta = 0.0376x^{0.8}$
The velocity profile in the boundary layer is given by

$$\frac{u}{U} = \left(\frac{y}{\delta}\right)^{1/7}$$

$\rho = \text{constant} = 2.38 \times 10^{-3} \text{ slug/ft}^3$

$U_0 = 16.4 \text{ ft/sec.}$ Uniform plate width, w.

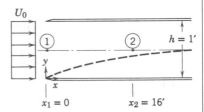

FIND:

Pressure drop, $p_1 - p_2$.

SOLUTION:

Let us first check to see if the boundary layer has reached the channel centerline at station ②. At $x = 16$ ft, $\delta = 0.0376x^{0.8} = 0.0376(16)^{0.8} = 0.0376(9.2) = 0.346$ ft. Since $\delta < h/2$, we can apply the Bernoulli equation along the center streamline.

Basic equation: $\dfrac{p_1}{\rho} + \dfrac{V_1^2}{2} + \cancel{gz_1} = \dfrac{p_2}{\rho} + \dfrac{V_2^2}{2} + \cancel{gz_2}$

Assumptions: (1) Steady flow
(2) Incompressible flow
(3) Flow along a streamline
(4) No friction outside boundary layer

displacement thickness

EXAMPLE PROBLEM 8.12 (continued)

(5) $z_1 = z_2$

Then

$$p_1 - p_2 = \frac{\rho}{2}(V_2^2 - V_1^2) = \frac{\rho V_1^2}{2}\left[\left(\frac{V_2}{V_1}\right)^2 - 1\right] = \frac{\rho U_0^2}{2}\left[\left\{\frac{U(x_2)}{U_0}\right\}^2 - 1\right]$$

We recognize that in order to satisfy continuity, the centerline velocity must increase in the direction of flow.

In order to determine U_2 we employ the notion of the displacement thickness. The continuity equation can be written

$$\rho U_1 A_1 = \rho U_2 A_2$$

For a plate width, w, then $A_1 = wh$, and $A_2 = w(h - 2\delta_2^*)$. Thus

$$\rho U_0 hw = \rho U_2 (h - 2\delta_2^*)w$$

so that

$$\frac{U_2}{U_0} = \frac{h}{h - 2\delta_2^*}$$

The task now is to calculate δ_2^*.

$$\delta_2^* = \int_0^\delta \left(1 - \frac{u}{U}\right) dy$$

Letting $\eta = y/\delta$, then $dy = \delta \, d\eta$, and

$$\delta_2^* = \delta_2 \int_0^1 \left(1 - \frac{u}{U}\right) d\eta = \delta_2 \int_0^1 (1 - \eta^{1/7}) \, d\eta = \delta_2[\eta - \tfrac{7}{8}\eta^{8/7}]_0^1 = \frac{\delta_2}{8}$$

$$\delta_2^* = \frac{0.346 \text{ ft}}{8} = 0.043 \text{ ft}$$

$$\frac{U_2}{U_0} = \frac{h}{h - 2\delta_2^*} = \frac{1.0}{1.0 - 0.086} = \frac{1.0}{0.914} = 1.095$$

$$p_1 - p_2 = \frac{\rho U_0^2}{2}\left[\left(\frac{U_2}{U_0}\right)^2 - 1\right] = \frac{2.38 \times 10^{-3}\text{slug}}{2} \times \frac{(16.4)^2 \text{ ft}^2}{\text{sec}^2} \times [(1.095)^2 - 1] \times \frac{\text{lbf-sec}^2}{\text{ft-slug}}$$

$$p_1 - p_2 = 0.064 \text{ lbf/ft}^2 \qquad\qquad\qquad (p_1 - p_2)$$

{This example illustrates the use of the displacement thickness concept.}

8–12 Momentum Integral Equation

Since all real fluids are viscous, a boundary layer will develop on all solid surfaces when there is flow over a body. Thus adjacent to any solid surface there is a region in which viscous forces are important. The natural question that arises is "how large is this viscous region?" In this section we shall develop an analysis that will enable us to determine a good approximation for the boundary-layer thickness as a function of distance along a body. Rather than solving the complete differential equations of motion for the boundary layer, we shall again apply the integral equations to a differential control volume. Our aim is to develop an equation that will enable us to predict (at least approximately) the manner in which the boundary layer grows as a function of distance along a body. We shall derive a relation that may be applied to both laminar and turbulent flow.

8–12.1 APPLICATION OF THE BASIC EQUATIONS

Consider the incompressible, steady flow over a solid surface as shown in Fig. 8.23. The boundary-layer thickness, δ, grows in some manner with increasing distance, x. For our analysis we choose a differential control volume of length, dx, width, dz, and height, δ, as shown in Fig. 8.23.

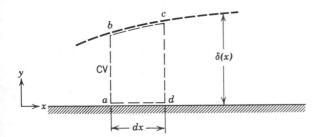

Fig. 8.23 *Differential control volume in a boundary layer.*

We wish to determine the boundary-layer thickness, δ, as a function of x. In looking at the differential control volume, *abcd*, we see that there will be mass flow across surfaces *ab* and *cd*. What about surface *bc*? Will there be a mass flow across this surface? In our earlier discussion of boundary layers (Chapter 2), we found that the edge of the boundary layer is not a streamline and, hence, it is reasonable to expect that there will be mass flow across the surface *bc*. Since the control surface *ad* is adjacent to a solid boundary, there will not be a flow across *ad*. Before proceeding to consider the forces acting on the control volume and the momentum fluxes through the control surface, let us apply the continuity equation to determine the mass flux through each portion of the control surface.

a. Continuity Equation

Basic equation:

$$0 = \overset{= \, 0(1)}{\underbrace{\frac{\partial}{\partial t} \iiint_{CV} \rho \, d\forall}} + \iint_{CS} \rho \vec{V} \cdot d\vec{A}$$

Assumptions: (1) Steady flow
(2) Two-dimensional flow

Then

$$0 = \iint_{CS} \rho \vec{V} \cdot d\vec{A} = \dot{m}_{ab} + \dot{m}_{bc} + \dot{m}_{cd}$$

or

$$\dot{m}_{bc} = -\dot{m}_{ab} - \dot{m}_{cd}$$

Now let us evaluate these terms:

Surface   *Mass Flux*

Surface *ab* is located at $x = x$. Since the flow is two-dimensional (no variation with *z*) the mass flux through *ab* is

ab

$$\dot{m}_{ab} = -\left\{ \int_0^\delta \rho u \, dy \right\} dz$$

Surface *cd* is located at $x + dx$. Expanding $\dot{m}$ in a Taylor series about the location $x = x$, we obtain

cd

$$\dot{m}_{x+dx} = \dot{m}_x + \frac{\partial \dot{m}}{\partial x}\bigg]_x dx,$$

and hence

$$\dot{m}_{cd} = \left\{ \int_0^\delta \rho u \, dy + \frac{\partial}{\partial x}\left[\int_0^\delta \rho u \, dy \right] dx \right\} dz$$

Thus, for surface *bc*, we obtain

bc

$$\dot{m}_{bc} = -\left\{ \frac{\partial}{\partial x}\left[\int_0^\delta \rho u \, dy \right] dx \right\} dz$$

b. Momentum Equation

Let us apply the x component of the momentum equation to the control volume *abcd*:

Basic equation:

$$F_{S_x} + \overset{= \, 0(3)}{\cancel{F_{B_x}}} = \overset{= \, 0(1)}{\cancel{\frac{\partial}{\partial t}}} \iiint_{CV} u\rho \, d\forall + \iint_{CS} u\rho \vec{V} \cdot d\vec{A}$$

Assumption: (3) $F_{B_x} = 0$

Then

$$F_{S_x} = \text{mf}_{ab} + \text{mf}_{bc} + \text{mf}_{cd}$$

where mf is the momentum flux.

In applying this equation to our differential control volume, *abcd*, we must obtain expressions for the momentum flux through the control surface and also the surface forces acting on the control volume. Let us consider the momentum flux first and again consider each segment of the control surface.

| *Surface* | *Momentum Flux* (mf) |

Surface *ab* is located at $x = x$. Since the flow is two-dimensional, the x momentum flux through *ab* is

ab

$$\text{mf}_{ab} = -\left\{ \int_0^\delta u\rho u \, dy \right\} dz$$

Surface *cd* is located at $x + dx$. Expanding the x momentum flux (mf) in a Taylor series about the location $x = x$, we obtain

cd

$$\text{mf}_{x+dx} = \text{mf}_x + \frac{\partial \text{mf}}{\partial x} \bigg]_x dx$$

or

$$\text{mf}_{cd} = \left\{ \int_0^\delta u\rho u \, dy + \frac{\partial}{\partial x} \left[\int_0^\delta u\rho u \, dy \right] dx \right\} dz$$

Since the mass crossing surface *bc* has velocity U, the momentum flux across *bc* is given by

bc

$$\text{mf}_{bc} = \dot{m}_{bc} U$$

$$\text{mf}_{bc} = -U \left\{ \frac{\partial}{\partial x} \left[\int_0^\delta \rho u \, dy \right] dx \right\} dz$$

From the above we can evaluate the integral over the control surface,

$$\iint_{CS} u\rho\vec{V} \cdot d\vec{A} = -\left\{\int_0^\delta u\rho u\, dy\right\} dz + \left\{\int_0^\delta u\rho u\, dy\right\} dz$$

$$+ \left\{\frac{\partial}{\partial x}\left[\int_0^\delta u\rho u\, dy\right] dx\right\} dz - U\left\{\frac{\partial}{\partial x}\left[\int_0^\delta \rho u\, dy\right] dx\right\} dz$$

Collecting terms

$$\iint_{CS} u\rho\vec{V} \cdot d\vec{A} = \left\{\frac{\partial}{\partial x}\left[\int_0^\delta u\rho u\, dy\right] dx - U\frac{\partial}{\partial x}\left[\int_0^\delta \rho u\, dy\right] dx\right\} dz$$

Now that we have a suitable expression for the integral of the x momentum flux, let's turn our attention to obtaining an expression for the surface forces acting on the control volume in the x direction. For convenience the differential control volume has been redrawn in Fig. 8.24. In analyzing the x component forces acting

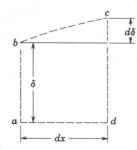

Fig. 8.24 *Differential control volume.*

on the control volume we recognize that normal forces act on three surfaces of the control surface. There is clearly a shear force acting on surface *ad*. Since the velocity gradient goes to zero at the edge of the boundary layer, there is no shear force acting along the surface *bc*.

Surface   *Force*

If the pressure at $x = x$ is p, then the force acting on surface *ab* is given by

ab $F_{ab} = p\delta\, dz$

(The boundary layer is very thin; its thickness has been greatly exaggerated in all sketches we have made. Because it is thin, pressure variations in the y direction may be neglected.)

Expanding the pressure, p, in a Taylor series, the pressure at $x + dx$ is given by

$$p_{x+dx} = p + \frac{\partial p}{\partial x}\bigg]_x dx$$

The force on surface cd is then given by

cd
$$F_{cd} = -\left(p + \frac{\partial p}{\partial x}\bigg]_x dx\right)(\delta + d\delta)\, dz$$

The average pressure acting over the surface bc is

$$p + \frac{1}{2}\frac{\partial p}{\partial x}\bigg]_x dx$$

Then the x component of the normal force acting over bc is given by

bc
$$F_{bc} = \left(p + \frac{1}{2}\frac{\partial p}{\partial x}\bigg]_x dx\right) d\delta\, dz$$

The shear force acting on ad is given by

ad
$$F_{ad} = -\tau_w\, dx\, dz$$

Summing the x component of each force acting on the control volume, we obtain

$$F_{S_x} = \left\{-\frac{\partial p}{\partial x}\delta\, dx - \overbrace{\frac{1}{2}\frac{\partial p}{\partial x}dx\, d\delta}^{\simeq 0} - \tau_w\, dx\right\} dz$$

where we note $dx\, d\delta \ll \delta\, dx$, and so neglect the second term.

On substituting the expressions for $\iint_{CS} u\rho\vec{V}\cdot d\vec{A}$ and F_{S_x} back into the momentum equation, we obtain

$$\left\{-\frac{\partial p}{\partial x}\delta\, dx - \tau_w\, dx\right\} dz = \left\{\frac{\partial}{\partial x}\left[\int_0^\delta u\rho u\, dy\right] dx - U\frac{\partial}{\partial x}\left[\int_0^\delta \rho u\, dy\right] dx\right\} dz$$

Dividing this equation through by $dx\, dz$, we obtain

$$-\delta\frac{\partial p}{\partial x} - \tau_w = \frac{\partial}{\partial x}\int_0^\delta u\rho u\, dy - U\frac{\partial}{\partial x}\int_0^\delta \rho u\, dy \qquad (8.37)$$

Equation 8.37 is a "momentum integral" equation that gives a relation between the

momentum integral equation

x components of the forces acting in a boundary layer and the momentum flux. In order to use this equation to estimate the boundary-layer thickness as a function of x we must:

1. Determine a first approximation to the pressure gradient, $\partial p/\partial x$. This is determined from inviscid flow theory, that is, the pressure gradient that would exist in the absence of a boundary layer. The velocity outside the boundary layer, the free stream velocity, U, is related to the pressure via Bernoulli's equation.
2. Assume some reasonable velocity profile shape inside the boundary layer.
3. Relate the wall shear stress to the velocity.

8–12.2 SPECIAL CASE—FLOW OVER A FLAT PLATE

To illustrate the method of using Eq. 8.37 we consider the special case of flow over a flat plate, for which $U = $ constant. From Bernoulli's equation we see that for this case, $p = $ constant, and thus $\partial p/\partial x = 0$.

The momentum integral equation then reduces to

$$\tau_w = U\frac{\partial}{\partial x}\int_0^\delta \rho u\, dy - \frac{\partial}{\partial x}\int_0^\delta u\rho u\, dy$$

Since $U = $ constant, we can write

$$\tau_w = \frac{\partial}{\partial x}\int_0^\delta U\rho u\, dy - \frac{\partial}{\partial x}\int_0^\delta u\rho u\, dy$$

The right-hand side can be combined into a single term

$$\tau_w = \frac{\partial}{\partial x}\int_0^\delta \rho u(U - u)\, dy$$

or

$$\tau_w = U^2\frac{\partial}{\partial x}\int_0^\delta \rho \frac{u}{U}\left(1 - \frac{u}{U}\right) dy$$

For incompressible flow,

$$\tau_w = \rho U^2\frac{\partial}{\partial x}\int_0^\delta \frac{u}{U}\left(1 - \frac{u}{U}\right) dy \qquad (8.38)$$

The velocity distribution, u/U, in the boundary layer is normally obtained as a function of y/δ. (Note that u/U is dimensionless and δ is a function of x only.) Consequently, it is convenient to change the variable of integration from y to

y/δ. Defining

$$\eta = \frac{y}{\delta}$$

then

$$dy = \delta \, d\eta$$

and the momentum integral equation for a zero pressure gradient is written

$$\tau_w = \rho U^2 \frac{d\delta}{dx} \int_0^1 \frac{u}{U}\left(1 - \frac{u}{U}\right) d\eta \qquad (8.39)$$

Equation 8.39 (or Eq. 8.38) was obtained by applying the basic equations (i.e. continuity and momentum) to a differential control volume. Reviewing the assumptions we made in the derivation, we see that these equations are subject to the restrictions:

1. Steady flow
2. Incompressible flow
3. Two-dimensional flow
4. No body forces
5. Flat plate flow (i.e. $\partial p / \partial x = 0$)

However, we have not made any specific assumption relating the wall shear stress, τ_w, to the velocity field. Thus, Eqs. 8.38 and 8.39 are valid for either a laminar or turbulent boundary layer flow on a flat plate.

8–13 Use of the Momentum Integral Equation for Zero Pressure Gradient Flow

The momentum integral equation for zero pressure gradient flow is given by

$$\tau_w = \rho U^2 \frac{\partial}{\partial x} \int_0^\delta \frac{u}{U}\left(1 - \frac{u}{U}\right) dy \qquad (8.38)$$

or

$$\tau_w = \rho U^2 \frac{d\delta}{dx} \int_0^1 \frac{u}{U}\left(1 - \frac{u}{U}\right) d\eta, \qquad \text{where } \eta = \frac{y}{\delta} \qquad (8.39)$$

We wish to solve this equation for the boundary-layer thickness as a function of x. In order to do this we must:

1. Assume a velocity distribution in the boundary layer, that is, a functional relationship

$$\frac{u}{U} = f\left(\frac{y}{\delta}\right)$$

use of the momentum integral equation for zero pressure gradient flow

(a) The assumed velocity distribution should satisfy certain physical boundary conditions:

$$\text{at} \quad y = 0, \qquad u = 0$$

$$\text{at} \quad y = \delta, \qquad u = U$$

$$\text{at} \quad y = \delta, \qquad \frac{\partial u}{\partial y} = 0$$

(b) Note that once the velocity distribution has been assumed, then the numerical value of the integral in Eq. 8.39 is simply

$$\int_0^1 \frac{u}{U}\left(1 - \frac{u}{U}\right) d\eta = \text{a constant} = \beta$$

and the momentum integral equation becomes

$$\tau_w = \rho U^2 \frac{d\delta}{dx} \beta$$

2. Obtain an expression for τ_w in terms of δ. This will then permit us to solve for $\delta(x)$.

8–13.1 LAMINAR FLOW

For laminar flow over a flat plate, a reasonable assumption for the velocity profile is a polynomial in y:

$$u = a + by + cy^2$$

The physical boundary conditions are:

$$\text{at} \quad y = 0, \qquad u = 0$$

$$\text{at} \quad y = \delta, \qquad u = U$$

$$\text{at} \quad y = \delta, \qquad \frac{\partial u}{\partial y} = 0$$

Evaluating the constants, a, b, and c gives

$$\frac{u}{U} = 2\left(\frac{y}{\delta}\right) - \left(\frac{y}{\delta}\right)^2 = 2\eta - \eta^2 \tag{8.40}$$

Since the wall shear stress for laminar flow is given by

$$\tau_w = \mu \left.\frac{\partial u}{\partial y}\right|_{y=0}$$

then on substituting the assumed velocity profile, Eq. 8.40, into this expression for τ_w we obtain

$$\tau_w = \mu \frac{\partial u}{\partial y}\bigg]_{y=0} = \mu \frac{U\partial(u/U)}{\delta\partial(y/\delta)}\bigg]_{y/\delta=0} = \frac{\mu U}{\delta} \frac{d(u/U)}{d\eta}\bigg]_{\eta=0}$$

or

$$\tau_w = \frac{2\mu U}{\delta}$$

We are now in a position to apply the momentum integral equation

$$\tau_w = \rho U^2 \frac{d\delta}{dx} \int_0^1 \frac{u}{U}\left(1 - \frac{u}{U}\right) d\eta \tag{8.39}$$

On substituting for τ_w and u/U we obtain

$$\frac{2\mu U}{\delta} = \rho U^2 \frac{d\delta}{dx} \int_0^1 (2\eta - \eta^2)(1 - 2\eta + \eta^2) \, d\eta$$

or

$$\frac{2\mu U}{\delta\rho U^2} = \frac{d\delta}{dx} \int_0^1 (2\eta - 5\eta^2 + 4\eta^3 - \eta^4) \, d\eta$$

Carrying out the integration and substituting limits, we obtain

$$\frac{2\mu}{\delta\rho U} = \frac{2}{15} \frac{d\delta}{dx}$$

or

$$\delta \, d\delta = \frac{15\mu}{\rho U} dx$$

which is a differential equation for δ. Integrating

$$\frac{\delta^2}{2} = \frac{15\mu}{\rho U} x + c$$

Assuming $\delta = 0$ at $x = 0$, then $c = 0$ and thus

$$\delta = \sqrt{\frac{30\mu}{\rho U}} x$$

or

$$\frac{\delta}{x} = \sqrt{\frac{30\mu}{\rho U x}} = \frac{5.48}{\sqrt{Re_x}} \tag{8.41}$$

use of the momentum integral equation for zero pressure gradient flow

Equation 8.41 expresses the laminar boundary-layer thickness on a flat plate as a function of distance along the plate. It has the same form as an exact solution derived from the complete differential equations of motion by H. Blasius in 1908.[9] Remarkably, Eq. 8.41 is only in error (the constant is too large) by about 10 percent.

Once we know the rate of boundary layer growth, we can compute all properties of the flow. The wall shear stress, or "skin friction," coefficient is defined as

$$C_f \equiv \frac{\tau_w}{\frac{1}{2}\rho U^2}$$

Substituting from the velocity profile and Eq. 8.41,

$$C_f = \frac{\tau_w}{\frac{1}{2}\rho U^2} = \frac{2\mu(U/\delta)}{\frac{1}{2}\rho U^2} = \frac{4\mu}{\rho U\delta} = 4\frac{\mu}{\rho Ux}\frac{x}{\delta} = 4\frac{1}{Re_x}\frac{\sqrt{Re_x}}{5.48}$$

Finally

$$C_f = \frac{0.730}{\sqrt{Re_x}} \tag{8.42}$$

With knowledge of τ_w, the viscous drag on the surface can be evaluated by integration over the area of the flat plate.

From Eq. 8.41 we can also compute the thickness of the laminar boundary layer at transition (Section 8–10). At $Re_x = 3 \times 10^6$, $x = 1.9$ ft for air at standard conditions, with $U = 250$ ft/sec. Thus

$$\frac{\delta}{x} = \frac{5.48}{\sqrt{Re_x}} = \frac{5.48}{\sqrt{3 \times 10^6}} = \frac{5.48}{1.73 \times 10^3} = 0.00316; \qquad \delta = 0.0060 \text{ ft}$$

At $Re_x = 4 \times 10^6$, $x = 2.6$ ft and

$$\frac{\delta}{x} = \frac{5.48}{\sqrt{Re_x}} = \frac{5.48}{\sqrt{4 \times 10^6}} = \frac{5.48}{2.0 \times 10^3} = 0.00274; \qquad \delta = 0.0071 \text{ ft}$$

In both of these cases, we have seen that the boundary-layer thickness was less than 1 percent of x. These calculations confirm that viscous effects are confined to a very narrow layer near the surface of a body.

8–13.2 TURBULENT FLOW

For turbulent flow over a flat plate, a reasonable assumption for the velocity profile is

$$\frac{u}{U} = \left(\frac{y}{\delta}\right)^{1/7} = \eta^{1/7}$$

[9] H. Blasius, "Boundary layers in fluids with small friction (in German)", Z. Math. u. Phys., **56**, 1, 1908. An English translation is available as NACA Technical Memorandum No. 1256.

However, this profile does not hold in the immediate vicinity of the wall, since at the wall it predicts $du/dy = \infty$. Consequently we cannot use this profile in the definition of τ_w to obtain an expression for τ_w in terms of δ as we did for laminar flow. For turbulent flow we will use the experimentally determined result (Ref. 14, p. 361):

$$\tau_w = 0.0225\rho U^2 \left(\frac{\nu}{U\delta}\right)^{1/4} \tag{8.43}$$

We are now in a position to apply the momentum integral equation

$$\tau_w = \rho U^2 \frac{d\delta}{dx} \int_0^1 \frac{u}{U}\left(1 - \frac{u}{U}\right) d\eta \tag{8.39}$$

On substituting for τ_w and u/U we obtain

$$0.0225\left(\frac{\nu}{U\delta}\right)^{1/4} = \frac{d\delta}{dx} \int_0^1 \eta^{1/7}(1 - \eta^{1/7})\, d\eta = \frac{7}{72} \frac{d\delta}{dx}$$

Thus we obtain a differential equation for δ:

$$\delta^{1/4}\, d\delta = 0.232 \left(\frac{\nu}{U}\right)^{1/4} dx$$

Integrating,

$$\tfrac{4}{5}\delta^{5/4} = 0.232 \left(\frac{\nu}{U}\right)^{1/4} x + c$$

Assuming $\delta \simeq 0$ at $x = 0$ (this is a reasonable assumption since δ has been shown to be so small), then $c = 0$ and

$$\delta = 0.376 \left(\frac{\nu}{U}\right)^{1/5} x^{4/5} \tag{8.44}$$

or

$$\frac{\delta}{x} = 0.376 \left(\frac{\nu}{Ux}\right)^{1/5} = \frac{0.376}{Re_x^{1/5}} \tag{8.45}$$

Using Eq. 8.45, we obtain the skin friction coefficient in terms of δ:

$$C_f = \frac{\tau_w}{\tfrac{1}{2}\rho U^2} = 0.0450 \left(\frac{\nu}{U\delta}\right)^{1/4}$$

On substituting for δ, we obtain

$$C_f = \frac{\tau_w}{\tfrac{1}{2}\rho U^2} = \frac{0.0576}{Re_x^{1/5}} \tag{8.46}$$

391   **use of the momentum integral equation for zero pressure gradient flow**

Experiments show that Eq. 8.46 predicts turbulent skin friction on a flat plate within about 3 percent. This agreement is remarkable in view of the approximate nature of our analysis.

In concluding this section, let us point out that use of the momentum integral equation is an approximate technique for predicting boundary layer development. The agreement we have obtained with experimental results shows that it is an effective method, which gives us considerable insight into the general behavior of boundary layers.

Example 8.13

Consider two-dimensional, laminar flow along a flat plate. The velocity profile in the boundary layer is assumed to be sinusoidal, that is,

$$\frac{u}{U} = \sin\left(\frac{\pi}{2}\frac{y}{\delta}\right)$$

Find an expression for:
(a) The rate of growth of δ as a function of x.
(b) The displacement thickness, δ^*, as a function of x.
(c) The total friction force on a plate of length, L, and width, b.

EXAMPLE PROBLEM 8.13

GIVEN:

Two-dimensional, laminar flow along a flat plate. The boundary-layer velocity profile is

$$\frac{u}{U} = \sin\left(\frac{\pi}{2}\frac{y}{\delta}\right) \qquad \text{for } 0 \le y \le \delta$$

and

$$\frac{u}{U} = 1 \qquad \text{for } y > \delta$$

FIND:

(a) $\delta(x)$
(b) $\delta^*(x)$
(c) Total friction force on a plate of length, L, and width, b

SOLUTION:

For flat plate flow, $\partial p/\partial x = 0$ and

$$\tau_w = \rho U^2 \frac{d\delta}{dx} \int_0^1 \frac{u}{U}\left(1 - \frac{u}{U}\right) d\eta \qquad (8.39)$$

Substituting $\dfrac{u}{U} = \sin\dfrac{\pi}{2}\eta$, then

$$\tau_w = \rho U^2 \frac{d\delta}{dx} \int_0^1 \sin\frac{\pi}{2}\eta\left(1 - \sin\frac{\pi}{2}\eta\right) d\eta$$

$$= \rho U^2 \frac{d\delta}{dx} \int_0^1 \left(\sin\frac{\pi}{2}\eta - \sin^2\frac{\pi}{2}\eta\right) d\eta$$

$$= \rho U^2 \frac{d\delta}{dx} \frac{2}{\pi}\left[-\cos\frac{\pi}{2}\eta - \frac{1}{2}\frac{\pi}{2}\eta + \frac{1}{4}\sin\pi\eta\right]_0^1$$

$$= \rho U^2 \frac{d\delta}{dx} \frac{2}{\pi}\left[0 + 1 - \frac{\pi}{4} + 0 + 0 - 0\right]$$

$$\tau_w = \rho U^2 \frac{d\delta}{dx} \frac{2}{\pi}\left[1 - \frac{\pi}{4}\right] = \frac{0.43}{\pi}\rho U^2\frac{d\delta}{dx} = \beta\rho U^2\frac{d\delta}{dx}$$

Now

$$\tau_w = \mu\frac{\partial u}{\partial y}\bigg|_{y=0} = \mu\frac{U}{\delta}\frac{\partial(u/U)}{\partial(y/\delta)}\bigg|_{y=0} = \mu\frac{U}{\delta}\frac{\pi}{2}\cos\frac{\pi}{2}\eta\bigg|_{\eta=0} = \frac{\pi\mu U}{2\delta}$$

Therefore

$$\tau_w = \frac{\pi\mu U}{2\delta} = \frac{0.43}{\pi}\rho U^2\frac{d\delta}{dx}$$

Separating variables

$$\frac{0.86}{\pi^2}\frac{\rho U}{\mu}\delta\, d\delta = dx$$

Integrating

$$\frac{0.86}{\pi^2}\frac{\rho U}{\mu}\frac{\delta^2}{2} = x + c$$

But $c = 0$, since $\delta = 0$ at $x = 0$, so

$$\delta = \sqrt{\frac{2\pi^2}{0.86}}\sqrt{\frac{x\mu}{\rho U}}$$

or

$$\frac{\delta}{x} = \sqrt{\frac{\pi^2}{0.86}}\sqrt{\frac{\mu}{\rho Ux}} = \frac{4.77}{\sqrt{Re_x}} \qquad \longleftarrow \qquad \delta(x)$$

The displacement thickness, δ^*, is given by

$$\delta^* = \delta \int_0^1 \left(1 - \frac{u}{U}\right) d\eta$$

$$= \delta \int_0^1 \left(1 - \sin\frac{\pi}{2}\eta\right) d\eta = \delta \left[\eta + \frac{2}{\pi}\cos\frac{\pi}{2}\eta\right]_0^1$$

$$\delta^* = \delta \left[1 - 0 + 0 - \frac{2}{\pi}\right] = \delta \left[1 - \frac{2}{\pi}\right]$$

Since, from part (a),

$$\frac{\delta}{x} = \frac{4.77}{\sqrt{Re_x}}$$

then

$$\frac{\delta^*}{x} = \left(1 - \frac{2}{\pi}\right)\frac{4.77}{\sqrt{Re_x}} \longleftarrow \qquad \delta^*(x)$$

The total friction force on one side of the plate is given by

$$F = \int_{A_p} \tau_w \, dA$$

Since $dA = b\,dx$ and $0 \le x \le L$, then

$$F = \int_0^L \tau_w b \, dx$$

Now

$$\tau_w = \rho U^2 \frac{d\delta}{dx} \int_0^1 \frac{u}{U}\left(1 - \frac{u}{U}\right) d\eta = \rho U^2 \frac{d\delta}{dx}\beta$$

and

$$F = \int_0^L \tau_w b \, dx = \int_0^L \rho U^2 \beta \frac{d\delta}{dx} b \, dx = \rho U^2 b\beta \int_0^L \frac{d\delta}{dx} dx = \rho U^2 \beta b \delta_L$$

From part (a), $\beta = \dfrac{0.43}{\pi}$ and $\delta_L = \dfrac{4.77L}{\sqrt{Re_L}}$, so

$$F = \frac{0.653\rho U^2 bL}{\sqrt{Re_L}} \longleftarrow \qquad F$$

$\left\{ \begin{array}{l} \text{This problem illustrates the application of the momentum integral equation to a flat} \\ \text{plate boundary layer flow.} \end{array} \right\}$

8–14 Pressure Gradients in Boundary-Layer Flow

We have restricted our discussion of boundary-layer flows to flow over a flat plate, that is, to flow in which the pressure gradient is zero. The momentum integral equation for this case was given as

$$\tau_w = \rho U^2 \frac{\partial}{\partial x} \int_0^\delta \frac{u}{U}\left(1 - \frac{u}{U}\right) dy \tag{8.38}$$

Recall that in deriving this equation, no assumption was made regarding the flow in the boundary layer; the equation is valid for both laminar and turbulent boundary layers.

Let us again consider the flow over a flat plate (Fig. 8.25). Equation 8.38 indicates that the wall shear stress is balanced by a decrease in fluid momentum. Thus the velocity profiles change as we move along the plate. The boundary layer thickness continues to increase and the fluid close to the wall is continually being slowed down (i.e. losing momentum).

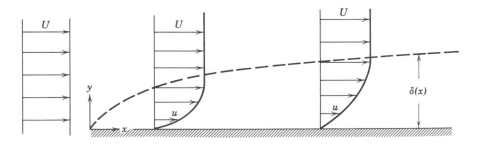

Fig. 8.25 *Laminar boundary layer on a flat plate.*

One question of interest is, "Will the fluid close to the wall ever be brought to rest?" That is, for the case of $\partial p/\partial x = 0$, is it possible that $\partial u/\partial y)_{y=0} = 0$?[10]

If we consider the wall shear stress distributions obtained for flat plates, we found that for laminar flow

$$\frac{\tau_w(x)}{\rho U^2} = \frac{\text{constant}}{\sqrt{Re_x}}$$

[10] Note that if $\dfrac{\partial u}{\partial y}\Big|_{y=0} = 0$, then the fluid layer near the wall will have a zero velocity since

$$u\Big)_{0+dy} = u_0 + \frac{\partial u}{\partial y}\Big|_{y=0} dy$$

and

$$u_0 = 0.$$

and for turbulent flow

$$\frac{\tau_w(x)}{\rho U^2} = \frac{\text{constant}}{(Re_x)^{1/5}}$$

Recalling that $\tau_w = \mu(\partial u/\partial y)_{y=0}$, we can then say that for any finite length plate, $\partial u/\partial y)_{y=0}$ will never be zero. The point on a solid boundary at which $\partial u/\partial y = 0$ is defined as the point of separation. Consequently we can conclude that for $\partial p/\partial x = 0$, the flow will not separate, that is, the fluid layer in the neighborhood of a solid surface cannot be brought to zero velocity.

8–14.1 EFFECT OF PRESSURE GRADIENT ON FLOW: SEPARATION

The pressure gradient is said to be adverse if the pressure increases in the direction of flow (i.e. if $\partial p/\partial x > 0$). When $\partial p/\partial x < 0$—that is, when the pressure decreases in the direction of flow—the pressure gradient is said to be favorable.

Consider the flow through a channel of variable cross section as shown in Fig. 8.26. To simplify our discussion we consider the flow along the straight wall.

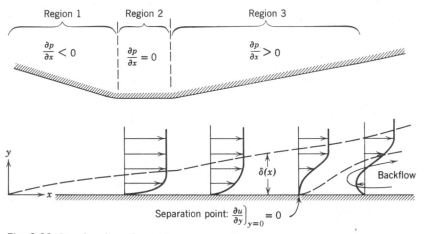

Fig. 8.26 *Boundary-layer flow with pressure gradient.*

If we consider the forces acting on a fluid particle close to the solid boundary, we see that there is a net retarding shear force on the particle no matter what the sign of the pressure gradient. For $\partial p/\partial x = 0$, the result is a decrease in momentum, but as we have already shown it is not sufficient to bring the particle to rest. Since $\partial p/\partial x < 0$ in Region ①, the pressure behind the particle (aiding its motion) is greater than that opposing its motion; hence the particle is "sliding down a pressure hill," without danger of being slowed to zero velocity. However in attempting to flow through Region ③, the particle encounters an adverse pressure gradient, $\partial p/\partial x > 0$, and the particle must "climb a pressure hill." The fluid particle could be brought to rest, thus causing the neighboring fluid to be deflected out away from the boundary; the flow is said to separate from the surface. The point on the boundary where $\partial u/\partial y)_{y=0} = 0$ is called the point of separation. Just downstream from the point of separation, the flow direction in the separated region is actually opposite to the main flow direction. The low energy fluid in the separated region is forced back upstream by the increased pressure downstream.

Thus, we see that an adverse pressure gradient, $\partial p/\partial x > 0$, is a necessary condition for separation. Does this mean that if $\partial p/\partial x > 0$, we will have separation? No, it does not. We have not shown that $\partial p/\partial x > 0$ will always lead to separation, but rather have reasoned that we cannot have separation unless $\partial p/\partial x > 0$. This conclusion can be shown rigorously using the complete differential equations for boundary-layer flow.

The turbulent boundary layer has a much fuller, or more blunt velocity profile than the laminar boundary layer, Fig. 8.21. Therefore the turbulent velocity profile at the same freestream speed contains much more momentum. Separation occurs when the momentum of fluid layers near the surface is reduced to zero by the combined action of pressure and viscous forces. Because the turbulent layer has more momentum, it is better able to resist separation in an adverse pressure gradient. We will discuss some consequences of this behavior in Section 8–15.3.

8–14.2 DETERMINATION OF PRESSURE GRADIENT

Since the pressure gradient has such a pronounced effect on the flow behavior, it is important that we be able to determine analytically the magnitude of the pressure gradient for a given flow situation. In fact, to calculate the growth and predict the behavior of a boundary layer, we must have an expression for the pressure gradient.

For both external and internal flows we can obtain a first approximation for the pressure gradient from ideal flow theory, that is, from the pressure variation in the flow of a frictionless (inviscid) fluid under the same conditions. As pointed out in Chapter 5, for frictionless irrotational flow (i.e. potential flow), the stream function, ψ, and the velocity potential, ϕ, satisfy Laplace's equation. These together with the Euler equations provide the basis for determining the pressure distribution. However, a detailed discussion of potential flow is beyond the scope of this text.[11]

[11] An introduction to potential flow is presented in many fluid mechanics texts, for example: Hansen, A. G., *Fluid Mechanics* (New York: Wiley, 1967); Li, W. H. and Lam, S. H., *Principles of Fluid Mechanics* (Reading, Mass.: Addison-Wesley, 1964); Sabersky, R. H., Acosta, A. J., and Hauptmann, E. G., *Fluid Flow*, 2nd ed. (New York: Macmillan, 1971).

Anyone interested in a detailed study of potential flow theory may find the following books of interest: Streeter, V. L., *Fluid Dynamics* (New York: McGraw-Hill, 1948); Valentine, H. R., *Applied Hydrodynamics* (London: Butterworths, 1959); Robertson, J. M., *Hydrodynamics in Theory and Application* (Englewood Cliffs, N.J.: Prentice-Hall, 1965).

Since we have had some experience with internal flows, let us consider the incompressible flow through the plane wall diffuser shown in Fig. 8.27. We are interested in determining the pressure distribution along the diffuser, and as a first approximation we will use the pressure distribution obtained for the flow of an inviscid fluid.

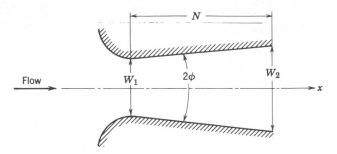

Fig. 8.27 *Two-dimensional plane wall diffuser.*

Let us apply the x component of the Euler equations to this flow:

Basic equation:

$$\rho \overset{=\,0(2)}{\cancel{B_x}} - \frac{\partial p}{\partial x} = \rho \left[\overset{=\,0(3)}{\cancel{\frac{\partial u}{\partial t}}} + u\frac{\partial u}{\partial x} + \overset{=\,0(4)}{v\cancel{\frac{\partial u}{\partial y}}} + \overset{=\,0(4)}{w\cancel{\frac{\partial u}{\partial z}}} \right] \tag{6.1a}$$

Assumptions: (1) Inviscid flow
(2) $B_x = 0$
(3) Steady flow
(4) One-dimensional flow: $u = V(x)$, $\partial/\partial y = \partial/\partial z = 0$
(5) Incompressible flow

Then

$$\frac{\partial p}{\partial x} = -\rho V \frac{\partial V}{\partial x}$$

or

$$\frac{dp}{dx} = -\rho V \frac{dV}{dx} \tag{8.47}$$

since all properties are functions of x only. Under these assumptions, the continuity

8/incompressible viscous flow

equation becomes

$$0 = \underbrace{\frac{\partial}{\partial t} \iiint\limits_{CV} \rho \, d\forall}_{= \; 0(3)} + \iint\limits_{CS} \rho \vec{V} \cdot d\vec{A}$$

or

$$\rho V A = \dot{m} = \text{constant}$$

Then

$$V A = \frac{\dot{m}}{\rho} = \text{constant} \tag{8.48}$$

Differentiating with respect to x,

$$V \frac{dA}{dx} + A \frac{dV}{dx} = 0$$

or

$$V \frac{dV}{dx} = -\frac{V^2}{A} \frac{dA}{dx} \tag{8.49}$$

Substituting Eq. 8.49 into Eq. 8.47, we obtain

$$\frac{dp}{dx} = \rho \frac{V^2}{A} \frac{dA}{dx}$$

and replacing V from Eq. 8.48 gives

$$\frac{dp}{dx} = \frac{\dot{m}^2}{\rho A^3} \frac{dA}{dx} \tag{8.50}$$

For a two-dimensional, plane wall diffuser of uniform depth, b, the cross section area varies linearly with x. From the geometry of Fig. 8.27,

$$A = Wb = (W_1 + 2x \tan \phi)b$$

Hence

$$\frac{dA}{dx} = 2b \tan \phi$$

which is constant for any given diffuser geometry. As a result, the pressure gradient is proportional to $1/A^3$, as shown in Fig. 8.28. Thus the pressure rises steeply near the diffuser inlet, but less steeply near the outlet section.

399 **pressure gradients in boundary-layer flow**

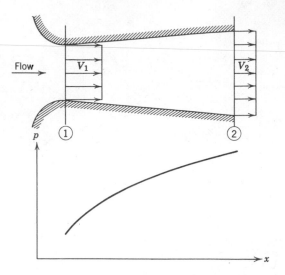

Fig. 8.28 *Pressure distribution for steady, inviscid incompressible flow in a plane wall diffuser.*

As discussed in Section 8-8.2, it is convenient to express the pressure rise through a diffuser in terms of a dimensionless pressure recovery coefficient, C_p. where

$$C_p = \frac{p_2 - p_1}{\frac{1}{2}\rho \bar{V}_1^2} \tag{8.31}$$

Applying the Bernoulli equation to the diffuser flow of Fig. 8.27,

$$\frac{p_1}{\rho} + \frac{V_1^2}{2} = \frac{p_2}{\rho} + \frac{V_2^2}{2}$$

or

$$p_2 - p_1 = \frac{\rho V_1^2}{2} - \frac{\rho V_2^2}{2} = \frac{\rho V_1^2}{2}\left[1 - \left(\frac{V_2}{V_1}\right)^2\right]$$

and

$$C_p = 1 - \left(\frac{V_2}{V_1}\right)^2$$

The velocity ratio can be eliminated, using Eq. 8.48, to obtain

$$C_p = 1 - \left(\frac{A_1}{A_2}\right)^2 = 1 - \frac{1}{AR^2} \tag{8.51}$$

where AR is the area ratio of the diffuser. As Eq. 8.51 shows, the recovery coefficient for steady, incompressible, inviscid flow through a diffuser is solely a function of the diffuser geometry. However, in practice, frictional effects are not negligible, and experimental data, such as those shown in Fig. 8.17, must be used for design.

Part E. Fluid Flow about Immersed Bodies

Whenever there is relative motion between a solid body and the fluid in which it is immersed, then the body experiences a net force, $\vec{F}$, due to the action of the fluid. In general, the infinitesimal force, $d\vec{F}$, acting on an element of surface area will be neither normal to nor in the plane of the surface. This can be seen clearly when one considers the nature of the surface forces that contribute to the force, $\vec{F}$. If the body is moving through a viscous fluid, then both shear and pressure forces act on the body, that is,

$$\vec{F} = \iint_{\substack{\text{Body} \\ \text{surface}}} d\vec{F} = \iint_{\substack{\text{Body} \\ \text{surface}}} d\vec{F}_{\text{shear}} + \iint_{\substack{\text{Body} \\ \text{surface}}} d\vec{F}_{\text{pressure}}$$

The resultant force, $\vec{F}$, can be resolved into components parallel and perpendicular to the direction of motion. The component of force parallel to the direction of motion is called the drag force, F_D, and the force component perpendicular to the direction of motion is called the lift force, F_L.

Recognizing that

$$d\vec{F}_{\text{shear}} = \vec{\tau}_w \, dA$$

and

$$d\vec{F}_{\text{pressure}} = -p \, d\vec{A}$$

one might be inclined to think that drag and lift could be evaluated analytically. This proves not to be so; there are very few cases in which the lift and drag can be determined without recourse to experimental results. As we have seen, the presence of an adverse pressure gradient often leads to separation; flow separation prohibits the analytical determination of the force acting on a body. Therefore for most shapes of interest, we must resort to the use of experimentally measured coefficients for lift and drag computations.

8–15 Drag

The drag force is the component of force on a body acting parallel to the direction of motion. In discussing the need for experimental results in fluid mechanics (Chapter 7, Dimensional Analysis and Similitude) we considered the problem of

401   **drag**

determining the drag force, F_D, on a smooth sphere of diameter, d, moving through a viscous, incompressible fluid with velocity, V; the fluid density and viscosity were ρ and μ, respectively. The drag force, F_D, was written as

$$F_D = f_1(d, V, \mu, \rho)$$

Application of the Buckingham Pi theorem resulted in two dimensionless Π parameters that can be written in functional form as

$$\frac{F_D}{\rho V^2 d^2} = f_2\left(\frac{\rho V d}{\mu}\right)$$

Note that d^2 is proportional to the cross-section area ($A = \pi d^2/4$) and, hence, we could write

$$\frac{F_D}{\rho V^2 A} = f_3\left(\frac{\rho V d}{\mu}\right) = f_3(Re) \tag{8.52}$$

While Eq. 8.52 was obtained for a sphere, the form of the equation is valid for incompressible flow over any body; the characteristic length employed in the Reynolds number depends on the particular body shape.

The drag coefficient, C_D, is defined as

$$C_D \equiv \frac{F_D}{\frac{1}{2}\rho V^2 A} \tag{8.53}$$

The number $\frac{1}{2}$ has been inserted (as was done in the defining equation for the friction factor) so as to form the familiar dynamic pressure. Then, Eq. 8.52 can be written as

$$C_D = f(Re) \tag{8.54}$$

Note that we have not considered compressibility or free surface effects in this discussion of the drag force. Had these been included we would have obtained the functional form

$$C_D = f(Re, Fr, M)$$

At this point we shall consider the drag force on several bodies for which Eq. 8.54 is valid. The total drag force is the sum of the friction drag and the pressure drag. However, the drag coefficient is a function only of the Reynolds number.

8–15.1 FLOW OVER A FLAT PLATE PARALLEL TO THE FLOW: FRICTION DRAG

This flow situation has been considered in detail in Section 8–13. Since the pressure gradient is zero, the total drag is equal to the friction drag. Thus here

$$\text{Drag} = \iint_{\substack{\text{Plate} \\ \text{surface}}} \tau_w \, dA$$

and

$$C_D = \frac{F_D}{\frac{1}{2}\rho V^2 A} = \frac{\displaystyle\iint_{PS} \tau_w \, dA}{\frac{1}{2}\rho V^2 A} \qquad (8.55)$$

The drag coefficient for a flat plate parallel to the flow depends on the shear stress distribution along the plate. Where the boundary layer is entirely laminar or entirely turbulent, the drag coefficient may be calculated using suitable expressions for τ_w (see Section 8–13). Approximate relations have been developed for a boundary layer that is initially laminar and undergoes transition to a turbulent boundary layer along the length of the plate.

The variation in drag coefficient for a flat plate parallel to the flow is shown qualitatively in Fig. 8.29. In Fig. 8.29, a range of transition curves is shown. The

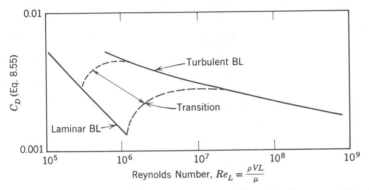

Fig. 8.29 *Variation of drag coefficient with Reynolds number for a smooth flat plate parallel to the flow.*

exact Reynolds number at which transition occurs depends on a combination of factors, such as surface roughness and freestream turbulence, as pointed out in Section 8–10. Transition tends to occur earlier (i.e. at lower Reynolds number) as surface roughness or freestream turbulence are increased. Figure 8.29 shows that drag is less, for a given length of plate, when laminar flow is maintained over the longest possible distance.

8–15.2 FLOW OVER A FLAT PLATE NORMAL TO THE FLOW: PRESSURE DRAG

In flow over a flat plate normal to the flow (Fig. 8.30), we see that the wall shear stress does not contribute to the drag force. The drag is given by

$$F_D = \iint_{Surface} p \, dA$$

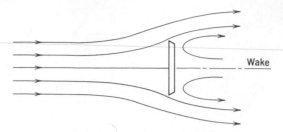

Fig. 8.30 *Flow over a flat plate normal to the flow.*

For this geometry the flow separates from the corners of the plate; there is backflow in the low energy wake of the plate. While the pressure over the rear surface of the plate is essentially constant, its magnitude cannot be determined analytically. Consequently, we must resort to experiments to determine the drag coefficient.

Because of end effects, the drag coefficient for a finite plate normal to the flow depends on the ratio of plate width to height and on the Reynolds number. For values of *Re* (based on height) greater than about 1000, the drag coefficient is independent of the Reynolds number. The variation of C_D with the ratio of plate width to height (b/h) is shown in Fig. 8.31. (The ratio b/h is defined as the

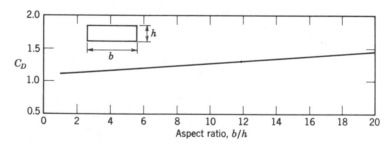

Fig. 8.31 *Variation of drag coefficient with aspect ratio for a flat plate of finite width normal to the flow with $Re_h > 1000$ (data from Ref. 11).*

aspect ratio of the plate.) For $b/h = 1.0$, the drag coefficient is a minimum at $C_D = 1.18$; this value is just slightly higher than that for a circular disc ($C_D = 1.17$) at large Reynolds number.

The drag coefficient for all objects with sharp corners is essentially independent of Reynolds number because the separation points are fixed by the geometry of the object. Additional drag data for a few selected objects are given in Table 8.6.

table 8.6 *Drag Coefficient Data for Selected Objects (Re $\gtrsim$ 1000)**

OBJECT	DIAGRAM		$C_D(Re \gtrsim 10^3)$
Square Cylinder		$b/h = \infty$	2.05
		$b/h = 1$	1.05
Disc			1.17
Ring			1.20†
Hemisphere (open end facing flow)			1.42
Hemisphere (open end facing downstream)			0.38
C-section (open)			2.30
			1.20

* Data from Ref. 11.
† Based on ring area.

8–15.3 FLOW OVER A SPHERE AND CYLINDER: FRICTION AND PRESSURE DRAG

We have looked at two special flow cases in which either friction or pressure drag was the sole form of drag present. In the former case the drag coefficient was a strong function of the Reynolds number while in the latter case C_D was independent of Reynolds number for $Re \gtrsim 1000$.

In the case of flow over a sphere, both friction drag and pressure drag contribute to the total drag. The drag coefficient for flow over a sphere is shown in Fig. 8.32 as a function of Reynolds number.

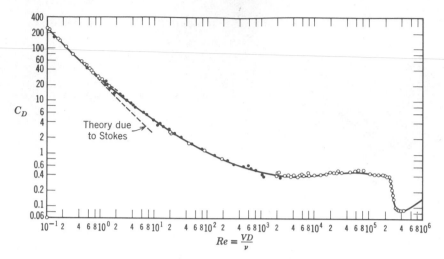

Fig. 8.32 *Drag coefficient of a sphere as a function of Reynolds number (data from Ref. 11).*

At very low Reynolds number,[12] $Re \leq 1$, there will be no flow separation from a sphere; the wake is laminar and the drag is predominantly friction drag. Stokes has shown analytically for very low Reynolds number flows, where inertia forces may be neglected, that drag force on a sphere of diameter, d, moving at velocity, V, through a fluid of viscosity, μ, is given by

$$F_D = 3\pi\mu V d$$

The drag coefficient, C_D, defined by Eq. 8.53 is then

$$C_D = \frac{24}{Re}$$

As shown in Fig. 8.32 this expression agrees with experimental values at low Re, but begins to deviate significantly from the experimental data for $Re > 1.0$.

As the Reynolds number is increased up to about 1000, the drag coefficient drops continuously. As a result of flow separation the drag is a combination of friction and pressure drag. The contribution of friction drag decreases with increasing Reynolds number; at $Re \simeq 1000$, the friction drag is approximately 5 percent of the total drag.

In the range of Reynolds number, $10^3 < Re < 2 \times 10^5$, the drag coefficient curve is relatively flat. The drag coefficient undergoes a rather sharp drop at a

---

[12] See the film, *The Fluid Mechanics of Drag*, A. H. Shapiro, principal, for a good discussion of drag on spheres and other shapes. Another excellent film is *Low Reynolds Number Flows*, principal, Sir G. I. Taylor. See also Reference 12.

Reynolds number of approximately 2×10^5. Experiments show that for $Re < 2 \times 10^5$ the boundary layer on the forward portion of the sphere is laminar. Separation of the boundary layer occurs just upstream of the midsection of the sphere; a relatively wide turbulent wake is present downstream of the sphere. In the separated region behind the sphere the pressure is essentially constant and lower than the pressure over the forward portion of the sphere (Fig. 8.33). It is this pressure difference that is the main contributor to the drag.

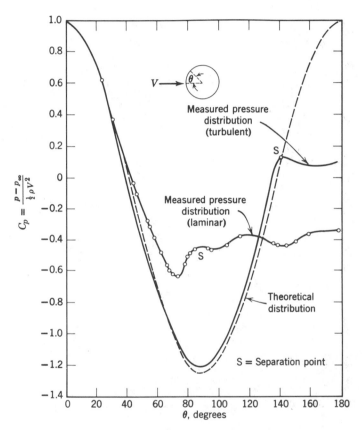

Fig. 8.33 *Pressure distribution around a sphere for laminar and turbulent boundary layer flow, compared to inviscid flow (data from Ref. 13).*

For Reynolds numbers larger than about 2×10^5, the boundary layer on the forward portion of the sphere becomes turbulent, that is, transition occurs in the boundary layer. The point of separation moves downstream of the center of the sphere and the size of the wake is decreased. The net pressure force on the sphere is reduced (Fig. 8.33), and the drag coefficient decreases abruptly.

A turbulent boundary layer, since it has more momentum than a laminar boundary layer, can travel farther against an adverse pressure gradient, as discussed in Section 8–14.1. Consequently, turbulent boundary layer flow is desirable on a blunt body, because it delays separation and, thus, reduces the pressure drag.

Transition in the boundary layer may be affected by roughness of the sphere surface and turbulence in the flow stream. Therefore the reduction in drag associated with turbulence in the boundary layer does not occur at a unique value of Reynolds number. Experiments with smooth spheres in a flow with low turbulence level show that transition may be delayed to a critical Reynolds number, Re_D, of about 4×10^5. For rough surfaces and/or turbulent flow streams, transition can occur at a critical Reynolds number as low as 1×10^5.

The drag coefficient with turbulent boundary layer flow is about 5 times less than that for laminar flow near the critical Reynolds number. The corresponding reduction in drag force can appreciably affect the range of a sphere (e.g. a golf ball). The "dimples" on a golf ball are designed to "trip" the boundary layer and, thus, to guarantee turbulent boundary layer flow, and minimum drag. To illustrate this effect more graphically, we obtained samples of golf balls without dimples a few years ago. One of our students volunteered to hit some drives with the smooth balls. In 50 tries with each type of ball, the average distance with the standard balls was 215 yards; the average with the smooth balls was only 125 yards!

Adding roughness elements to a sphere can also suppress local oscillations in location of the transition between laminar and turbulent flow in the boundary

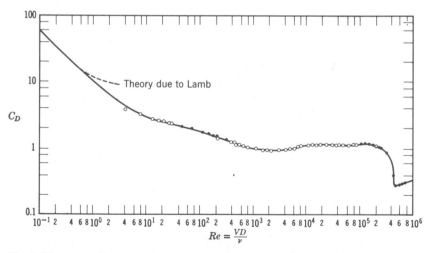

Fig. 8.34 *Drag coefficient for circular cylinders as a function of the Reynolds number.*

layer. These oscillations can lead to variations in drag, and to random fluctuations in lift (see Section 8–16). In baseball, the "knuckle ball" pitch is intended to behave erratically, to keep the batter confused. By throwing the ball with almost no spin, the pitcher relies on the seams to cause transition in an unpredictable fashion, as the ball moves on its way to the batter. This causes the desired variation in the flight path of the ball.

The drag coefficient for flow over a circular cylinder is shown in Fig. 8.34. The variation of C_D with Reynolds number shows the same characteristics as observed in the flow over a sphere, but the values of C_D are about twice as large.

Example 8.14

A cylindrical chimney 3 ft in diameter and 75 ft tall is exposed to a uniform 35 mph wind. Atmospheric conditions are 59 F and 14.7 psia. End effects and gusts may be neglected. Estimate the bending moment at the base of the chimney due to wind forces.

EXAMPLE PROBLEM 8.14

GIVEN:

Cylindrical chimney, $D = 3$ ft, $L = 75$ ft in uniform flow with

$$V = 35 \text{ mph} \qquad p = 14.7 \text{ psia} \qquad T = 59 \text{ F}$$

Neglect end effects.

FIND:

Bending moment at bottom of chimney.

SOLUTION:

The drag coefficient is given by

$$C_D = \frac{F_D}{\frac{1}{2}\rho V^2 A}, \qquad \text{and thus } F_D = \tfrac{1}{2}\rho V^2 A C_D$$

Since the force per unit length is uniform over the entire length, the resultant force, F_D, will act at the midpoint of the chimney. Hence

$$M_0 = F_D \frac{L}{2} = \frac{L}{4}\rho V^2 A C_D$$

$$V = \frac{35 \text{ mi}}{\text{hr}} \times \frac{5280 \text{ ft}}{\text{mi}} \times \frac{\text{hr}}{3600 \text{ sec}} = 51.4 \text{ ft/sec}$$

$$\rho = \frac{p}{RT} = \frac{14.7 \text{ lbf}}{\text{in.}^2} \times \frac{144 \text{ in.}^2}{\text{ft}^2} \times \frac{\text{lbm-R}}{53.3 \text{ ft-lbf}} \times \frac{1}{519 \text{ R}} \times \frac{\text{slug}}{32.2 \text{ lbm}} = 2.38 \times 10^{-3} \text{ slug/ft}^3$$

$$Re = \frac{\rho V d}{\mu} = \frac{V d}{\nu} = \frac{51.4 \text{ ft}}{\text{sec}} \times 3 \text{ ft} \times \frac{\text{sec}}{1.58 \times 10^{-4} \text{ ft}^2} = 9.75 \times 10^5$$

From Fig. 8.34, $C_D = 0.35$.
    Thus

$$M_0 = \frac{L}{4} \rho V^2 A C_D = \frac{L}{4} \rho V^2 L \, d C_D = \frac{L^2}{4} \rho V^2 \, d C_D$$

$$= \frac{(75)^2 \text{ ft}^2}{4} \times \frac{2.38 \times 10^{-3} \text{ slug}}{\text{ft}^3} \times \frac{(51.4)^2 \text{ ft}^2}{\text{sec}^2} \times 3 \text{ ft} \times 0.35 \times \frac{\text{lbf-sec}^2}{\text{slug-ft}}$$

$$M_0 = 9.27 \times 10^3 \text{ ft-lbf} \qquad\qquad\qquad M_0$$

Example 8.15

A fuel burning dragster weighing 2000 lbf attains a speed of 200 mph in the quarter mile. Immediately after passing through the traps, the driver opens the drag chute. The chute area is 25 ft², and it has a constant drag coefficient of 1.2. Air and rolling resistance of the car may be neglected. The local air density is 0.0024 slug/ft³.

    Find the time required for the machine to decelerate to 100 mph.

EXAMPLE PROBLEM 8.15

GIVEN:

Dragster weighing 2000 lbf, moving with speed, $V = 200$ mph, is slowed by the drag force on a chute.

  Chute area, $A = 25 \text{ ft}^2$ Drag coefficient, $C_D = 1.2$ $\rho_{air} = 0.0024 \text{ slug/ft}^3$

Air and rolling resistance of the car may be neglected.

FIND:

Time required for machine to decelerate to 100 mph.

EXAMPLE PROBLEM 8.15 (*continued*)

SOLUTION:

Taking the car as a system and writing Newton's 2nd law in the direction of motion

$$-F_D = ma = m\frac{dV}{dt}$$

$V_0 = 200 \text{ mph}$
$V_f = 100 \text{ mph}$

Since

$$C_D = \frac{F_D}{\frac{1}{2}\rho V^2 A}, \quad \text{then} \quad F_D = \tfrac{1}{2}C_D\rho V^2 A$$

Therefore

$$-\frac{1}{2}C_D\rho V^2 A = m\frac{dV}{dt}$$

$$-\frac{1}{2}C_D\rho\frac{A}{m}\int_0^t dt = \int_{V_0}^{V_f}\frac{dV}{V^2}$$

$$-\frac{1}{2}C_D\rho\frac{A}{m}t = -V^{-1}\Big]_{V_0}^{V_f} = -\frac{1}{V}\Big]_{V_0}^{V_f} = -\frac{1}{V_f}+\frac{1}{V_0} = -\frac{(V_0 - V_f)}{V_f V_0}$$

Finally,

$$t = \frac{V_0 - V_f}{V_f V_0}\frac{2m}{C_D\rho A}$$

$$= \frac{100\text{ mph}}{100\text{ mph}} \times \frac{hr}{200\text{ mi}} \times \frac{3600\text{ sec}}{hr} \times \frac{mi}{5280\text{ ft}} \times \frac{2}{1.2} \times 2000\text{ lbm} \times \frac{ft^3}{0.0024\text{ slug}}$$

$$\times \frac{1}{25\text{ ft}^2} \times \frac{slug}{32.2\text{ lbm}}$$

$$t = 5.9\text{ sec} \qquad\qquad t$$

8–15.4 STREAMLINING

The extent of the separated flow region behind many of the objects discussed in the previous section can be reduced by streamlining, or fairing, the body shape. The objective of streamlining is to reduce the adverse pressure gradient that occurs behind the point of maximum thickness on the body, as pointed out in Section 2–5.1. This delays boundary layer separation and, thus, reduces the pressure drag. However, addition of a faired tail section increases the surface area of the body; this causes skin friction drag to increase. The optimum streamlined shape is thus the shape that gives minimum total drag. These effects are discussed at length in the film series, *The Fluid Dynamics of Drag.*[13]

Even the most advanced theories are as yet unable to predict the location of separation analytically. Thus it is not possible to determine optimum bodies by purely analytical means, and we must rely on experimental data or on wind tunnel tests of proposed designs. Let us look at two general classes of results before going on to discuss lift.

The pressure gradient around a "teardrop" shape (i.e. a "streamlined" cylinder) is less severe than that around a cylinder of circular section. The trade-off between pressure and friction drag for this case is illustrated by the results presented in Fig. 8.35, for tests at $Re_c = 4 \times 10^5$. From the figure, the minimum drag coefficient

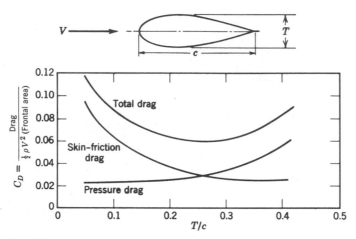

Fig. 8.35 *Drag coefficient on a streamlined strut as a function of thickness ratio, showing contribution of skin friction and pressure to total drag (data from Ref. 14).*

is $C_D \simeq 0.06$, which occurs when $T/c \simeq 0.25$. This value is approximately 20 percent of the minimum drag coefficient for a cylinder of the same thickness! Consequently a streamlined strut about 5 times the thickness of a cylindrical strut could be used with no penalty in aerodynamic drag.

Pressure distribution and drag data[14] for two symmetric airfoils of infinite span and 12 percent thickness at zero angle of attack are presented in Fig. 8.36. These

[13] A. H. Shapiro, principal.

[14] Note that drag coefficients for airfoils are based on the planform area, that is

$$C_D = \frac{F_D}{\frac{1}{2}\rho V^2 A_p}$$

where A_p is the maximum projected wing area.

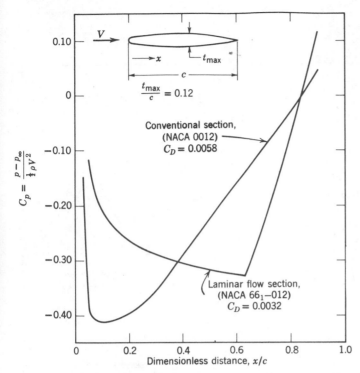

Fig. 8.36 *Pressure distributions at zero angle of attack for conventional (NACA 0012) and laminar flow (NACA 66₁–012) airfoils (data from Ref. 15).*

data illustrate the effect of boundary layer transition on the drag of well stream-lined objects, which is primarily due to skin friction.

Transition on the conventional (NACA 0012) airfoil takes place where the pressure gradient becomes adverse, at $x/c = 0.13$, that is, near the point of maximum thickness. Thus most of the airfoil surface is covered with a turbulent boundary layer; the drag coefficient is $C_D = 0.0058$. The point of maximum thickness has been moved aft on the airfoil (NACA 66₁–012) designed for laminar flow. The boundary layer is maintained in the laminar state by the favorable pressure gradient to $x/c = 0.63$. Thus the bulk of the flow is laminar; $C_D = 0.0032$ for this section. These figures show that the drag for the laminar flow section is only about half that for the conventional airfoil.

Tests in special wind tunnels have shown that laminar flow can be maintained up to length Reynolds numbers as high as $Re_x = 30 \times 10^6$ by appropriate profile shaping. Because they have favorable drag characteristics, laminar flow airfoils are used in the design of most modern subsonic aircraft.

413   **drag**

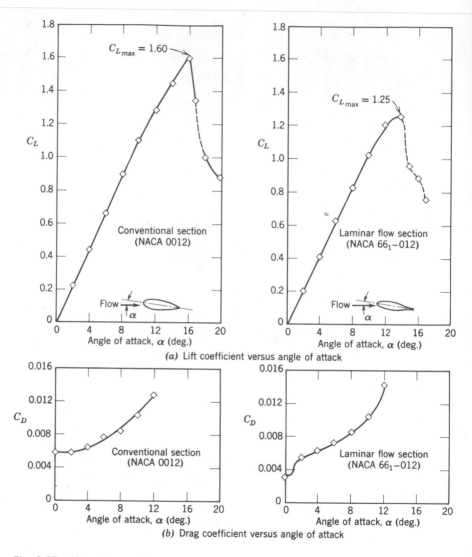

Fig. 8.37 *Lift and drag coefficient versus angle of attack for conventional (NACA 0012) and laminar flow (NACA 66_1–012) airfoil sections of 12 percent thickness (data from Ref. 15).*

As previously defined, lift is the component of the resultant force perpendicular to the direction of fluid motion. One of the most commonly observed examples of dynamic lift is the flow over an airfoil. The lift coefficient, C_L, is defined as

$$C_L \equiv \frac{F_L}{\frac{1}{2}\rho V^2 A_p} \tag{8.56}$$

The lift and drag coefficients for an airfoil are functions of both the Reynolds number and the angle of attack; the angle of attack, α, is the angle between the airfoil chord and the freestream velocity vector. For an airfoil the maximum cross-section area at right angles to the flow changes with angle of attack. Consequently, in defining the lift and drag coefficients of an airfoil, the area, A_p, the maximum projected area of the wing is used.

Lift and drag coefficient data for the conventional and laminar flow profiles discussed in Section 8–15.4 are plotted in Fig. 8.37, for a Reynolds number of 9×10^6, based on chord length. Since both of these sections are symmetric, the lift coefficient at zero angle of attack is zero. As the angle of attack is increased, the lift coefficient increases smoothly until a maximum value is reached. Further increases in angle of attack produce a sudden decrease in C_L. The airfoil is said to have *stalled* when C_L drops in this fashion.

Airfoil stall results when flow separation occurs over a major portion of the upper surface of the airfoil. As the angle of attack is increased, the stagnation point moves back along the lower surface of the airfoil, as shown schematically in Fig. 8.38. The flow on the upper surface then must accelerate sharply to round the nose of the airfoil. The minimum pressure becomes lower, and it moves farther forward on the upper surface. The adverse pressure gradient following the point of minimum pressure becomes more severe. Finally it causes the flow to separate completely from the upper surface; the airfoil stalls.

Movement of the minimum pressure point and accentuation of the adverse gradient are also responsible for the sudden increase in C_D at an angle of attack of about 1° for the laminar section, which is apparent in Fig. 8.37. The sudden rise in C_D is due to early transition from laminar to turbulent boundary-layer flow on the upper surface. Aircraft with laminar flow sections are designed to operate in the low drag region only.

Because laminar flow sections have very sharp leading edges, all of the effects we have described are exaggerated, and they stall at lower angles of attack than conventional sections, as shown in Fig. 8.37. The maximum possible lift coefficient, $C_{L_{max}}$, is also less for laminar flow sections.

Polar plots (i.e. plots of C_L versus C_D) are often used to present airfoil data. Such a plot is given in Fig. 8.39 for the two sections we have been discussing. The value of the lift/drag ratio, C_L/C_D, at $C_L = 0.1$ is shown for both sections.

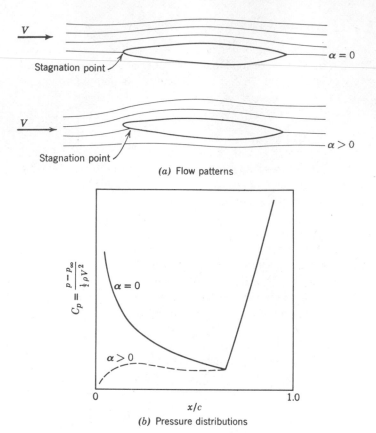

(a) Flow patterns

(b) Pressure distributions

Fig. 8.38 *Effect of angle of attack on flow pattern and pressure distribution around an airfoil (NACA 66_1–012 Section; data from Ref. 15).*

For an aircraft of given weight, the power required for level flight at fixed speed is inversely proportional to the lift/drag ratio. Thus the advantages of laminar flow designs may be seen clearly.

Note that all of the airfoil data presented have been for two-dimensional sections, that is, slices from airfoils of infinite span. End effects, which occur on airfoils of finite span, reduce the lift coefficient and cause the drag coefficient to increase. Thus the lift/drag ratio (L/D) values that we can achieve in practice are less than those which we obtain from two-dimensional test results.

The parameter used to define the effective span of an airfoil is the aspect ratio, defined as

$$ar = \frac{A_p}{c^2}$$

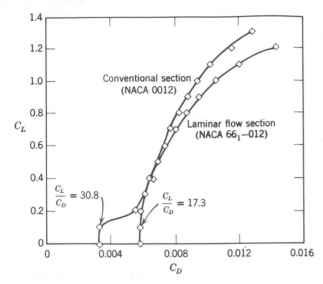

Fig. 8.39 *Lift-drag polar for conventional (NACA 0012) and laminar flow (NACA 66₁–012) airfoil sections.*

where A_p is the planform area, and c is the chord. Thus for a rectangular planform of span S,

$$ar = \frac{Sc}{c^2} = \frac{S}{c}$$

The maximum lift/drag ratio for a low-drag section may be as high as 80 for infinite aspect ratio. A sailplane (glider) with $ar = 40$ might have $L/D = 40$, and a typical light plane ($ar \sim 12$) might have $L/D \sim 20$ or so. Two examples of rather poor shapes are lifting bodies used for reentry from the upper atmosphere, and water skis, which are *hydro* foils of low aspect ratio. For both of these shapes, L/D values are less than unity.

Mother nature is well aware of the effects of aspect ratio on aerodynamic performance. Soaring birds, such as the albatross or California condor, have thin wings of long span. Birds that must maneuver quickly to catch their prey, such as owls, have wings of relatively short span, but large area, which gives low wing loading and thus high maneuverability.

For finite airfoils, lift reduction results from changes in the flow pattern arising from end effects. Increased drag results from *downwash* velocities induced by *trailing vortices*, and the energy which is left behind in these vortices. These effects are shown schematically in Fig. 8.40. The downwash velocities tend to reduce lift and increase drag by reducing the effective angle of attack. (They are

417   **lift**

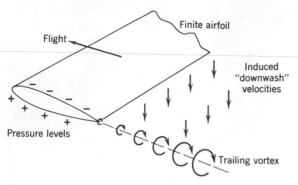

Fig. 8.40 *Schematic representation of the trailing vortex system of a finite airfoil.*

suppressed when an airplane, especially a low-wing craft, is close to the ground.) The trailing vortices are a result of leakage flows around the wingtips, from the high pressure area on the bottom surface to the low pressure region above. They may be very strong and persistent, especially for large aircraft at low speeds (and consequently high angles of attack). For example, the trailing vortices behind C-5A or 747 class aircraft present a hazard to light planes for distances of 5–10 *miles* behind the larger plane. Velocities exceeding 200 mph have been measured in these vortices.[15]

As we have seen, aircraft can be fitted with low drag airfoils to give excellent performance at cruise conditions. However, since the maximum lift coefficient is low for thin airfoils, some additional effort must be expended to obtain acceptably low landing speeds. In steady state flight conditions, the lift must equal the aircraft weight. Thus, from Eq. 8.56,

$$W = F_L = C_L \tfrac{1}{2}\rho V^2 A$$

The minimum flight velocity is obtained when $C_L = C_{L_{max}}$. Solving for V_{min},

$$V_{min} = \sqrt{\frac{2W}{\rho C_{L_{max}} A}} \tag{8.57}$$

According to Eq. 8.57, the minimum landing speed can be reduced by increasing either $C_{L_{max}}$ or the wing area. Two basic techniques are available for controlling these variables: variable geometry wing sections (e.g. through the use of flaps) or boundary layer control techniques.

Flaps are movable portions of a wing surface that may be extended during landing and takeoff to increase the effective wing area. The effect on lift and drag

[15] Sforza, P. M., "Aircraft Vortices: Benign or Baleful?" *Space/Aeronautics*, **53**, 4, April 1970, pp. 42–49. See also the University of Iowa film, *Form Drag, Lift and Propulsion*.

of two typical flap configurations is shown in Fig. 8.41, as applied to an NACA 23012 airfoil section. The maximum lift coefficient for this section is increased from 1.52 in the "clean" condition to 3.48 with double-slotted flaps. According to Eq. 8.57, the corresponding reduction in landing speed would be 34 percent.

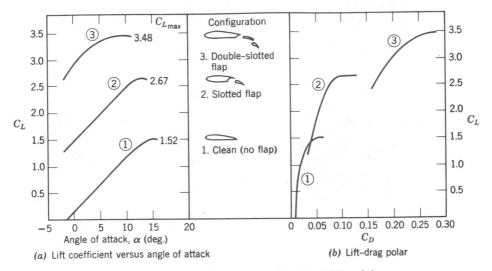

Fig. 8.41 *Effect of flaps on aerodynamic characteristics of NACA 23012 airfoil section (data from Ref. 15).*

Figure 8.41 shows that drag is increased substantially by high lift devices. From Fig. 8.41*b*, the drag at $C_{L_{max}}$ ($C_D \simeq 0.28$) with double slotted flaps is about five times larger than the drag at $C_{L_{max}}$ ($C_D \simeq 0.055$) for the clean airfoil.

Although the details of boundary layer control techniques[16] are beyond the scope of this text, the basic purpose of all of them is to delay separation or reduce drag, by adding energy to the boundary layer through blowing, or removing low energy boundary layer fluid by suction. Many examples of practical boundary layer control systems may be seen on commercial transport aircraft at your local airport. One of the more sophisticated systems, on the Boeing 727 transport, is shown in Fig. 8.42. On this aircraft, leading edge devices are used in conjunction with triple-slotted trailing edge flaps, to achieve a value of $C_{L_{max}}$ in excess of 3.6.

Another method of boundary layer control, which we shall discuss briefly, is the use of moving surfaces to reduce skin friction effects on the boundary layer. This method is hard to apply to practical devices, because of geometric and weight

---

[16] See the excellent film, *Boundary Layer Control*, D. C. Hazen, principal, for a review of these techniques.

Fig. 8.42 *Application of high-lift boundary layer control devices to jet transport aircraft for reduction of landing speed (Photographs courtesy of Boeing Airplane Company). (a) The wing of the Boeing 727 is one of the most mechanized of those in commercial service today. During the approach to landing, huge triple-slotted, trailing edge flaps roll out from under the wing and deflect downward to increase lift. About the same time, a section of the leading edge near the outboard end of the wing slides forward to open a slot that keeps the air flow close to the wing's upper surface. And at the leading edge of the wing near the root, a Kruger flap drops down from under the wing, increasing the radius of the leading edge to prevent flow separation. After touchdown, spoilers (not shown) pop up in front of each flap to kill the lift and ensure that the plane remains on the ground despite the lift-augmenting devices.*

complications, but it is very important in recreation. Most golfers, tennis players, ping-pong enthusiasts, and baseball pitchers can attest to this!

It has long been known that a spinning projectile in flight is affected by a force perpendicular to the direction of motion and to the spin axis. This effect, known as the Magnus effect, is responsible for the systematic drift of artillery shells. Tennis and ping-pong players use spin to control the trajectory and bounce of a shot. In golf, a drive can leave the tee at 275 ft/sec or more, with a backspin of 5000 rpm! This spin provides significant aerodynamic lift, that substantially increases the carry of a drive. It is also largely responsible for hooking and slicing, when shots are not hit squarely. The baseball pitcher uses spin to throw a curve ball.

(b)

Fig. 8.42 (b) *The view from the passenger's seat in a modern airliner—this one, a Boeing 707—includes a remarkably good view of the increasingly complex machinery needed to make air behave as it passes over the wing. Both the inboard and outboard trailing edge flaps (shown here fully deflected) are equipped with double slots near the pivot. The slots duct air from the under surface to the upper surface of the flap, creating a streamwise jet that causes the air flow to cling more closely to the upper surface. The two rows of short blades in front of the flaps are vortex generators whose purpose is to churn up the slow-moving layers of air next to the surface. This encourages the flow to follow the wing surface more closely. The raised panels just above the flaps are lift-destroying devices called spoilers. Normally, they are used to control lift for landing and maneuvering. They are raised here to help the plane descend from high altitude.*

421   **lift**

Flow about a spinning sphere is shown in Fig. 8.43a. The spin affects boundary layer separation and, thus, influences the pressure distribution. Separation is delayed on the upper surface of the sphere in Fig. 8.43a, and it occurs earlier on the lower surface. Thus the pressure is reduced on the upper surface and increased on the lower surface; the wake is deflected downward as shown. The pressure forces cause a lift in the direction shown; spin in the opposite direction would produce a negative lift, that is, a downward force. The force is directed perpendicular to both V and the spin axis.

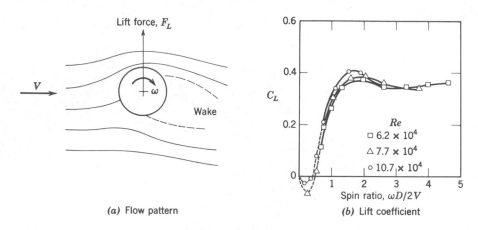

(a) Flow pattern   (b) Lift coefficient

Fig. 8.43 *Flow pattern and lift coefficient for a spinning sphere in uniform flow (data from Ref. 14).*

Example 8.16

An airplane is designed according to the following specifications.

$$\text{Weight} = 3000 \text{ lbf}$$
$$\text{Wing area} = 300 \text{ ft}^2$$
$$\text{Takeoff speed} = 100 \text{ ft/sec}$$

Model tests show that the lift and drag coefficients vary with the angle of attack of the wing according to the approximate relations:

$$C_L = 0.35(1 + 0.2\alpha)$$
$$C_D = 0.008(1 + \alpha)$$

for small α, where α is the angle of attack measured in degrees. The atmospheric density is 0.00238 slug/ft^3.

Find:

(a) The angle of attack that insures takeoff at the desired speed.
(b) The power required for takeoff.

EXAMPLE PROBLEM 8.16

GIVEN:

Aircraft with specifications:

$$\text{Weight} = 3000 \text{ lbf} \qquad\qquad C_L = 0.35(1 + 0.2\alpha)$$
$$\text{Wing area} = 300 \text{ ft}^2 \qquad\quad C_D = 0.008(1 + \alpha)$$
$$\text{Takeoff speed} = 100 \text{ ft/sec} \qquad\qquad \text{where } \alpha \text{ is angle of attack in degrees}$$
$$\rho_{air} = 0.00238 \text{ slug/ft}^3$$

FIND:

(a) α for takeoff at 100 ft/sec.
(b) Power required for takeoff.

SOLUTION:

To ensure takeoff, lift force must be equal to weight.

$$C_L = \frac{F_L}{\frac{1}{2}\rho V^2 A} = \frac{3000 \text{ lbf}}{\frac{1}{2}} \times \frac{\text{ft}^3}{0.00238 \text{ slug}} \times \frac{\text{sec}^2}{(100)^2 \text{ ft}^2} \times \frac{1}{300 \text{ ft}^2} \times \frac{\text{slug-ft}}{\text{lbf-sec}^2}$$

$$C_L = 0.84$$

Since

$$C_L = 0.35(1 + 0.2\alpha)$$

$$\alpha = \left[\frac{C_L}{0.35} - 1\right]\frac{1}{0.2} = \left[\frac{0.84}{0.35} - 1\right]\frac{1}{0.2} = \frac{1.4}{0.2} = 7° \longleftarrow \qquad\qquad \alpha$$

Power required for takeoff = $F_D V$

$$C_D = \frac{F_D}{\frac{1}{2}\rho V^2 A} \qquad F_D = \tfrac{1}{2}C_D \rho V^2 A$$

For $\alpha = 7°$,

$$C_D = 0.008(1 + 7) = 0.064$$

EXAMPLE PROBLEM 8.16 (continued)

$$\text{Power} = F_D V = \tfrac{1}{2}C_D \rho V^2 A V$$

$$= \tfrac{1}{2}(0.064) \times \frac{0.00238 \text{ slug}}{ft^3} \times \frac{(100)^2 \text{ ft}^2}{\sec^2} \times 300 \text{ ft}^2 \times \frac{100 \text{ ft}}{\sec} \times \frac{\text{lbf-sec}^2}{\text{slug-ft}} \times \frac{\text{hp-sec}}{550 \text{ ft-lbf}}$$

$$\text{Power} = 41.5 \text{ hp} \qquad\qquad\qquad\qquad\qquad\qquad\qquad\qquad\qquad\qquad \text{Power}$$

Part F. Flow Measurement

The choice of a flow metering device is influenced by the accuracy required, cost, complication, ease of reading or data reduction, and service life. Ordinarily, the simplest and cheapest device for the desired accuracy should be chosen. Accordingly, a few simple methods of flow measurement are discussed first.

8–17 Simple Methods

With liquid flows, simple tanks can be used to determine flowrate by measuring the volume or weight of liquid collected during a known time interval. If the time interval is long enough to be measured accurately, extremely precise values of flowrate may be determined in this way.

For gas flows, compressibility must be considered in making volumetric measurements. Specific weights of gases are also generally too small to permit accurate determination of weight flowrate directly. However, a volumetric sample can often be collected by displacing a "bell," or inverted jar over water or other liquid (the pressure is held constant by means of counterweights). If volumetric or weight determinations are set up carefully, no calibration is required; this is a great advantage of these methods.

Another simple method, but one that does require calibration, makes use of the radial pressure gradient caused by streamline curvature. Design of a simple "elbow flowmeter" is illustrated in Example Problem 8.17.

Example 8.17

The flowrate of air in a flat duct is to be determined by installing pressure taps across a bend. The duct is 12 in. deep and 4 in. wide. The inner radius of the bend is 10 in. The velocity profile is assumed uniform. If the measured pressure difference between the taps is 1.54 in. of water, compute the approximate flowrate in ft³/sec.

GIVEN:

Flow through elbow, as shown.

$$p_2 - p_1 = 1.54 \text{ in. } H_2O$$

Flow uniform. Air at STP.

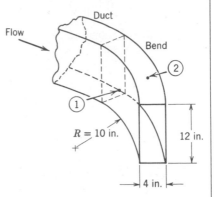

FIND:

Q in ft^3/sec.

SOLUTION:

Apply Euler's n component equation across the flow streamlines.

Basic equation:

$$\frac{\partial p}{\partial r} = \frac{\rho V^2}{r}$$

Plan view of bend

Assumptions: (1) Frictionless flow
  (2) Incompressible flow
  (3) Uniform flow at measurement section

For this flow, $p = p(r)$, so

$$\frac{\partial p}{\partial r} = \frac{dp}{dr} = \frac{\rho V^2}{r}$$

or

$$dp = \rho V^2 \frac{dr}{r}$$

Integrating

$$p_2 - p_1 = \rho V^2 \ln r \Big]_{r_1}^{r_2} = \rho V^2 \ln \frac{r_2}{r_1}$$

Solving,

$$V = \left[\frac{p_2 - p_1}{\rho \ln (r_2/r_1)} \right]^{1/2}$$

But

$$\Delta p = p_2 - p_1 = \gamma \Delta h = 62.4 \frac{\text{lbf}}{\text{ft}^3} \times 1.54 \text{ in.} \times \frac{\text{ft}}{12 \text{ in.}} = 8 \frac{\text{lbf}}{\text{ft}^2}$$

EXAMPLE PROBLEM 8.17 (continued)

For air at STP, $\rho = 0.00238$ slug/ft³. Substituting,

$$V = \left[\frac{8 \text{ lbf}}{\text{ft}^2} \times \frac{\text{ft}^3}{0.00238 \text{ slug}} \times \frac{1}{\ln(1.4)} \right]^{1/2} = 100 \text{ ft/sec}$$

For uniform flow

$$Q = VA = \frac{100 \text{ ft}}{\text{sec}} \times 4 \text{ in.} \times 12 \text{ in.} \times \frac{\text{ft}^2}{144 \text{ in.}^2} = 33.3 \text{ ft}^3/\text{sec} \qquad \xleftarrow{\quad} Q$$

$\left\{ \begin{array}{c} \text{In actual applications, the velocity profile in a channel bend will not be uniform.} \\ \text{Thus, elbow flowmeters require calibration.} \end{array} \right\}$

8–18 Flow Meters for Internal Flows

All nonmechanical flowmeters for internal flow are (except the laminar flow element, Section 8–18.4) based on acceleration of a fluid stream through some form of nozzle, as shown schematically in Fig. 8.44. The pressure drop is related to the flowrate by applying the Bernoulli and continuity equations.

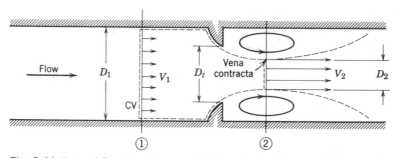

Fig. 8.44 *Internal flow through a generalized nozzle, showing control volume used for analysis.*

Basic equations:

$$\frac{p_1}{\rho} + \frac{V_1^2}{2} + \cancel{gz_1}^{(7)} = \frac{p_2}{\rho} + \frac{V_2^2}{2} + \cancel{gz_2}^{(7)}$$

$$0 = \cancel{\frac{\partial}{\partial t} \int\!\!\int\!\!\int_{CV} \rho \, d\forall}^{= 0(1)} + \int\!\!\int_{CS} \rho \vec{V} \cdot d\vec{A}$$

Assumptions: (1) Steady flow
(2) Incompressible flow
(3) Flow along a streamline
(4) No friction
(5) Uniform velocity at sections ① and ②
(6) No streamline curvature at section ① or ②, so pressure is uniform across those sections.
(7) $z_1 = z_2$

Then from the Bernoulli equation,

$$p_1 - p_2 = \frac{\rho}{2}(V_2^2 - V_1^2) = \frac{\rho V_2^2}{2}\left[1 - \left(\frac{V_1}{V_2}\right)^2\right]$$

and from continuity

$$0 = \{-|\rho V_1 A_1|\} + \{|\rho V_2 A_2|\}$$

or

$$V_1 A_1 = V_2 A_2 \quad \text{or} \quad \left(\frac{V_1}{V_2}\right)^2 = \left(\frac{A_2}{A_1}\right)^2$$

Substituting,

$$p_1 - p_2 = \frac{\rho V_2^2}{2}\left[1 - \left(\frac{A_2}{A_1}\right)^2\right] = \frac{\rho V_2^2}{2}\left[1 - \left(\frac{D_2}{D_1}\right)^4\right] = \frac{\rho V_2^2}{2}(1 - \beta^4)$$

where $\beta = D_2/D_1$. Solving for V_2,

$$V_2 = \sqrt{\frac{2(p_1 - p_2)}{\rho(1 - \beta^4)}} \tag{8.58a}$$

For uniform flow, $\dot{m} = \rho V A = \rho V_2 A_2$, so that

$$\dot{m} = A_2\sqrt{\frac{2\rho(p_1 - p_2)}{1 - \beta^4}} \tag{8.58b}$$

Equations 8.58 would give the correct velocity and mass flowrate if all seven of our assumptions were exactly true. In any real flow, there will be some frictional effects resulting from turbulent mixing. Therefore assumption (4) is questionable. Frictional effects are taken into account by defining a velocity coefficient, C_v, as

$$C_v \equiv \frac{\text{actual velocity}}{\text{ideal velocity}} \tag{8.59}$$

where the ideal velocity is the value computed from Eq. 8.58a, for frictionless flow. Boundary layer development and streamline curvature effects also reduce the effective flow area. Thus, as shown in Fig. 8.44, the flow area at the vena contracta, or point of minimum area, where the flow streamlines are essentially straight, is considerably smaller than at the nozzle throat. These effects, which reduce the effective flow area, are taken into account by defining a contraction coefficient, C_c, where

$$C_c \equiv \frac{\text{effective flow area}}{\text{geometric area of throat}} \tag{8.60}$$

Thus

$$V_{2\,\text{actual}} = C_v V_{2\,\text{ideal}} = C_v V_2$$

$$A_{2\,\text{effective}} = C_c A_{t\,\text{geometric}} = C_c A_t$$

and

$$\dot{m}_{\text{actual}} = \rho V_{2\,\text{actual}} A_{2\,\text{effective}} = C_c C_v \rho V_2 A_t$$

Substituting from Eq. 8.58b,

$$\dot{m}_{\text{actual}} = C_v C_c A_t \sqrt{\frac{2\rho(p_1 - p_2)}{1 - \beta^4}} = \frac{C_v C_c}{\sqrt{1 - \beta^4}} A_t \sqrt{2\rho(p_1 - p_2)} \tag{8.61}$$

The factor $1/\sqrt{1 - \beta^4}$, often called the "velocity of approach" factor, is a function of geometry alone. It is often combined with the velocity and contraction coefficients into a single coefficient, C_d, the "discharge" coefficient. Thus

$$C_d \equiv \frac{C_v C_c}{\sqrt{1 - \beta^4}} \tag{8.62}$$

In terms of the discharge coefficient, the actual mass flowrate, given by Eq. 8.61, may then be written

$$\dot{m}_{\text{actual}} = C_d A_t \sqrt{2\rho(p_1 - p_2)} \tag{8.63}$$

As we have noted, selection of a flowmeter depends on factors such as cost, accuracy, need for calibration, and ease of installation and maintenance. Some of these factors are compared for orifice plate, flow nozzle and venturi meters in Table 8.7.

To apply any of these flowmeters, they must be installed in a section of straight pipe, with sufficient length upstream from the meter to provide a fully-developed turbulent velocity distribution at section ①. If a flowmeter is installed downstream from a valve, elbow or other disturbance, a straight section of pipe must be placed in front of the meter. Approximately 10 diameters of straight pipe are required for venturi meters, and up to 40 diameters for orifice plate or flow nozzle meters. When a meter has been properly installed, the flowrate may be computed from Eq. 8.63, after choosing an appropriate value for the empirical discharge coefficient defined in Eq. 8.62. Some design data for incompressible flow are given in the next few sections. The same basic methods can be extended to compressible flows, but these will not be treated here. For complete details, see Reference 16.

table 8.7 *Configurations and Properties of Orifice, Flow Nozzle and Venturi Flowmeters*

FLOWMETER TYPE	DIAGRAM	HEAD LOSS	COST
Orifice		High	Low
Flow nozzle		Intermediate	Intermediate
Venturi		Low	High

8–18.1 THE ORIFICE PLATE

The orifice plate is a thin plate that may be clamped between pipe flanges. Since its geometry is simple, it is low in cost and easy to install or replace. The sharp edge of the orifice will not foul with scale or suspended matter. However, suspended matter can build up at the inlet side of a concentric orifice in a horizontal pipe; an eccentric orifice may be placed flush with the bottom of the pipe to avoid this difficulty. The primary disadvantages of the orifice are its limited capacity, and the high head loss due to the uncontrolled expansion downstream from the metering element.

Pressure taps may be placed at the vena contracta, or 1/2 diameter downstream from the orifice plate, or in the flanges (see Ref. 16 for details). Some discharge coefficient values for concentric orifices with taps 1 pipe diameter upstream and 1/2 diameter downstream are given in Fig. 8.45 as a function of Reynolds number and diameter ratio, β. Recommended design practice is to use orifice diameters such that $0.25 < \beta < 0.90$.

8–18.2 THE FLOW NOZZLE

Flow nozzles may be used as metering elements in either plenums or ducts, as shown in Fig. 8.46. As shown in Fig. 8.46, the nozzle section is approximately

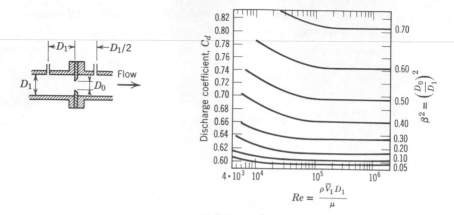

Fig. 8.45 *Discharge coefficients for concentric orifices with pressure taps as shown (data from Ref. 16).*

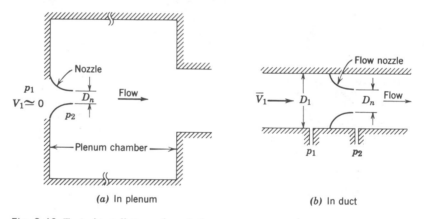

(a) In plenum

(b) In duct

Fig. 8.46 *Typical installations of nozzle flowmeters.*

a quarter ellipse. Design details and locations for pressure taps may be found in Reference 16.

a. Plenum Installations

For plenum installation, the approach velocity, V_1, is essentially zero, so $D_1 \to \infty$. and $\beta \simeq 0$. Under these conditions, C_c is nearly unity, and Eq. 8.62 shows that C_d equals C_v. Consequently, Eq. 8.63 becomes

$$\dot{m}_{\text{actual}} = C_v A_t \sqrt{2\rho(p_1 - p_2)} \qquad (8.64)$$

For plenum installation, nozzles may be fabricated from spun aluminum, molded fiberglass or other cheap materials. Thus, they are simple and cheap to make and install. Since the plenum pressure is equal to p_2, the location of the downstream pressure tap is not critical. Meters suitable for a wide range of flowrates may be made by installing several nozzles in a plenum. At low flowrates, most of them may be plugged with rubber balls or other handy objects. For larger flowrates, more nozzles may be used.

For plenum nozzles, typical velocity coefficient values are in the range, $0.95 < C_v < 0.99$, generally increasing for larger Reynolds numbers. Thus the mass rate of flow can be computed within approximately plus or minus 2 percent using Eq. 8.64 with $C_v = 0.97$.

b. Pipe Installation

Equation 8.63, repeated below, must be used with an experimental value for C_d to compute the mass flowrate through a flow nozzle in a pipe.

$$\dot{m}_{\text{actual}} = C_d A_t \sqrt{2\rho(p_1 - p_2)} \tag{8.63}$$

Experimentally, the discharge coefficient is a function of both the Reynolds number and diameter ratio, as shown in Fig. 8.47.

Figure 8.47 shows that C_d is essentially constant for large Reynolds number $(Re_{D_1} > 2 \times 10^5)$. Thus at high flowrates, the flowrate may be computed directly.

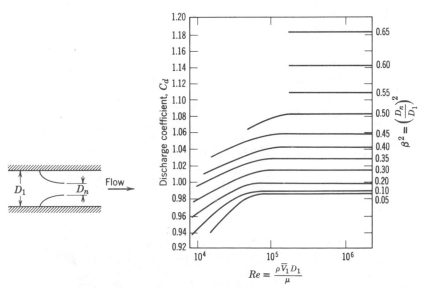

Fig. 8.47 *Discharge coefficients for ASME long radius flow nozzles (data from Ref. 16).*

flow meters for internal flow

At lower flowrates, where C_d is a weak function of the Reynolds number, iteration may be required.

For some configurations, C_d values larger than unity are possible. The reason is that the discharge coefficient includes the velocity of approach factor, $1/\sqrt{1 - \beta^4}$, which is always greater than one.

Flow nozzles are intermediate between orifice plates and venturis in cost and ease of installation. Their head loss is slightly less than that of an orifice of the same diameter ratio, because no vena contracta is present.

8–18.3 THE VENTURI

The venturi meter was sketched in Table 8.7. It is generally made from a casting that is then machined to close tolerances to assure performance that duplicates that of the standard design. As a result, venturis are bulky, heavy, and expensive. However, the conical diffuser section downstream from the throat gives excellent pressure recovery; the overall head loss is therefore low. The venturi is also self-cleaning due to its smooth internal contour.

Experiments show that the discharge coefficient for venturi meters is independent of diameter ratio, over the range $0.25 < \beta < 0.75$. The variation of C_d with Reynolds number is shown in Fig. 8.48. Note that C_d is constant for Reynolds numbers larger than about 2×10^5.

8–18.4 THE LAMINAR FLOW ELEMENT

The orifice, flow nozzle, and venturi all produce a pressure drop proportional to the square of the flowrate, according to Eq. 8.63. In practice, a meter size must

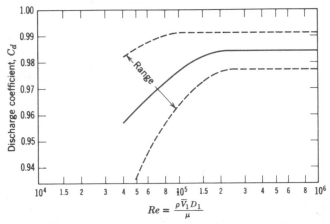

Fig. 8.48 *Discharge coefficients for Venturi meters, $0.25 < \beta < 0.75$ (data from Ref. 16).*

432   **8/incompressible viscous flow**

be chosen to accommodate the largest flowrate expected. Because the pressure drop versus flowrate relationship is nonlinear, the range of flowrate that can be measured accurately is limited.

The laminar flow element[17] is designed to produce a pressure drop directly proportional to flowrate. The laminar flow element (LFE) contains a section subdivided into many passages, each small enough in diameter to assure fully-developed laminar flow. As shown in Section 8–4, pressure drop in laminar duct flow is directly proportional to flowrate. However, the relationship between pressure drop and flowrate also depends upon the fluid viscosity, which is a strong function of temperature. Therefore the fluid temperature must be known to obtain accurate metering with a LFE.

The LFE costs approximately as much as a venturi, but it is much lighter and smaller. Thus, the LFE is becoming widely used in applications where compactness and extended range are important.

Example 8.18

An air flowrate of 30 ft^3/sec at standard conditions is expected in a 10 in. diameter duct. An orifice meter is to be installed to measure the rate of flow. The manometer available to make the measurement has a maximum range of 6 in. of water. What diameter orifice plate should be used with taps placed 1 diameter upstream and 1/2 diameter downstream from the plate? Approximately what head loss will result if no energy is recovered and $C_c \simeq 0.60$?

EXAMPLE PROBLEM 8.18

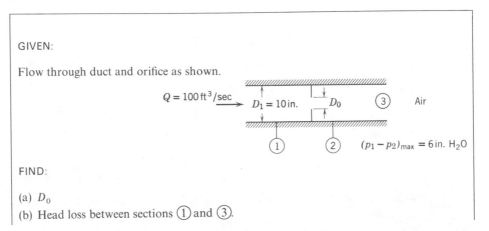

GIVEN:

Flow through duct and orifice as shown.

$Q = 100 \, ft^3/sec$ $D_1 = 10 \, in.$ D_0 ③ Air

① ② $(p_1 - p_2)_{max} = 6 \, in. \, H_2O$

FIND:

(a) D_0
(b) Head loss between sections ① and ③.

[17] Patented and manufactured by Meriam Instrument Co., 10920 Madison Ave., Cleveland, Ohio, 44102.

EXAMPLE PROBLEM 8.18 (continued)

SOLUTION:

The orifice plate may be designed using Eq. 8.63 and data from Fig. 8.45.

Computing equation : $\quad \dot{m}_{\text{actual}} = C_d A_t \sqrt{2\rho(p_1 - p_2)}$ (8.63)

But

$$A_t = \frac{\pi}{4}D_t^2 = \frac{\pi}{4}D_1^2\beta^2 = A_1\beta^2$$

Substituting and solving for $\beta^2 C_d$,

$$\beta^2 C_d = \frac{\dot{m}_{\text{actual}}}{A_1\sqrt{2\rho(p_1 - p_2)}} = \frac{\rho Q}{A_1\sqrt{2\rho(p_1 - p_2)}}$$

For the water manometer,

$$p_1 - p_2 = \gamma_{\text{H}_2\text{O}}\Delta h = \frac{62.4 \text{ lbf}}{\text{ft}^3} \times 6 \text{ in.} \times \frac{\text{ft}}{12 \text{ in.}} = 31.2 \text{ lbf/ft}^2$$

$$A_1 = \frac{\pi}{4}D_1^2 = \frac{\pi}{4}\left[10 \text{ in.} \times \frac{\text{ft}}{12 \text{ in.}}\right]^2 = 0.52 \text{ ft}^2$$

Then

$$\beta^2 C_d = \frac{\dfrac{0.00238 \text{ slug}}{\text{ft}^3} \times \dfrac{30 \text{ ft}^3}{\text{sec}}}{0.52 \text{ ft}^2 \left[\dfrac{2(0.00238) \text{ slug}}{\text{ft}^3} \times \dfrac{31.2 \text{ lbf}}{\text{ft}^2} \times \dfrac{\text{slug-ft}}{\text{lbf-sec}^2}\right]^{1/2}} = 0.356$$

The duct Reynolds number is

$$Re = \frac{\rho \bar{V}_1 D_1}{\mu} = \frac{\rho(Q/A_1)D_1}{\mu} = \frac{4Q}{\pi\nu D_1}$$

$$Re = \frac{4}{\pi} \times \frac{30 \text{ ft}^3}{\text{sec}} \times \frac{\text{sec}}{2 \times 10^{-4} \text{ ft}^2} \times \frac{1}{10 \text{ in.}} \times \frac{12 \text{ in.}}{\text{ft}} = 2.3 \times 10^5$$

Since in Fig. 8.45, C_d is a function of both Re and β, we must iterate. Guessing $\beta = 0.7$, we obtain

$$C_d = \frac{0.356}{\beta^2} = \frac{0.356}{0.49} = 0.73$$

From Fig. 8.45, $C_d = 0.69$ at $\beta = 0.7$ and $Re = 2.3 \times 10^5$. Therefore, a value of $\beta = 0.71$, which gives

$$C_d = \frac{0.356}{(0.71)^2} = 0.708$$

is a better choice. Thus $\beta = 0.71$, and

$$D_o = (0.71)(10 \text{ in.}) = 7.1 \text{ in.}$$

To evaluate the head loss, apply Eq. 8.24 between stations ① and ③. Thus

Computing equation:
$$\left[\frac{p_1}{\rho} + \frac{\overline{V}_1^2}{2} + gz_1 \right] - \left[\frac{p_3}{\rho} + \frac{\overline{V}_3^2}{2} + gz_2 \right] = h_{l_T} \qquad (8.24)$$

Assumptions: (1) $\overline{V}_1 = \overline{V}_3$
(2) Neglect Δz

Then

$$h_{l_T} = \frac{p_1 - p_3}{\rho}$$

The pressure at ③ may be found by applying the momentum equation to a control volume between sections ② and ③.

Basic equation:
$$F_{S_x} + F_{B_x} \overset{= 0(3)}{=} \frac{\partial}{\partial t} \overset{= 0(4)}{\int\int\int_{CV}} u \rho \, d\forall + \int\int_{CS} u \rho \overline{V} \cdot d\overline{A}$$

Assumptions: (3) $F_{B_x} = 0$
(4) Steady flow
(5) Uniform flow at sections ② and ③
(6) Pressure uniform across duct at section ②
(7) No friction

Then

$$(p_2 - p_3)A_1 = u_2\{-|\rho \overline{V}_2 A_2|\} + u_3\{|\rho \overline{V}_3 A_3|\}$$
$$= (u_3 - u_2)\rho Q = (\overline{V}_3 - \overline{V}_2)\rho Q$$

or

$$p_3 - p_2 = (\overline{V}_2 - \overline{V}_3)\frac{\rho Q}{A_1}$$

Now $\overline{V}_3 = Q/A_1$, and

$$\overline{V}_2 = \frac{Q}{A_2} = \frac{Q}{C_c A_t} = \frac{Q}{C_c \beta^2 A_1}$$

Thus

$$p_3 - p_2 = \frac{\rho Q^2}{A_1^2}\left[\frac{1}{C_c \beta^2} - 1 \right]$$

EXAMPLE PROBLEM 8.18 (continued)

Assuming $C_c \simeq 0.6$,

$$p_3 - p_2 = \frac{0.00238 \text{ slug}}{\text{ft}^3} \times \frac{(30)^2 \text{ ft}^6}{\text{sec}^2} \times \frac{1}{(0.52)^2 \text{ ft}^4}\left[\frac{1}{(0.6)(0.71)^2} - 1\right] = 18.3 \text{ lbf/ft}^2$$

Then

$$h_{l_T} = \frac{p_1 - p_3}{\rho} = \frac{p_1 - p_2 - (p_3 - p_2)}{\rho}$$

$$= \frac{(31.2 - 18.3) \text{ lbf}}{\text{ft}^2} \times \frac{\text{ft}^3}{0.00238 \text{ slug}} \times \frac{\text{slug-ft}}{\text{lbf-sec}^2}$$

$$h_{l_T} = 5420 \frac{\text{ft}^2}{\text{sec}^2}\left[\text{or } \frac{\text{ft-lbf}}{\text{slug}}\right] \longleftarrow \qquad h_{l_T}$$

$\left\{\begin{array}{l}\text{This problem illustrates flowmeter calculations, and shows use of the momentum} \\ \text{equation to compute the pressure rise in a sudden expansion.}\end{array}\right\}$

8–19 Mechanical Flowmeters

In specialized applications, particularly for remote use or recording, mechanical flowmeters may be specified. Common examples include household water and natural gas meters, which are generally calibrated to read directly in units of the product used, or gasoline metering pumps at the local filling station, which measure

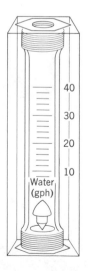

Fig. 8.49 *Float-type mechanical flowmeter (Courtesy F. W. Dwyer Instrument Co., Michigan City, Indiana).*

total flow and automatically compute the dollar value. A large variety of such meters is available commercially. Manufacturers' literature should be consulted for design and installation details.

Float-type meters may be used to give a direct indication of flowrate for both liquids and gases. An example is shown in Fig. 8.49. In operation, the ball or float is carried upward in the tapered clear tube by the flowing fluid until the drag force and float weight are in equilibrium. Such meters are available with factory calibration for a number of common fluids and ranges of flowrate.

The final type of meter that we shall consider is the turbine flowmeter. In this form of flowmeter, a free-running turbine or vaned impeller is mounted in a cylindrical section of tube (Fig. 8.50). The rate of rotation of the impeller is closely proportional to the fluid flowrate over a wide range. This linearity is a principal advantage of the turbine flowmeter.

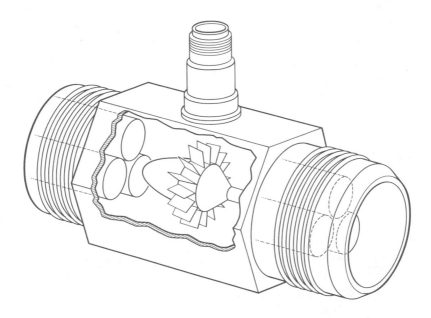

Fig. 8.50 *Turbine-type mechanical flowmeter (Courtesy Potter Aeronautical Corp., Union, New Jersey).*

Rotational speed of the turbine element can be sensed using a magnetic pickup external to the meter. This sensing method thus requires no penetrations or seals in the duct. Thus turbine flowmeters can be used safely to measure flow-rates in corrosive or toxic fluids. The electrical signal can be displayed or recorded easily, or integrated to provide total flow.

In some practical situations, for example in air handling or refrigeration equipment, it is either impractical or impossible to install a fixed flowmeter. In many such cases, it is possible to obtain flowrate data using a traversing technique.

To make a flowrate measurement by traverse, the duct cross section is subdivided into segments of equal area. The fluid velocity is measured at the center of each such segment using a pitot or total head tube or a suitable anemometer. Then the volume flowrate is approximated for each segment by the product of velocity and area. The flowrate through the entire duct is the sum of these segmental flowrates.

problems

8.1 Approximately how far from the entrance of a 1 in. diameter pipe would flow be fully-developed if the Reynolds number is 1500, based on average velocity?

8.2 Standard air enters a 12 in. diameter duct. The volume flowrate is 60 ft^3/min. Determine whether the flow is laminar or turbulent. Estimate the entrance length required for fully-developed flow to be established.

8.3 The velocity profile for fluid flow between stationary parallel plates is given by

$$u = ay(h - y)$$

where a is a constant, h is the total gap width between plates, y is distance measured upward from the lower plate. Determine the ratio $\bar{V}/u_{max}$.

8.4 An incompressible fluid flows between two infinite stationary parallel plates. The velocity profile is given by

$$u = u_{max}(Ay^2 + By + C)$$

where A, B, and C are constants, and y is measured from the center of the gap. The total gap width is h units.

Use appropriate boundary conditions to express the constants in terms of h. Develop an expression for volumetric flowrate per unit depth.

8.5 A viscous oil flows steadily between parallel plates. The flow is laminar and fully-developed. The velocity profile is given by

$$u = -\frac{2h^2}{\mu}\frac{\partial p}{\partial x}\left[1 - \left(\frac{2y}{h}\right)^2\right]$$

where the total gap width between the plates, $h = 0.01$ ft, and y is measured from the centerline of the gap. The oil viscosity is 0.01 lbf-sec/ft^2, and the pressure gradient is -8 lbf/ft^3.

Find:

(a) The magnitude and direction of the shear stress on the upper plate.
(b) The volumetric flowrate through the channel, per foot of width.

8.6　　A fluid flows steadily between two parallel plates. The distance between the plates is h. The velocity profile for fully-developed laminar flow is given by

$$u = -\frac{1}{2\mu}\frac{\partial p}{\partial x}\left[y^2 - \left(\frac{h}{2}\right)^2\right]$$

(a) Derive an equation for the shear stress as a function of y. Plot this function.
(b) For $\mu = 2.4 \times 10^{-5}$ lbf-sec/ft^2, $\partial p/\partial x = -4.0$ lbf/ft^3, and $h = 0.05$ in., calculate the maximum shear stress in lbf/ft^2.

8.7　　A block that weighs 10 lbf slides down a plane inclined at 30 degrees from the horizontal. The block rides on an oil film 0.001 in. thick. The oil viscosity is 0.0002 lbf-sec/ft^2, and the terminal velocity of the block is 1.0 ft/sec.
　　　Find the area of the block surface in contact with the oil film.

8.8　　A sealed journal bearing is formed from concentric cylinders. The inner and outer radii are 1.00 and 1.05 in., respectively; the journal length is 4 in., and it turns at 240 rpm. The gap is filled with oil in laminar motion. The velocity profile is linear across the gap. The torque needed to turn the journal is 2.0 in.-lbf.
　　　Calculate the viscosity of the oil. Will the torque increase or decrease with time? Why?

8.9　　Water at 60 F flows between parallel plates with gap width $b = 0.10$ ft. The upper plate moves with velocity $U = 10$ ft/sec in the positive x direction. The pressure gradient is $\partial p/\partial x = -0.06$ lbf/ft^3.
　　　Locate the point of maximum velocity and determine its magnitude (let $y = 0$ at the bottom plate). Sketch the velocity and shear stress distributions. Determine the volume of flow that passes a given cross section ($x = $ constant) in 10 sec.

8.10　A fluid flows at 140 F flows between two large flat plates. The lower plate moves to the left at a speed of 1.0 ft/sec. The plate spacing is 0.01 ft, and the flow is laminar.
　　　Determine the pressure gradient required to produce zero net flow at a cross section.

8.11　The velocity distribution for fully-developed laminar flow in a pipe is given by

$$u = -\frac{R^2}{4\mu}\frac{\partial p}{\partial z}\left[1 - \left(\frac{r}{R}\right)^2\right]$$

Determine the radial distance from the pipe axis at which the velocity equals the average velocity.

439　　**problems**

8.12   Consider fully-developed laminar flow in a circular pipe. Use a cylindrical control volume as shown in Fig. 8.51. Indicate the forces acting on the control volume. Using the momentum equation, develop an expression for the velocity distribution.

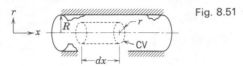

Fig. 8.51

8.13   Consider first water and then SAE 10 lubricating oil flowing at 100 F in a 1/4 in. diameter tube. Determine the maximum flowrate (and the corresponding pressure gradient, $\partial p/\partial x$) for each fluid at which laminar flow would be expected.

8.14   Consider fully-developed laminar flow in the annulus between two concentric pipes (Fig. 8.52). The inner pipe is stationary, and the outer pipe moves in the x direction with velocity V_0. Assume the axial pressure gradient to be zero ($\partial p/\partial x = 0$).

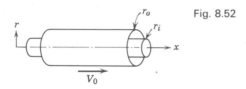

Fig. 8.52

Obtain a general expression for the shear stress, τ, as a function of the radius, r, in terms of a constant, C_1. Obtain a general expression for the velocity profile, $V(r)$, in terms of two constants, C_1 and C_2. Evaluate the constants, C_1 and C_2.

8.15   A tube 17.6 in. long, with an inside diameter of 0.030 in., is used as a capillary viscometer. Calibration tests are made using water at 60 F, and a flowrate of 1 cm^3/sec is measured for an applied pressure drop of 10 psi. (Assume the pressure drop in the entrance length is twice that for the same length of fully-developed flow.)

Determine the percentage error in viscosity that would result if Eq. 8.13c were used directly to compute it, without considering the entrance length.

8.16   A continuous belt (Fig. 8.53) passing upward through a chemical bath at velocity, U_0, picks up a liquid film of thickness, h, density, ρ, and viscosity, μ. Gravity tends to make the liquid drain down, but the movement of the belt keeps the fluid from running off completely. Assume that the flow is fully-developed laminar flow with

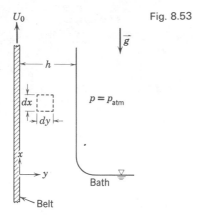

Fig. 8.53

zero pressure gradient, and that the atmosphere produces no shear at the outer surface of the film.

State clearly the boundary conditions to be satisfied by the velocity at $y = 0$ and $y = h$. Determine an expression for the velocity profile.

8.17   A film of water (at 60 F) in steady, laminar motion runs down a long slope, inclined 30° below the horizontal. The thickness of the film is 0.03 in. Assume that flow is fully-developed, and at zero pressure gradient.

Determine the surface shear stress, and the volume flowrate per unit width.

8.18   The velocity profile for turbulent flow through smooth pipes is often represented by the empirical equation

$$\frac{u}{U} = \left[1 - \frac{r}{R} \right]^{1/n}$$

Show that the ratio of average to centerline velocities is given by

$$\frac{\bar{V}}{U} = \frac{2n^2}{(n + 1)(2n + 1)}$$

Evaluate the ratio for $n = 7$ (corresponding to a Reynolds number of 110,000) and for $n = 8.8$ ($Re \simeq 1.1 \times 10^6$). Using the results of Section 8–5, plot $\bar{V}/U$ as a function of Reynolds number.

8.19   Consider the velocity profile for fully-developed laminar pipe flow, Eq. 8.14, and the empirical "power law" profile for turbulent pipe flow, Eq. 8.16. Assume $n = 7$ for the turbulent profile.

Determine the value of r/R for each profile at which u is equal to the average velocity, $\bar{V}$.

441   **problems**

8.20   Kerosene at 70 F flows in a smooth tube with an inside diameter of 1 in. The flow
        Reynolds number is 4000. For laminar flow, the pressure gradient is found to be
        $\partial p/\partial x = -0.18$ lbf/ft^3, while for turbulent flow, $\partial p/\partial x = -0.45$ lbf/ft^3.
        Plot the variation of shear stress as a function of radius for both flow conditions.

8.21   The kinetic energy flux coefficient, α, is defined by Eq. 8.21. Using the "power
        law" turbulent velocity profile, show that

$$\alpha = \left[\frac{U}{\overline{V}}\right]^3 \frac{2n^2}{(3 + n)(3 + 2n)}$$

        Evaluate α for $n = 7$.

8.22   A momentum flux coefficient, β, may be defined as

$$\int_A u\rho u\, dA = \beta \int_A \overline{V}\rho u\, dA = \beta \dot{m}\overline{V}$$

        Evaluate β for a laminar velocity profile, Eq. 8.14, and for a "power law" turbulent
        velocity profile, Eq. 8.16 (choose $n = 7$).

8.23   Water flows in a constant area pipeline; the pipe diameter is 2 in. and the average
        velocity of the flow is 5 ft/sec. At the pipe inlet the pressure is 85 psig. The outlet
        of the pipe is at an elevation 80 ft higher than the inlet; the outlet pressure is
        atmospheric. Determine the head loss between inlet and outlet of the pipe.

8.24   The pipe of Problem 8.23 is placed on a horizontal surface. The flowrate and outlet
        pressure are to remain the same.
        Compute the inlet pressure for this new condition.

8.25   A 50 ft length of new wrought iron pipe of 1.0 in. inside diameter is to be used
        in a horizontal position to convey water at 60 F. The inlet pressure is 80 psia and
        the exit pressure is 60 psia. The flow is expected to be turbulent with a friction
        factor of 0.025. Calculate the average velocity of the water in the pipe.

8.26   Resistance to fluid flow can be defined by analogy to Ohm's law for electric
        current. Thus, resistance to flow is given by the ratio of pressure drop (driving
        potential) to volume flowrate (current). Show that resistance to laminar flow is
        given by

$$\text{Resistance} = \frac{128\mu L}{\pi D^4}$$

        which is a function of geometry only, and is independent of flowrate. Find the
        maximum pressure drop for which this relation is valid for a tube 2 in. long and
        0.010 in. inside diameter for both ethyl alcohol and castor oil at 100 F.

8.27 An empirical correlation for friction factor in turbulent flow in smooth pipes was developed by H. Blasius in 1911. He found that

$$f = \frac{0.316}{Re^{1/4}}, \qquad Re \leq 100{,}000$$

correlated data well. Show that in turbulent flow, the predicted pressure drop is proportional to $(\bar{V})^{7/4}$ when the Blasius correlation is used. How does pressure drop depend on the tube diameter at a given flowrate?

8.28 Water at 78 F flows in a pipe whose inside diameter is 1.2 in. The flow Reynolds number is 100,000, and the friction factor is 0.04.

Determine the slope that the pipe must have to maintain constant pressure along its length. If the temperature remains constant, determine the heat transfer per 100 ft of pipe.

8.29 A smooth, 3 in. diameter pipe carries water horizontally at 150 F at a mass flowrate of 0.006 slug/sec. The pressure drop is observed to be 0.065 lbf/ft² per 100 ft of pipe. From the Moody chart, the friction factor could be chosen as 0.021 or 0.042. Which is correct?

8.30 A correlating equation for flow in the fully rough flow regime is

$$f = \frac{1}{4[0.57 - \log_{10}(e/D)]^2}$$

as originally obtained by von Kármán. Compare values obtained from this equation with those from the fully rough region of the Moody chart for values of e/D equal to 0.01, 0.001, and 0.0001.

8.31 Air at standard conditions flows through a sudden expansion in a circular duct. The upstream and downstream duct diameters are 3 and 9 in., respectively. The pressure downstream is 0.25 in. of water *higher* than that upstream.

Determine the average velocity of the air approaching the expansion, and the volume flowrate.

8.32 Water flows through a 2 in. diameter tube that suddenly contracts to a 1 in. diameter. The pressure drop across the contraction is 0.5 psi. Determine the flowrate in gallons per minute.

8.33 Air flows out of a clean room test chamber through a 6 in. diameter duct. The original duct had a square-edged entrance, but this has been replaced with a well-rounded one. The pressure in the chamber is 0.10 in. of water above ambient. Losses due to friction are negligible compared to the entrance and exit losses.

Determine the increase in volume flowrate that results from the change in entrance contour.

8.34 Space has been found for a conical diffuser 18 in. long in the clean room ventilation system described in Problem 8.33. The best diffuser of this size is to be used. Assume data from Fig. 8.17 may be used.

Determine the appropriate diffuser angle and area ratio for this installation, and predict the flowrate that will be found after it is installed.

8.35 Water flows through a galvanized iron pipe at a flowrate of 6 ft³/sec. The inside diameter of the pipe is 6 in., and the water temperature is 75 F.
Compute the friction factor for the flow.

8.36 Water at 70 F flows through a 4 in. (internal diameter) concrete drainage pipe at a rate of 1 slug/sec.
Determine the pressure drop in psi per 100 ft of horizontal pipe.

8.37 Water flows from a large reservoir as shown in Fig. 8.54. The pipe is of cast iron, with an inside diameter of 8 in. The flowrate is 5 ft³/sec, and the discharge

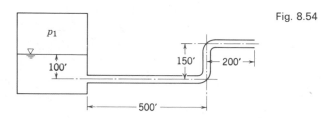

Fig. 8.54

is to atmospheric pressure. The mean temperature for the flow is 50 F; the entire system is insulated.
Determine the gage pressure, p_1, required to produce the flow. Calculate the temperature rise between the liquid surface and the exit.

8.38 Air at 60 F flows through a straight, 3 ft diameter, smooth duct 800 ft long. The flowrate is 12,000 ft³/min, and the pressure is the same at both ends of the duct.
Determine the change in elevation between the inlet and outlet.

8.39 Water is to flow by gravity from one reservoir to a lower one through a straight, inclined pipe. The flowrate required is 0.25 ft³/sec, the pipe diameter is 2 in. and the total length is 800 ft. Each reservoir is open to the atmosphere. Neglecting minor losses, calculate the difference in level required to maintain this flow.

8.40 Two reservoirs are connected by three clean cast-iron pipes in series, $L_1 = 2000$ ft, $D_1 = 12$ in.; $L_2 = 3000$ ft, $D_2 = 16$ in.; $L_3 = 5000$ ft, $D_3 = 18$ in. When the discharge is 4 cfs of water at 60 F, determine the difference in elevation between the reservoirs.

8.41 Shown in Fig. 8.55 is a water flow system with a variable elevation reservoir, B. Determine the water level in reservoir B so that no water flows into or out of the reservoir. The velocity in the 12 in. diameter pipe is 10 ft/sec. Neglect all minor losses.

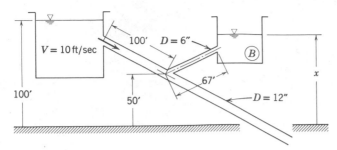

Fig. 8.55

8.42 Lightweight crude oil ($SG = 0.855$) is pumped horizontally through a 1 mile length of 12 in. diameter pipe. The average roughness size is 0.01 in. The flowrate is 1500 gpm.

Calculate the horsepower required to drive the pump if it is 75 percent efficient.

8.43 Water from a pump flows through a 10 in. diameter pipe for a distance of 15,000 ft from the pump discharge to a reservoir open to the atmosphere (Fig. 8.56). The level of the water in the reservoir is 20 ft above the pump discharge, and the average velocity of the fluid in the pipe is 4.5 ft/sec.

Calculate the pressure at the pump discharge.

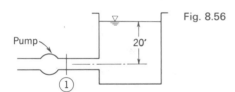

Fig. 8.56

8.44 A fire nozzle is supplied through 300 ft of 3 in. diameter hose, with a relative roughness of 0.001. The nozzle outlet is 1 in. in diameter. Water from the nozzle must reach the top of a building 80 ft tall and 30 ft across the street from the firemen. Water for the nozzle comes from a fire hydrant at 6 psig, and the pressure is boosted by a pump on board the pumper truck.

Neglecting minor losses, determine the minimum power that must be delivered by the pump.

8.45 Six hundred gpm of kerosene at 140 F flow in a pipe system in a refinery (Fig. 8.57). The pipe is commercial steel, with an inside diameter of 6 in. Determine the total length, L, of the straight pipe in the system.

8.46 Heavy crude oil ($SG = 0.925$) is pumped through a pipeline laid on flat ground. The line is made from steel pipe with 24 in. inside diameter and has a wall thickness

445 **problems**

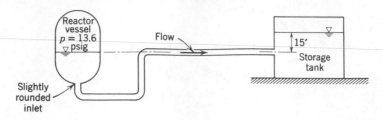

Fig. 8.57

of 1/2 in. The allowable tensile stress in the pipe wall is limited to 40,000 psi by corrosion considerations. At the same time, it is important to keep the oil under pressure to ensure that gases remain in solution. The minimum recommended pressure is 75 psia. The pipeline carries a flow of 400,000 barrels (a barrel holds 55 gal) per day.

Determine the maximum spacing between pumping stations. Compute the power added to the oil at each pumping station.

8.47   Gasoline flows in a long, underground pipeline at a constant temperature of 60 F. Two pumping stations at the same elevation are located 8 miles apart. The pressure drop between the stations is 210 psi. The pipeline is made from 24 in. pipe. Although made from commercial steel, age and corrosion have raised the pipe roughness to approximately that for galvanized iron. The specific gravity of the gasoline is 0.68.

Compute the volume rate of flow of gasoline through the pipe, in ft^3/sec.

8.48   A siphon is shown in Fig. 8.58. Assume that the only head loss occurs at the tube inlet which may be considered a reentrant entrance. The tube inside diameter is 3 in. The fluid is water at 60 F.

Determine the volume flowrate through the siphon. Compute the pressure at point A.

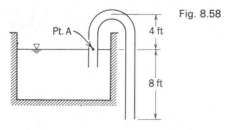

Fig. 8.58

8.49   Two reservoirs containing water are connected by a constant area, galvanized iron pipe that has one right-angle bend. The surface pressure at the upper reservoir is atmospheric whereas the gage pressure at the lower reservoir surface is 10 psig.

The pipe diameter is 3 in. Assume that the only significant losses occur in the pipe and bend (Fig. 8.59).

Determine the magnitude and direction of the volumetric flowrate.

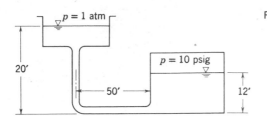

Fig. 8.59

8.50   You are watering your lawn with an *old* hose. Due to the buildup of lime deposits over the years, the 0.75 in. i.d. hose now has an average roughness height of 0.022 in. One 50 ft length of the hose, attached to your spigot, delivers 20 gallons of water (60 F) per minute.

Compute the pressure at the spigot, in psi. Assuming the pressure at the spigot remains constant, estimate the delivery if two 50 ft lengths of the hose are connected together.

8.51   A circular tank 5 ft in diameter is filled to a depth of 15 ft with water at 60 F. A smooth tube 20 ft long is attached to a well-rounded entrance at the tank bottom. The 2 in. diameter tube discharges to atmosphere 15 ft below the tank bottom.

Estimate the time required for the tank level to drop 5 ft. (You may wish to use the computer to solve this problem. If so, the Blasius correlation for friction factor, for turbulent flow,

$$f = \frac{0.316}{Re^{1/4}}$$

might come in handy.)

8.52   A spray system for a sewage treatment plant is shown in Fig. 8.60. Water at 60 F is pumped through a spray arm. The effective flow area of each nozzle is 0.25 in.^{2} The pipe inside diameter is 1 in., and the material is galvanized iron. Determine the flowrate of water through the spray arm.

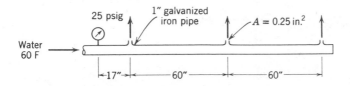

Fig. 8.60

447   **problems**

8.53 A hydraulic press is powered by a remote high pressure pump. The pump outlet pressure is 3000 psig, whereas the pressure required for the press is 2800 psig, at a flowrate of 8.5 gpm. The press and pump are to be connected by 150 ft of smooth, drawn steel tubing.

Determine the minimum tubing diameter that may be used.

8.54 A pump is located 15 ft to one side and 12 ft above a reservoir (Fig. 8.61). The pump is designed for a flowrate of 100 gpm. For satisfactory operation, the suction head at the pump inlet must not be lower than -20 feet of water.

Determine the smallest standard commercial steel pipe which will give the required performance.

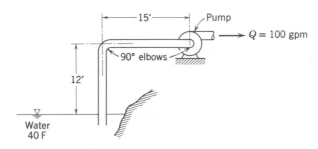

Fig. 8.61

8.55 The tank in Fig. 8.62 has two lengths of 3 in. galvanized pipe attached at the levels shown. The flow may be assumed to be in the fully rough region, and the entrances are slightly rounded.

Determine the ratio of h_2 to h_1 that will cause the same flowrate in each pipe line. Compute the minimum value of h_1 that will produce flow in the fully rough region.

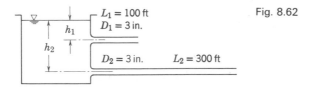

Fig. 8.62

8.56 A swimming pool has a partial flow filtration system. Water at 75 F is pumped from the pool through the system shown in Fig. 8.63. The pump delivers 30 gpm at 60 psig. The pipe is galvanized iron ($e/D \simeq 0.009$). The pressure loss through

448   **8/incompressible viscous flow**

the filter is given approximately by

$$\Delta p = 0.6Q^2$$

where Δp is in psi and Q is in gallons per minute.
　　Determine the flow through each branch of the system.

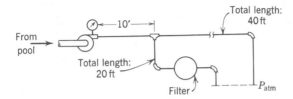

Fig. 8.63

8.57　The results of Example Problem 8.11 indicated that the pipes used were oversized. Rework the problem using the same geometry, but with 4 in. commercial steel pipes.

8.58　Changes in an industrial process require a water flow of 1500 gpm into an industrial plant. The pressure in the water main, located in the street 150 ft from the plant, is 120 psig. The present water line is a 2 in. galvanized pipe, but it is old enough so that its e/D value should be doubled to account for corrosion. Both old and new lines will require installation of 4 elbows in their total length of 200 ft. The water pressure required in the plant is 80 psig.
　　What size line should be installed?

8.59　In a certain air-conditioning installation, a flowrate of 1200 ft³/min of air at 70 F and 14.7 psia is required. A smooth sheet metal duct 1 ft square is to be used.
　　Determine the pressure drop in in. of water for a 100 ft horizontal duct run.

8.60　Consider flow of standard air at 1200 ft³/min. Compare the pressure drop per unit length of a round duct with that for rectangular ducts of aspect ratio 1, 2, and 3. Assume all ducts are smooth, with a cross-section area of 1 ft².

8.61　Determine the minimum size smooth rectangular duct with an aspect ratio of 2 that will pass 2400 ft³/min of standard air with a head loss of 1 in. of water per 100 ft of duct.

8.62　The head versus capacity curve for a certain fan may be approximated by the equation

$$h = 30 - 10^{-7}(Q^2)$$

where h is the output head in in. of water and Q is the air flowrate in ft³/min. The fan outlet dimensions are 8 × 16 in.

449　**problems**

Determine the air flowrate delivered by the fan into a 200 ft straight length of 8 × 16 in. rectangular duct.

8.63　A submarine moves at 20 knots through sea water at 45 F. Assume that near the bow, the boundary layer behaves as though on a flat plate.

Determine the range of distance from the bow that transition from laminar to turbulent flow in the boundary layer might be expected.

8.64　An airplane cruises at 300 mph at an altitude of 30,000 ft on a standard day. Assume that the boundary layers on the wing surfaces behave as on a flat plate.

Determine the expected extent of laminar flow in the wing boundary layers.

8.65　Velocity profiles in laminar boundary layers are often approximated by the equations

$$\text{Linear} \quad \frac{u}{U} = \frac{y}{\delta}$$

$$\text{Parabolic} \quad \frac{u}{U} = 2\left(\frac{y}{\delta}\right) - \left(\frac{y}{\delta}\right)^2$$

$$\text{Cubic} \quad \frac{u}{U} = \frac{3}{2}\left(\frac{y}{\delta}\right) - \frac{1}{2}\left(\frac{y}{\delta}\right)^3$$

$$\text{Sinusoidal} \quad \frac{u}{U} = \sin\frac{\pi}{2}\left(\frac{y}{\delta}\right)$$

Compare the shapes of these velocity profiles by plotting y/δ (on the ordinate) versus u/U (on the abscissa).

8.66　Evaluate the ratio δ^*/δ for each of the laminar boundary-layer velocity profiles given in Problem 8.65.

8.67　Consider a laminar boundary layer on a flat plate with a velocity profile given by

$$\frac{u}{U} = \frac{3}{2}\eta - \frac{\eta^3}{2}; \quad \eta = y/\delta$$

For this profile

$$\frac{\delta}{x} = \frac{4.65}{\sqrt{Re_x}}$$

Determine an expression for δ^*/x.

8.68　The velocity profile in a turbulent boundary layer is often approximated by the "power law" equation

$$\frac{u}{U} = \left(\frac{y}{\delta}\right)^{1/7}$$

Compare the shape of this profile with the parabolic laminar boundary layer profile (Problem 8.65) by plotting y/δ (on the ordinate) versus u/U (on the abscissa) for both profiles.

8.69 Evaluate the ratio δ^*/δ for the "power law" form used to represent the turbulent velocity profile, that is,

$$\frac{u}{U} = \left(\frac{y}{\delta}\right)^{1/7}$$

Compare with the value for the cubic laminar boundary layer velocity profile given in Problem 8.67.

8.70 Air flows in the entrance region of a square duct, as shown in Fig. 8.64. The velocity at the inlet is uniform, $U_0 = 100$ ft/sec, and the duct is 3.2 in. square. At a section 1 ft downstream from the entrance, the displacement thickness, δ^*, on each wall measures 0.040 in.

Determine the pressure change between sections ① and ②.

Fig. 8.64

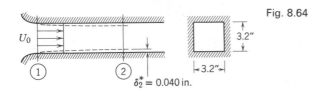

U_0

3.2"

|←3.2"→|

① ②

$\delta_2^* = 0.040$ in.

8.71 Air is flowing over a thin flat plate which is 3 ft long and 1 ft wide. The flow is uniform at the leading edge of the plate. The velocity profile in the boundary layer is assumed to be linear, that is,

$$\frac{u}{U} = \frac{y}{\delta} = \eta$$

$$U = 100 \text{ ft/sec}$$

$$\rho = 0.0024 \text{ slug/ft}^3$$

Treat the flow as two-dimensional, that is, assume flow conditions are independent of z (Fig. 8.65).

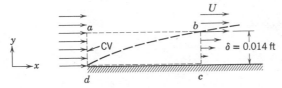

U

a b

y CV

$\delta = 0.014$ ft

d c

Fig. 8.65

Using the control volume *abcd*, shown by the dashed lines, compute the mass flowrate across surface *ab*. Determine the magnitude and direction of the *x* component of the force required to hold the plate stationary.

8.72   Rework Problem 8.71 with the same information, except assume

$$\frac{u}{U} = 2\left(\frac{y}{\delta}\right) - \left(\frac{y}{\delta}\right)^2$$

at section ②.

8.73   The most general sinusoidal velocity profile for laminar boundary-layer flow on a flat plate is

$$u = A \sin (By) + C$$

State three boundary conditions applicable to the laminar boundary-layer velocity profile. Evaluate the constants A, B, and C.

8.74   The velocity profile in a laminar boundary-layer flow at zero pressure gradient is to be approximated by the linear expression,

$$\frac{u}{U} = \eta; \qquad \eta = \frac{y}{\delta}$$

Use the momentum integral equation with this profile to obtain an expression for the ratio δ/x.

8.75   Water flows over a flat plate at a freestream speed of 0.5 ft/sec. There is no pressure gradient, and the laminar boundary layer is 0.25 in. thick. Assume a sinusoidal velocity profile,

$$\frac{u}{U} = \sin \frac{\pi}{2}\left(\frac{y}{\delta}\right)$$

Derive an equation for the shear stress at any location within the boundary layer. For the flow conditions given, compute the local wall shear stress and skin friction coefficient.

8.76   When a cubic velocity profile for the laminar boundary layer,

$$\frac{u}{U} = \frac{3}{2}\eta - \frac{1}{2}\eta^3; \qquad \eta = \frac{y}{\delta}$$

is used in the momentum integral equation, the variation of δ with *x* is found to be

$$\frac{\delta}{x} = \frac{4.64}{\sqrt{Re_x}}$$

For a laminar boundary layer with cubic profile, obtain an expression for the local skin friction coefficient,

$$C_f = \frac{\tau_w}{\frac{1}{2}\rho U^2}$$

in terms of distance and flow properties.

8.77 Air at standard conditions flows over a flat plate. The free-stream speed is 50 ft/sec.
Find δ and τ_w two feet from the leading edge for (a) completely laminar flow (assume a parabolic velocity profile), and (b) completely turbulent flow (assume a "1/7 power" velocity profile).

8.78 Transition from laminar to turbulent boundary-layer flow actually occurs over a finite length of surface, during which the velocity profile and wall shear stress adjust from laminar to turbulent forms. A useful approximation during transition is that the momentum flux within the boundary layer remains constant.
Assuming constant momentum flux, show that

$$\delta_{turbulent} = \frac{72}{105}\delta_{laminar}$$

for transition from a parabolic laminar velocity profile to a "1/7 power" turbulent velocity profile.

8.79 Figure 8.66 shows two hypothetical boundary-layer velocity profiles. Calculate the momentum flux of each profile. If the two profiles were subjected to the same pressure gradient conditions, which would be most likely to separate first? Why?

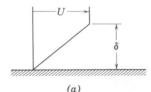

(a)   (b) $u = \frac{2}{3} U$

Fig. 8.66

8.80 A plane-wall diffuser is sketched in Fig. 8.67. We wish to compare the flow of an ideal fluid (recall $\mu = 0$) and a real fluid in such a passage.
Consider first the case where $\phi = 0$, that is, a straight channel. What can be said of the pressure gradient for the real and ideal fluids? Which fluid gives the higher value of p_2?
Now consider a case where ϕ is not equal to zero, but is small enough to avoid separation. Again, what can be said of the pressure gradient for real and ideal fluids? Which case results in the highest exit pressure?

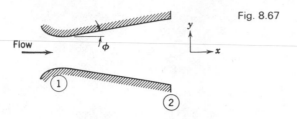

Fig. 8.67

Flow

τ_ϕ

8.81   A uniform flow of air at 200 ft/sec enters a plane wall diffuser with negligible boundary-layer thickness. The inlet width is 3 in., and the air density is 0.0024 slug/ft³. The diffuser walls diverge slightly to accommodate the boundary-layer growth, so that the pressure gradient is negligible. Flat plate boundary-layer behavior may be assumed.

      Explain why the Bernoulli equation is applicable to this flow. Estimate the diffuser width 3.6 ft downstream from the entrance.

8.82   Water at 80 F flows over a flat plate at a speed of 3 ft/sec. The plate is 1.2 ft long and 3 ft wide. The boundary layer on each surface of the plate is laminar. Assume the velocity profile is approximated by a linear expression, for which

$$\frac{\delta}{x} = \frac{3.46}{\sqrt{Re_x}}$$

Determine the drag force on the plate.

8.83   Equation 8.46 gives the local shear stress, τ_w, in terms of Re_x for turbulent boundary-layer flow at zero pressure gradient. The total skin friction coefficient is defined as

$$\bar{C}_f = \frac{\text{Drag}}{\frac{1}{2}\rho U^2 bL}$$

where b is the width, and L is the length of a flat plate.

      Show, using Eq. 8.46, that

$$\bar{C}_f = \frac{0.072}{(UL/v)^{1/5}}$$

for turbulent boundary-layer flow with the "1/7 power" velocity profile. (Experiments show that the constant should be "adjusted" to a value of 0.074.)

8.84   A flat-bottomed barge, which is 75 ft long, 25 ft wide, and submerged to a depth of 5 ft, is to be pushed up a river at the rate of 5 mph. Estimate the power required to overcome skin friction when the water temperature is 60 F.

8.85   A sheet of plastic material 1/8 in. thick, with specific gravity, $SG = 1.5$, is dropped into a large tank containing water. The sheet is 2 ft high and 3 ft wide. It falls vertically.

Estimate the terminal velocity of the sheet, assuming the only drag is due to skin friction, and that the boundary layers are turbulent from the leading edge.

8.86 The vertical component of the landing speed of a parachute is 20 ft/sec. The parachute may be treated as an open hemisphere. The total weight of chute and jumper is 250 lbf. Assume a standard day.

Determine the diameter of the open parachute.

8.87 A rotary mixer is constructed from two circular discs, as shown in Fig. 8.68. The mixer is rotated in a large vessel containing a brine solution ($\rho = 2.10$ slug/ft^3) at 60 rpm. The drag on the rods, and motion induced in the liquid may be neglected.

Calculate the torque and horsepower required to drive the mixer.

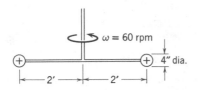

Fig. 8.68

8.88 A vehicle is built to try for the land speed record at the Bonneville Salt Flats, elevation 4400 ft. The engine delivers 500 hp to the rear wheels, and careful streamlining has resulted in a drag coefficient of 0.3, based on a 15 ft^2 frontal area.

Compute the theoretical maximum ground speed of the car (a) in still air and (b) with a 20 mph head wind.

8.89 An F-4 aircraft is slowed after landing by dual parachutes deployed from the rear. Each parachute is 20 ft in diameter, and may be assumed to have the same drag coefficient as an open hemisphere facing upstream. The F-4 weighs 32,000 lbf, and lands at 160 knots.

Estimate the time required to decelerate the aircraft to 100 knots, assuming the brakes are not used, and drag of the aircraft is negligible.

8.90 A typical full-size American sedan (1973 Ford LTD) has a frontal area of 23.4 ft^2, and a drag coefficient of 0.5. Plot a curve of power required to overcome drag (in hp) versus road speed in standard air. If rolling resistance is 1.5 percent of curb weight (4500 lbf), determine the speed at which aerodynamic forces exceed frictional resistance. How much power is required to cruise at 75 mph, and at 60 mph?

8.91 At low Reynolds number ($Re < 1$), drag of a sphere can be predicted analytically from the complete equations of motion. The result is called Stokes' flow, and the drag is given by

$$\text{Drag} = 3\pi\mu D V$$

where μ is the fluid viscosity, and D and V are the sphere diameter and relative velocity, respectively. A small sphere ($D = 0.25$ in.) is observed to fall through castor oil at a terminal velocity of 0.2 ft/sec. The temperature is 70 F.

Show that for Stokes flow

$$C_D = \frac{24}{Re}; \qquad Re \leq 1$$

and compute the drag coefficient for the sphere.

Determine the density of the sphere. If dropped in water, would the sphere fall slower or faster? Why?

8.92   A spherical hydrogen-filled balloon 2 ft in diameter exerts an upward force on a restraining string of 0.30 lbf when held stationary in standard air at zero velocity. With a wind speed of 10 ft/sec, the string holding the balloon makes an angle of 60° with the horizontal. Calculate the drag coefficient of the balloon under these conditions, neglecting the weight of the string.

8.93   A certain light airplane carries 40 ft of external guy wires stretched normal to the direction of motion. The diameter of the wires is 0.25 in. Basing calculations on two-dimensional flow around the wires, what power saving may be effected by removing the wires if the speed is 180 mph in standard air at sea level?

8.94   A water tower consists of a 40 ft diameter sphere on top of a vertical tower 50 ft tall. Estimate the bending moment exerted on the base of the tower due to the aerodynamic force imposed by a 60 mph wind on a standard day. Interference at the joint between the sphere and tower and drag of tower may be neglected.

8.95   The radio antenna on a car is about 0.25 in. in diameter and 4 ft long. Estimate the torque that tends to snap it off, if the car is driven at 75 mph on a standard day.

8.96   A spherical balloon contains helium and ascends through air at 14 psia, 40 F. The balloon and its payload together weigh 300 lbf. Determine the required diameter if it is to ascend at 10 ft/sec.

8.97   The guy wires on the light plane in Problem 8.93 are to be streamlined rather than removed. Data from Fig. 8.35 may be used to select an optimum faired shape for the guys.

Evaluate the maximum power saving that results from fairing the wires.

8.98   A light plane has a 30 ft effective wingspan and a 5 ft chord. It was originally designed to use a conventional (NACA 0012) airfoil section. With this airfoil, its cruising speed on a standard day near sea level is 135 mph. A conversion to a laminar flow (NACA 66_1–012 section) airfoil is proposed.

Determine the cruising speed that can be achieved with the new airfoil section for the same power.

8.99 An aircraft is flying in level flight at a speed of 120 ft/sec through air at standard conditions. The lift coefficient at this speed is 1.0 and the drag coefficient is 0.12. The aircraft weighs 2000 lbf.

Calculate the effective lift area for the craft.

8.100 Consider a kite weighing 3 lbf as a flat plate with an area of 10 ft². It is flown in standard air moving horizontally at 29 ft/sec. The kite makes an angle of 8° with the horizontal. Assume the lift coefficient is given by the equation

$$C_L = 2\pi \sin \alpha$$

where α is the angle of attack.

If the string makes an angle of 45° with the horizontal, determine the tension in the string.

8.101 The foils of a hydrofoil type watercraft have an effective area of 7 ft². Their co-efficients of lift and drag are 1.6 and 0.5, respectively. The total weight of the craft in running trim is 4000 lbf.

Determine the minimum velocity at which the craft is supported by the hydro-foils. At this velocity, find the power required to overcome water resistance. If the craft is fitted with a 150 hp engine, estimate its top speed.

8.102 An airplane with an effective lift area of 250 ft² is fitted with airfoils of NACA 23012 section (Fig. 8.41). The maximum flap setting that can be used on takeoff corres-ponds to configuration ② in Fig. 8.41.

Determine the maximum gross weight possible for the airplane if its takeoff speed is 80 mph (neglect added lift due to ground effect). Calculate the thrust required for takeoff.

8.103 An airplane that weighs 10,000 lbf is flown at constant elevation and speed on a circular path at 150 mph. The flight circle has a radius of 3000 ft, and produces a radial acceleration of 0.5 "G's" on the aircraft. The plane has an effective lifting area of 220 ft², and is fitted with NACA 0012 section airfoils.

Determine the drag on the aircraft, and the power required, in hp.

8.104 Aircraft and missiles flying at high altitude may have regions of laminar flow that become turbulent at lower altitudes at the same speed.

Explain a possible mechanism for this effect. Support your answer with calcula-tions based on standard atmosphere data.

8.105 Turbojet powered aircraft operate most efficiently at high altitudes because the air temperature is low, which increases the Carnot efficiency of the engine thermo-dynamic cycle. Because the density is low at high altitude, higher lift coefficients are needed to support the aircraft weight. The stalling speed at altitude is also high.

Consider an aircraft fitted with NACA 66_1–012 airfoil section wings. At sea level and 400 knots air speed, the required lift coefficient is 0.2. Boundary layer control devices cannot be used at air speeds above 200 knots.

Determine the stalling speed of this aircraft when flown at 30,000 ft on a standard day.

8.106   A tennis ball weighs 2 oz and is 2.5 in. in diameter. A typical shot is hit at a speed of 60 ft/sec, with a spin of 5500 rpm.

Determine the aerodynamic lift acting on the ball, and compute the radius of curvature of the path that is due to this force, if it acts horizontally.

8.107   The weight flowrate in a water flow system is determined by weighing the discharge collected over a timed interval to be 0.6 lbf/sec. The scales used can be read to the nearest 0.1 lbf, and the stopwatch is accurate to 1/5 sec.

Determine the precision with which the flowrate can be calculated for time intervals of (a) 10 sec, and (b) 1 min.

8.108   An "ergometer" is used to measure the rate at which air is consumed by a research subject running on a treadmill. The atmospheric temperature and pressure are 72 F and 29.6 in. of mercury, respectively. An inverted bell, which is used to measure airflow, is 18 in. in diameter. During a 30 sec test run, the bell rises 43 mm.

Determine the rate at which the subject consumes oxygen.

8.109   Water at 150 F flows through a 3 in. diameter orifice installed in a 6 in. inside diameter pipe. The pressure taps are located 1 pipe diameter upstream and $\frac{1}{2}$ pipe diameter downstream from the orifice plate. The flowrate is 300 gpm.

Determine the pressure drop across the pressure taps.

8.110   Ethyl alcohol ($SG = 0.75$) flows through a 12 in. line in a distillery. The flowrate is not expected to exceed 250 lbm/sec. A manometer with a range of 36 in. of water is available for use with an orifice meter.

Specify a recommended orifice diameter for use with this system. What minimum rate of flow could be measured within 10 percent accuracy if the manometer least count is 0.1 in. of water?

8.111   Airflow in a test of an internal combustion engine is to be measured using a flow nozzle installed in a plenum. The maximum flowrate required is 700 ft^3 of standard air per minute. The maximum pressure drop should be 10 in. of water or less, to avoid loading the engine. The manometer can be read accurately down to 1 in. of water pressure difference.

Determine the flow nozzle diameter that should be specified. Find the minimum rate of airflow (cfm) that can be metered accurately using this setup.

8.112   Consider a flow nozzle installation in a pipe (Fig. 8.69). Apply the basic equations to the control volume shown.

Evaluate the head loss from section ① to section ③. Express the result as a fraction of the quantity $(p_1 - p_2)/\rho$.

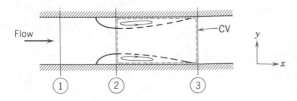

Fig. 8.69

8.113 A venturi meter with a 3 in. diameter throat is placed in a 6 in. line carrying water at 80 F. The pressure drop between the upstream tap and the venturi throat is 12 in. of mercury.

Compute the rate of flow in gallons per minute.

8.114 Air flows through the venturi meter described in Problem 8.113. Assume, for convenience, that the upstream pressure is 60 psia, and that the temperature is everywhere constant at 70 F.

Determine the maximum possible rate of air flow (in cfm) for which the assumption of incompressible flow is a valid engineering approximation. Compute the corresponding pressure reading on a mercury manometer.

references

1. L. F. Moody, "Friction Factors for Pipe Flow," *Trans. ASME*, **66**, 8, November 1944, pp. 671–684.
2. Crane Co., *Flow of Fluids through Valves, Fittings and Pipe*, Technical Paper No. **410**, Crane, Industrial Products Group, 4100 S. Kedzie Ave., Chicago, Illinois, 60632, 1957.
3. **ASHRAE**, *ASHRAE Guide and Data Book—Equipment Volume*. New York: American Society of Heating, Refrigeration and Air Conditioning Engineers, 1969.
4. C. I. Bradley, and D. J. Cockrell, "The Response of Diffusers to Flow Conditions at Their Inlet," Paper No. **5**, *Symposium on Internal Flows*, University of Salford, Salford, England, April 1971, pp. A32–A41.
5. G. Sovran, and E. D. Klomp, "Experimentally Determined Optimum Geometries for Rectilinear Diffusers with Rectangular, Conical or Annular Cross-Section," in *Fluid Mechanics of Internal Flow*, ed. by G. Sovran. Amsterdam: Elsevier, 1967, pp. 270–319.
6. W. J. Feiereisen, "An Experimental Investigation of Incompressible Flow without Swirl in R-Radial Diffusers," Ph.D. Thesis, School of Mechanical Engineering, Purdue University, June 1971.

7. L. R. Reneau, J. P. Johnston, and S. J. Kline, "Performance and Design of Straight, Two-Dimensional Diffusers," *Trans. ASME, J. Basic Engrg.,* **89D**, 1, March 1967, pp. 141–150.

8. H. Cross, "Analysis of Flow in Networks of Conduits or Conductors," University of Illinois, *Bulletin 286,* November 1946.

9. *Trane Air Conditioning Manual.* La Crosse, Wisconsin: The Trane Co., 1965.

10. SAE, *Aerospace Applied Thermodynamics Manual,* 2nd ed. New York: Society of Automotive Engineers, 1969.

11. S. F. Hoerner, *The Fluid Dynamics of Drag,* 2nd ed. Midland Park, New Jersey: Published by the author, 1965.

12. A. H. Shapiro, *Shape and Flow, The Fluid Dynamics of Drag.* New York: Anchor, 1961 (paperback).

13. A. Fage, "Experiments on a Sphere at Critical Reynolds Numbers," Great Britain, Aeronautical Research Council, *Reports and Memoranda No. 1766,* 1937.

14. S. Goldstein, ed., *Modern Developments in Fluid Dynamics,* Vols. I and II. Oxford: Clarendon Press, 1938. (Also reprinted in paperback by Dover, New York, 1967.)

15. I. H. Abbott, and A. E. von Doenhoff, *Theory of Wing Sections, Including a Summary of Airfoil Data.* New York: Dover, 1959 (paperback).

16. ASME, *Flowmeters: Their Theory and Application.* New York: American Society of Mechanical Engineers, 1959.

introduction to compressible flow

In Chapter 4, we developed control volume formulations from the basic equations expressed in system form. The basic equations included conservation of mass and linear momentum, and the first and second laws of thermodynamics. In our work thus far, we have applied the continuity and momentum formulations to a variety of problems to solve for the pressure and velocity, but we have not considered changes in fluid density. The energy equation was required only to solve for the head loss due to friction in duct flows.

"Compressible" flow implies appreciable variations in density throughout a flow field. Compressibility becomes important at high flow velocities; large changes in velocity bring about corresponding changes in pressure. For gas flows, these pressure changes are accompanied by variations in both density and temperature. Since two additional variables are encountered in treating compressible flow, two additional equations are needed. Both the energy equation and an equation of state must be applied to solve problems in compressible flow.

In our study of compressible fluid flow, we will deal primarily with the steady, one-dimensional flow of an ideal gas. Although many real flows of interest are more complex, these restrictions will allow us to concentrate on the effects of basic flow processes.

461

A review of the thermodynamics necessary for the study of compressible flows, including the equation of state and "$T\,ds$" equations, is presented in the next section.

9–1 Review of Thermodynamics

The pressure, density, and temperature of a substance may be related by an *equation of state*. Although many substances are quite complex in behavior, experience shows that most gases of engineering interest, at moderate pressure and temperature, are well represented by the *ideal gas* equation of state,

$$p = \rho R T \qquad (9.1)$$

where R is a constant for each gas.[1] Actually,

$$R = \frac{R_u}{M_w}$$

where R_u is the universal gas constant ($R_u = 1544$ ft-lbf/lbmole-R) and M_w is the molecular weight of the gas. Although no real substance behaves exactly as an ideal gas, it is well to point out[2] that Eq. 9.1 is in error by less than 1 percent for air at room temperature for pressures as high as 400 psia. For air at 1 atm, the equation is less than 1 percent in error for temperatures as low as -200 F.

The ideal gas has other features which are simple and useful. In general, the internal energy of a substance may be expressed as $u = u(v, T)$. Then

$$du = \left(\frac{\partial u}{\partial T}\right)_v dT + \left(\frac{\partial u}{\partial v}\right)_T dv$$

where $v = 1/\rho$, the specific volume. The specific heat at constant volume is defined as

$$c_v \equiv \left(\frac{\partial u}{\partial T}\right)_v$$

so that

$$du = c_v\, dT + \left(\frac{\partial u}{\partial v}\right)_T dv$$

You may recall from your earlier course in thermodynamics that for any substance that follows the equation of state, $p = \rho R T$,

$$\left(\frac{\partial u}{\partial v}\right)_T = 0 \quad \text{and hence } u = u(T) \qquad (9.2)$$

---

[1] For air, $R = 53.3$ ft-lbf/lbm-R.

[2] See, for example, *Engineering Thermodynamics* by J. B. Jones and G. A. Hawkins (New York: John Wiley, 1960), Chapter 4.

Consequently

$$du = c_v \, dT \tag{9.3}$$

for an ideal gas, which means that internal energy and temperature changes may be related if c_v is known. Furthermore, since $u = u(T)$ by Eq. 9.2, then $c_v = c_v(T)$ by Eq. 9.3.

The enthalpy of a substance is defined as $h \equiv u + p/\rho$. For an ideal gas, $p = \rho RT$ and, hence, $h = u + RT$. Since, for an ideal gas $u = u(T)$, then h also must be a function of temperature alone.

To obtain a relation between h and T, we express h in its most general form as

$$h = h(p, T)$$

or

$$dh = \left(\frac{\partial h}{\partial T}\right)_p dT + \left(\frac{\partial h}{\partial p}\right)_T dp$$

and

$$dh = c_p \, dT + \left(\frac{\partial h}{\partial p}\right)_T dp$$

from the definition, $c_p \equiv (\partial h/\partial T)_p$. We have shown that for an ideal gas h is a function of T only. Consequently $(\partial h/\partial p)_T = 0$ and

$$dh = c_p \, dT \tag{9.4}$$

Again, since h is a function of T alone, Eq. 9.4 requires that c_p be a function of T only.

The specific heats for an ideal gas have been shown to be functions of temperature only. We should also note that their difference is a constant. From

$$h = u + RT$$

we can write

$$dh = du + R \, dT$$

Combining this with Eq. 9.4, we have

$$dh = c_p \, dT = du + R \, dT = c_v \, dT + R \, dT$$

and

$$c_p - c_v = R \tag{9.5}$$

The ratio of specific heats is defined as

$$k \equiv \frac{c_p}{c_v} \tag{9.6}$$

Using the definition of k, Eq. 9.5 can be solved for either c_p or c_v in terms of k and R. Thus

$$c_p = \frac{kR}{k - 1}$$

and

$$c_v = \frac{R}{k - 1}$$

In all of the equations we have developed so far, the specific heats have been assumed to be functions of temperature. Within reasonable temperature ranges, the specific heats of an ideal gas may be treated as constants for calculations of engineering accuracy. Under these conditions

$$u_2 - u_1 = \int_{u_1}^{u_2} du = \int_{T_1}^{T_2} c_v\, dT = c_v(T_2 - T_1) \tag{9.7a}$$

$$h_2 - h_1 = \int_{h_1}^{h_2} dh = \int_{T_1}^{T_2} c_p\, dT = c_p(T_2 - T_1) \tag{9.7b}$$

These equations obviously may be used to advantage in simplifying analyses.

Values of M_w, c_p, c_v, R, and k are given in Appendix A, Table A.6.

The property entropy, introduced as a consequence of the second law of thermodynamics, is an extremely useful property in the analysis of compressible flows. State diagrams, particularly temperature-entropy (Ts) diagrams, are a valuable aid in the physical interpretation of analytical results. Since we shall make extensive use of the Ts diagram in solving compressible flow problems, it is worthwhile to review briefly some of the useful relationships involving the property entropy.[3]

Entropy is defined by the equation

$$\Delta S \equiv \int_{\text{rev}} \frac{\delta Q}{T} \quad \text{or} \quad dS = \left(\frac{\delta Q}{T}\right)_{\text{rev}} \tag{9.8}$$

The inequality of Clausius, deduced from the second law, states that

$$\oint \frac{\delta Q}{T} \leq 0$$

As a consequence of the second law these results can be extended to

$$dS \geq \frac{\delta Q}{T} \quad \text{or} \quad T\, dS \geq \delta Q \tag{9.9a}$$

---

[3] For a detailed discussion see, for example, J. B. Jones and G. A. Hawkins, *Engineering Thermodynamics* (New York: John Wiley, 1960), Chapter 9.

For *reversible* processes, the equality holds, that is

$$T\,ds = \delta Q/dm \quad \text{(reversible process)} \tag{9.9b}$$

while the inequality holds for *irreversible* processes

$$T\,ds > \delta Q/dm \quad \text{(irreversible process)} \tag{9.9c}$$

For an *adiabatic* process, $\delta Q/dm \equiv 0$. Thus

$$ds = 0 \quad \text{(reversible adiabatic process)}$$

and

$$ds > 0 \quad \text{(irreversible adiabatic process)}$$

A useful relationship among properties (p, v, T, s, u) can be obtained from a joint consideration of the first and second laws. The result is the familiar $T\,ds$ equation

$$T\,ds = du + p\,dv \tag{9.10a}$$

This is a relationship among properties, valid for all processes between equilibrium states. While it is based on the first and second laws, in itself it is a statement of neither.

An alternate form of Eq. 9.10a can be obtained by substituting

$$du = d(h - pv) = dh - p\,dv - v\,dp$$

Then

$$T\,ds = dh - v\,dp \tag{9.10b}$$

For an ideal gas, the entropy change can be readily evaluated from the $T\,ds$ equations

$$ds = \frac{du}{T} + \frac{p}{T}\,dv = c_v\frac{dT}{T} + R\frac{dv}{v}$$

$$ds = \frac{dh}{T} - \frac{v}{T}\,dv = c_p\frac{dT}{T} - R\frac{dp}{p}$$

For constant specific heats, the equations may be integrated to yield

$$s_2 - s_1 = c_v\ln\frac{T_2}{T_1} + R\ln\frac{v_2}{v_1}$$

$$s_2 - s_1 = c_p\ln\frac{T_2}{T_1} - R\ln\frac{p_2}{p_1}$$

These equations are so easy to derive when needed, that special case formulas are unnecessary.

Example 9.1

Air flows through a long duct of constant area at a rate of 0.3 lbm/sec. A short section of the duct is cooled by liquid nitrogen that surrounds the duct. The rate of heat loss in this section is 14.3 Btu/sec from the air. The pressure, temperature, and velocity at the entrance to the cooled section are 27.2 psia, 790 R, and 690 ft/sec, respectively. At the outlet, the pressure and temperature are 30 psia and 615 R. Compute the duct cross-section area, and the changes in enthalpy, internal energy, and entropy for this flow.

EXAMPLE PROBLEM 9.1

GIVEN:

Air flows steadily through a short section of constant area duct which is cooled by liquid nitrogen.

$$T_1 = 790 \text{ R} \qquad T_2 = 615 \text{ R}$$

$$p_1 = 27.2 \text{ psia} \qquad p_2 = 30 \text{ psia}$$

$$V_1 = 690 \text{ ft/sec}$$

FIND:

(a) The duct area
(b) Δh
(c) Δu
(d) Δs

SOLUTION:

The duct area may be found from the continuity equation.

Basic equation:
$$0 = \overset{= \, 0(1)}{\cancel{\frac{\partial}{\partial t} \iiint_{CV} \rho \, d\forall}} + \iint_{CS} \rho \vec{V} \cdot d\vec{A}$$

Assumptions: (1) Steady flow
(2) Uniform flow at each section
(3) Ideal gas

Then

$$0 = \{-|\rho_1 V_1 A_1|\} + \{|\rho_2 V_2 A_2|\}$$

EXAMPLE PROBLEM 9.1 (continued)

or

$$\dot{m} = \rho_1 V_1 A = \rho_2 V_2 A$$

since $A = A_1 = A_2 = $ constant. Using the ideal gas relation, $p = \rho RT$, we find

$$\rho_1 = \frac{p_1}{RT_1} = \frac{27.2 \text{ lbf}}{\text{in.}^2} \times \frac{144 \text{ in.}^2}{\text{ft}^2} \times \frac{\text{lbm-R}}{53.3 \text{ ft-lbf}} \times \frac{1}{790 \text{ R}} = 0.093 \text{ lbm/ft}^3$$

From continuity

$$A = \frac{\dot{m}}{\rho_1 V_1} = \frac{0.3 \text{ lbm}}{\text{sec}} \times \frac{\text{ft}^3}{0.093 \text{ lbm}} \times \frac{\text{sec}}{690 \text{ ft}} = 0.00468 \text{ ft}^2 \qquad \longleftarrow \qquad A$$

For an ideal gas, $dh = c_p \, dT$, so

$$\Delta h = h_2 - h_1 = \int_{T_1}^{T_2} c_p \, dT = c_p(T_2 - T_1)$$

$$\Delta h = 0.240 \frac{\text{Btu}}{\text{lbm-R}} \times (615 - 790) \text{ R} = -42.0 \frac{\text{Btu}}{\text{lbm}} \qquad \longleftarrow \qquad \Delta h$$

Also

$$du = c_v \, dT,$$

so

$$\Delta u = u_2 - u_1 = \int_{T_1}^{T_2} c_v \, dT = c_v(T_2 - T_1)$$

$$\Delta u = 0.171 \frac{\text{Btu}}{\text{lbm-R}} \times (615 - 790) \text{ R} = -29.9 \frac{\text{Btu}}{\text{lbm}} \qquad \longleftarrow \qquad \Delta u$$

The entropy change may be obtained from the $T \, ds$ equation

$$T \, ds = dh - v \, dp$$

$$ds = \frac{dh}{T} - \frac{v \, dp}{T} = c_p \frac{dT}{T} - R \frac{dp}{p}$$

or

$$\Delta s = s_2 - s_1 = \int_{T_1}^{T_2} c_p \frac{dT}{T} - \int_{p_1}^{p_2} R \frac{dp}{p} = c_p \ln \frac{T_2}{T_1} - R \ln \frac{p_2}{p_1}$$

$$= 0.240 \frac{\text{Btu}}{\text{lbm-R}} \ln \left(\frac{615}{790}\right) - \frac{53.3 \text{ ft-lbf}}{\text{lbm-R}} \times \frac{\text{Btu}}{778 \text{ ft-lbf}} \ln \left(\frac{30}{27.2}\right)$$

$$\Delta s = (0.240)(-0.250) - (0.0685)(0.0980) = -0.0667 \frac{\text{Btu}}{\text{lbm-R}} \qquad \longleftarrow \qquad \Delta s$$

review of thermodynamics

For the special case of an isentropic process, $ds = 0$, and the $T\,ds$ equations reduce to

$$0 = du + p\,dv$$
$$0 = dh - v\,dp$$

For an ideal gas, we have

$$0 = c_v\,dT + p\,dv$$
$$0 = c_p\,dT - v\,dp$$

Solving for dT,

$$dT = \frac{v\,dp}{c_p}$$

$$dT = -\frac{p\,dv}{c_v}$$

Combining and rearranging

$$\frac{v\,dp}{c_p} = -\frac{p\,dv}{c_v}$$

or

$$\frac{dp}{p} + \frac{c_p}{c_v}\frac{dv}{v} = \frac{dp}{p} + k\frac{dv}{v} = 0$$

Integrating (for $k = $ constant)

$$\ln p + k \ln v = \ln c$$

or

$$\ln p + \ln v^k = \ln c$$

Taking antilogarithms, this equation reduces to

$$pv^k = \text{constant} \qquad\qquad (9.11a)$$

or

$$\frac{p}{\rho^k} = \text{constant} \qquad\qquad (9.11b)$$

Either form of Eq. 9.11 is valid for any isentropic flow of an ideal gas.

Qualitative information that is useful in drawing state diagrams also can be obtained from the $T\,ds$ equations. The slopes of lines of constant pressure and of constant volume on the Ts diagram are evaluated in Example Problem 9.2 to complete our review of thermodynamic fundamentals.

468   **9/introduction to compressible flow**

Example 9.2

For an ideal gas, determine the slope of (a) a constant volume line and (b) a constant pressure line, in the *Ts* plane.

EXAMPLE PROBLEM 9.2

GIVEN:

An ideal gas.

FIND:

(a) Slope of constant volume line in *Ts* plane.
(b) Slope of constant pressure line in *Ts* plane.

SOLUTION:

The *T ds* equations, Eqs. 9.10a and 9.10b may be applied.

$$T\,ds = du + p\,dv \qquad (9.10a)$$

$$T\,ds = dh - v\,dp \qquad (9.10b)$$

Substituting for an ideal gas, $du = c_v\,dT$ and $dh = c_p\,dT$, we obtain

$$T\,ds = c_v\,dT + p\,dv$$

$$T\,ds = c_p\,dT - v\,dp$$

For a constant volume process, $dv = 0$. From the first equation

$$\left.\frac{dT}{ds}\right|_{\substack{\text{Constant} \\ \text{volume}}} = \left.\frac{\partial T}{\partial s}\right|_v = \frac{T}{c_v} \qquad \longleftarrow \underline{\text{Constant volume slope}}$$

which is the slope of a line of constant volume. For a constant pressure process, $dp = 0$. From the second equation

$$\left.\frac{dT}{ds}\right|_{\substack{\text{Constant} \\ \text{pressure}}} = \left.\frac{\partial T}{\partial s}\right|_p = \frac{T}{c_p} \qquad \longleftarrow \underline{\text{Constant pressure slope}}$$

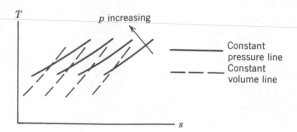

Note that the slope of each line is proportional at any point to the absolute temperature. Furthermore, at any point a constant volume line has a slope $c_p/c_v = k$ times
larger than a line of constant pressure.

9–2 Propagation of Sound Waves

9–2.1 SPEED OF SOUND

The terms supersonic and subsonic are familiar terms; they refer to velocities that
are respectively greater than and less than the speed of sound. The speed of sound
(or of a pressure wave of infinitesimal strength) is an important characteristic
parameter in a compressible flow. We have previously (Chapters 2 and 7) introduced the Mach number, $M = V/c$, the ratio of the local flow speed to the local
speed of sound, as an important nondimensional parameter characterizing compressible flows. Before proceeding with our study of compressible flows it is
logical to obtain a general relation for calculating the sonic speed.

Consider the propagation of a sound wave of infinitesimal strength into an
undisturbed medium as shown in Fig. 9.1a. We are interested in relating the speed
of propagation of the wave, c, to fluid property changes across the wave. If the
pressure and density in the undisturbed medium ahead of the wave are denoted
by p and ρ, the passage of the wave will cause them to undergo an infinitesimal
change and become $p + dp$ and $\rho + d\rho$. Since the wave is propagating into a
stationary fluid, the velocity ahead of the wave is zero. The velocity behind the
wave will be dV_x to the left.

Unfortunately the flow of Fig. 9.1a appears unsteady to a stationary observer,
observing the motion of the wave from a fixed point on the ground. However, we
can put a moving control volume around a segment of the wave, as shown in
Fig. 9.1b; the flow will then appear steady to an observer located *on* this inertial
control volume. The velocity approaching the control volume will be c and the
velocity leaving, $c - dV_x$.

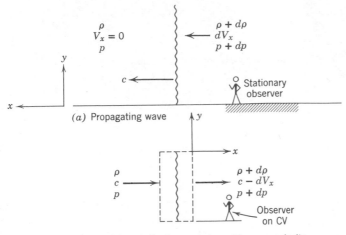

Fig. 9.1 *Propagating sound wave showing control volume chosen for analysis.*

The basic equations may be applied to the differential control volume shown in Fig. 9.1b (we shall hereafter use V_x for the x component of velocity to avoid confusion with internal energy, u):

a. Continuity Equation

Basic equation:

$$0 = \overset{=\,0(1)}{\frac{\partial}{\partial t}\iiint\limits_{CV} \rho\, d\forall} + \iint\limits_{CS} \rho\vec{V}\cdot d\vec{A} \qquad (4.14)$$

Assumptions: (1) Steady flow
(2) Uniform flow

Then

$$0 = \{-|\rho c A|\} + \{|(\rho + d\rho)(c - d\dot{V}_x)A|\} \qquad (9.12a)$$

or

$$0 = -\rho\!\!\!\diagup\!\!\! c A + \rho\!\!\!\diagup\!\!\! c A - \rho\, dV_x A + d\rho c A - \overset{\simeq\,0}{d\rho\, d\!\!\!\diagup\!\!\! V_x A}$$

471   **propagation of sound waves**

Neglecting the product of differentials, $d\rho\, dV_x$, compared to either $d\rho$ or dV_x, we obtain

$$0 = -\rho A\, dV_x + cA\, d\rho$$

or

$$dV_x = \frac{c}{\rho}\, d\rho \qquad\qquad (9.12b)$$

b. Momentum Equation

Basic equation:

$$F_{S_x} + \overset{=0(3)}{\cancel{F_{B_x}}} - \underset{\text{CV}}{\int\int\int} \overset{=0(4)}{\cancel{a_{rf_x}}} \rho\, d\forall = \overset{=0(1)}{\cancel{\frac{\partial}{\partial t}}} \underset{\text{CV}}{\int\int\int} V_x \rho\, d\forall + \underset{\text{CS}}{\int\int} V_x \rho \vec{V} \cdot d\vec{A} \quad (4.29a)$$

Assumptions: (3) $F_{B_x} = 0$
(4) $a_{rf_x} = 0$

The surface forces acting on the infinitesimal control volume are

$$F_{S_x} = dR_x + pA - (p + dp)A$$

where dR_x represents all forces applied to the horizontal portions of the control surface shown in Fig. 9.1b. However, since we consider only a portion of a moving sound wave, $dR_x = 0$, because there is no relative motion along the wave. Thus the surface force simplifies to

$$F_{S_x} = -A\, dp$$

Substituting into the basic equation

$$-A\, dp = c\{-|\rho cA|\} + (c - dV_x)\{|(\rho + dp)(c - dV_x)A|\}$$

Using the continuity equation in the form of Eq. 9.12a, this reduces to

$$-A\, dp = c\{-|\rho cA|\} + (c - dV_x)\{|\rho cA|\}$$
$$= (-c + c - dV_x)\{|\rho cA|\}$$
$$-A\, dp = -\rho cA\, dV_x$$

or

$$dV_x = \frac{1}{\rho c} dp \qquad (9.12c)$$

Combining Eqs. 9.12b and 9.12c, we obtain

$$dV_x = \frac{c}{\rho} d\rho = \frac{1}{\rho c} dp$$

$$dp = c^2 d\rho$$

or

$$c^2 = \frac{dp}{d\rho}$$

To evaluate the derivative of a thermodynamic property, we must specify the property to be held constant during differentiation. For the present case, the limit at vanishing strength of a sound wave will be

$$\lim_{\text{strength} \to 0} \frac{dp}{d\rho} = \frac{\partial p}{\partial \rho}\bigg|_{s = \text{constant}}$$

Perhaps a more physical justification for the assumption of isentropic propagation is that an infinitesimal pressure change is reversible. Hence, since there is too little time for heat transfer, the process is reversible and adiabatic. Thus the speed of propagation of a sound wave is given by

$$c = \sqrt{\frac{\partial p}{\partial \rho}\bigg|_s}$$

For an ideal gas, the pressure and density in isentropic flow are related by

$$\frac{p}{\rho^k} = \text{constant} \qquad (9.11b)$$

as shown previously. Taking logarithms,

$$\ln p - k \ln \rho = \ln c$$

and differentiating

$$\frac{dp}{p} - k\frac{d\rho}{\rho} = 0$$

$$\left.\frac{\partial p}{\partial \rho}\right|_s = k\frac{p}{\rho}$$

But $p/\rho = RT$, so finally

$$c = \sqrt{kRT} \tag{9.13}$$

for an ideal gas.

The important feature of sound propagation in an ideal gas, as shown by Eq. 9.13, is that the speed of sound is a function of temperature only. The variation in atmospheric temperature, T, with altitude on a standard day has been discussed in Chapter 3. The corresponding variation in c is computed in Example Problem 9.3, and sketched as a function of altitude.

Example 9.3

Compute the speed of sound at sea level in standard air. Evaluate the speed of sound and plot for altitudes to 40,000 ft.

EXAMPLE PROBLEM 9.3

GIVEN:

Air under standard atmospheric conditions.

FIND:

(a) The speed of sound at sea level.
(b) Plot the speed of sound for altitudes to 40,000 ft.

SOLUTION:

Assume an ideal gas.

Computing equation: $c = \sqrt{kRT}$

From Fig. 3.3, the temperature at sea level on a standard day is 59 F. Thus

$$c = \left(1.4 \times \frac{53.3 \text{ ft-lbf}}{\text{lbm-R}} \times (460 + 59)\,\text{R} \times \frac{32.2 \text{ lbm}}{\text{slug}} \times \frac{\text{slug-ft}}{\text{lbf-sec}^2}\right)^{1/2} = 1117 \text{ ft/sec} \quad \underleftarrow{\quad c}$$

Temperatures at various altitudes may also be found from Fig. 3.3. The resulting sound speeds are plotted below:

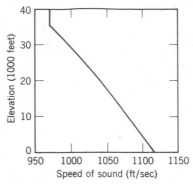

From the plot we see that the speed of sound in air on a standard day varies from 1117 ft/sec at sea level to 971 ft/sec at altitudes above 36,000 ft.

9–2.2 TYPES OF FLOW—THE MACH CONE

Flows for which $M < 1$ are termed *subsonic*, while those for which $M > 1$ are *supersonic*. If there are both regions where $M < 1$ and $M > 1$ in a flow, the flow-field is called *transonic*. Although most flows within our experience are subsonic, there are important practical cases where $M \geq 1$ occurs in a flowfield. Perhaps the most obvious are supersonic aircraft and the current trend toward transonic flows in aircraft compressors and fans, and in steam turbines. Yet another flow regime, *hypersonic* flow ($M \gtrsim 5$), is of interest in missile and reentry vehicle design. Some important qualitative differences between subsonic and supersonic flows can be deduced from the properties of a simple moving sound source.

Consider a point source capable of emitting an instantaneous infinitesimal disturbance. These disturbances propagate in all directions with speed, c. At any time, t, the location of the wave front from the disturbance emitted at time, t_0, will be represented by a sphere with radius, $c(t - t_0)$, whose center is coincident with the location of the disturbance at time, t_0.

We are interested in determining the nature of the disturbance propagation for different velocities of the moving source. Four cases are shown in Fig. 9.2:

1. $V = 0$. The sound pattern propagates uniformly in all directions. Any given sound pulse is located at the instant Δt after emission at radius $c\Delta t$ from the source. At the instant $2\Delta t$, of course the radius is $c(2\Delta t)$. Any given wave front is spherical in shape; all wave fronts are concentric spheres.

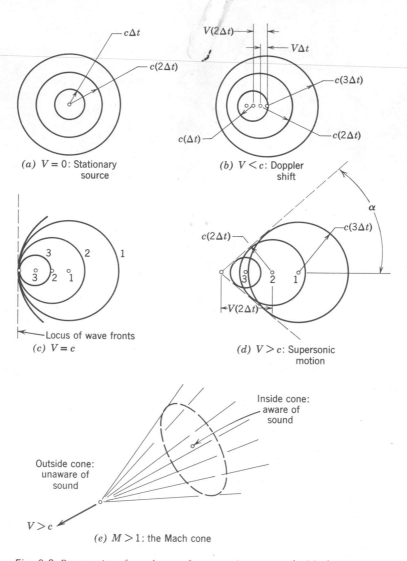

(a) $V = 0$: Stationary source

(b) $V < c$: Doppler shift

(c) $V = c$

Locus of wave fronts

(d) $V > c$: Supersonic motion

Outside cone: unaware of sound

Inside cone: aware of sound

$V > c$

(e) $M > 1$: the Mach cone

Fig. 9.2 *Propagation of sound waves from a moving source: the Mach cone.*

2. $0 < V < c$. The concentricity of the wave pattern is lost. Any individual wave front is spherical, but each successive sound is emitted from a different position, $V\Delta t$ distant from the previous position.

    If you imagine the circles shown to be the amplitude peaks of a sinusoidal tone, the same qualitative picture holds for a moving source of continuous sound. If this source moves with the constant velocity, V, we can imagine the whole pattern shown in Fig. 9.2b carried along with the emitter. Thus a stationary observer would hear more peaks per unit time as the source approaches him than after it passes. This is known as the "Doppler effect." (Have you ever heard a fast-moving train whistle through a crossing?)

3. $V = c$. The locus of the leading surfaces of all waves will be a plane at the source, perpendicular to the path of motion. No sound wave will travel in front of the source. Therefore no observer in front of the source will hear it approaching.

4. $V > c$. In this case, the locus of the leading surfaces of the sound waves will be a cone. Again, no sound will be heard in front of this cone.

The cone angle can be related to the Mach number at which the source moves. From the geometry of Fig. 9.2d,

$$\sin \alpha = \frac{c}{V} = \frac{1}{M}$$

or

$$\alpha = \sin^{-1}\left(\frac{1}{M}\right)$$

The cone depicted in Fig. 9.2e is termed the Mach cone,[4] and angle, α, is referred to as the Mach angle. The regions inside and outside the cone are sometimes called the *zone of action* and the *zone of silence*, respectively.

9–3 *Reference State: Stagnation Properties*

If we wish to describe the state of a fluid at any point in a flow field, we must specify two independent thermodynamic properties (usually pressure and temperature), plus the fluid velocity at the point in question.[5]

In our discussion of compressible flow, we shall find it very convenient to employ the stagnation state as a reference state. The stagnation state is the state characterized by a condition of zero velocity. Consider a point in a flow field at which the temperature is T, the pressure, p, and the velocity, V. The stagnation state at that point in the flow field would be characterized by the stagnation pressure, p_0, stagnation temperature, T_0, and zero velocity. Thus the stagnation properties at any point in a flow field are those properties that would exist at that point in the flow if the velocity at the point were reduced to zero.

Before we can proceed to calculate the stagnation properties, we must specify the process by which the fluid is imagined to undergo this deceleration to zero velocity.

---

[4] See the Shell film, *Transonic Flight,* for a well-presented discussion of the Mach cone.

[5] You will recall from your thermodynamics course that the state of a pure substance, in the absence of motion, gravity, and surface, magnetic, or electrical effects, is defined by two independent thermodynamic properties.

Recall in the discussion of incompressible flow in Chapter 6, integration of Euler's equation for frictionless, incompressible flow along a streamline led to the Bernoulli equation

$$\frac{p}{\rho} + \frac{V^2}{2} + gz = \text{constant} \tag{6.10}$$

Then for incompressible flow, a frictionless deceleration leads to a stagnation pressure, p_0, given by

$$p_0 = p + \rho\frac{V^2}{2} \quad \text{or} \quad \frac{p_0}{p} = 1 + \frac{\rho V^2}{2p} \tag{9.14}$$

For compressible flow we again employ a frictionless deceleration process; in addition we specify that the deceleration process be adiabatic. In short we specify an isentropic deceleration process. This then leads us to the definition of local isentropic stagnation properties:

> *Local isentropic stagnation properties* are those properties that would be obtained at any point in a flow field if the fluid at that point were decelerated from local conditions to zero velocity following a frictionless, adiabatic, that is, *isentropic,* process.

Isentropic stagnation properties are reference properties that may be determined at any point in a fluid flow. The variations in these properties from point to point in a flow field give information about the flow process between points. This will become apparent in our discussion of one-dimensional flow cases. However, before proceeding let us first develop equations for computing the local isentropic stagnation properties for the flow of an ideal gas.

9–3.1 LOCAL ISENTROPIC STAGNATION PROPERTIES FOR THE FLOW OF AN IDEAL GAS

The stagnation properties at a point are the properties that would be obtained if the fluid at that point were decelerated isentropically to zero velocity. In order to calculate the stagnation properties we need to look at the deceleration process. At the beginning of the process the conditions are those corresponding to the actual flow at the point (velocity, V, pressure, p, and temperature, T, etc.); at the end of the process the velocity is zero and the conditions are the corresponding stagnation properties (stagnation pressure, p_0, and stagnation temperature, T_0, etc.).

We need to develop an expression describing the relationship among the fluid properties during the process. Since both the initial and final properties are specified, it makes sense to develop the relationship among properties in differential form. This can then be integrated to obtain expressions for the stagnation conditions in terms of the initial conditions (the conditions corresponding to the actual flow at the point).

*The nature of the deceleration process is shown in Fig. 9.3. We are interested in determining the stagnation properties at point ①. To determine a relationship among the fluid properties during the deceleration process, we shall apply the continuity and momentum equations to the differential stream tube control volume shown.

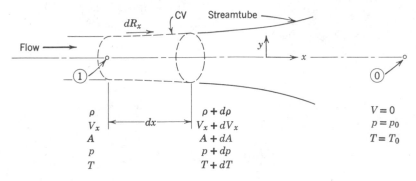

Fig. 9.3 *Compressible flow in an infinitesimal streamtube.*

a. *Continuity Equation*

Basic equation:

$$0 = \overset{= \ 0(1)}{\cancel{\frac{\partial}{\partial t} \iiint_{\mathrm{CV}}}} \rho \, d\forall + \iint_{\mathrm{CS}} \rho \vec{V} \cdot d\vec{A} \qquad (4.14)$$

Assumptions: (1) Steady flow
(2) Uniform flow

Then

$$0 = \{ -|\rho V_x A| \} + \{ |(\rho + d\rho)(V_x + dV_x)(A + dA)| \}$$

or

$$\rho V_x A = (\rho + d\rho)(V_x + dV_x)(A + dA) \qquad (9.15a)$$

b. *Momentum Equation*

Basic equation:

$$F_{S_x} + \overset{= \ 0(3)}{\cancel{F_{B_x}}} - \overset{= \ 0(4)}{\cancel{\iiint_{\mathrm{CV}} a_{rf_x} \rho \, d\forall}} = \overset{= \ 0(1)}{\cancel{\frac{\partial}{\partial t} \iiint_{\mathrm{CV}}}} V_x \rho \, d\forall + \iint_{\mathrm{CS}} V_x \rho \vec{V} \cdot d\vec{A} \quad (4.29a)$$

Assumptions: (3) $F_{B_x} = 0$
(4) $a_{rf_x} = 0$
(5) Frictionless flow

The surface forces acting on the infinitesimal control volume are

$$F_{S_x} = dR_x + pA - (p + dp)(A + dA)$$

The force dR_x is applied along the streamtube boundary as shown in Fig. 9.3. There the average pressure is $p + dp/2$, and the area component in the x direction is dA. There is no friction. Thus

$$F_{S_x} = \left(p + \frac{dp}{2}\right) dA + pA - (p + dp)(A + dA)$$

$$F_{S_x} = p\,dA + \frac{dp\,dA}{2}^{\;\simeq\,0} + pA - pA - dpA - p\,dA - dp\,dA^{\;\simeq\,0}$$

Substituting this result into the momentum equation,

$$-dpA = V_x\{-|\rho V_x A|\} + (V_x + dV_x)\{|(\rho + d\rho)(V_x + dV_x)(A + dA)|\}$$

which may be simplified using Eq. 9.15a, to obtain

$$-dpA = (-V_x + V_x + dV_x)(\rho V_x A)$$

Finally

$$dp = -\rho V_x\,dV_x = -\rho d\left(\frac{V_x^2}{2}\right)$$

or

$$\frac{dp}{\rho} + d\left(\frac{V_x^2}{2}\right) = 0 \qquad\qquad (9.15b)$$

Equation 9.15b is a relation among properties during the deceleration process. In developing this relation we have specified a frictionless deceleration process. Before we can integrate this relation between the initial and final (stagnation) states we must specify the relation that exists between the pressure, p, and the density, ρ, along the process path.

Since the deceleration process is isentropic (i.e. both frictionless and adiabatic), then during the process p and ρ are related by the expression

$$\frac{p}{\rho^k} = \text{constant}$$

Our task now is to integrate Eq. 9.15b subject to this relation. Along the stagnation streamline there is only a single component of velocity; V_x is the magnitude of the velocity and, hence, we can drop the subscript and write

$$\frac{dp}{\rho} + d\left(\frac{V^2}{2}\right) = 0 \qquad (9.15c)$$

From $p/\rho^k = c$, we can write

$$p = c\rho^k \quad \text{and} \quad \rho = p^{1/k}c^{-1/k}$$

Then, from Eq. 9.15c

$$-d\left(\frac{V^2}{2}\right) = \frac{dp}{\rho} = p^{-1/k}c^{1/k}\,dp$$

We can integrate this equation

$$-\int_V^0 d\left(\frac{V^2}{2}\right) = c^{1/k}\int_p^{p_0} p^{-1/k}\,dp$$

to obtain

$$\frac{V^2}{2} = c^{1/k}\frac{k}{k-1}[p^{(k-1)/k}]_p^{p_0} = c^{1/k}\frac{k}{k-1}[p_0^{(k-1)/k} - p^{(k-1)/k}]$$

$$\frac{V^2}{2} = c^{1/k}\frac{k}{k-1}p^{(k-1)/k}\left[\left(\frac{p_0}{p}\right)^{(k-1)/k} - 1\right]$$

Since $c^{1/k} = p^{1/k}/\rho$, then

$$\frac{V^2}{2} = \frac{k}{k-1}\frac{p^{1/k}}{\rho}p^{(k-1)/k}\left[\left(\frac{p_0}{p}\right)^{(k-1)/k} - 1\right] = \frac{k}{k-1}\frac{p}{\rho}\left[\left(\frac{p_0}{p}\right)^{(k-1)/k} - 1\right]$$

Since we are seeking an expression for the stagnation pressure, we can rewrite this equation as

$$\left(\frac{p_0}{p}\right)^{(k-1)/k} = 1 + \frac{k-1}{k}\frac{\rho}{p}\frac{V^2}{2}$$

and

$$\frac{p_0}{p} = \left[1 + \frac{k-1}{k}\frac{\rho V^2}{2p}\right]^{k/(k-1)}$$

For an ideal gas $p = \rho RT$ and, hence,

$$\frac{p_0}{p} = \left[1 + \frac{k-1}{2}\frac{V^2}{kRT}\right]^{k/(k-1)}$$

481   **reference state: stagnation properties**

Also, for an ideal gas $c = \sqrt{kRT}$ and thus

$$\frac{p_0}{p} = \left[1 + \frac{k-1}{2}\frac{V^2}{c^2}\right]^{k/(k-1)}$$

$$\frac{p_0}{p} = \left[1 + \frac{k-1}{2}M^2\right]^{k/(k-1)} \tag{9.16a}$$

Equation 9.16a enables us to calculate the stagnation pressure at any point in a flow field provided we know the static pressure and the Mach number at that point.

Since the process of decelerating the flow at a point to determine the corresponding stagnation properties was an isentropic process, we can readily obtain expressions for other stagnation properties by applying the relation

$$\frac{p}{\rho^k} = \text{constant}$$

between end points of the process. Thus

$$\frac{p_0}{p} = \left(\frac{\rho_0}{\rho}\right)^k$$

and

$$\frac{\rho_0}{\rho} = \left(\frac{p_0}{p}\right)^{1/k}$$

From the ideal gas equation of state, $p = \rho RT$. Then

$$\frac{T_0}{T} = \frac{p_0}{p}\frac{\rho}{\rho_0} = \frac{p_0}{p}\left(\frac{p_0}{p}\right)^{-1/k} = \left(\frac{p_0}{p}\right)^{(k-1)/k}$$

Utilizing Eq. 9.16a we can summarize the equations for determining the stagnation properties as follows:

$$\frac{p_0}{p} = \left[1 + \frac{k-1}{2}M^2\right]^{k/(k-1)} \tag{9.16a}$$

$$\frac{T_0}{T} = 1 + \frac{k-1}{2}M^2 \tag{9.16b}$$

$$\frac{\rho_0}{\rho} = \left[1 + \frac{k-1}{2}M^2\right]^{1/(k-1)} \tag{9.16c}$$

From Eqs. 9.16 the ratio of the stagnation property to the corresponding static property at any point in a flow field can be found if the local Mach number is known. In fact these ratios could be calculated once and for all and tabulated. The calculation procedure is illustrated in Example Problem 9.4; ratios of local isentropic stagnation properties to the corresponding properties are tabulated in Table B.1 of Appendix B.

In Example Problem 9.5, the range of validity of the assumption of incompressible flow in terms of Mach number is investigated.

Example 9.4

Air flows steadily through the duct shown in the sketch, from 50 psia, 140 F and 600 ft/sec at the initial state to a Mach number of 1.3 at the final state. Also at the outlet, stagnation conditions are known to be 55 psia and 630 R. Compute values of the stagnation pressure and temperature at the inlet and of the local pressure and temperature at the duct outlet. Locate the inlet and outlet static state points on a *Ts* diagram, and sketch the stagnation processes.

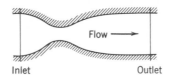

Flow →

Inlet   Outlet

EXAMPLE PROBLEM 9.4

GIVEN:

Steady flow of air through a duct as shown in the sketch.

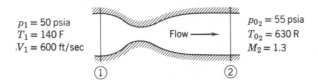

$p_1 = 50$ psia
$T_1 = 140$ F
$V_1 = 600$ ft/sec

Flow →

$p_{0_2} = 55$ psia
$T_{0_2} = 630$ R
$M_2 = 1.3$

① ②

FIND:

(a) p_{0_1} (b) T_{0_1} (c) p_2 (d) T_2
(e) Locate state points ① and ② on a *Ts* diagram and sketch the stagnation processes.

483   **reference state: stagnation properties**

EXAMPLE PROBLEM 9.4 (continued)

SOLUTION:

To obtain stagnation conditions at station ①, we must calculate the Mach number, $M_1 = V_1/c_1$.

Computing equation: $c = \sqrt{kRT}$

Assumption: (1) Ideal gas

Then

$$c_1 = \left[1.4 \times \frac{53.3 \text{ ft-lbf}}{\text{lbm-R}} \times (460 + 140) \text{ R} \times \frac{32.2 \text{ lbm}}{\text{slug}} \times \frac{\text{slug-ft}}{\text{lbf-sec}^2} \right]^{1/2} = 1200 \text{ ft/sec}$$

and

$$M_1 = \frac{V_1}{c_1} = \frac{600}{1200} = 0.5$$

Stagnation conditions may be evaluated from Eqs. 9.16. Thus

$$\frac{p_{0_1}}{p_1} = \left[1 + \frac{k-1}{2} M_1^2 \right]^{k/(k-1)} = [1 + 0.2(0.5)^2]^{3.5} = 1.186$$

and

$$p_{0_1} = 1.186\, p_1 = (1.186)(50) = 59.3 \text{ psia} \qquad\qquad\qquad\qquad\longleftarrow p_{0_1}$$

$$\frac{T_{0_1}}{T_1} = 1 + \frac{k-1}{2} M_1^2 = 1.05$$

and

$$T_{0_1} = 1.05\, T_1 = (1.05)(600) = 630 \text{ R} \qquad\qquad\qquad\qquad\longleftarrow T_{0_1}$$

At section ②, Eqs. 9.16 may be applied again. Thus

$$\frac{p_{0_2}}{p_2} = \left[1 + \frac{k-1}{2} M_2^2 \right]^{k/(k-1)} = [1 + 0.2(1.3)^2]^{3.5} = (1.338)^{3.5} = 2.76$$

and

$$p_2 = \frac{p_{0_2}}{2.76} = \frac{55}{2.76} = 19.9 \text{ psia} \qquad\qquad\qquad\qquad\qquad\longleftarrow p_2$$

$$\frac{T_{0_2}}{T_2} = 1 + \frac{k-1}{2} M_2^2 = 1.338$$

and

$$T_2 = \frac{T_{0_2}}{1.338} = \frac{630}{1.338} = 471 \text{ R} \qquad\qquad\qquad\qquad\qquad\longleftarrow T_2$$

The entropy change must be evaluated to locate state ② with respect to state ①. Using the $T\,ds$ equation

$$T\,ds = dh - v\,dp$$

or

$$ds = \frac{dh}{T} - \frac{v\,dp}{T} = c_p\frac{dT}{T} - R\frac{dp}{p}$$

Integrating

$$s_2 - s_1 = \int_{T_1}^{T_2} c_p\frac{dT}{T} - \int_{p_1}^{p_2} R\frac{dp}{p} = c_p\ln\frac{T_2}{T_1} - R\ln\frac{p_2}{p_1}$$

$$= \frac{0.240\text{ Btu}}{\text{lbm-R}}\ln\left(\frac{471}{600}\right) - \frac{53.3\text{ ft-lbf}}{\text{lbm-R}} \times \frac{\text{Btu}}{778\text{ ft-lbf}}\ln\left(\frac{19.9}{50}\right)$$

or

$$s_2 - s_1 = (0.240)(-0.242) - (0.0685)(-0.921) = 0.0050\text{ Btu/lbm-R}$$

Therefore state ② lies to the right of state ① on the Ts plane, as shown in the sketch below:

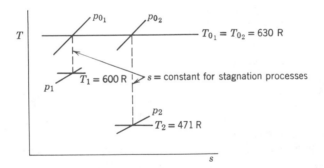

p_{0_1} p_{0_2}

$T_{0_1} = T_{0_2} = 630$ R

$T_1 = 600$ R

s = constant for stagnation processes

p_1

p_2

$T_2 = 471$ R

s

$\left\{\begin{array}{l}\text{The process the fluid follows between states ① and ② is not specified. However, it}\\\text{need not be known. A unique isentropic stagnation process is defined at each state}\\\text{point.}\end{array}\right\}$

Example 9.5

We have derived equations for stagnation pressure ratio, p_0/p, for both compressible and incompressible flow. By writing both equations in terms of Mach number, compare their behavior. Find the Mach number below which the two equations agree within 5 percent.

485 **reference state: stagnation properties**

GIVEN:

The incompressible and compressible forms of the equations for stagnation pressure ratio, p_0/p.

Incompressible

$$\frac{p_0}{p} = 1 + \frac{\rho V^2}{2p} \qquad (9.14)$$

Compressible (isentropic)

$$\frac{p_0}{p} = \left[1 + \frac{k-1}{2}M^2\right]^{k/(k-1)} \qquad (9.16a)$$

FIND:

(a) Behavior of both equations as a function of Mach number.
(b) Mach number below which they agree within 5 percent.

SOLUTION:

First let us write Eq. 9.14 in terms of Mach number. Using the ideal gas equation of state, and $c^2 = kRT$,

$$\frac{p_0}{p} = 1 + \frac{\rho V^2}{2p} = 1 + \frac{V^2}{2RT} = 1 + \frac{kV^2}{2kRT} = 1 + \frac{k}{2}\frac{V^2}{c^2}$$

Thus

$$\frac{p_0}{p} = 1 + \frac{k}{2}M^2 \qquad (1)$$

for incompressible flow.

Eq. 9.16a, for isentropic stagnation, may be expanded using the binomial theorem,

$$(1+x)^n = 1 + nx + \frac{n(n-1)}{2!}x^2 + \cdots, |x| < 1$$

For Eq. 9.16a, $x = [(k-1)/2]M^2$, and $n = k/(k-1)$. Thus the series converges for $[(k-1)/2]M^2 < 1$, and

$$\frac{p_0}{p} = 1 + \left(\frac{k}{k-1}\right)\left[\frac{k-1}{2}M^2\right] + \left(\frac{k}{k-1}\right)\left(\frac{k}{k-1}-1\right)\frac{1}{2!}\left[\frac{k-1}{2}M^2\right]^2$$

$$+ \left(\frac{k}{k-1}\right)\left(\frac{k}{k-1}-1\right)\left(\frac{k}{k-1}-2\right)\frac{1}{3!}\left[\frac{k-1}{2}M^2\right]^3 + \cdots$$

$$\frac{p_0}{p} = 1 + \frac{k}{2}M^2 + \frac{k}{8}M^4 + \frac{k(2-k)}{48}M^6 + \cdots$$

$$\frac{p_0}{p} = 1 + \frac{k}{2}M^2\left[1 + \frac{1}{4}M^2 + \frac{(2-k)}{24}M^4 + \cdots\right] \tag{2}$$

for isentropic compressible stagnation.

In the limit, as $M \to 0$, the term in brackets in Eq. (2) approaches 1.0. Thus, for flow at low Mach number, the incompressible and compressible equations give the same result. The variation with Mach number is shown in the sketch below. As Mach number is increased, the compressible equation gives larger values for the stagnation pressure ratio, p_0/p.

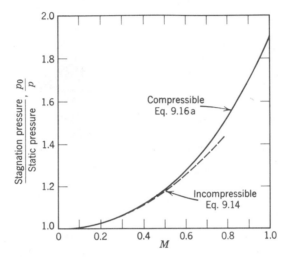

Equations (1) and (2) may be compared quantitatively most simply by writing

$$\frac{p_0}{p} - 1 = \frac{k}{2}M^2 \quad \text{(incompressible)}$$

$$\frac{p_0}{p} - 1 = \frac{k}{2}M^2\left[1 + \frac{1}{4}M^2 + \frac{(2-k)}{24}M^4 + \cdots\right] \quad \text{(compressible)}$$

The term in brackets is approximately equal to 1.02 at $M = 0.3$, and to 1.04 at $M = 0.4$. Thus for calculations of engineering accuracy, flow may be considered incompressible if $M < 0.3$. The two agree within 5 percent for $M \gtrsim 0.45$.

9–3.2 CRITICAL CONDITIONS

Stagnation conditions are extremely useful as reference conditions for the thermodynamic properties; this is not true for velocity, since $V = 0$ by definition at stagnation. A more useful reference value for velocity is the critical speed, that is, the speed at a Mach number of unity. Even if there is no point in a given flow field where the Mach number is equal to unity, such a condition is still useful as a reference condition.

Using an asterisk to denote conditions at $M = 1$, then by definition

$$V^* = M^*c^* \equiv c^* \qquad (9.17)$$

At critical conditions, Eqs. 9.16 for stagnation properties become (for $k = 1.4$)

$$\frac{p_0^*}{p^*} = \left[1 + \frac{k-1}{2}\right]^{k/(k-1)} = 1.893 \qquad (9.18a)$$

$$\frac{T_0^*}{T^*} = 1 + \frac{k-1}{2} = 1.200 \qquad (9.18b)$$

$$\frac{\rho_0^*}{\rho^*} = \left[1 + \frac{k-1}{2}\right]^{1/(k-1)} = 1.577 \qquad (9.18c)$$

The critical velocity may be written in terms of both the critical temperature, T^*, and the critical stagnation temperature, T_0^*.

For an ideal gas, $c^* = \sqrt{kRT^*}$, and thus, $V^* = \sqrt{kRT^*}$. From Eq. 9.18b

$$T^* = \frac{T_0^*}{1 + (k-1)/2} = \frac{2}{k+1} T_0^*$$

Substituting

$$V^* = c^* = \sqrt{\frac{2k}{k+1} RT_0^*} \qquad (9.19)$$

We shall employ both the stagnation conditions and the critical conditions as reference conditions in the next chapter in considering a variety of one-dimensional compressible flows.

problems

9.1   Ten pounds of air in a closed system expands reversibly with constant entropy from 40 psia, 140 F, to 20 psia. Calculate the air temperature after the expansion.

9.2   Ten pounds of air is cooled in a closed tank from 500 to 100 F. The initial pressure is 400 psia. Compute the changes in entropy, internal energy, and enthalpy.

9.3   A turbine operating on air has inlet conditions of 30 psia, 200 F, and an exhaust pressure of 15 psia. If the expansion is adiabatic and $\Delta KE = 0$, is an exhaust temperature of 40 F possible? Is it possible if $\Delta KE \neq 0$?

9.4   Air is compressed irreversibly from 15 psia, 40 F, to 30 psia, 240 F. Calculate the change in entropy of the air.

488   **9/introduction to compressible flow**

9.5 A fluid passing through a steady-flow system can be heated reversibly from state ① to state ② in either of two ways. In one case, $T = a + bs$; in the other case, $T = c + es^2$. a, b, c, and e are constants. In which case is the heat transfer greater?

9.6 Is an adiabatic expansion of air from 40 psia, 140 F, to 20 psia, 80 F, possible? Justify your answer.

9.7 In a closed system, a gas undergoes a cycle made up of the following processes: 1–2 reversible isothermal compression; 2–3 reversible constant volume heating; 3–4 reversible constant pressure expansion; 4–1 reversible adiabatic expansion.

(a) Sketch pv and Ts diagrams.
(b) State whether each of the following quantities is positive, zero, negative, or indeterminate in sign:

$$\oint \delta W, \quad \oint \delta Q, \quad \oint dS, \quad \oint dU, \quad \oint dH$$

9.8 An ideal gas is heated at constant volume from state ① to state ②, expanded isothermally to state ③, expanded adiabatically to state ④ which is at the same pressure as state ①, and then restored to state ① by a constant-pressure process. All four processes are reversible.

(a) Sketch pv and Ts diagrams of the cycle.
(b) State whether each of the following quantities is greater than zero, less than zero, equal to zero, or of indeterminate sign:

$$\oint \delta Q, \quad \oint \delta W, \quad \oint du, \quad \oint ds$$

9.9 An aircraft is flying at a speed of 600 mph through air at a pressure of 12 psia and temperature of 0 F. Calculate the Mach number of the craft.

9.10 Air at a temperature of 60 F is flowing with a velocity of 500 ft/sec. A bullet is fired into the air stream with a velocity of 2500 ft/sec. (Direction of bullet is opposite to that of the air.) Calculate

(a) The Mach number of the air flow.
(b) The Mach number of the bullet if it were fired in still air.
(c) The Mach number of the bullet with respect to the moving air.

9.11 What is the speed of sound in carbon dioxide at 300 F? What is the acoustic speed in water at 60 F?

9.12 A photograph of a bullet shows a Mach angle of 28°. Determine the speed of the bullet for standard air.

9.13 Air at 80 F is flowing at a Mach number of 1.9.

(a) Find the air velocity in feet per second.
(b) What is the Mach angle?

489   **problems**

9.14 A projectile is fired into a gas in which the pressure is 50 psia and the density is 0.27 lbm/ft³. It is observed experimentally that a Mach cone emanates from the projectile with a total angle of 20°. What is the speed of the projectile with respect to the gas?

9.15 A supersonic aircraft flies at an altitude of 10,000 ft at a speed of 3000 ft/sec. How long after passing directly above a ground observer will it be before the sound of the aircraft is heard? Assume the properties of the atmosphere to be $p = 12.2$ psia and $T = 40$ F. The sound disturbance may be assumed to propagate along a Mach line.

9.16 Consider an air flow with a speed of 1900 ft/sec at 14.7 psia and 60 F. What is the stagnation pressure? The stagnation enthalpy? The stagnation temperature?

9.17 The pressure at the nose of an aircraft in flight was found to be 27.8 psi (the velocity of air relative to the craft was zero at this point). Estimate the speed and Mach number of the craft. The pressure and temperature of the undisturbed air were 14.7 psi and 70 F respectively.

9.18 A body moves through standard air at 600 ft/sec. What is the pressure at a point on the body where the velocity of the air relative to the body is zero? Assume

(a) Compressible flow
(b) Incompressible flow

9.19 Air flows in a duct where the static temperature is 50 F and the static pressure is 10 psia. Calculate the stagnation pressure if the air speed is

(a) 200 ft/sec
(b) 2000 ft/sec.

9.20 A plane is flying at 10,000 ft where the air temperature is 20 F. At a point on the plane where the relative air velocity is zero the temperature is found to be 120 F. Determine the speed of the plane.

9.21 An airplane is flying at a speed of 500 fps and at an altitude of 1000 ft where the temperature is 60 F. The plane climbs to 40,000 ft where the temperature is −68 F and levels off at a speed of 1000 fps. Calculate the Mach number of flight in both cases.

9.22 Air flows steady through a section of an insulated constant area duct as shown. Properties change along the duct as a result of friction.

(a) Beginning with the general form of the first law of thermodynamics, show that the equation can be reduced to

$$h_1 + \frac{V_1^2}{2} = h_2 + \frac{V_2^2}{2} = \text{constant}$$

(b) Denoting the constant by h_0 (the stagnation enthalpy), show that for an adiabatic flow with friction

$$\frac{T_0}{T} = 1 + \frac{k-1}{2}M^2$$

(c) For this flow does $T_{0_1} = T_{0_2}$? $p_{0_1} = p_{0_2}$?

9.23   Consider the steady adiabatic flow of air through a long straight pipe with a cross section area 0.5 ft². At the inlet (section ①) the air is at 30 psia, 140 F, and has a velocity of 500 ft/sec. At a section downstream the air is at 11.3 psia and has a velocity of 985 ft/sec. Determine p_{0_1}, p_{0_2}, T_{0_1}, T_{0_2}.

9.24   Air flows steadily through a constant area duct. At section ①, the air is at 60 psia, 600 R, with velocity 500 ft/sec. As a result of heat transfer and friction, the air at section ② downstream is at 40 psia, 800 R. Determine p_{0_1}, p_{0_2}, T_{0_1}, T_{0_2}.

one-dimensional compressible flow

We shall begin our study of steady, one-dimensional compressible flows by considering isentropic flow. We can learn much about the general behavior of compressible flows by studying this simple flow case. In addition, the assumption of isentropic flow is a reasonable first approximation to the actual flow in a nozzle.

Following consideration of the one-dimensional, steady isentropic flow of any compressible fluid, we shall look in greater detail at the flow of an ideal gas. Once we are familiar with the procedure for the calculation of isentropic flows, we shall consider isentropic flows through nozzles. Following isentropic flow, we shall study adiabatic flow in a constant area duct with friction and frictionless flow in a constant area duct with heat transfer. A discussion of normal shocks will conclude the chapter.

We can summarize the contents of this chapter by saying that we shall study a variety of one-dimensional compressible flows, each of which involves a separate effect.

10–1 Basic Equations for Isentropic Flow

Consider the steady, one-dimensional isentropic flow of any compressible fluid through a channel of arbitrary cross section; a portion of such a duct is shown in

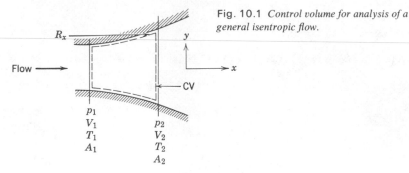

Fig. 10.1 *Control volume for analysis of a general isentropic flow.*

Fig. 10.1. In developing the governing equations for this flow, we shall apply the basic equations, derived in Chapter 4, to the finite, fixed control volume of Fig. 10.1. The properties at sections ① and ② are labeled with the appropriate subscripts; R_x is the x component of the surface force acting on the control volume.

a. Continuity Equation

Basic equation:

$$0 = \overset{= 0(1)}{\cancel{\frac{\partial}{\partial t} \iiint_{CV} \rho \, d\forall}} + \iint_{CS} \rho \vec{V} \cdot d\vec{A} \tag{4.14}$$

Assumptions: (1) Steady flow
 (2) One-dimensional flow

$$0 = \{-|\rho_1 V_1 A_1|\} + \{|\rho_2 V_2 A_2|\}$$

Using scalar magnitudes and dropping absolute value signs gives the familiar form

$$\rho_1 V_1 A_1 = \rho_2 V_2 A_2 = \rho V A = \dot{m} = \text{constant} \tag{10.1a}$$

b. Momentum Equation

Basic equation:

$$F_{S_x} + \overset{= 0(3)}{\cancel{F_{B_x}}} = \overset{= 0(1)}{\cancel{\frac{\partial}{\partial t} \iiint_{CV} V_x \rho \, d\forall}} + \iint_{CS} V_x \rho \vec{V} \cdot d\vec{A} \tag{4.19a}$$

Assumptions: (3) $F_{B_x} = 0$

The surface force will be due to pressure forces at the surfaces ① and ②, and to the distributed pressure force along the channel walls, R_x. Substituting,

$$R_x + p_1 A_1 - p_2 A_2 = V_1\{-|\rho_1 V_1 A_1|\} + V_2\{|\rho_2 V_2 A_2|\}$$

Using scalar magnitudes and dropping absolute value signs, we obtain

$$R_x + p_1 A_1 - p_2 A_2 = \dot{m} V_2 - \dot{m} V_1 \qquad (10.1b)$$

c. First Law of Thermodynamics

Basic equation:

$$\overset{=0(4)}{\cancel{\dot{Q}}} \quad \overset{=0(5)}{\cancel{\dot{W}_s}} \quad \overset{=0(6)}{\cancel{\dot{W}_{shear}}} \quad = \overset{=0(1)}{\frac{\cancel{\partial}}{\partial t}} \iiint_{CV} e\rho \, d\forall + \iint_{CS} (e + pv)\rho \vec{V} \cdot d\vec{A} \qquad (4.45)$$

where

$$e = u + \frac{V^2}{2} + \overset{\simeq 0(7)}{\cancel{gz}}$$

Assumptions: (4) $\dot{Q} = 0$ (isentropic flow)
(5) $\dot{W}_s = 0$
(6) $\dot{W}_{shear} = 0$
(7) Effects of gravity are negligible.

Under the assumptions, the first law reduces to

$$0 = \left(u_1 + p_1 v_1 + \frac{V_1^2}{2}\right)\{-|\rho_1 V_1 A_1|\} + \left(u_2 + p_2 v_2 + \frac{V_2^2}{2}\right)\{|\rho_2 V_2 A_2|\}$$

But we know from continuity that the mass flowrate terms in brackets are equal, so they may be canceled. We may also substitute $h \equiv u + pv$, to obtain

$$h_1 + \frac{V_1^2}{2} = h_2 + \frac{V_2^2}{2} = h + \frac{V^2}{2} = \text{constant} \qquad (10.1c)$$

The combination $h + V^2/2$ occurs often in compressible flow problems. It is convenient to introduce the definition of stagnation enthalpy, h_0, and to note

$$h_0 \equiv h + \frac{V^2}{2}$$

495 **basic equations for isentropic flow**

Physically, the stagnation enthalpy is that enthalpy which would be reached if the fluid were decelerated adiabatically to zero velocity. We note that the stagnation enthalpy is constant throughout an adiabatic flow field.

d. Second Law of Thermodynamics

Basic equation:

$$\iint_{CS} \frac{1}{T} \overset{=\,0(4)}{\cancel{\frac{\dot{Q}}{A}dA}} \leq \overset{=\,0(1)}{\cancel{\frac{\partial}{\partial t}}} \iiint_{CV} s\rho\, d\forall + \iint_{CS} s\rho \vec{V} \cdot d\vec{A} \tag{4.46}$$

Then

$$0 = s_1\{-|\rho_1 V_1 A_1|\} + s_2\{|\rho_2 V_2 A_2|\}$$

Since the mass flowrate terms { } are equal by continuity,

$$s_1 = s_2 = s = \text{constant} \tag{10.1d}$$

e. Equation of State

Equations of state are relations among thermodynamic properties. These may be expressed in tables, charts, or algebraic equations. Since, for a pure substance, it is possible to specify any thermodynamic property in terms of any other two thermodynamic properties, we can write

$$h = h(s, p) \tag{10.1e}$$

and

$$\rho = \rho(s, p) \tag{10.1f}$$

as equations of state.

Before summarizing the simplified forms of the basic equations for steady, one-dimensional, isentropic flow of any compressible fluid, let us turn to a representation of an isentropic flow on an hs diagram. At some point in the isentropic flow field (call it state ①) the flow properties are h_1, s_1, ρ_1, V_1, and so on. Clearly, in isentropic flow, the flow may proceed to state ② (with properties h_2, $s_2 = s_1$, ρ_2, V_2, etc.) or to state ③ (with properties h_3, $s_3 = s_1$, ρ_3, V_3, etc.) as shown in Fig. 10.2.

How are the isentropic stagnation properties at states ①, ②, and ③ related? To answer this question, consider the results obtained from the first law analysis for isentropic flow. From Eq. 10.1c and the definition of stagnation enthalpy, we have

$$h_1 + \frac{V_1^2}{2} = h_2 + \frac{V_2^2}{2} = h + \frac{V^2}{2} = h_0 = \text{constant}$$

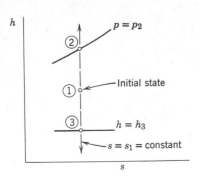

Fig. 10.2 *Representation of isentropic flow in the hs plane.*

Thus all states in an isentropic flow have the same stagnation enthalpy. In addition, by definition, all states in an isentropic flow (including the stagnation state) have the same entropy. Thus all stagnation states in an isentropic flow have the same stagnation enthalpy and stagnation entropy. Since at the stagnation state the velocity, is zero, it follows that the stagnation properties are constant for all points in an isentropic flow.

For isentropic flow, the first law, in the form

$$h + \frac{V^2}{2} = \text{constant} = h_0$$

suggests another interpretation for the stagnation enthalpy. In the absence of potential energy, h_0 represents the total energy per unit mass. The kinetic energy per unit mass of the flow is then represented by the enthalpy difference, $h_0 - h$. This is illustrated graphically in Fig. 10.3.

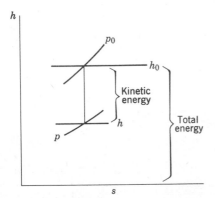

Fig. 10.3 *Schematic hs diagram illustrating interpretation of energy per unit mass in a flow.*

basic equations for isentropic flow

Equations 10.1a–f are the simplified forms of the basic equations that describe steady, one-dimensional isentropic flow of any compressible fluid. There are 6 independent equations. If all the properties of the flow are known at station ①, then we have a total of seven unknowns ($p_2, A_2, V_2, \rho_2, h_2, s_2$, and R_x) in these 6 equations. Consequently, an isentropic flow can proceed to a variety of states from state ①. Each such state must have $s_2 = s_1$, satisfying Eq. 10.1d. This leaves, in effect, 6 unknowns and 5 equations. Thus the problem at this point is indeterminate. In order to determine conditions at state ②, one of the six unknowns must be specified.

What causes fluid properties to change in an isentropic flow? You undoubtedly recognize that it is the effect of area variation. In the next section we shall look at the effect of area variation on flow properties in isentropic flow.

10–2 *Effect of Area Variation on Flow Properties in Isentropic Flow*

In considering the effect of area variation on flow properties in isentropic flow, we shall concern ourselves primarily with velocity and pressure. We wish to determine the effect of a change, dA, in the area, A, on the velocity, V, and the pressure, p; that is, for a change, dA, in the area, A, are dV and dp positive or negative?

To answer these questions, it is convenient to work with the differential forms of the governing equations. These were derived for the differential control volume of Fig. 9.3 (Section 9–3.1). The differential momentum equation for isentropic flow reduces to

$$\frac{dp}{\rho} + d\left(\frac{V^2}{2}\right) = 0 \qquad (9.15c)$$

or

$$dp = -\rho V\, dV$$

Dividing by ρV^2, we obtain

$$\frac{dp}{\rho V^2} = -\frac{dV}{V} \qquad (10.2)$$

A convenient differential form of the continuity equation can be obtained from Eq. 10.1a

$$\rho A V = \text{constant}$$

Taking the natural logarithm of both sides

$$\ln \rho + \ln A + \ln V = \ln c$$

Differentiating

$$\frac{d\rho}{\rho} + \frac{dA}{A} + \frac{dV}{V} = 0 \qquad (10.3)$$

Solving Eq. 10.3 for dA/A gives

$$\frac{dA}{A} = -\frac{dV}{V} - \frac{d\rho}{\rho}$$

Substituting from Eq. 10.2,

$$\frac{dA}{A} = \frac{dp}{\rho V^2} - \frac{d\rho}{\rho}$$

or

$$\frac{dA}{A} = \frac{dp}{\rho V^2}\left[1 - \frac{V^2}{dp/d\rho}\right]$$

Now recall that for an isentropic process $dp/d\rho = \partial p/\partial\rho)_s = c^2$, so

$$\frac{dA}{A} = \frac{dp}{\rho V^2}\left[1 - \frac{V^2}{c^2}\right] = \frac{dp}{\rho V^2}[1 - M^2] \qquad (10.4)$$

From Eq. 10.4 we see that for $M < 1$ an area change causes a pressure change of the same sign (positive dA means positive dp for $M < 1$); for $M > 1$ an area change causes a pressure change of opposite sign.

Substituting from Eq. 10.2 into Eq. 10.4 we obtain

$$\frac{dA}{A} = \frac{-dV}{V}[1 - M^2] \qquad (10.5)$$

From Eq. 10.5 we see that for $M < 1$ an area change causes a velocity change of opposite sign (positive dA means negative dV for $M < 1$); for $M > 1$ an area change causes a velocity change of the same sign.

These results are summarized in Fig. 10.4. For subsonic flows ($M < 1$) flow acceleration in a *nozzle* requires a passage of diminishing cross section, that is, the area must decrease to cause a velocity increase. This produces a passage shaped like that shown in the upper left of Fig. 10.4, and this result is in accord with our experience. A subsonic *diffuser* requires that the passage area increase to cause a velocity decrease. Again this result agrees with our experience.

In supersonic flows ($M > 1$), these effects are different. According to Eq. 10.5, a *supersonic nozzle* must be built with an area increase in the flow direction. A *supersonic diffuser* must be a converging channel. Although these predictions are contrary to the bulk of our experience, laboratory experiments show that they are valid. We can also recall seeing divergent nozzles designed to produce super-sonic flow on missiles and launch vehicles.

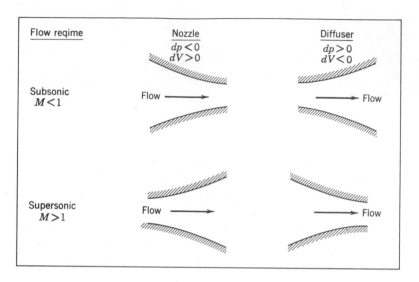

Fig. 10.4 *Nozzle and diffuser shapes as a function of initial Mach number.*

What of the remaining case, $M = 1$? Further inspection of Eq. 10.5 shows that at $M = 1$ then $dA/dV = 0$. The fact that $dA/dV = 0$ means that the passage area must pass through a minimum or maximum at $M = 1$. Inspection of Fig. 10.4 shows that $M = 1$ can be reached only in a throat or section of minimum area.

To accelerate flow from rest to supersonic speed ($M > 1$) requires first a subsonic or converging nozzle. Under the proper conditions, the flow will be at $M = 1$ at the throat where the area is a minimum. Further acceleration is possible if a supersonic (diverging) nozzle segment is added downstream from a throat. Isentropic flow in converging nozzles will be treated in Section 10–3.4, and isentropic flow in converging-diverging nozzles will be covered in Section 10–3.5.

To decelerate a supersonic flow isentropically from $M > 1$ to subsonic speed also requires a throat. Initially, deceleration is accomplished in a converging, supersonic diffuser section. The throat, at which $M = 1$, is then followed by a subsonic (diverging) diffuser section. Although a diffuser is by nature less efficient than a nozzle (due primarily to the adverse pressure gradients and attendant possibility of separation), in principle the same piece of hardware could be used as a converging-diverging (C-D) nozzle to produce supersonic flow, or as a diffuser to reduce a supersonic inlet stream to a subsonic outlet stream.

10–3 Isentropic Flow of an Ideal Gas

10–3.1 BASIC EQUATIONS

In Section 10–1 we applied the basic equations to a finite control volume for the steady, one-dimensional, isentropic flow of any compressible fluid. The only modification required in restricting our discussion to an ideal gas is in the equation of state. For an ideal gas the equation of state is simplified to $p = \rho RT$. In addition, for the isentropic flow of an ideal gas we have the process equation, p/ρ^k = constant. Then, for the isentropic flow of an ideal gas we can summarize the basic equations as follows:

$$\text{Continuity} \qquad \rho_1 V_1 A_1 = \rho_2 V_2 A_2 = \rho V A = \dot{m} \qquad (10.1a)$$

$$\text{Momentum} \qquad R_x + p_1 A_1 - p_2 A_2 = \dot{m} V_2 - \dot{m} V_1 \qquad (10.1b)$$

$$\text{First law} \qquad h_1 + \frac{V_1^2}{2} = h_2 + \frac{V_2^2}{2} = h + \frac{V^2}{2} \qquad (10.1c)$$

$$\text{Second law} \qquad s_1 = s_2 = s \qquad (10.1d)$$

$$\text{Equation of state} \quad p = \rho RT \qquad (9.1)$$

$$\text{Process equation} \quad p/\rho^k = \text{constant} \qquad (9.11)$$

These are the governing equations for steady, one-dimensional, isentropic flow of an ideal gas. If all the properties at state ① are known, then we have 8 unknowns $(\rho_2, A_2, V_2, p_2, h_2, s_2, T_2,$ and $R_x)$ in these 6 equations. However, since we have counted both h and T as unknowns, it is legitimate to include the known relationship between h and T for an ideal gas with constant specific heats, namely

$$\Delta h = h_2 - h_1 = c_p \Delta T = c_p (T_2 - T_1) \qquad (9.7b)$$

Thus, as in the general case (Section 10–1), the problem is indeterminate. One condition (other than s_2) must be specified at state ② before conditions at state ② can be completely determined.

10–3.2 REFERENCE CONDITIONS FOR ISENTROPIC FLOW OF AN IDEAL GAS

Expressions for stagnation properties in the flow of an ideal gas were developed in Chapter 9 (Section 9–3.1). For completeness these expressions are repeated here.

$$\text{Stagnation pressure:} \qquad \frac{p_0}{p} = \left[1 + \frac{k-1}{2} M^2\right]^{k/(k-1)} \qquad (9.16a)$$

$$\text{Stagnation temperature:} \qquad \frac{T_0}{T} = 1 + \frac{k-1}{2} M^2 \qquad (9.16b)$$

Stagnation density:
$$\frac{\rho_0}{\rho} = \left[1 + \frac{k-1}{2}M^2\right]^{1/(k-1)}$$
(9.16c)

The stagnation enthalpy, h_0, was defined by Eq. 10.1c as

$$h_0 \equiv h + \frac{V^2}{2}$$
(10.1c)

As shown in Section 10–1, the stagnation properties are constant throughout an isentropic flow field.

The critical conditions, that is, the values of the flow properties at which the Mach number is unity, were introduced in Section 9–3.2. Since the stagnation properties are constant in an isentropic flow, then from Eqs. 9.16 (or from Eqs. 9.18) we can write, for $k = 1.4$

$$\frac{p_0}{p^*} = \left[1 + \frac{k-1}{2}\right]^{k/(k-1)} = 1.893$$

$$\frac{T_0}{T^*} = 1 + \frac{k-1}{2} = 1.200$$

$$\frac{\rho_0}{\rho^*} = \left[1 + \frac{k-1}{2}\right]^{1/(k-1)} = 1.577$$

In addition, from Eq. 9.19 we have

$$V^* = c^* = \sqrt{\frac{2k}{k+1}RT_0}$$

In Section 10–2 we saw that it was necessary for a passage to have a point of minimum area (a throat) to accelerate a flow isentropically from rest to a Mach number greater than unity. Furthermore, in such a flow the Mach number is unity at the throat. If the area at which the Mach number is unity is designated by A^*, then it is possible to express the contour of a passage in terms of the area ratio, A/A^*.

Since for steady, one-dimensional flow, the continuity equation can be written as

$$\rho A V = \text{constant} = \rho^* A^* V^*$$

then

$$\frac{A}{A^*} = \frac{\rho^*}{\rho}\frac{V^*}{V} = \frac{\rho^*}{\rho}\frac{c^*}{Mc} = \frac{1}{M}\frac{\rho^*}{\rho}\sqrt{\frac{T^*}{T}}$$

$$\frac{A}{A^*} = \frac{1}{M}\frac{\rho^*}{\rho_0}\frac{\rho_0}{\rho}\sqrt{\frac{T^*/T_0}{T/T_0}}$$

$$\frac{A}{A^*} = \frac{1}{M} \frac{\left[1 + \dfrac{k-1}{2}M^2\right]^{1/(k-1)}}{\left[1 + \dfrac{k-1}{2}\right]^{1/(k-1)}} \left[\frac{1 + \dfrac{k-1}{2}M^2}{1 + \dfrac{k-1}{2}}\right]^{1/2}$$

$$\frac{A}{A^*} = \frac{1}{M} \left[\frac{1 + \dfrac{k-1}{2}M^2}{1 + \dfrac{k-1}{2}}\right]^{(k+1)/2(k-1)} \tag{10.6}$$

From Eq. 10.6 we see that a choice of M gives a unique value of A/A^*. The variation of A/A^* with M is shown in Fig. 10.5. Note that the curve is double

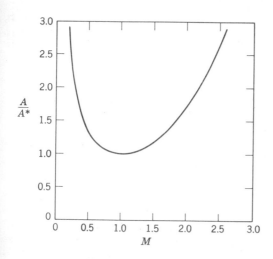

Fig. 10.5 *Variation of A/A^* with Mach number in isentropic flow for $k = 1.4$.*

valued, that is, for a given value of A/A^*, there are two possible values of Mach number other than unity. This is consistent with the results of Section 10–2 (see Fig. 10.4), where it was found that a converging-diverging passage with a point of minimum area is required to accelerate a flow from subsonic to supersonic speed.

Example 10.1

Air flows isentropically in a channel. At section ①, the Mach number is 0.3, the area is 0.01 ft², and the pressure and temperature are 100 psia and 140 F, respectively. At section ②, the Mach number is 0.8. Plot a Ts diagram for the process, evaluate the properties at section ②, and sketch the channel shape.

GIVEN:

Isentropic flow of air in a channel. At stations ① and ②, the following data are given:

$$T_1 = 140\ F$$
$$p_1 = 100\ psia$$
$$M_1 = 0.3 \qquad M_2 = 0.8$$
$$A_1 = 0.01\ ft^2$$

FIND:

(a) Plot a Ts diagram for the process.
(b) Properties at station ②.
(c) Sketch the channel shape.

SOLUTION:

On the Ts plane, the process follows a line $s = $ constant. Stagnation conditions remain fixed for isentropic flow. Consequently, the stagnation temperature at section ② can be calculated (for air, $k = 1.4$) from

$$T_{0_2} = T_{0_1} = T_1\left[1 + \frac{k-1}{2}M_1^2\right]$$

$$= (460 + 140)\,R\,[1 + 0.2(0.3)^2]$$

$$T_{0_2} = T_{0_1} = (600)(1.018)\,R = 611\ R$$

and

$$T_2 = \frac{T_{0_2}}{\left[1 + \dfrac{k-1}{2}M_2^2\right]} = \frac{611\ R}{[1 + 0.2(0.8)^2]} = \frac{611\ R}{1.128} = 541\ R \text{ or } 81\ F$$

Also, for an ideal gas

$$c_2 = \sqrt{kRT_2} = \left[1.4 \times \frac{53.3\ \text{ft-lbf}}{\text{lbm-R}} \times 541\ R \times \frac{32.2\ \text{lbm}}{\text{slug}} \times \frac{\text{slug-ft}}{\text{lbf-sec}^2}\right]^{1/2} = 1141\ \text{ft/sec}$$

T diagram showing: $p_{0_1} = p_{0_2}$, $T_{0_1} = T_{0_2}$, ①, p_1, T_1, ②, p_2, T_2, s, T_{0_1}, T_{0_2}, T_2, c_2

EXAMPLE PROBLEM 10.1 (continued)

From the definition of Mach number,

$$V_2 = M_2 c_2 = (0.8)\,1141\text{ ft/sec} = 913\text{ ft/sec} \qquad V_2$$

Using the isentropic relation, $p/\rho^k = \text{constant}$,

$$\frac{p_2}{p_1} = \left(\frac{T_2}{T_1}\right)^{k/(k-1)} = \left(\frac{541}{600}\right)^{3.5} = (0.902)^{3.5} = 0.698$$

and

$$p_2 = 0.698p_1 = (0.698)\,100\text{ psia} = 69.8\text{ psia} \qquad p_2$$

Also

$$\frac{\rho_2}{\rho_1} = \left(\frac{T_2}{T_1}\right)^{1/(k-1)} = (0.902)^{2.5} = 0.774$$

$$\rho_1 = \frac{p_1}{RT_1} = \frac{100\text{ lbf}}{\text{in.}^2} \times \frac{144\text{ in.}^2}{\text{ft}^2} \times \frac{\text{lbm-R}}{53.3\text{ ft-lbf}} \times \frac{1}{(460+140)\text{ R}} = 0.450\text{ lbm/ft}^3 \qquad \rho_1$$

Thus

$$\rho_2 = 0.774\rho_1 = (0.774)\,0.450\text{ lbm/ft}^3 = 0.348\text{ lbm/ft}^3 \qquad \rho_2$$

From continuity,

$$\rho_1 A_1 V_1 = \rho_2 A_2 V_2$$

so that

$$A_2 = A_1 \frac{\rho_1}{\rho_2}\frac{V_1}{V_2}$$

But $V_1 = M_1 c_1$; $c_1 = \sqrt{kRT_1} = 1200\text{ ft/sec}$; $V_1 = 360\text{ ft/sec}$, so

$$A_2 = 0.01\text{ ft}^2 \left(\frac{1}{0.774}\right)\left(\frac{360}{913}\right) = 0.00510\text{ ft}^2 \qquad A_2$$

Finally,

$$p_{0_2} = p_2 \left(\frac{T_{0_2}}{T_2}\right)^{k/(k-1)} = 69.8\text{ psia}\,(1.128)^{3.5} = 69.8\text{ psia}\,(1.52) = 106\text{ psia} \qquad p_{0_2}$$

Note that for isentropic flow, $T_{0_2} = T_{0_1}$ and $p_{0_2} = p_{0_1}$. Since the flow is subsonic and accelerates from ① to ②, the passage must be a nozzle, shaped as shown:

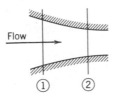

Flow

①   ②

In the previous section we saw (Eqs. 9.16a, b, c and 10.6) that the properties at a point in a compressible flow may be related to appropriate reference conditions by functions of the local Mach number. This makes it possible to compute all property ratios once, and to tabulate (or plot) them as functions of Mach number, M.

Table B.1 of Appendix B lists values of T/T_0, p/p_0, ρ/ρ_0, and A/A^* as functions of M for isentropic flow. The use of the tables significantly reduces the labor of calculations.

Since the reference conditions remain constant in an isentropic flow, the ratio of properties at two points in a flow may readily be found from the tables. Use of the tables is illustrated in Example Problem 10.2.

You are encouraged to use the tables to check problem solutions.

Example 10.2

Air flows isentropically in a channel. At section ①, the Mach number is 0.3, the area is 0.01 ft², and the pressure and temperature are 140 F and 100 psia, respectively. Evaluate the properties at section ②, where the Mach number is 0.8, using the isentropic flow tables. (Note these are the data of Example 10.1.)

EXAMPLE PROBLEM 10.2

GIVEN:

Isentropic flow of air in a channel. At stations ① and ②, the following data are given:

$$T_1 = 140 \, F$$
$$p_1 = 100 \, \text{psia}$$
$$M_1 = 0.3 \qquad M_2 = 0.8$$
$$A_1 = 0.01 \, \text{ft}^2$$

FIND:

Properties at station ②, using isentropic flow tables.

** This section may be omitted without loss of continuity in the text material.

10/one-dimensional compressible flow

EXAMPLE PROBLEM 10.2 (*continued*)

SOLUTION:

The *Ts* diagram was sketched in Example Problem 10.1. From Table B.1, Appendix B, we find

M	T/T_0	p/p_0	ρ/ρ_0	A/A^*
0.3	0.982	0.940	0.956	2.035
0.8	0.887	0.656	0.740	1.038

For isentropic flow, $T_{0_1} = T_{0_2} = T_0$. Thus

$$\frac{T_2}{T_1} = \frac{T_2}{T_0}\frac{T_0}{T_1} = \frac{0.887}{0.982} = 0.902; \qquad T_2 = 0.902T_1 = 541 \text{ R} \qquad \underleftarrow{\quad T_2 \quad}$$

Also $p_{0_2} = p_{0_1} = p_0$, and $\rho_{0_2} = \rho_{0_1} = \rho_0$, so

$$\frac{p_2}{p_1} = \frac{p_2}{p_0}\frac{p_0}{p_1} = \frac{0.656}{0.940} = 0.698; \qquad p_2 = 0.698p_1 = 69.8 \text{ psia} \qquad \underleftarrow{\quad p_2 \quad}$$

and

$$\frac{\rho_2}{\rho_1} = \frac{\rho_2}{\rho_0}\frac{\rho_0}{\rho_1} = \frac{0.740}{0.956} = 0.774$$

$$\rho_1 = \frac{p_1}{RT_1} = \frac{100 \text{ lbf}}{\text{in.}^2} \times \frac{144 \text{ in.}^2}{\text{ft}^2} \times \frac{\text{lbm-R}}{53.3 \text{ ft-lbf}} \times \frac{1}{(460 + 140)\text{R}} = 0.450 \text{ lbm/ft}^3$$

$$\rho_2 = 0.774\,\rho_1 = 0.348 \text{ lbm/ft}^3 \qquad \underleftarrow{\quad \rho_2 \quad}$$

Also

$$T_0 = \frac{T_1}{(T_1/T_0)} = \frac{600 \text{ R}}{0.982} = 611 \text{ R} \qquad \underleftarrow{\quad T_0 \quad}$$

and

$$p_0 = \frac{p_1}{(p_1/p_0)} = \frac{100 \text{ psia}}{0.940} = 106 \text{ psia} \qquad \underleftarrow{\quad p_0 \quad}$$

The area may be computed using the ratio A/A^*. Thus

$$\frac{A_2}{A_1} = \frac{A_2}{A^*}\frac{A^*}{A_1} = \frac{1.038}{2.035} = 0.510$$

or

$$A_2 = 0.510A_1 = 0.00510 \text{ ft}^2 \qquad \underleftarrow{\quad A_3 \quad}$$

The velocity may be computed from continuity, $\rho_1 V_1 A_1 = \rho_2 V_2 A_2$.

10–3.4 ISENTROPIC FLOW IN A CONVERGING NOZZLE

In this section we are interested in investigating the operation of a converging nozzle under various back pressures. Flow through the converging nozzle shown in Fig. 10.6 is supplied from a large plenum chamber, where conditions are assumed to be stagnation conditions; the flow is induced by a vacuum pump downstream and is controlled by the valve shown.

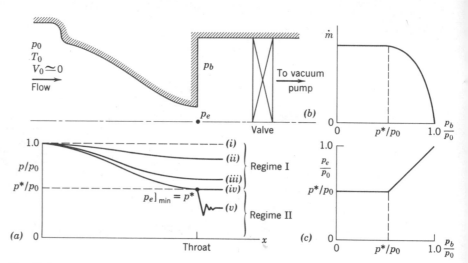

Fig. 10.6 *Converging nozzle operating at various back pressures.*

The back pressure, p_b, to which the nozzle discharges is controlled by the valve. The upstream stagnation conditions (p_0, T_0, etc.) are maintained constant. The pressure in the exit plane of the nozzle is denoted as p_e. We are interested in investigating the effect of variations in back pressure on the pressure distribution through the nozzle, on the mass flowrate and on the exit plane pressure. The results are illustrated graphically in Fig. 10.6. Let us look at each of the cases shown.

When the valve is closed, there is clearly no flow through the nozzle. The pressure is p_0 throughout as shown by condition (i) in Fig. 10.6a.

If the back pressure, p_b, is now reduced to a value slightly less than p_0, there will be flow through the nozzle with a decrease in pressure in the direction of flow as shown by condition (ii). Flow at the exit plane will be subsonic with the exit plane pressure equal to the back pressure.

What happens as we continue to decrease the back pressure? The flow rate will continue to increase and the exit plane pressure will continue to decrease as shown by condition (*iii*) in Fig. 10.6. Will this continue indefinitely as the back pressure is lowered?

Recall from our previous discussion that in a converging section the Mach number cannot increase beyond unity in isentropic flow. Thus with continued decrease of the back pressure, the flow at the exit plane of the nozzle will eventually reach a Mach number of unity. The corresponding pressure is the critical pressure, p^*. Condition (*iv*) illustrates the condition where M_e equals unity and p_b/p_0 equals p^*/p_0.

From Eq. 9.16a, the critical pressure ratio is given by

$$\frac{p^*}{p_0} = 1 \bigg/ \left[1 + \frac{k-1}{2} M^{*2}\right]^{k/(k-1)} = \left(\frac{2}{k+1}\right)^{k/(k-1)}$$

For $k = 1.4$, $p^*/p_0 = 0.528$.

What happens when the back pressure is reduced further to a value below p^*, to condition (*v*)? Since the Mach number at the throat is unity ($V_e = c_e$), information about conditions in the exhaust duct cannot be transmitted upstream. Consequently, reductions in p_b below a value equal to p^* have no effect on flow conditions in the nozzle; thus neither the pressure distribution through the nozzle, the nozzle exit pressure, nor the mass flowrate are affected by lowering p_b below p^*. When p_b is equal to p^*, the nozzle is said to be *choked*.

For p_b less than p^*, the flow leaving the nozzle will expand suddenly to match the lower back pressure as shown for condition (*v*) in Fig. 10.6a. This sudden expansion process is not isentropic, and the pressure distribution cannot be predicted by one-dimensional theory.

Fig. 10.7 *Schematic Ts diagram for choked flow through a converging nozzle.*

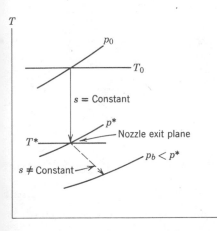

isentropic flow of an ideal gas

Flow through a converging nozzle may be divided into two regimes:

1. In Regime I, $1 \geq p_b/p_0 \geq p*/p_0$. The flow to and through the throat is isentropic.
2. In Regime II, $p_b/p_0 < p*/p_0$. The flow to the throat is isentropic, but a non-isentropic sudden expansion occurs in the flow leaving the nozzle.

The flow processes corresponding to Regime II are shown on a Ts diagram in Fig. 10.7. Two problems involving converging nozzles are solved in Example Problems 10.3 and 10.4.

   Although isentropic flow is an idealization, it is often a very good approximation for the actual behavior of nozzles. Since a nozzle is a device that accelerates a flow, the internal pressure gradient is favorable. This tends to keep the wall boundary layers thin and to minimize the effects of friction.

Example 10.3

A converging nozzle with a throat area of 0.01 ft^2 is operated with air at a back pressure of 59.1 psia. The nozzle is fed from a large plenum chamber where the stagnation pressure and temperature are 100 psia and 600 R, respectively. The exit Mach number and mass flowrate are to be determined using (i) isentropic flow relations, and (ii) tables for isentropic flow.

EXAMPLE PROBLEM 10.3

GIVEN:

Air flow through a converging nozzle at the conditions shown:

Flow is isentropic.

FIND:

(a) M_e
(b) $\dot{m}$

$p_0 = 100$ psia $p_b = 59.1$ psia
$T_0 = 600$ R

p_e

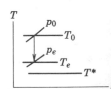

SOLUTION:

The pressure ratio is

$$\frac{p_b}{p_0} = \frac{59.1}{100} = 0.591 > 0.528$$

so the flow is not choked. Thus, $p_b = p_e$, and the flow is isentropic, as sketched on the Ts diagram above.

(*i*) *Isentropic Flow Relations*

The Mach number, M_e, may be found from the pressure ratio. Thus

$$\frac{p_e}{p_0} = \frac{p_b}{p_0} = \left(\frac{T_e}{T_0}\right)^{k/(k-1)} = \left[\frac{1}{1 + \frac{k-1}{2}M_e^2}\right]^{k/(k-1)}$$

Solving for M_e,

$$1 + \frac{k-1}{2}M_e^2 = \left(\frac{p_0}{p_b}\right)^{(k-1)/k}$$

and

$$M_e = \left\{\left[\left(\frac{p_0}{p_b}\right)^{(k-1)/k} - 1\right]\frac{2}{k-1}\right\}^{1/2} = \left\{\left[\left(\frac{100}{59.1}\right)^{0.286} - 1\right](5)\right\}^{1/2}$$

or

$$M_e = \{[1.162 - 1](5)\}^{1/2} = \sqrt{0.81} = 0.90 \qquad\qquad M_e$$

$$\frac{T_e}{T_0} = \left(\frac{p_e}{p_0}\right)^{(k-1)/k} = \left(\frac{p_b}{p_0}\right)^{(k-1)/k} = (0.591)^{0.286} = 0.861$$

$$T_e = 0.861 T_0 = 0.861(600\ R) = 517\ R$$

$$V_e = M_e c_e = M_e\sqrt{kRT_e} = (0.9)\left[1.4 \times \frac{53.3\ \text{ft-lbf}}{\text{lbm-R}} \times 517\ R \times \frac{32.2\ \text{lbm}}{\text{slug}} \times \frac{\text{slug-ft}}{\text{lbf-sec}^2}\right]^{1/2}$$

$$= 0.9 \times \frac{1114\ \text{ft}}{\text{sec}} = 1000\ \text{ft/sec}$$

$$\rho_e = \frac{p_e}{RT_e} = \frac{p_b}{RT_e} = \frac{59.1\ \text{lbf}}{\text{in.}^2} \times \frac{144\ \text{in.}^2}{\text{ft}^2} \times \frac{\text{lbm-R}}{53.3\ \text{ft-lbf}} \times \frac{1}{517\ R} = 0.31\ \text{lbm/ft}^3$$

Finally,

$$\dot{m} = \rho_e V_e A_e = \frac{0.31\ \text{lbm}}{\text{ft}^3} \times \frac{1000 \cdot \text{ft}}{\text{sec}} \times 0.01\ \text{ft}^2$$

$$\dot{m} = 3.1\ \text{lbm/sec} \qquad\qquad \dot{m}$$

(*ii*) ***Tables for Isentropic Flow*

The pressure ratio is

$$\frac{p_b}{p_0} = \frac{59.1}{100} = 0.591 > 0.528$$

so the flow is not choked. Thus $p_e = p_b$, and the flow is isentropic. From Table B.1, Appendix

B, $p/p_0 = 0.591$ at $M = 0.90$. Thus

$$M_e = 0.90 \qquad\qquad\qquad\qquad\qquad\qquad M_e$$

Also from Table B.1, at $M = 0.90$, $T/T_0 = 0.861$. Thus $T_e = 517$ R. The mass flowrate may be evaluated as in the preceding solution.

Example 10.4

Air flows isentropically through a converging nozzle. At a section where the nozzle area is 0.013 ft², the local pressure, temperature and Mach number are 60 psia, 40 F and 0.52, respectively. The back pressure is 30 psia. The Mach number at the throat, the mass flowrate, and the throat area are to be determined, using (i) isentropic flow relations and (ii) tables for isentropic flow.

EXAMPLE PROBLEM 10.4

GIVEN:

Air flow through a converging nozzle at the conditions shown:

$p_1 = 60$ psia

$T_1 = 40$ F

$M_1 = 0.52$

$A_1 = 0.013$ ft²

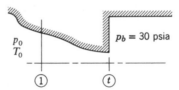

FIND:

(a) M_t
(b) $\dot{m}$
(c) A_t

SOLUTION:

(i) Isentropic Flow Relations

We must first check for choking, to determine if flow is isentropic down to p_b. To check, we must first evaluate the stagnation conditions.

$$\frac{T_0}{T_1} = 1 + \frac{k-1}{2}M_1^2 = 1 + 0.2(0.52)^2 = 1.054; \qquad T_0 = 1.054(500\ \text{R}) = 527\ \text{R}$$

$$\frac{p_0}{p_1} = \left(\frac{T_0}{T_1}\right)^{k/(k-1)} = (1.054)^{3.5} = 1.202$$

Thus $p_0 = 1.202p_1 = 1.202(60\ \text{psia}) = 72.1\ \text{psia}$

The back pressure ratio is

$$\frac{p_b}{p_0} = \frac{30}{72.1} = 0.416 < 0.528,$$

so flow is choked! For choked flow,

$$M_t = 1.0 \qquad\qquad\qquad\qquad\qquad\qquad\qquad\qquad M_t$$

The Ts diagram is

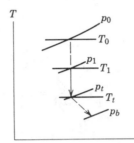

The mass flowrate may be found from conditions at section ①.

$$V_1 = M_1 c_1; \qquad c_1 = \sqrt{kRT_1} = 1096\ \text{ft/sec}; \qquad V_1 = 0.52(1096\ \text{ft/sec}) = 570\ \text{ft/sec}$$

$$\rho_1 = \frac{p_1}{RT_1} = \frac{60\ \text{lbf}}{\text{in.}^2} \times \frac{144\ \text{in.}^2}{\text{ft}^2} \times \frac{\text{lbm-R}}{53.3\ \text{ft-lbf}} \times \frac{1}{500\ \text{R}} = 0.324\ \text{lbm/ft}^3$$

$$\dot{m} = \rho_1 A_1 V_1 = \frac{0.324\ \text{lbm}}{\text{ft}^3} \times 0.013\ \text{ft}^2 \times \frac{570\ \text{ft}}{\text{sec}} = 2.4\ \text{lbm/sec} \qquad\qquad \dot{m}$$

The throat area may be computed by applying continuity between section ① and the throat, that is, $\dot{m} = \rho_1 V_1 A_1 = \rho_t V_t A_t$, so

$$A_t = A_1 \frac{\rho_1}{\rho_t} \frac{V_1}{V_t}$$

In isentropic flow, $T_0 = $ constant. At the throat, $M_t = 1$, so

$$\frac{T_0}{T_t} = 1 + \frac{k-1}{2}M_t^2 = 1 + \frac{k-1}{2} = 1.2$$

$$T_t = \frac{T_0}{1.2} = \frac{527\ \text{R}}{1.2} = 439\ \text{R}$$

$$V_t = M_t c_t = \sqrt{kRT_t} = 1027\ \text{ft/sec}$$

Flow is choked, so $p_t/p_0 = 0.528$; $p_t = 0.528p_0 = 0.528(72.1 \text{ psia}) = 38.1 \text{ psia}$. Thus

$$\rho_t = \frac{p_t}{RT_t} = \frac{38.1 \text{ lbf}}{\text{in.}^2} \times \frac{144 \text{ in.}^2}{\text{ft}^2} \times \frac{\text{lbm-R}}{53.3 \text{ ft-lbf}} \times \frac{1}{439 \text{ R}} = 0.234 \text{ lbm/ft}^3$$

and finally

$$A_t = A_1 \frac{\rho_1}{\rho_t} \frac{V_1}{V_t} = 0.013 \text{ ft}^2 \left(\frac{0.324}{0.234}\right)\left(\frac{570}{1027}\right) = 0.010 \text{ ft}^2 \quad\longleftarrow\quad A_t$$

(ii) **Tables for Isentropic Flow

We must first check for choking. From Table B.1, Appendix B, at $M_1 = 0.52$

$$\frac{T_1}{T_0} = 0.949; \qquad T_0 = \frac{T_1}{0.949} = \frac{500 \text{ R}}{0.949} = 527 \text{ R}$$

and

$$\frac{p_1}{p_0} = 0.832; \qquad p_0 = \frac{p_1}{0.832} = \frac{60 \text{ psia}}{0.832} = 72.1 \text{ psia}$$

The back pressure ratio is

$$\frac{p_b}{p_0} = \frac{30}{72.1} = 0.416 < 0.528$$

so flow is choked! For choked flow,

$$M_t = 1.0 \quad\longleftarrow\quad M_t$$

The Ts diagram was given above. The mass flowrate calculation is the same as in the previous solution. From Table B.1, at $M_1 = 0.52$, $A_1/A^* = 1.303$. For choked flow, $A_t = A^*$. Thus

$$A_t = A^* = \frac{A_1}{1.303} = \frac{0.013 \text{ ft}^2}{1.303} = 0.010 \text{ ft}^2 \quad\longleftarrow\quad A_t$$

10–3.5 ISENTROPIC FLOW IN A CONVERGING–DIVERGING NOZZLE

Having considered isentropic flow in a converging nozzle, we turn now to isentropic flow in a converging-diverging nozzle. As in the previous case, the flow through the converging-diverging passage of Fig. 10.8 is induced by a vacuum pump downstream, and is controlled by the valve shown. The upstream stagnation conditions are assumed constant; the pressure in the exit plane of the nozzle is denoted as p_e; the nozzle discharges to the back pressure, p_b. We are interested in investigating the effect of variations in back pressure on the pressure distribution through the nozzle. The results are illustrated graphically in Fig. 10.8. Let us consider each of the cases shown.

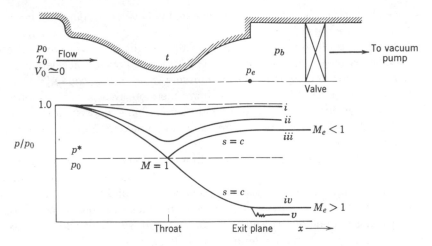

Fig. 10.8 *Pressure distributions for isentropic flow in a converging-diverging nozzle.*

With the valve initially closed, there is no flow through the nozzle; the pressure is constant at p_0 as shown by the dashed line in the figure. Opening the valve slightly (p_b slightly less than p_0) produces the pressure distribution curve (i). If the flowrate is low enough, at all points on this curve the flow will be subsonic and essentially incompressible. Under these conditions, the C–D nozzle will behave like a venturi, with flow accelerating in the converging portion, until a point of maximum velocity and minimum pressure is reached at the throat, then decelerating in the diverging portion to the nozzle exit.

As the valve is opened farther and the flowrate is increased, a more sharply defined pressure minimum occurs, as shown by curve (ii). Although compressibility effects become important, the flow is still subsonic everywhere, and deceleration takes place in the diverging section. Finally, as the valve is opened farther, curve (iii) results. At the point of minimum area the flow finally reaches $M = 1$ and the nozzle is choked, that is, the flowrate is the maximum possible for the given nozzle and stagnation conditions.

All of the flows giving rise to pressure distributions i, ii, and iii are isentropic; each curve is associated with a unique rate of mass flow. Finally when curve (iii) is reached, critical conditions are present at the throat. For this flowrate, the flow is choked, and

$$\dot{m} = \rho^* A^* V^*$$

where $A^* = A_t$.

In our discussion of the effect of area variation on isentropic flow we noted acceleration of a flow beyond Mach one at a throat is possible in a diverging section downstream of the throat. At this point, then, we ask the question, "What

515   **isentropic flow of an ideal gas**

back pressure, p_b, is necessary to accelerate the flow isentropically in the diverging portion of the nozzle?"

Accelerating flow in the diverging section results in a pressure decrease. This condition is illustrated by curve (iv) in Fig. 10.8. The flow will accelerate isentropically in the nozzle provided the back pressure is set at the value p_{iv}. Thus we see that with a throat Mach number of unity there are two possible isentropic flow conditions in the converging-diverging nozzle. This is consistent with the results of Fig. 10.5, where we found the two Mach numbers for a given value of A/A^* in isentropic flow.

Lowering the back pressure below that of condition (iv), say to condition (v), has no effect on the flow in the nozzle. The flow is isentropic from the plenum chamber to the nozzle exit (same as condition (iv)) and then undergoes a sudden expansion (which is irreversible) to meet the back pressure. A nozzle operating under these conditions is said to be *underexpanded*, since additional expansion takes place outside the nozzle to meet the back pressure.

A converging-diverging nozzle is generally intended to produce supersonic flow at the exit plane. If the back pressure is set at p_{iv}, flow will be isentropic through the nozzle, and supersonic at the nozzle exit. Nozzles operating at $p_b = p_{iv}$ (corresponding to curve (iv) in Fig. 10.8) are said to be at *design conditions*.

The only time flow leaving a C–D nozzle is supersonic is when the back pressure is at the design value or below. The exit Mach number is fixed by the area ratio A_e/A^*. Although C–D nozzles are inherently efficient at design conditions because they have favorable pressure gradients, it is difficult to design a nozzle for completely uniform supersonic flow. For rocket-propelled vehicles, other compromises in nozzle design must be made to match ambient pressure conditions, and to minimize structural weight. Thus, although most such vehicles use C–D nozzles to obtain the maximum possible thrust, few approach the theoretical optimum thrust. Because only a single supersonic Mach number can be obtained for a given area ratio, nozzles for supersonic wind tunnels are often built with variable geometry.

You have undoubtedly noticed that nothing has been said about the operation of converging-diverging nozzles when the back pressure is in the range $p_{iii} > p_b > p_{iv}$. The reason for this is simply that for such cases isentropic flow is not possible. Under these conditions a shock (which may be treated as an irreversible discontinuity involving entropy increase) occurs somewhere within the flow. Following a discussion of normal shocks in Section 10–6, we shall return to complete the discussion of converging-diverging nozzle flows. Nozzles operating with $p_{iii} > p_b > p_{iv}$ are said to be *overexpanded*, because the pressure at some point in the nozzle is less than the back pressure. Obviously an overexpanded nozzle could be made to operate at a new design condition by cutting off some of the diverging portion.

One other comment should be made at this point. Real compressible fluid flows are affected by friction, heating or cooling, and the possible presence (in supersonic flow) of shock waves. We have treated isentropic flow first because it is a useful idealized model for many real flow processes and because it gives us valuable insight into the behavior of fluids in compressible flow. In the next two sections we take up the effects of friction and heat transfer separately to gain insight into the effect of each factor on flow behavior. Following this we return to a discussion of the normal shock and complete our study of converging-diverging nozzle flows. In later courses it will be possible to explore real flows and the results of combining several of these effects.

Example 10.5

Air flows isentropically in a converging-diverging nozzle, with an exit area of 0.01 ft². The nozzle is fed from a large plenum where the stagnation pressure and temperature are 100 psia and 600 R, respectively. The exit pressure is 95.4 psia, and the Mach number at the throat is 0.68. Flow conditions at the throat and the exit Mach number are to be determined.

EXAMPLE PROBLEM 10.5

GIVEN:

Isentropic flow of air in C–D nozzle as shown:

$$p_0 = 100 \text{ psia}$$
$$T_0 = 600 \text{ R}$$
$$p_b = 95.4 \text{ psia}$$
$$M_t = 0.68 \qquad A_e = 0.01 \text{ ft}^2$$

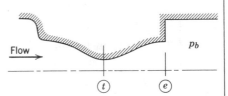

FIND:

(a) Properties at nozzle throat
(b) M_e

SOLUTION:

(*i*) *Isentropic Flow Relations*

The stagnation temperature is constant for isentropic flow. Thus, since

$$\frac{T_0}{T} = 1 + \frac{k-1}{2}M^2$$

isentropic flow of an ideal gas

$$T_t = \frac{T_0}{1 + \dfrac{k-1}{2}M_t^2} = \frac{600\ R}{1 + 0.2(0.68)^2} = \frac{600\ R}{1.0925} = 549\ R \longleftarrow \qquad T_t$$

Also

$$P_t = P_0\left(\frac{T_t}{T_0}\right)^{k/(k-1)} = (100\ \text{psia})\left(\frac{1}{1.0925}\right)^{3.5} = (100\ \text{psia})(0.734) = 73.4\ \text{psia} \longleftarrow \quad P_t$$

$$\rho_t = \frac{P_t}{RT_t} = \frac{73.4\ \text{lbf}}{\text{in.}^2} \times \frac{144\ \text{in.}^2}{\text{ft}^2} \times \frac{\text{lbm-R}}{53.3\ \text{ft-lbf}} \times \frac{1}{549\ R} = 0.361\ \text{lbm/ft}^3 \longleftarrow \quad \rho_t$$

$$c_t = \sqrt{kRT_t} = 1149\ \text{ft/sec} \longleftarrow \qquad c_t$$

and

$$V_t = M_t c_t = (0.68)(1149\ \text{ft/sec}) = 781\ \text{ft/sec} \longleftarrow \qquad V_t$$

Since the exit plane pressure is above the critical value, flow must be subsonic there. Consequently $p_e = p_b$. The temperature may be found from the stagnation conditions

$$T_e = T_0\left(\frac{p_e}{p_0}\right)^{(k-1)/k} = (600\ R)(0.954)^{0.286} = (600\ R)(0.987) = 592\ R$$

Also, since

$$\frac{T_0}{T_e} = 1 + \frac{k-1}{2}M_e^2,$$

$$M_e = \left[\left(\frac{2}{k-1}\right)\left(\frac{T_0}{T_e} - 1\right)\right]^{1/2} = \left[(5)\left\{\frac{600}{592} - 1\right\}\right]^{1/2} = (0.0676)^{1/2} = 0.26 \longleftarrow \quad M_e$$

The Ts diagram for this flow is:

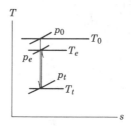

*(ii)*** *Tables for Isentropic Flow*

The stagnation temperature is constant for isentropic flow. From Table B.1, Appendix B, at $M = 0.68$,

EXAMPLE PROBLEM 10.5 (continued)

$$\frac{T}{T_0} = 0.915; \quad T_t = 0.915T_0 = (0.915)(600 \text{ R}) = 549 \text{ R} \qquad\qquad\qquad T_t$$

$$\frac{p}{p_0} = 0.734; \quad p_t = 0.734p_0 = (0.734)(100 \text{ psia}) = 73.4 \text{ psia} \qquad\qquad p_t$$

$$\frac{\rho}{\rho_0} = 0.802; \quad \rho_0 = \frac{p_0}{RT_0} = \frac{100 \text{ lbf}}{\text{in.}^2} \times \frac{144 \text{ in.}^2}{\text{ft}^2} \times \frac{\text{lbm-R}}{53.3 \text{ ft-lbf}} \times \frac{1}{600 \text{ R}} = 0.450 \text{ lbm/ft}^3$$

$$\rho_t = 0.802\rho_0 = (0.802)(0.450 \text{ lbm/ft}^3) = 0.361 \text{ lbm/ft}^3 \qquad\qquad \rho_t$$

and

$$\frac{A}{A^*} = 1.110; \quad A_t = 1.110A^*$$

but at this point, A^* is not known.

At the exit, $p_e = 95.4$ psia. Thus $p_e/p_0 = 0.954$, and from Table B.1,

$$M_e = 0.26 \qquad\qquad\qquad\qquad\qquad\qquad\qquad\qquad\qquad\qquad\qquad\qquad\qquad\qquad M_e$$

Since A_e is known, we can compute A^*. From Table B.1, at $M = 0.26$, $A/A^* = 2.32$. Thus

$$A^* = \frac{A_e}{2.32} = \frac{0.01 \text{ ft}^2}{2.32} = 0.00431 \text{ ft}^2$$

and

$$A_t = 1.110A^* = (1.110)(0.00431 \text{ ft}^2) = 0.00478 \text{ ft}^2 \qquad\qquad A_t$$

Note that the solution for A_t using the tables was relatively effortless. To find A_t using the isentropic relations, we could have applied continuity between the throat and exit planes. Since this would have required calculation of all properties at the exit, it would have been a lengthy process.

Example 10.6

The nozzle of Example 10.5 has a design back pressure of 8.7 psia, but is operated at a back pressure of 5 psia. Flow within the nozzle may be assumed isentropic. Determine the pressure and temperature at the nozzle throat, and the exit Mach number. Use (*i*) isentropic flow relations, and (*ii*) tables for isentropic flow.

519   **isentropic flow of an ideal gas**

GIVEN:

Air flow through C–D nozzle as shown:

$$p_0 = 100 \text{ psia}$$

$$T_0 = 600 \text{ R}$$

$$p_b = 5 \text{ psia}$$

$$A_t = 0.00478 \text{ ft}^2 \qquad A_e = 0.01 \text{ ft}^2$$

$$p_e (\text{design}) = 8.7 \text{ psia}$$

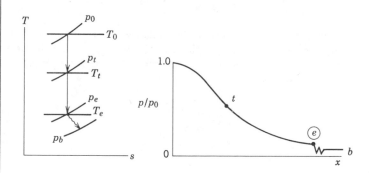

FIND:

(a) T_t
(b) p_t
(c) M_e

SOLUTION:

The operating back pressure is *below* the design value. Consequently the nozzle is under-expanded, and the Ts diagram and pressure distribution will be as shown below:

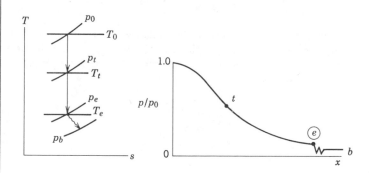

Flow *within* the nozzle will be isentropic, but the sudden expansion from p_e to p_b will be irreversible. Assume $p_e = p_e$ (design) $= 8.7$ psia.

(i) Isentropic Flow Relations

For isentropic flow at design conditions, the nozzle will be *choked*, that is, $M_t = 1$.

Then

$$\frac{T_t}{T_0} = \frac{1}{1 + \frac{k-1}{2}M_t^2} = \frac{1}{1.2}; \qquad T_t = \frac{T_0}{1.2} = 500 \text{ R} \qquad\qquad\qquad\qquad T_t$$

and

$$\frac{p_t}{p_0} = \left(\frac{T_t}{T_0}\right)^{k/(k-1)} = \left(\frac{1}{1.2}\right)^{3.5} = 0.528; \qquad p_t = 52.8 \text{ psia} \qquad\qquad\qquad p_t$$

The exit Mach number can be computed from the pressure ratio. Thus

$$\frac{p_e}{p_0} = \left(\frac{T_e}{T_0}\right)^{k/(k-1)} = \left[1 \middle/ 1 + \frac{k-1}{2}M_e^2\right]^{k/(k-1)}$$

Solving,

$$M_e = \left\{\left(\frac{2}{k-1}\right)\left[\left(\frac{p_0}{p_e}\right)^{(k-1)/k} - 1\right]\right\}^{1/2} = \left\{(5)\left[\left(\frac{100}{8.7}\right)^{0.286} - 1\right]\right\}^{1/2}$$

$$M_e = \{(5)[2.010 - 1]\}^{1/2} = \sqrt{5.052} = 2.25 \qquad\qquad\qquad\qquad M_e$$

(*ii*)** *Tables for Isentropic Flow*

From Table B.1, Appendix B, at $M = 1$,

$$\frac{T}{T_0} = 0.833; \qquad T_t = (0.833)(600 \text{ R}) = 500 \text{ R} \qquad\qquad\qquad\qquad T_t$$

and

$$\frac{p}{p_0} = 0.528; \qquad p_t = 52.8 \text{ psia} \qquad\qquad\qquad\qquad p_t$$

Again from Table B.1, at $p/p_0 = p_e/p_0 = 0.087$,

$$M_e \simeq 2.25 \quad \text{(by interpolation)} \qquad\qquad\qquad\qquad M_e$$

$\left.\begin{array}{l} \text{To check, note } A_e/A^* = 0.01/0.00478 = 2.09. \text{ For } M = 2.25, A/A^* \simeq 2.096. \text{ Therefore} \\ \text{the solution checks.} \end{array}\right\}$

10–4 Adiabatic Flow in a Constant Area Duct with Friction

To obtain an overall view of the problem of frictional adiabatic flow, apply the basic equations to the uniform flow of an ideal gas with constant specific heats through the finite control volume shown in Fig. 10.9.

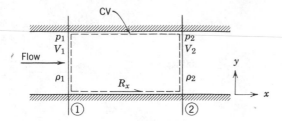

10–4.1 BASIC EQUATIONS

a. Continuity Equation

Basic equation:

$$0 = \overset{= \, 0(1)}{\frac{\partial}{\partial t}\iiint_{CV} \rho \, d\forall} + \iint_{CS} \rho \vec{V} \cdot d\vec{A} \tag{4.14}$$

Assumptions: (1) Steady flow
(2) Uniform flow at each section

Then

$$0 = \{-|\rho_1 V_1 A_1|\} + \{|\rho_2 V_2 A_2|\}$$

The area is constant, so

$$\rho_1 V_1 = \rho_2 V_2 \equiv G \tag{10.7a}$$

b. Momentum Equation

Basic equation:

$$F_{S_x} + \overset{= \, 0(3)}{\cancel{F_{B_x}}} = \overset{= \, 0(1)}{\frac{\partial}{\partial t}\iiint_{CV} V_x \rho \, d\forall} + \iint_{CS} V_x \rho \vec{V} \cdot d\vec{A} \tag{4.19a}$$

Assumption: (3) $F_{B_x} = 0$

The surface force is due to pressure forces at sections ① and ②, and to the friction force, R_x, of the duct wall *on* the flow. Substituting, recognizing that $A_2 = A_1 = A$, then

$$R_x + p_1 A - p_2 A = V_1\{-|\rho_1 V_1 A|\} + V_2\{|\rho_2 V_2 A|\}$$

Using scalar magnitudes and dropping absolute value signs, we obtain

$$R_x + p_1 A - p_2 A = \dot{m} V_2 - \dot{m} V_1 \qquad (10.7b)$$

c. First Law of Thermodynamics

Basic equation:

$$\overset{=\,0(4)}{\cancel{\dot{Q}}} - \overset{=\,0(5)}{\cancel{\dot{W}_s}} - \overset{=\,0(6)}{\cancel{\dot{W}_{shear}}} = \frac{\partial}{\partial t} \iiint_{CV} e\rho \, d\forall + \iint_{CS} (e + pv)\rho \vec{V} \cdot d\vec{A} \qquad (4.45)$$

where

$$e = u + \frac{V^2}{2} + \overset{\simeq\,0(7)}{\cancel{gz}}$$

Assumptions: (4) $\dot{Q} = 0$ (adiabatic flow)
(5) $\dot{W}_s = 0$
(6) $\dot{W}_{shear} = 0$
(7) Effects of gravity are negligible

With these restrictions, the equation becomes

$$0 = \left(u_1 + \frac{V_1^2}{2} + p_1 v_1\right)\{-|\rho_1 V_1 A|\} + \left(u_2 + \frac{V_2^2}{2} + p_2 v_2\right)\{|\rho_2 V_2 A|\}$$

Since the mass flowrate terms in { } are identical by continuity,

$$u_1 + \frac{V_1^2}{2} + p_1 v_1 = u_2 + \frac{V_2^2}{2} + p_2 v_2$$

or

$$h_1 + \frac{V_1^2}{2} = h_2 + \frac{V_2^2}{2} \qquad (10.7c)$$

We could also write

$$h_{0_1} = h_{0_2}$$

which is a physical consequence of our assumption of adiabatic flow.

d. Second Law of Thermodynamics

Basic equation:

$$\iint_{CS} \frac{1}{T} \overset{=\,0(4)}{\cancel{\frac{\dot{Q}}{A}}} dA \le \frac{\partial}{\partial t} \overset{=\,0(1)}{\cancel{\iiint_{CV} s\rho \, d\forall}} + \iint_{CS} s\rho \vec{V} \cdot d\vec{A} \qquad (4.46)$$

523   **adiabatic flow in a constant area duct with friction**

Then

$$0 \leq s_1\{-|\rho_1 V_1 A|\} + s_2\{|\rho_2 V_2 A|\} = \dot{m}(s_2 - s_1)$$

Because the flow is frictional and, hence, irreversible, the inequality in the above equation holds. The control volume form of the second law then tells us that $s_2 - s_1 > 0$.

This fact is of little help in calculating the actual entropy change between any two points in an adiabatic frictional flow. To calculate the entropy change we rely on the $T\,ds$ equations. Since

$$T\,ds = dh - v\,dp$$

for an ideal gas we can write

$$ds = c_p \frac{dT}{T} - R \frac{dp}{p}$$

For constant specific heats the equation can be integrated to give

$$s_2 - s_1 = c_p \ln \frac{T_2}{T_1} - R \ln \frac{p_2}{p_1} \tag{10.7d}$$

e. Equations of State

For an ideal gas, the equation of state is given by

$$p = \rho R T \tag{10.7e}$$

Equations 10.7a–e are the governing equations for the steady, one-dimensional, adiabatic, frictional flow of an ideal gas in a constant area duct. If all the properties at state ① are known, then we have 7 unknowns (ρ_2, V_2, p_2, h_2, s_2, T_2, and R_x) in these 5 equations. However, since we have counted h_2 and T_2 both as unknowns, it is legitimate to include the known relationship between h and T for an ideal gas with constant specific heats, namely

$$\Delta h = h_2 - h_1 = c_p \Delta T = c_p(T_2 - T_1) \tag{10.7f}$$

We thus have the situation of 6 equations and 7 unknowns.

Then if all conditions at point ① are known, how many possible states ② are there? The mathematics of the situation (6 equations and 7 unknowns) indicate there are an infinite number of possible states at ②.

With an infinite number of possible states at ② for a given state ①, what is to be expected if all possible states at ② are plotted on a Ts diagram? It follows that the locus of all possible states ② reachable from state ① is a continuous curve passing through state ①.

How might we determine this curve? Perhaps the simplest way is to assume different values of T_2. For an assumed value of T_2 we could then calculate the corresponding values of all other properties at state ② and also R_x.

10–4.2 THE FANNO LINE

The results of these calculations are shown qualitatively on the Ts plane of Fig. 10.10. The locus of all possible downstream states is referred to as the *Fanno line*.[1] In carrying out the calculations we discover some interesting features of Fanno line

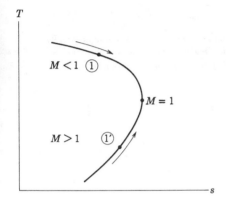

Fig. 10.10 *Schematic Ts diagram for frictional adiabatic (Fanno line) flow in a constant area duct.*

flow. At the point of maximum entropy, the Mach number is unity. On the upper branch of the curve the Mach number is always less than unity, and it increases monotonically as we proceed to the right along the curve. At every point on the lower portion of the curve the Mach number is greater than unity; furthermore the Mach number decreases monotonically as we move to the right along the curve.

For any initial state in a Fanno line flow any point on the Fanno line is a mathematically possible downstream state. Indeed we determined the locus of all possible downstream states by assuming values of T_2 and calculating the corresponding properties. While the Fanno line represents all mathematically possible downstream states, are they all physically attainable downstream states? A moment's reflection will indicate they are not. Why not? The second law requires that the entropy must increase. Consequently we must always move to the right along the Fanno line of Fig. 10.10. Indeed it is the effect of friction that causes the flow to change from the initial state. Referring again to Fig. 10.10, we see that for an initially subsonic flow (state ①), the effect of friction increases the Mach number toward unity. For a flow that is initially supersonic (state ①'), the effect of friction decreases the Mach number toward unity.

[1] A convenient way to remember this name is to think of *Fr*ictional *A*diabatic *N*... flow.

In developing the simplified form of the first law for Fanno line flow we found that the stagnation enthalpy remains constant for the flow. Consequently when the fluid is an ideal gas with constant specific heats, the stagnation temperature must also remain constant. What happens to the stagnation pressure? Friction causes the local isentropic stagnation pressure to decrease for all Fanno line flows as shown in Fig. 10.11. Recall that the second law requires $s_2 - s_1 > 0$. Since the entropy must increase in the direction of flow, the flow process must proceed to the right on the Ts diagram. In Fig. 10.11, a path from state ① to state ② is shown on the subsonic portion of the curve. The corresponding values of local isentropic stagnation pressure, p_{0_1} and p_{0_2}, show clearly that $p_{0_2} < p_{0_1}$. An

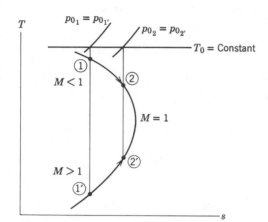

Fig. 10.11 *Schematic of Fanno line flow on Ts plane, showing reduction in local isentropic stagnation pressure caused by friction.*

identical result is obtained for flow on the supersonic branch of the curve from state ① to state ②. Again $p_{0_{2'}} < p_{0_{1'}}$. Thus p_0 decreases for any Fanno line flow.

At this point it may be beneficial to summarize the effects of friction on flow properties in Fanno line flow. The summary is presented in Table 10.1.

In deducing the effect of friction on flow properties for Fanno line flow, we have used the shape of the Fanno line on the Ts diagram and the basic governing equations (Eqs. 10.7a–f). The reader is encouraged to follow through the logic indicated in the right-hand column of the table.

We have noted that the entropy must increase in the direction of flow; it is the effect of friction that causes the change in flow properties along the Fanno line curve. From Fig. 10.10 we see that there is a maximum entropy point, corresponding to $M = 1$ for each Fanno line. The maximum entropy point is reached by increasing the amount of friction (through addition of duct length), just enough to produce a Mach number of unity (choked flow) at the exit. How do we compute this critical length of duct?

table 10.1   *Summary of Effects of Friction on Properties in Fanno Line Flow*

PROPERTY	SUBSONIC $M < 1$	SUPERSONIC $M > 1$	OBTAINED FROM:
Stagnation Temperature, T_0	Constant	Constant	Energy equation
Entropy, s	Increases	Increases	Second law
Stagnation Pressure, p_0	Decreases	Decreases	$T_0 = $ constant, s increases
Temperature, T	Decreases	Increases	Shape of Fanno line
Velocity, V	Increases	Decreases	Energy equation, and trend of T
Mach number, M	Increases	Decreases	Trends of V, T, and definition of M
Density, ρ	Decreases	Increases	Continuity equation, and effect on V
Pressure, p	Decreases	Increases	Equation of state, and effects on ρ, T

To compute the critical duct length, we must analyze the flow in detail, accounting for the effect of friction. This analysis requires that we begin with a differential control volume, develop expressions in terms of Mach number, and integrate along the flow to the point where $M = 1$. This analysis is completed in Section 10–4.3, where tables for Fanno line flow are developed. The detailed analysis is so complex algebraically that it tends to obscure the physics of the flow; the general trends in properties caused by friction can be demonstrated using finite control volumes and the basic governing equations. This approach is illustrated in Example Problem 10.7.

Example 10.7

Air flows in an insulated tube with an inside diameter of 0.25 in. At section ①, where the Mach number is 0.3, the stagnation pressure and temperature are 100 psia and 140 F. At section ②, downstream, the temperature is 40 F. Determine the mass flowrate, the properties at section ②, and the force exerted on the duct wall by the flowing gas.

EXAMPLE PROBLEM 10.7

GIVEN:

Air flow in insulated tube

FIND:

(a) $\dot{m}$
(b) Properties at station ②
(c) Force on duct wall

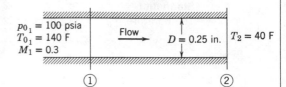

$p_{0_1} = 100$ psia
$T_{0_1} = 140$ F
$M_1 = 0.3$

Flow

$D = 0.25$ in.

$T_2 = 40$ F

① ②

SOLUTION:

The mass flowrate can be obtained from properties at section ①. From the definition of T_0,

$$\frac{T_{0_1}}{T_1} = 1 + \frac{k-1}{2}M_1^2 = 1 + 0.2(0.3)^2 = 1.018$$

$$T_1 = \frac{T_{0_1}}{1.018} = \frac{(460 + 140)\,R}{1.018} = 589\,R$$

$$\frac{p_{0_1}}{p_1} = \left(\frac{T_{0_1}}{T_1}\right)^{k/(k-1)} = (1.018)^{3.5} = 1.064$$

$$p_1 = \frac{p_{0_1}}{1.064} = \frac{100\,\text{psia}}{1.064} = 93.9\,\text{psia}$$

For an ideal gas,

$$\rho_1 = \frac{p_1}{RT_1} = \frac{93.9\,\text{lbf}}{\text{in.}^2} \times \frac{\text{lbm-R}}{53.3\,\text{ft-lbf}} \times \frac{1}{589\,R} \times \frac{144\,\text{in.}^2}{\text{ft}^2} = 0.431\,\text{lbm/ft}^3$$

and

$$c_1 = \sqrt{kRT_1} = \left[1.4 \times \frac{53.3\,\text{ft-lbf}}{\text{lbm-R}} \times 589\,R \times \frac{32.2\,\text{lbm}}{\text{slug}} \times \frac{\text{slug-ft}}{\text{lbf-sec}^2}\right]^{1/2} = 1190\,\text{ft/sec}$$

$$V_1 = M_1 c_1 = (0.3)\,1190\,\text{ft/sec} = 357\,\text{ft/sec}$$

$$A_1 = \frac{\pi D^2}{4} = \frac{\pi}{4}(0.25)^2\,\text{in.}^2 \times \frac{\text{ft}^2}{144\,\text{in.}^2} = 3.41 \times 10^{-4}\,\text{ft}^2$$

$$\dot{m} = \rho_1 A_1 V_1 = \frac{0.431\,\text{lbm}}{\text{ft}^3} \times 0.000341\,\text{ft}^2 \times \frac{357\,\text{ft}}{\text{sec}} = 0.0525\,\text{lbm/sec} \qquad \dot{m}$$

Flow is adiabatic and frictional and, therefore, a Fanno line flow. Hence T_0 is constant, and

$$T_{0_2} = T_{0_1} = 600 \text{ R} \qquad\qquad\qquad T_{0_2}$$

Also

$$\frac{T_{0_2}}{T_2} = 1 + \frac{k-1}{2}M_2^2$$

Solving for M_2,

$$M_2 = \left[\frac{2}{k-1}\left(\frac{T_{0_2}}{T_2} - 1\right)\right]^{1/2} = \left[5\left(\frac{600}{500} - 1\right)\right]^{1/2} = 1.0 \qquad\qquad M_2$$

$$c_2 = \sqrt{kRT_2} = 1096 \text{ ft/sec} \qquad\qquad\qquad c_2$$

$$V_2 = M_2 c_2 = (1.0)(1096 \text{ ft/sec}) = 1096 \text{ ft/sec} \qquad\qquad V_2$$

From continuity $\rho_1 V_1 = \rho_2 V_2$, so

$$\rho_2 = \rho_1 \frac{V_1}{V_2} = 0.431 \frac{\text{lbm}}{\text{ft}^3} \frac{357}{1096} = 0.140 \text{ lbm/ft}^3 \qquad\qquad \rho_2$$

$$p_2 = \rho_2 R T_2 = 0.140 \frac{\text{lbm}}{\text{ft}^3} \times \frac{53.3 \text{ ft-lbf}}{\text{lbm-R}} \times 500 \text{ R} \times \frac{\text{ft}^2}{144 \text{ in.}^2} = 25.9 \text{ psia} \qquad\qquad p_2$$

$$\frac{p_{0_2}}{p_2} = \left(\frac{T_{0_2}}{T_2}\right)^{k/(k-1)} = \left[1 + \frac{k-1}{2}M_2^2\right]^{k/(k-1)} = [1 + 0.2(1)^2]^{3.5} = (1.2)^{3.5} = 1.89$$

$$p_{0_2} = 1.89 p_2 = (1.89)(25.9 \text{ psia}) = 49.0 \text{ psia} \qquad\qquad\qquad p_{0_2}$$

The frictional force may be found by applying the momentum equation to the control volume below:

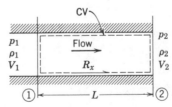

Basic equation:

$$\qquad\qquad = 0(1) \qquad = 0(2)$$

$$F_{S_x} + \cancel{F_{B_x}} = \cancel{\frac{\partial}{\partial t}} \iiint_{CV} V_x \rho \, d\forall + \iint_{CS} V_x \rho \vec{V} \cdot d\vec{A} \qquad\qquad (4.19a)$$

Assumptions: (1) $F_{B_x} = 0$
(2) Steady flow
(3) Uniform flow at each section

529   **adiabatic flow in a constant area duct with friction**

Then

$$p_1A - p_2A + R_x = V_1\{-|\rho_1 V_1 A|\} + V_2\{|\rho_2 V_2 A|\} = \dot{m}(V_2 - V_1)$$

and

$$R_x = (p_2 - p_1)A + \dot{m}(V_2 - V_1)$$

$$= (25.9 - 93.9)\frac{\text{lbf}}{\text{in.}^2} \times \frac{\pi}{4}(0.25)^2 \text{ in.}^2$$

$$+ \frac{0.0525 \text{ lbm}}{\text{sec}} \times (1096 - 357)\frac{\text{ft}}{\text{sec}} \times \frac{\text{slug}}{32.2 \text{ lbm}} \times \frac{\text{lbf-sec}^2}{\text{slug-ft}}$$

$$R_x = -3.34 + 1.20 = -2.14 \text{ lbf (i.e. to left)}$$

This is the force exerted on the control volume by the duct wall. The force of the *fluid* on the *duct* is

$$K_x = -R_x = 2.14 \text{ lbf} \quad \text{(to the right)} \qquad\qquad K_x$$

**10–4.3 TABLES FOR COMPUTATION OF FANNO LINE FLOW

The primary independent variable in Fanno line flow is the friction force, R_x. Knowledge of the total friction force between any two points in a Fanno line flow would enable us to predict downstream conditions when conditions upstream were known. The total friction force depends on the wall shear stress. Since the

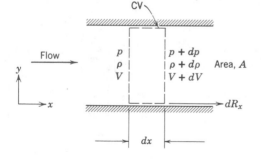

Fig. 10.12 *Differential control volume used for analysis of Fanno line flow.*

wall shear stress varies along the duct, we must develop a differential equation and then integrate to find property variations. In setting up the differential equation we use the differential control volume shown in Fig. 10.12 for our analysis.

** This section may be omitted without loss of continuity in the text material.

a. Continuity Equation

Basic equation:

$$0 = \overset{= \ 0(1)}{\frac{\partial}{\partial t} \iiint\limits_{CV} \rho \, d\forall} + \iint\limits_{CS} \rho \vec{V} \cdot d\vec{A} \qquad (4.14)$$

Assumptions: (1) Steady flow
(2) Uniform flow at each section

$$0 = \{-|\rho V A|\} + \{|(\rho + d\rho)(V + dV)A|\}$$

or

$$\rho V A = (\rho + d\rho)(V + dV)A$$

Simplifying and canceling A,

$$0 = \overset{}{-\rho V} + \rho V + \rho \, dV + V d\rho + \overset{\simeq \ 0}{d\rho \, dV}$$

which reduces to

$$\rho \, dV + V \, d\rho = 0 \qquad (10.8a)$$

since products of differentials are negligible.

b. Momentum Equation

Basic equation:

$$F_{S_x} + \overset{= \ 0(3)}{F_{B_x}} = \overset{= \ 0(1)}{\frac{\partial}{\partial t} \iiint\limits_{CV} V_x \rho \, d\forall} + \iint\limits_{CS} V_x \rho \vec{V} \cdot d\vec{A} \qquad (4.19a)$$

Assumption: (3) $F_{B_x} = 0$

The momentum equation becomes

$$dR_x + pA - (p + dp)A = V\{-|\rho V A|\} + (V + dV)\{|(\rho + d\rho)(V + dV)A|\}$$

which simplifies, using continuity, to give

$$\frac{dR_x}{A} - dp = \rho V \, dV \qquad (10.8b)$$

c. First Law of Thermodynamics

Basic equation:

$$\cancel{\dot{Q}}^{=\,0(4)} - \cancel{\dot{W}_s}^{=\,0(5)} - \cancel{\dot{W}_{shear}}^{=\,0(6)} = \cancel{\frac{\partial}{\partial t}\iiint_{CV} e\rho\,d\forall}^{=\,0(1)} + \iint_{CS} (e + pv)\rho\vec{V}\cdot d\vec{A} \quad (4.45)$$

where

$$e = u + \frac{V^2}{2} + \cancel{gz}^{\simeq\,0(7)}$$

Assumptions: (4) Adiabatic flow, $\dot{Q} = 0$
(5) $\dot{W}_s = 0$
(6) $\dot{W}_{shear} = 0$
(7) Effects of gravity are negligible

Under these restrictions, we obtain

$$0 = \left(u + \frac{V^2}{2} + pv\right)\{-|\rho VA|\} + \left[u + du + \frac{V^2}{2} + d\left(\frac{V^2}{2}\right) + pv + d(pv)\right]$$

$$\times \{|(\rho + d\rho)(V + dV)A|\}$$

Noting from continuity that the flowrate terms { } are equal, and substituting $h = u + pv$

$$dh + d\left(\frac{V^2}{2}\right) = 0 \quad (10.8c)$$

To carry out our analysis, we must relate dR_x to the flow variables. We note that $dR_x = -\tau_w P_w\,dx$, where P_w is the wetted perimeter of the duct. To account for ducts of other than circular shape, we introduce the hydraulic diameter

$$D_h \equiv \frac{4A}{P_w} \quad (8.35)$$

(The factor of 4 is introduced so that D_h reduces to D, the diameter, for circular ducts.) In terms of D_h,

$$dR_x = -\tau_w \frac{4A}{D_h}\,dx \quad (10.9)$$

To obtain an expression for τ_w in terms of flow variables, we assume that changes in flow variables with x are gradual, and use correlations for fully-developed duct flow. (Such an assumption should be checked against experimental data.)

In fully-developed duct flow, the net pressure force on an infinitesimal slice of fluid is $-A\,dp$, and the force due to wall friction is $-\tau_w P_w\,dx$. These two forces must sum to zero. Thus

$$-A\,dp - \tau_w P_w\,dx = 0$$

Solving for τ_w,

$$\tau_w = -\frac{A}{P_w}\frac{dp}{dx} \qquad (10.10)$$

But for duct flow,

$$\frac{dp}{dx} = -f\frac{1}{D_h}\frac{\rho V^2}{2}$$

Substituting into Eq. 10.10,

$$\tau_w = \frac{A}{P_w}\frac{f}{D_h}\frac{\rho V^2}{2}$$

or

$$\tau_{w_j} = \frac{f}{4}\frac{\rho V^2}{2} \qquad (10.11)$$

Combining Eqs. 10.9 and 10.11,

$$dR_x = -\frac{A}{D_h}f\frac{\rho V^2}{2}\,dx$$

or

$$\frac{dR_x}{A} = -\frac{f}{D_h}\frac{\rho V^2}{2}\,dx$$

Substituting this result into the momentum equation (Eq. 10.8b), we obtain

$$-f\frac{1}{D_h}\frac{\rho V^2}{2}\,dx - dp = \rho V\,dV$$

or

$$\frac{dp}{p} = -f\frac{1}{D_h}\frac{\rho V^2}{2p}\,dx - \frac{\rho V\,dV}{p}$$

Noting that $p/\rho = RT = c^2/k$, and $V\,dV = d(V^2/2)$, we obtain

$$\frac{dp}{p} = -f\frac{1}{D_h}\frac{kM^2}{2}\,dx - \frac{k}{c^2}d\left(\frac{V^2}{2}\right)$$

adiabatic flow in a constant area duct with friction

and finally

$$\frac{dp}{p} = -f \frac{1}{D_h} \frac{kM^2}{2} dx - \frac{kM^2}{2} \frac{d(V^2)}{V^2} \tag{10.12}$$

In order to obtain an equation relating M and x, we need to eliminate dp/p and $d(V^2)/V^2$ from Eq. 10.12.

From the ideal gas equation of state, $p = \rho RT$, we can write

$$\frac{dp}{p} = \frac{d\rho}{\rho} + \frac{dT}{T}$$

From the definition of Mach number, $M = V/c$ then $V^2 = M^2 c^2 = M^2 kRT$ and

$$\frac{d(V^2)}{V^2} = \frac{dT}{T} + \frac{d(M^2)}{M^2}$$

From the continuity equation $d\rho/\rho = -dV/V$

$$\frac{d\rho}{\rho} = -\frac{1}{2} \frac{d(V^2)}{V^2}$$

Combining these three equations we obtain

$$\frac{dp}{p} = \frac{1}{2} \frac{dT}{T} - \frac{1}{2} \frac{d(M^2)}{M^2} \tag{10.13a}$$

$$\frac{d(V^2)}{V^2} = \frac{dT}{T} + \frac{d(M^2)}{M^2} \tag{10.13b}$$

On substituting Eqs. 10.13 into Eq. 10.12, then

$$\frac{1}{2} \frac{dT}{T} - \frac{1}{2} \frac{d(M^2)}{M^2} = -f \frac{1}{D_h} \frac{kM^2}{2} dx - k \frac{M^2}{2} \frac{dT}{T} - \frac{kM^2}{2} \frac{d(M^2)}{M^2}$$

This equation can be simplified to

$$\left(\frac{1 + kM^2}{2} \right) \frac{dT}{T} = -f \frac{1}{D_h} \frac{kM^2}{2} dx + \left(\frac{1 - kM^2}{2} \right) \frac{d(M^2)}{M^2} \tag{10.14}$$

We have been successful in reducing the number of variables somewhat. However, in order to obtain an equation relating M and x, we must obtain an expression for dT/T in terms of M.

Such an expression can be most readily obtained from the stagnation temperature equation

$$\frac{T_0}{T} = 1 + \frac{k - 1}{2} M^2 \tag{9.16b}$$

Since the stagnation temperature is constant for Fanno line flow,

$$T\left(1 + \frac{k-1}{2}M^2\right) = \text{constant}$$

and

$$\frac{dT}{T} + \frac{M^2\dfrac{(k-1)}{2}}{\left(1 + \dfrac{k-1}{2}M^2\right)}\frac{d(M^2)}{M^2} = 0$$

Substituting for dT/T into Eq. 10.14

$$\frac{M^2\dfrac{(k-1)}{2}\left(\dfrac{1+kM^2}{2}\right)}{\left(1 + \dfrac{k-1}{2}M^2\right)}\frac{d(M^2)}{M^2} = +f\frac{1}{D_h}\frac{kM^2}{2}dx - \left(\frac{1-kM^2}{2}\right)\frac{d(M^2)}{M^2}$$

Combining terms,

$$\frac{(1-M^2)}{\left(1 + \dfrac{k-1}{2}M^2\right)}\frac{d(M^2)}{kM^4} = \frac{f}{D_h}dx \qquad (10.15)$$

We have succeeded in obtaining a differential equation that relates changes in M with x. All we have to do now is integrate the equation to find M as a function of x.

Integration of Eq. 10.15 between states ① and ② would produce a complicated function of both M_1 and M_2. The function would have to be evaluated numerically for each new combination of M_1 and M_2 encountered in a problem. There must be a simpler way. Calculations can be simplified considerably by use of the critical conditions (conditions where, by definition, $M = 1$). All Fanno line flows tend toward $M = 1$, so the integration is carried out between a station where the Mach number is M and the station where sonic conditions occur (the critical conditions). The Mach number will reach unity when the maximum possible length of duct is used as shown in Fig. 10.13.

The task then is to perform the integration

$$\int_M^1 \frac{(1-M^2)}{kM^4\left(1 + \dfrac{k-1}{2}M^2\right)}d(M^2) = \int_0^{L_{max}}\frac{f}{D_h}dx \qquad (10.16)$$

Integration of the left-hand side may be carried out using integration by parts. On the right-hand side the friction factor, f, may vary with x since the Reynolds

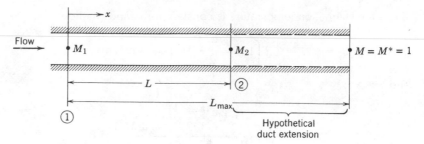

Fig. 10.13 *Coordinates and notation used for analysis of Fanno line flow.*

number will vary along the duct. Note, however, that since the ρV product is constant along the duct (from continuity) the variation in Reynolds number is brought about solely by variations in fluid viscosity.

If we define a mean friction factor, $\bar{f}$, over the duct length as

$$\bar{f} = \frac{1}{L_{max}} \int_0^{L_{max}} f \, dx$$

then integration of Eq. 10.16 leads to

$$\frac{1 - M^2}{kM^2} + \frac{k + 1}{2k} \ln \left[\frac{(k + 1)M^2}{2\left(1 + \frac{k - 1}{2}M^2\right)} \right] = \frac{\bar{f}L_{max}}{D_h} \qquad (10.17)$$

Equation 10.17 gives the maximum value of $\bar{f}L/D_h$ corresponding to any given initial Mach number. For any given Mach number we can compute the corresponding value of $\bar{f}L_{max}/D_h$. These values are tabulated in Table B.2 of Appendix B.

Since the parameter $\bar{f}L_{max}/D_h$ is a function only of M, the duct length, L, required for the flow Mach number to change from an initial value, M_1, to a final value, M_2 (as illustrated in Fig. 10.13), may be found from

$$\frac{\bar{f}L}{D_h} = \left(\frac{\bar{f}L_{max}}{D_h}\right)_{M_1} - \left(\frac{\bar{f}L_{max}}{D_h}\right)_{M_2}$$

The critical conditions are appropriate reference conditions to use in tabulating fluid properties as a function of local Mach number. Thus, for example, since T_0 is constant, we can write

$$\frac{T}{T^*} = \frac{T/T_0}{T^*/T_0} = \frac{1}{1 + \frac{k - 1}{2}M^2} \Bigg/ \frac{1}{1 + \frac{k - 1}{2}M^{*2}} = \frac{\left(\frac{k + 1}{2}\right)}{\left(1 + \frac{k - 1}{2}M^2\right)}$$

$$(10.18a)$$

Similarly

$$\frac{V}{V^*} = \frac{M\sqrt{kRT}}{M^*\sqrt{kRT^*}} = M\sqrt{\frac{T}{T^*}} = \left[\frac{\left(\dfrac{k+1}{2}\right)M^2}{1+\dfrac{k-1}{2}M^2}\right]^{1/2} \qquad (10.18b)$$

From continuity

$$\frac{\rho}{\rho^*} = \frac{V^*}{V} = \left[\frac{1+\dfrac{k-1}{2}M^2}{\left(\dfrac{k+1}{2}\right)M^2}\right]^{1/2} \qquad (10.18c)$$

From the ideal gas equation of state

$$\frac{p}{p^*} = \frac{\rho}{\rho^*}\frac{T}{T^*} = \frac{1}{M}\left[\frac{\left(\dfrac{k+1}{2}\right)}{1+\dfrac{k-1}{2}M^2}\right]^{1/2} \qquad (10.18d)$$

The ratio of local stagnation pressure to the reference stagnation pressure is given by

$$\frac{p_0}{p_0^*} = \frac{p_0}{p}\frac{p}{p^*}\frac{p^*}{p_0^*} = \frac{\left(1+\dfrac{k-1}{2}M^2\right)^{k/(k-1)}\dfrac{1}{M}\left[\left(\dfrac{k+1}{2}\right)\middle/\left(1+\dfrac{k-1}{2}M^2\right)\right]^{1/2}}{\left(1+\dfrac{k-1}{2}M^{*2}\right)^{k/(k-1)}}$$

$$\frac{p_0}{p_0^*} = \frac{1}{M}\left[\left(\frac{2}{k+1}\right)\left(1+\frac{k-1}{2}M^2\right)\right]^{(k+1)/2(k-1)} \qquad (10.18e)$$

The ratios in Eqs. 10.18 are tabulated as functions of Mach number in Table B.2 of Appendix B.

Example 10.8

Air flows in an insulated tube with an inside diameter of 0.25 in. At section ①, where the Mach number is 0.3, the stagnation pressure and temperature are 100 psia and 140 F. At section ②, downstream, the Mach number is 0.6. At section ③, farther downstream, the temperature is 40 F. The duct walls are smooth; the average friction factor, $\bar{f}$, may be taken as the value at section ①. Determine the length of duct between section ① and sections ② and ③, and the Mach number at section ③.

537   **adiabatic flow in a constant area duct with friction**

GIVEN:

Air flow (with friction) in an insulated tube.

$$M_1 = 0.3 \qquad M_2 = 0.6 \qquad T_3 = 40\,\text{F}$$

$$p_{0_1} = 100\,\text{psia}$$

$$T_{0_1} = 140\,\text{F} \qquad \bar{f} = f_1$$

FIND:

(a) L_{12}
(b) M_3
(c) L_{13}

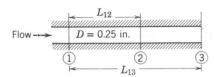

Flow →   D = 0.25 in.

SOLUTION:

Flow is frictional and adiabatic, that is, a Fanno line flow. To evaluate f_1, we need Re at station ①. Since stagnation properties are given, local properties may be found using isentropic relations. From Table B.1, Appendix B, at $M = 0.3$,

$$\frac{p}{p_0} = \frac{p_1}{p_{0_1}} = 0.940; \qquad p_1 = 94.0\,\text{psia}$$

$$\frac{T}{T_0} = \frac{T_1}{T_{0_1}} = 0.982; \qquad T_1 = (0.982)(600\,\text{R}) = 589\,\text{R}$$

For an ideal gas,

$$\rho_1 = \frac{p_1}{RT_1} = \frac{94\,\text{lbf}}{\text{in.}^2} \times \frac{144\,\text{in.}^2}{\text{ft}^2} \times \frac{1\text{bm-R}}{53.3\,\text{ft-lbf}} \times \frac{1}{589\,\text{R}} = 0.431\,\text{lbm/ft}^3$$

and

$$V_1 = M_1 c_1 = M_1 \sqrt{kRT_1} = (0.3)(1190\,\text{ft/sec}) = 357\,\text{ft/sec}$$

At $T_1 = 129\,\text{F}$, the viscosity is $\mu_1 = 4.8 \times 10^{-7}\,\text{lbf-sec/ft}^2$, from Fig. A.1, Appendix A. Thus

$$Re_1 = \frac{\rho_1 V_1 D}{\mu_1} = \frac{0.431\,\text{lbm}}{\text{ft}^3} \times \frac{357\,\text{ft}}{\text{sec}} \times 0.25\,\text{in.} \times \frac{\text{ft}}{12\,\text{in.}} \times \frac{\text{ft}^2}{4.8 \times 10^{-7}\,\text{lbf-sec}}$$

$$\times \frac{\text{slug}}{32.2\,\text{lbm}} \times \frac{\text{lbf-sec}^2}{\text{slug-ft}} = 2.07 \times 10^5$$

From Fig. 8.12, $f_1 = 0.0157$.

From Table B.2, at $M = 0.3, \bar{f}L_{max}/D_h]_1 = 1.32$. For $M = 0.6, \bar{f}L_{max}/D_h]_2 = 0.123$. For a circular pipe, $D_h = D$, so

$$\frac{\bar{f}L_{max}}{D_h}\bigg]_1 - \frac{\bar{f}L_{max}}{D_h}\bigg]_2 = 1.32 - 0.123 = 1.20$$

$$L_{12} = L_{max}]_1 - L_{max}]_2 = 1.20\frac{D}{\bar{f}} = 1.20 \times \frac{0.25 \text{ in.}}{0.0157} \times \frac{\text{ft}}{12 \text{ in.}}$$

$$L_{12} = 1.59 \text{ ft} \qquad\qquad\qquad\qquad\qquad\qquad\qquad\qquad L_{12}$$

Again from Table B.2, at $M = 0.3$,

$$\frac{T}{T^*} = \frac{T_1}{T^*} = 1.179; \qquad T^* = \frac{T_1}{1.179} = \frac{589 \text{ R}}{1.179} = 500 \text{ R}$$

Then

$$\frac{T_3}{T^*} = \frac{500}{500} = 1.0, \qquad \text{so } M_3 = 1 \qquad\qquad\qquad\qquad\qquad M_3$$

Finally, then

$$L_{13} = L_{max}]_1 = 1.32\frac{D}{\bar{f}} = 1.32 \times \frac{0.25 \text{ in.}}{0.0157} \times \frac{\text{ft}}{12 \text{ in.}}$$

$$L_{13} = 1.75 \text{ ft} \qquad\qquad\qquad\qquad\qquad\qquad\qquad\qquad L_{13}$$

10–5 Frictionless Flow in a Constant Area Duct with Heat Transfer

To explore the effects of heat transfer on a compressible flow, let us apply the basic equations to the steady, uniform, frictionless flow of an ideal gas with constant specific heats through the finite control volume shown in Fig. 10.14.

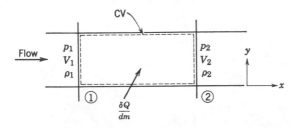

Fig. 10.14 *Control volume used for integral analysis of frictionless flow with heat transfer.*

10-5.1 BASIC EQUATIONS

a. Continuity Equation

Basic equation:

$$0 = \overset{= \ 0(1)}{\cancel{\frac{\partial}{\partial t} \iiint_{CV} \rho \, d\forall}} + \iint_{CS} \rho \vec{V} \cdot d\vec{A} \tag{4.14}$$

Assumptions: (1) Steady flow
(2) Uniform flow at each section

$$0 = \{-|\rho_1 V_1 A_1|\} + \{|\rho_2 V_2 A_2|\}$$

The area is constant, so

$$\rho_1 V_1 = \rho_2 V_2 = G \tag{10.19a}$$

b. Momentum Equation

Basic equation:

$$F_{S_x} + \overset{= \ 0(3)}{\cancel{F_{B_x}}} = \overset{= \ 0(1)}{\cancel{\frac{\partial}{\partial t} \iiint_{CV} V_x \rho \, d\forall}} + \iint_{CS} V_x \rho \vec{V} \cdot d\vec{A} \tag{4.19a}$$

Assumption: (3) $F_{B_x} = 0$

Since there is no friction between the duct walls and the flow, and $A_2 = A_1 = A$, then

$$p_1 A - p_2 A = V_1\{-|\rho_1 V_1 A|\} + V_2\{|\rho_2 V_2 A|\}$$

This can be written as

$$p_1 A - p_2 A = \dot{m} V_2 - \dot{m} V_1 \tag{10.19b}$$

or

$$p_1 + \rho_1 V_1^2 = p_2 + \rho_2 V_2^2$$

c. First Law of Thermodynamics

Basic equation:

$$\dot{Q} - \overset{= \ 0(4)}{\cancel{\dot{W}_s}} - \overset{= \ 0(5)}{\cancel{\dot{W}_{shear}}} = \overset{= \ 0(1)}{\cancel{\frac{\partial}{\partial t} \iiint_{CV} e\rho \, d\forall}} + \iint_{CS} (e + pv)\rho \vec{V} \cdot d\vec{A} \tag{4.45}$$

where

$$e = u + \frac{V^2}{2} + \underset{\nearrow}{g\cancel{z}} \overset{\simeq \; 0(6)}{}$$

Assumptions: (4) $\dot{W}_s = 0$
(5) $\dot{W}_{\text{shear}} = 0$
(6) Effects of gravity are negligible

With these restrictions

$$\dot{Q} = \left(u_1 + \frac{V_1^2}{2} + p_1 v_1 \right) \{-|\rho_1 V_1 A|\} + \left(u_2 + \frac{V_2^2}{2} + p_2 v_2 \right) \{|\rho_2 V_2 A|\}$$

or

$$\dot{Q} = \dot{m} \left(h_2 + \frac{V_2^2}{2} - h_1 - \frac{V_1^2}{2} \right)$$

But

$$\frac{\delta Q}{dm} = \frac{1}{\dot{m}} \dot{Q}$$

so

$$\frac{\delta Q}{dm} + h_1 + \frac{V_1^2}{2} = h_2 + \frac{V_2^2}{2} \qquad (10.19c)$$

or

$$\frac{\delta Q}{dm} + h_{0_1} = h_{0_2}$$

We see that heat transfer to the flow causes the stagnation enthalpy and, hence, the stagnation temperature to change.

d. Second Law of Thermodynamics

Basic equation:

$$\iint_{\text{CS}} \frac{1}{T} \frac{\dot{Q}}{A} dA \leq \underset{\nearrow}{\frac{\partial}{\partial t} \iiint_{\text{CV}}} sp \, d\mathrm{V} + \iint_{\text{CS}} sp\vec{V} \cdot d\vec{A} \overset{= \; 0(1)}{} \qquad (4.46)$$

$$\iint_{\text{CS}} \frac{1}{T} \frac{\dot{Q}}{A} dA \leq \dot{m}(s_2 - s_1)$$

541   **frictionless flow in a constant area duct with heat transfer**

The flow is frictionless, and the heat transfer may be carried out over an arbitrarily small difference in temperature. Since under these conditions the process could be considered reversible, the equality in Eq. 4.46 would hold. However, even with heat transfer uniform over the control surface, we would need to know the temperature distribution over the area to evaluate the integral on the left-hand side of Eq. 4.46. Consequently the second law does not enable us to calculate the actual entropy change between any two points in the flow. Note, however, that the rate of heat transfer, $\dot{Q}$, may be either positive (heat addition to the flow) or negative (heat rejection from the flow). Consequently the entropy change in a frictionless flow with heat transfer may be positive or negative.

To calculate the entropy change we rely on the Tds equations. Since

$$Tds = dh - v\,dp$$

for an ideal gas we can write

$$ds = c_p \frac{dT}{T} - R \frac{dp}{p}$$

For constant specific heats the equation can be integrated to give

$$s_2 - s_1 = c_p \ln \frac{T_2}{T_1} - R \ln \frac{p_2}{p_1} \qquad (10.19d)$$

e. *Equations of State*

For an ideal gas, the equation of state is given by

$$p = \rho RT \qquad (10.19e)$$

Equations 10.19a–e are the governing equations for the steady, one-dimensional, frictionless flow of an ideal gas in a constant area duct with heat transfer. If all properties at state ① are known, then we have 7 unknowns (ρ_2, V_2, p_2, h_2, s_2, T_2, and $\delta Q/dm$) in these 5 equations. However, since we have counted both h_2 and T_2 as unknowns, we can legitimately include the known relationship between h and T for an ideal gas with constant specific heats, namely

$$\Delta h = h_2 - h_1 = c_p \Delta T = c_p(T_2 - T_1) \qquad (10.19f)$$

We thus have the situation of 6 equations and 7 unknowns.

10–5.2 THE RAYLEIGH LINE

If all conditions at state ① are known, how many possible states ② are there? The mathematics of the situation (6 equations and 7 unknowns) indicates there are an infinite number of possible states at ②.

With an infinite number of possible states at ② for a given state ①, what is to be expected if all possible states at ② are plotted on a Ts diagram? It follows that the locus of all possible states ②, reachable from state ①, is a continuous curve passing through state ①.

How can we determine this curve? Perhaps the simplest way is to assume different values of T_2. For an assumed value of T_2 we could then calculate the corresponding values of all other properties at state ② and also $\delta Q/dm$.

The results of these calculations are shown qualitatively on the Ts plane of Fig. 10.15. The locus of all possible downstream states is referred to as the

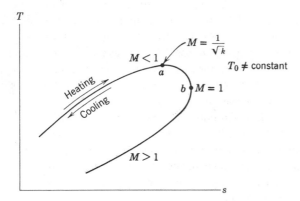

Fig. 10.15 *Schematic Ts diagram for frictionless flow in a constant area duct with heat transfer (Rayleigh line flow).*

Rayleigh line. In carrying out the calculations we would discover some interesting features of Rayleigh line flow. At the point of maximum temperature (point a of Fig. 10.15) the value of the Mach number is $1/\sqrt{k}$. At the point of maximum entropy, the Mach number is unity. On the upper branch of the curve the Mach number is always less than unity, and it increases monotonically as we proceed to the right along the curve. At every point on the lower portion of the curve the Mach number is greater than unity, and it decreases monotonically as we move to the right along the curve. Note also that regardless of the initial Mach number, heat addition causes the flow to proceed to the right and heat rejection causes the flow to proceed to the left along the Rayleigh line.

For any initial state in a Rayleigh line flow any point on the Rayleigh line is a mathematically possible downstream state. Indeed we determined the locus of all possible downstream states by assuming values of T_2 and proceeding to calculate the corresponding properties. While the Rayleigh line represents all mathematically possible states, are they all physically attainable downstream states? A moment's reflection will indicate they are. Since we are considering a flow with heat transfer, the second law does not impose any restrictions on the sign of the entropy change.

The effects of heat transfer on properties in compressible fluid flow may be determined by studying the basic equations, Eqs. 10.19a–f, and by using the Rayleigh line, Fig. 10.15. These effects are summarized in Table 10.2, and the development of each effect is discussed in the next few paragraphs.

Since the heating along the Rayleigh line is reversible, $Tds = \delta Q/dm$. Thus the direction of entropy change is always determined by the heat transfer; entropy increases with heating and decreases with cooling. Similarly, the first law, Eq. 10.19c, shows that heating increases the stagnation enthalpy and cooling decreases it; since $\Delta h_0 = c_p \Delta T_0$, the same effect is found on the stagnation temperature.

The effect on temperature of heating and cooling may be deduced from the shape of the Rayleigh line in Fig. 10.15. We see that for $M < 1/\sqrt{k}$ (for air, $1/\sqrt{k} = 0.85$) or for $M > 1$, heating causes T to increase, and in the same regions cooling causes T to decrease. However, we also see the most unexpected result that for $1/\sqrt{k} < M < 1$, *heat addition* acts to *reduce* the stream temperature, and *heat rejection* causes the stream temperature to *increase* !

For subsonic flow, the Mach number increases monotonically with heating, until the value $M = 1$ is reached. For a given set of inlet conditions, all possible downstream states lie on a single Rayleigh line. Therefore the point $M = 1$ corresponds to the maximum possible heat addition. If the flow is initially supersonic, heating will reduce the Mach number. Again the maximum possible heat addition is that which reduces the Mach number to $M = 1.0$.

The effect of heat transfer on static pressure is obtained from the shape of the Rayleigh line and of constant pressure lines on the Ts plane (see Fig. 10.16). As

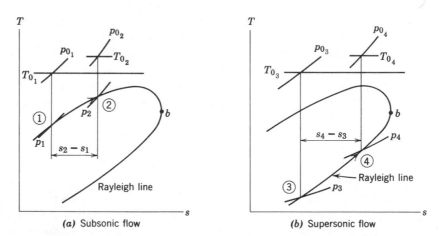

(a) Subsonic flow   (b) Supersonic flow

Fig. 10.16 *Reduction in stagnation pressure due to heating for two flow cases.*

table 10.2 *Summary of Heat Transfer Effects on Fluid Properties*

PROPERTY	HEATING		COOLING		OBTAINED FROM:
	$M < 1$	$M > 1$	$M < 1$	$M > 1$	
Entropy, s	Increase	Increase	Decrease	Decrease	Second law
Stagnation Temperature, T_0	Increase	Increase	Decrease	Decrease	First law, and $\Delta h_0 = c_p\,\Delta T_0$
Temperature, T	$\left(M < \frac{1}{\sqrt{k}}\right)$ Increase $\left(\frac{1}{\sqrt{k}} < M < 1\right)$ Decrease	Increase	$\left(M < \frac{1}{\sqrt{k}}\right)$ Decrease $\left(\frac{1}{\sqrt{k}} < M < 1\right)$ Increase	Decrease	Shape of Rayleigh line
Mach Number, M	Increase	Decrease	Decrease	Increase	Trend on Rayleigh line
Pressure, p	Decrease	Increase	Increase	Decrease	Trend on Rayleigh line
Velocity, V	Increase	Decrease	Decrease	Increase	Momentum equation, and effect on p
Density, ρ	Decrease	Increase	Increase	Decrease	Continuity, and effect on V
Stagnation Pressure, p_0	Decrease	Decrease	Increase	Increase	Fig. 10.16

heating occurs for $M < 1$, the pressure falls. For $M > 1$, pressure increases with heating, due to the shapes of the lines p = constant. Once the pressure variation has been found, the effect on velocity may be found from the momentum equation

$$p_1 A - p_2 A = \dot{m} V_2 - \dot{m} V_1 \qquad (10.19b)$$

or

$$p + \left(\frac{\dot{m}}{A}\right) V = \text{constant}$$

Thus, since $\dot{m}/A$ = constant, trends in p and V must be opposite. From the continuity equation the trend in ρ is opposite that in V.

The local isentropic stagnation pressure always decreases with heating. This is illustrated schematically in Fig. 10.16. A reduction in stagnation pressure has obvious practical implications for heating processes, such as combustion chambers. Note also that the same amount of heating (same change in T_0) causes a larger change in p_0 for supersonic flow; because the heating occurs at a lower temperature in supersonic flow, the entropy increase is larger.

Example 10.9

Air flows with negligible friction through a duct with cross section area 0.25 ft². At section ①, the flow properties are p_1 = 20 psia, T_1 = 600 R, and V_1 = 360 ft/sec. At section ②, the pressure is 10 psia. Heat is added to the flow between sections ① and ②. Determine the properties at section ②, the heat added in Btu/lbm, and the entropy change. Finally, plot the process on a Ts diagram.

EXAMPLE PROBLEM 10.9

GIVEN:

Frictionless flow of air in duct shown:

$p_1 = 20 \, \text{psia} \qquad p_2 = 10 \, \text{psia}$

$T_1 = 600 \, \text{R}$

$V_1 = 360 \, \text{ft/sec}$

$A_1 = A_2 = A = 0.25 \, \text{ft}^2$

FIND:

(a) Properties at ②
(b) $\delta Q/dm$ in Btu/lbm
(c) $s_2 - s_1$
(d) Sketch on Ts diagram

SOLUTION:

Apply the x component of the momentum equation, using the coordinates and control volume shown.
Basic equation:

$$F_{S_x} + \overset{= 0(1)}{\cancel{F_{B_x}}} = \overset{= 0(2)}{\cancel{\frac{\partial}{\partial t} \int\int\int_{CV} V_x \rho \, d\forall}} + \int\int_{CS} V_x \rho \vec{V} \cdot d\vec{A} \qquad (4.19a)$$

Assumptions: (1) $F_{B_x} = 0$
(2) Steady flow
(3) Uniform flow at each section

Then

$$p_1 A - p_2 A = V_1\{-|\rho_1 V_1 A|\} + V_2\{|\rho_2 V_2 A|\} = \dot{m}(V_2 - V_1)$$

or

$$p_1 - p_2 = \frac{\dot{m}}{A}(V_2 - V_1) = \rho_1 V_1(V_2 - V_1)$$

Solving for V_2,

$$V_2 = \frac{p_1 - p_2}{\rho_1 V_1} + V_1$$

For an ideal gas,

$$\rho_1 = \frac{p_1}{RT_1} = \frac{20 \text{ lbf}}{\text{in.}^2} \times \frac{144 \text{ in.}^2}{\text{ft}^2} \times \frac{\text{lbm-R}}{53.3 \text{ ft-lbf}} \times \frac{1}{600 \text{ R}} = 0.090 \text{ lbm/ft}^3$$

$$V_2 = \frac{(20 - 10) \text{ lbf}}{\text{in.}^2} \times \frac{144 \text{ in.}^2}{\text{ft}^2} \times \frac{\text{ft}^3}{0.090 \text{ lbm}} \times \frac{\text{sec}}{360 \text{ ft}} \times \frac{32.2 \text{ lbm}}{\text{slug}} \times \frac{\text{slug-ft}}{\text{lbf-sec}^2} + 360 \text{ ft/sec}$$

$$V_2 = 1790 \text{ ft/sec} \qquad\qquad\qquad\qquad\qquad\qquad\qquad\qquad \underline{V_2}$$

From continuity, $G = \rho_1 V_1 = \rho_2 V_2$, so

$$\rho_2 = \rho_1 \frac{V_1}{V_2} = 0.090 \frac{\text{lbm}}{\text{ft}^3}\left(\frac{360}{1790}\right) = 0.0181 \text{ lbm/ft}^3 \qquad\qquad \underline{\rho_2}$$

Solving for T_2,

$$T_2 = \frac{p_2}{\rho_2 R} = \frac{10 \text{ lbf}}{\text{in.}^2} \times \frac{144 \text{ in.}^2}{\text{ft}^2} \times \frac{\text{ft}^3}{0.0181 \text{ lbm}} \times \frac{\text{lbm-R}}{53.3 \text{ ft-lbf}} = 1490 \text{ R} \qquad \underset{\longleftarrow}{T_2}$$

$$T_{0_2} = T_2\left(1 + \frac{k-1}{2}M_2^2\right)$$

$$c_2 = \sqrt{kRT_2} = 1890 \frac{\text{ft}}{\text{sec}}; \qquad M_2 = \frac{V_2}{c_2} = \frac{1790}{1890} = 0.947$$

$$T_{0_2} = 1490 \text{ R}[1 + 0.2(0.947)^2] = (1490 \text{ R})(1.179) = 1760 \text{ R} \qquad \underset{\longleftarrow}{T_{0_2}}$$

and

$$p_{0_2} = p_2\left(\frac{T_{0_2}}{T_2}\right)^{k/(k-1)} = (10 \text{ psia})(1.179)^{3.5} = 17.8 \text{ psia} \qquad \underset{\longleftarrow}{p_{0_2}}$$

The heat transfer is obtained from the energy equation.
Basic equation:

$$\dot{Q} - \overset{= \, 0(4)}{\cancel{\dot{W}_s}} - \overset{= \, 0(5)}{\cancel{\dot{W}_{shear}}} = \overset{= \, 0(2)}{\cancel{\frac{\partial}{\partial t}\iiint_{CV} e\rho \, d\mathcal{V}}} + \iint_{CS} (e + pv)\rho \vec{V} \cdot d\vec{A} \qquad (4.45)$$

where

$$e = u + \frac{V^2}{2} + \overset{\approx \, 0(6)}{\cancel{gz}}$$

Assumptions: (4) $\dot{W}_s = 0$
(5) $\dot{W}_{shear} = 0$
(6) Neglect changes in z

Then

$$\dot{Q} = \left(u_1 + p_1 v_1 + \frac{V_1^2}{2}\right)\{-|\rho_1 V_1 A|\} + \left(u_2 + p_2 v_2 + \frac{V_2^2}{2}\right)\{|\rho_2 V_2 A|\}$$

$$\dot{Q} = \dot{m}\left(h_2 + \frac{V_2^2}{2} - h_1 - \frac{V_1^2}{2}\right) = \dot{m}(h_{0_2} - h_{0_1}) = \dot{m}c_p(T_{0_2} - T_{0_1})$$

and

$$\frac{\delta Q}{dm} = \frac{1}{\dot{m}}\dot{Q} = c_p(T_{0_2} - T_{0_1})$$

$$T_{0_1} = T_1\left(1 + \frac{k-1}{2}M_1^2\right)$$

$$c_1 = \sqrt{kRT_1} = 1200 \frac{ft}{sec}; \qquad M_1 = \frac{V_1}{c_1} = \frac{360}{1200} = 0.3$$

$$T_{0_1} = (600 \text{ R})[1 + 0.2(0.3)^2] = (600 \text{ R})(1.018) = 611 \text{ R}$$

so

$$\frac{\delta Q}{dm} = \frac{0.240 \text{ Btu}}{\text{lbm-R}}(1760 - 611) \text{ R} = 276 \text{ Btu/lbm} \qquad\qquad\qquad \overleftarrow{\qquad} \quad \delta Q/dm$$

Using the $T ds$ equation,

$$T ds = dh - v dp,$$

we obtain for an ideal gas

$$ds = c_p \frac{dT}{T} - R\frac{dp}{p}$$

Integrating

$$s_2 - s_1 = c_p \ln \frac{T_2}{T_1} - R \ln \frac{p_2}{p_1}$$

But $R = c_p - c_v$, so we obtain

$$s_2 - s_1 = 0.240 \frac{\text{Btu}}{\text{lbm-R}} \times \ln \left(\frac{1490}{600}\right) - (0.240 - 0.171)\frac{\text{Btu}}{\text{lbm-R}} \times \ln \left(\frac{10}{20}\right)$$

$$= (0.240)(0.910) - (0.069)(-0.693)$$

or

$$s_2 - s_1 = 0.218 + 0.048 = 0.266 \text{ Btu/lbm-R} \qquad\qquad \overleftarrow{\qquad} \quad s_2 - s_1$$

The process follows a Rayleigh line:

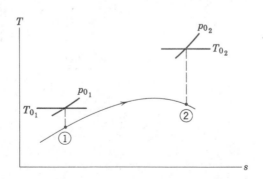

frictionless flow in a constant area duct with heat transfer

EXAMPLE PROBLEM 10.9 (continued)

To complete our analysis, let us examine the change in p_0 by comparing p_{0_2} with p_{0_1}.

$$\frac{p_{0_1}}{p_1} = \left(\frac{T_{0_1}}{T_1}\right)^{k/(k-1)} = (1.018)^{3.5} = 1.064$$

$$p_{0_1} = 1.064 p_1 = (1.064)(20 \text{ psia}) = 21.3 \text{ psia} \qquad\qquad\qquad p_{0_1}$$

Comparing the two values, we see that p_{0_2} is less than p_{0_1}.

$\left\{\begin{array}{l}\text{In general, the stagnation pressure is decreased by heating and increased by cooling in}\\ \text{Rayleigh line flow.}\end{array}\right\}$

**10–5.3 TABLES FOR COMPUTATION OF RAYLEIGH LINE FLOW

In Section 10–5.1 we wrote the basic equations for Rayleigh line flow between two arbitrary states ① and ② in the flow. To facilitate the solution of problems it is convenient to tabulate dimensionless properties in terms of local Mach number as we did for Fanno line flow. The reference state is again taken as the critical condition, that is, the state at which the Mach number is unity; the properties at the critical condition are denoted by (*).

The dimensionless properties (such as T/T^* and p/p^*) may be obtained by writing the basic equations between a point in the flow where the properties are M, T, p, and so on (unsubscripted) and the critical state ($M = 1$ and properties denoted, for example, as T^* and p^*).

The pressure ratio p/p^* may be obtained from the momentum equation

$$pA - p^*A = \dot{m}V^* - \dot{m}V \tag{10.19b}$$

or

$$p + \rho V^2 = p^* + \rho^* V^{*2}$$

Substituting $\rho = p/RT$ and factoring out pressures,

$$p\left[1 + \frac{V^2}{RT}\right] = p^*\left[1 + \frac{V^{*2}}{RT^*}\right]$$

Noting that $V^2/RT = k(V^2/kRT) = kM^2$, then

$$p[1 + kM^2] = p^*[1 + kM^{*2}]$$

and finally

$$\frac{p}{p^*} = \frac{1+k}{1+kM^2} \tag{10.20a}$$

** This section may be omitted without loss of continuity in the text material.

From the ideal gas equation of state

$$\frac{T}{T^*} = \frac{p}{p^*}\frac{\rho^*}{\rho}$$

From the continuity equation

$$\frac{\rho^*}{\rho} = \frac{V}{V^*} = \frac{M}{M^*}\frac{c}{c^*} = \frac{M}{M^*}\sqrt{\frac{T}{T^*}}$$

Then on substituting for ρ^*/ρ we obtain

$$\frac{T}{T^*} = \frac{p}{p^*}\frac{M}{M^*}\sqrt{\frac{T}{T^*}}$$

and

$$\frac{T}{T^*} = \left[\frac{p}{p^*}\frac{M}{M^*}\right]^2 = \left[M\left(\frac{1+k}{1+kM^2}\right)\right]^2 \tag{10.20b}$$

Also from continuity, using Eq. 10.20b

$$\frac{\rho^*}{\rho} = \frac{V}{V^*} = \frac{M^2(1+k)}{1+kM^2} \tag{10.20c}$$

The dimensionless stagnation temperature, T_0/T_0^*, can be determined from

$$\frac{T_0}{T_0^*} = \frac{T_0}{T}\frac{T}{T^*}\frac{T^*}{T_0^*} = \left(1 + \frac{k-1}{2}M^2\right)\left[M\left(\frac{1+k}{1+kM^2}\right)\right]^2\left[\frac{1}{\left(1 + \dfrac{k-1}{2}\right)}\right]$$

$$\frac{T_0}{T_0^*} = \frac{2(k+1)M^2\left(1 + \dfrac{k-1}{2}M^2\right)}{(1+kM^2)^2} \tag{10.20d}$$

Similarily

$$\frac{p_0}{p_0^*} = \frac{p_0}{p}\frac{p}{p^*}\frac{p^*}{p_0} = \left(1 + \frac{k-1}{2}M^2\right)^{k/(k-1)}\left(\frac{1+k}{1+kM^2}\right)\frac{1}{\left(1 + \dfrac{k-1}{2}\right)^{k/(k-1)}}$$

$$\frac{p_0}{p_0^*} = \frac{1+k}{1+kM^2}\left[\left(\frac{2}{k+1}\right)\left(1 + \frac{k-1}{2}M^2\right)\right]^{k/(k-1)} \tag{10.20e}$$

Since the ratios in Eqs. 10.20a–e are functions of Mach number only, they may be computed once and presented in tabular form. These ratios are tabulated as functions of Mach number in Table B.3 of Appendix B.

frictionless flow in a constant area duct with heat transfer

Example 10.10

Air flows with negligible friction in a constant area duct. At section ①, the flow properties are $V_1 = 2400$ ft/sec, $p_1 = 20$ psia, and $T_1 = 600$ R. Heat is added to the flow between section ① and section ②, where the Mach number is 1.2. Determine the flow properties at section ②, the heat added in Btu/lbm, the entropy change, and sketch the process on a Ts diagram. Use tables.

EXAMPLE PROBLEM 10.10

GIVEN:

Frictionless flow of air as shown:

$V_1 = 2400$ ft/sec $\quad M_2 = 1.2$

$p_1 = 20$ psia

$T_1 = 600$ R

FIND:

(a) Properties at station ②
(b) $\delta Q/dm$, in Btu/lbm
(c) $s_2 - s_1$
(d) Sketch on a Ts diagram

SOLUTION:

For an ideal gas,

$$\rho_1 = \frac{p_1}{RT_1} = \frac{20 \text{ lbf}}{\text{in.}^2} \times \frac{144 \text{ in.}^2}{\text{ft}^2} \times \frac{\text{lbm-R}}{53.3 \text{ ft-lbf}} \times \frac{1}{600 \text{ R}} = 0.09 \text{ lbm/ft}^3$$

To obtain property ratios from tables, we need both Mach numbers.

$$c_1 = \sqrt{kRT_1} = 1200 \frac{\text{ft}}{\text{sec}}; \qquad M_1 = \frac{V_1}{c_1} = \frac{2400}{1200} = 2.0$$

From Table B.3, Appendix B,

M	T/T^*	p/p^*	T_0/T_0^*	p_0/p_0^*
2.0	0.529	0.364	0.793	1.503
1.2	0.912	0.796	0.979	1.019

EXAMPLE PROBLEM 10.10 (continued)

Using these data, since critical properties are constant,

$$\frac{T_2}{T_1} = \frac{T_2/T^*}{T_1/T^*} = \frac{0.912}{0.529} = 1.72; \qquad T_2 = 1.72 T_1 = 1030 \text{ R} \qquad\qquad \underleftarrow{\quad T_2 \quad}$$

$$\frac{p_2}{p_1} = \frac{p_2/p^*}{p_1/p^*} = \frac{0.796}{0.364} = 2.19; \qquad p_2 = 2.19 p_1 = 43.8 \text{ psia} \qquad\qquad \underleftarrow{\quad p_2 \quad}$$

$$\rho_2 = \frac{p_2}{RT_2} = \frac{43.8 \text{ lbf}}{\text{in.}^2} \times \frac{144 \text{ in.}^2}{\text{ft}^2} \times \frac{\text{lbm-R}}{53.3 \text{ ft-lbf}} \times \frac{1}{1030 \text{ R}} = 0.115 \text{ lbm/ft}^3 \qquad \underleftarrow{\quad \rho_2 \quad}$$

From continuity, $G = \rho_1 V_1 = \rho_2 V_2$, so

$$V_2 = V_1 \frac{\rho_1}{\rho_2} = (2400 \text{ ft/sec}) \left(\frac{0.090}{0.115} \right) = 1880 \text{ ft/sec} \qquad\qquad \underleftarrow{\quad V_2 \quad}$$

The heat added may be obtained from the energy equation, which reduces to (see Example Problem 10.9)

$$\frac{\delta Q}{dm} = h_{0_2} - h_{0_1} = c_p (T_{0_2} - T_{0_1})$$

From the isentropic flow tables (Table B.1), at $M = 2.0$

$$\frac{T}{T_0} = \frac{T_1}{T_{0_1}} = 0.556; \qquad T_{0_1} = \frac{T_1}{0.556} = \frac{600 \text{ R}}{0.556} = 1080 \text{ R} \qquad\qquad \underleftarrow{\quad T_{0_1} \quad}$$

and at $M = 1.2$

$$\frac{T}{T_0} = \frac{T_2}{T_{0_2}} = 0.776; \qquad T_{0_2} = \frac{T_2}{0.776} = \frac{1030 \text{ R}}{0.776} = 1330 \text{ R} \qquad\qquad \underleftarrow{\quad T_{0_2} \quad}$$

Substituting,

$$\frac{\delta Q}{dm} = \frac{0.240 \text{ Btu}}{\text{lbm-R}} \times (1330 - 1080) \text{ R} = 60.0 \text{ Btu/lbm} \qquad\qquad \underleftarrow{\quad \delta Q/dm \quad}$$

The entropy change may be found from the $T ds$ equation,

$$T ds = dh - v dp$$

For an ideal gas,

$$ds = c_p \frac{dT}{T} - R \frac{dp}{p}$$

Integrating,

$$s_2 - s_1 = c_p \ln \frac{T_2}{T_1} - R \ln \frac{p_2}{p_1}$$

But $R = c_p - c_v$, so

$$s_2 - s_1 = \frac{0.240 \text{ Btu}}{\text{lbm-R}} \ln\left(\frac{1030}{600}\right) - (0.240 - 0.171)\frac{\text{Btu}}{\text{lbm-R}} \ln\left(\frac{43.8}{20}\right)$$

$$s_2 - s_1 = 0.240(0.540) - (0.069)(0.784) = 0.075 \text{ Btu/lbm-R} \qquad \longleftarrow \quad s_2 - s_1$$

Finally, let us check the effect on p_0. From Table B.1, at $M = 2.0$

$$\frac{p}{p_0} = \frac{p_1}{p_{0_1}} = 0.128; \qquad p_{0_1} = \frac{p_1}{0.128} = \frac{20 \text{ psia}}{0.128} = 156 \text{ psia}$$

and at $M = 1.2$

$$\frac{p}{p_0} = \frac{p_2}{p_{0_2}} = 0.412; \qquad p_{0_2} = \frac{p_2}{0.412} = \frac{43.8 \text{ psia}}{0.412} = 106 \text{ psia}$$

Thus, $p_{0_2} < p_{0_1}$, as expected for a heating process.
  The process follows the supersonic branch of a Rayleigh line:

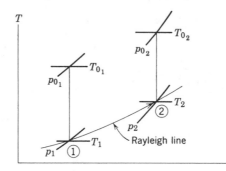

10–6 Normal Shocks

We have previously mentioned normal shocks in the section on nozzle flow. In practice, these irreversible discontinuities can occur in any supersonic flow field, in either internal flow or external flow[2] Knowledge of property changes across shocks and of shock behavior is also important in understanding the design of supersonic diffusers, for example, for inlets on high performance aircraft, and supersonic wind tunnels. Accordingly, the purpose of the present section is to analyze the normal shock process and to develop equations to predict shock behavior.

[2] The Shell film *Schlieren* shows several examples of shock formation in external flow.

Before applying the basic equations to normal shocks, it is important that we have in mind a clear physical picture of the shock itself. Although it is physically impossible to have jump discontinuities in fluid properties, the normal shock is nearly discontinuous. The thickness of a shock is in the order of 10^{-5} in., or approximately twice the mean free path of the gas molecules. Across this small distance, large changes in pressure, temperature, and other properties occur. Local fluid accelerations can reach tens of millions of "G's"! These considerations justify treating the normal shock as an abrupt discontinuity, since we are interested in changes occurring across the shock rather than in the details of its structure.

10–6.1 BASIC EQUATIONS

To begin our analysis, let us apply the basic equations to the thin control volume shown in Fig. 10.17, where for generality we have depicted a shock standing in a passage of arbitrary shape.

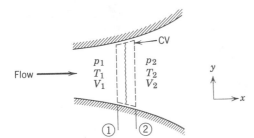

Fig. 10.17 *Control volume used for analysis of normal shock.*

a. Continuity Equation

Basic equation:

$$0 = \overset{= \; 0(1)}{\cancel{\frac{\partial}{\partial t} \iiint_{CV} \rho \, d\mathbb{V}}} + \iint_{CS} \rho \vec{V} \cdot d\vec{A} \qquad (4.14)$$

Assumptions: (1) Steady flow
                 (2) Uniform flow at each section
                 (3) $A_1 = A_2 = A$, because the shock is so extraordinarily thin

$$0 = \{-|\rho_1 V_1 A|\} + \{|\rho_2 V_2 A|\}$$

Writing the result in terms of scalar magnitudes, we obtain

$$\rho_1 V_1 = \rho_2 V_2 = G \qquad (10.21a)$$

b. Momentum Equation

Basic equation:

$$= 0(5) \quad = 0(1)$$

$$F_{S_x} + F_{B_x} = \frac{\partial}{\partial t} \iiint_{CV} V_x \rho \, d\forall + \iint_{CS} V_x \rho \vec{V} \cdot d\vec{A} \qquad (4.19a)$$

Assumptions: (4) Negligible friction at duct walls because the shock is so thin
(5) $F_{B_x} = 0$

Under these conditions,

$$F_{S_x} = p_1 A - p_2 A = V_1\{-|\rho_1 V_1 A|\} + V_2\{|\rho_2 V_2 A|\}$$

Using scalar magnitudes and dropping absolute value signs, we obtain

$$p_1 A - p_2 A = \dot{m} V_2 - \dot{m} V_1 \qquad (10.21b)$$

or

$$p_1 + \rho_1 V_1^2 = p_2 + \rho_2 V_2^2$$

c. First Law of Thermodynamics

Basic equation:

$$= 0(6) \quad = 0(7) \quad = 0(8) \qquad = 0(1)$$

$$\dot{Q} - \dot{W}_s - \dot{W}_{shear} = \frac{\partial}{\partial t} \iiint_{CV} e \rho \, d\forall + \iint_{CS} (e + pv) \rho \vec{V} \cdot d\vec{A} \qquad (4.45)$$

where

$$\simeq 0(9)$$

$$e = u + \frac{V^2}{2} + gz$$

Assumptions: (6) $\dot{Q} = 0$ (adiabatic flow)
(7) $\dot{W}_s = 0$
(8) $\dot{W}_{shear} = 0$
(9) Effects of gravity are negligible

Then

$$0 = \left(u_1 + p_1 v_1 + \frac{V_1^2}{2}\right)\{-|\rho_1 V_1 A|\} + \left(u_2 + p_2 v_2 + \frac{V_2^2}{2}\right)\{|\rho_2 V_2 A|\}$$

However, from continuity, Eq. 10.21a, the mass flowrate terms in brackets are equal. We may also substitute $h \equiv u + pv$, to obtain

$$h_1 + \frac{V_1^2}{2} = h_2 + \frac{V_2^2}{2} \tag{10.21c}$$

or, in terms of stagnation enthalpy,

$$h_{0_1} = h_{0_2}$$

Physically, we should expect the total energy of the flow to remain constant, since there is no energy addition.

d. Second Law of Thermodynamics

Basic equation:

$$\overset{=\,0(6)}{} \quad \overset{=\,0(1)}{}$$

$$\iint\limits_{CS} \frac{1}{T} \frac{\dot{Q}}{A} dA \leq \frac{\partial}{\partial t} \iiint\limits_{CV} s\rho \, d\forall + \iint\limits_{CS} s\rho \vec{V} \cdot d\vec{A} \tag{4.46}$$

Then

$$0 \leq s_1\{-|\rho_1 V_1 A|\} + s_2\{|\rho_2 V_2 A|\}$$

Flow through the normal shock is irreversible because of the almost discontinuous property changes across the shock. Consequently, the inequality in the above equation holds. The control volume form of the second law then tells us that $s_2 - s_1 > 0$.

This fact is of little help in calculating the actual entropy change across the shock. To calculate the entropy change we rely on the $T\,ds$ equations. Since

$$T\,ds = dh - v\,dp$$

then for an ideal gas we can write

$$ds = c_p \frac{dT}{T} - R \frac{dp}{p}$$

For constant specific heats this equation can be integrated to give

$$s_2 - s_1 = c_p \ln \frac{T_2}{T_1} - R \ln \frac{p_2}{p_1} \tag{10.21d}$$

d. Equation of State

For an ideal gas, the equation of state is given by

$$p = \rho RT \tag{10.21e}$$

Equations 10.21a–e are the governing equations for the flow of an ideal gas through a normal shock. If all the properties at state ① (immediately upstream of the shock) are known, then we have 6 unknowns (ρ_2, V_2, p_2, h_2, s_2, and T_2) in these 5 equations. However, since we have counted both h_2 and T_2 as unknowns, it is legitimate to include the known relationship between h and T for an ideal gas with constant specific heats, namely

$$\Delta h = h_2 - h_1 = c_p \Delta T = c_p(T_2 - T_1) \qquad (10.21f)$$

We thus have the situation of 6 equations and 6 unknowns.

Then if all conditions at state ① (immediately ahead of the shock) are known, how many possible states ② (immediately behind the shock) are there? The mathematics of the situation (6 equations and 6 unknowns) indicates that there is a unique state ② for a given state ①.

We can obtain a physical picture of the flow through a normal shock by employing some of the notions developed in the consideration of Fanno line and Rayleigh line flows. For convenience let us first rewrite the governing equations for a normal shock.

$$\rho_1 V_1 = \rho_2 V_2 = G \qquad (10.21a)$$

$$p_1 A - p_2 A = \dot{m} V_2 - \dot{m} V_1 \qquad (10.21b)$$

$$h_1 + \frac{V_1^2}{2} = h_2 + \frac{V_2^2}{2} \qquad (10.21c)$$

$$s_2 - s_1 = c_p \ln \frac{T_2}{T_1} - R \ln \frac{p_2}{p_1} \qquad (10.21d)$$

$$p = \rho RT \qquad (10.21e)$$

$$h_2 - h_1 = c_p(T_2 - T_1) \qquad (10.21f)$$

Flow through a normal shock must satisfy Eqs. 10.21a–f. Since all conditions at state ① are known, we can locate state ① on a Ts diagram. If we were to draw a Fanno line curve through state ①, we would have a locus of mathematical states that satisfy Eqs. 10.21a, c, d, e, and f. (The Fanno line curve does not satisfy Eq. 10.21b.) Drawing a Rayleigh line curve through state ① gives a locus of mathematical states that satisfy Eqs. 10.21a, b, d, e, and f. (The Rayleigh line curve does not satisfy Eq. 10.21c.) These curves are shown in Fig. 10.18.

The normal shock must satisfy all 6 of Eqs. 10.21a–f. Consequently, for a given state ①, the end state (state ②) of the normal shock must lie on both the Fanno line and the Rayleigh line passing through state ①. Hence the intersection of the two lines at state ② represents the conditions downstream from the shock, corresponding to the given upstream conditions at state ①. In Fig. 10.18 the flow

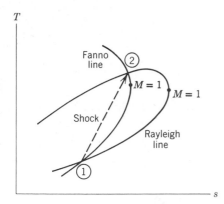

T

Fanno
line

②

$M = 1$

$M = 1$

Shock

①

Rayleigh
line

s

Fig. 10.18 *Intersection of Fanno line and*
Rayleigh line as a solution of the normal shock
equations.

through the shock has been indicated as occurring from state ① to state ②. This is the only possible direction of the shock process as dictated by the second law ($s_2 > s_1$).

From Fig. 10.18 we note also that the flow through a normal shock involves a change from supersonic to subsonic speeds. Normal shocks can occur only in a flow that is initially supersonic.

As an aid in summarizing the effects of a normal shock on the flow properties, a schematic of the normal shock process is illustrated on the Ts plane of Fig. 10.19.

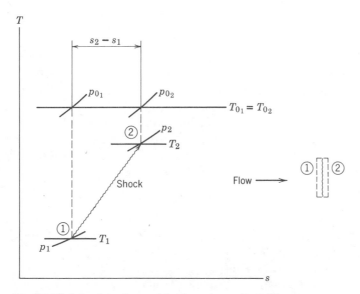

T

$s_2 - s_1$

p_{0_1}

p_{0_2}

$T_{0_1} = T_{0_2}$

②

p_2

T_2

Shock

Flow

① ②

①

T_1

p_1

s

Fig. 10.19 *Schematic of normal shock process on the Ts plane.*

559   **normal shocks**

This figure, together with the governing basic equations, is the basis for Table 10.3. The reader is encouraged to follow through the logic indicated in the right-hand column of the table.

table 10.3 *Summary of Property Changes Across a Normal Shock*

PROPERTY	EFFECT	OBTAINED FROM:
Stagnation Temperature, T_0	Constant	Energy equation
Entropy, s	Increase	Second law
Stagnation Pressure, p_0	Decrease	Ts diagram
Temperature, T	Increase	Ts diagram
Velocity, V	Decrease	Energy equation, and effect on T
Density, ρ	Increase	Continuity equation, and effect on V
Pressure, p	Increase	Momentum equation, and effect on V
Mach Number, M	Decrease	$M = V/c$, and effect on V and T

In theory the solution of 6 equations in 6 unknowns poses no difficulty, but in practice the algebra becomes quite involved. It makes sense to do the algebra once and recast the equations in a form suitable for tabulating property ratios across the shock. This we shall do in the next Section. To illustrate the direct application of the control volume equations to the solution of a normal shock, consider Example Problem 10.11, in which the problem is overspecified (one property downstream of the shock is given).

Example 10.11

A normal shock stands in a duct. The fluid is air, which may be considered an ideal gas. Properties upstream from the shock are $p_1 = 10$ psia, $T_1 = 40$ F, and $V_1 = 2190$ ft/sec. The temperature at station ②, downstream from the shock, is $T_2 = 384$ F. Determine the property values at station ②, and compare them with the upstream values. Sketch the process on a Ts diagram.

EXAMPLE PROBLEM 10.11

GIVEN:

Normal shock in a duct as shown:

$$T_1 = 40\ F \qquad T_2 = 384\ F$$

$$p_1 = 10\ \text{psia}$$

$$V_1 = 2190\ \text{ft/sec}$$

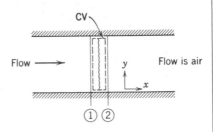

FIND:

(a) Properties at ②
(b) Sketch on a Ts diagram

SOLUTION:

Let us first compute the remaining properties at section ①. For an ideal gas

$$\rho_1 = \frac{p_1}{RT_1} = \frac{10\ \text{lbf}}{\text{in.}^2} \times \frac{144\ \text{in.}^2}{\text{ft}^2} \times \frac{1\text{bm-R}}{53.3\ \text{ft-lbf}} \times \frac{1}{(460 + 40)\ R} = 0.054\ \text{lbm/ft}^3$$

$$c_1 = \sqrt{kRT_1} = 1095\ \frac{\text{ft}}{\text{sec}}; \qquad M_1 = \frac{V_1}{c_1} = \frac{2190}{1095} = 2.0$$

$$\frac{T_{0_1}}{T_1} = 1 + \frac{k-1}{2}M_1^2 = 1 + 0.2(2)^2 = 1.8; \qquad T_{0_1} = 1.8 T_1 = 900\ R$$

$$\frac{p_{0_1}}{p_1} = \left(\frac{T_{0_1}}{T_1}\right)^{k/(k-1)} = (1.8)^{3.5} = 7.82; \qquad p_{0_1} = 7.82 p_1 = 78.2\ \text{psia}$$

V_2 may be evaluated by applying the energy equation to the control volume shown. Basic equation:

$$\require{cancel}\begin{array}{cccc} = 0(1) & = 0(2) & = 0(3) & = 0(4) \end{array}$$

$$\cancel{\dot{Q}} - \cancel{\dot{W}_s} - \cancel{\dot{W}_{shear}} = \frac{\partial}{\partial t}\iiint_{CV} e\rho\, d\forall + \iint_{CS} (e + pv)\rho \vec{V} . d\vec{A} \qquad (4.45)$$

where

$$e = u + \frac{V^2}{2} + \cancel{gz}^{\ \simeq\ 0(6)}$$

Assumptions: (1) $\dot{Q} = 0$
(2) $\dot{W}_s = 0$
(3) $\dot{W}_{shear} = 0$
(4) Steady flow
(5) Uniform flow at each section
(6) Neglect gravity term
(7) $A_1 = A_2 = A$

Then

$$0 = \left(u_1 + p_1 v_1 + \frac{V_1^2}{2}\right)\{-|\rho_1 V_1 A|\} + \left(u_2 + p_2 v_2 + \frac{V_2^2}{2}\right)\{|\rho_2 V_2 A|\}$$

$$0 = \dot{m}\left(u_2 + p_2 v_2 + \frac{V_2^2}{2} - u_1 - p_1 v_1 - \frac{V_1^2}{2}\right)$$

or

$$h_1 + \frac{V_1^2}{2} = h_2 + \frac{V_2^2}{2}$$

Solving for V_2,

$$V_2 = [V_1^2 + 2(h_1 - h_2)]^{1/2} = [V_1^2 + 2c_p(T_1 - T_2)]^{1/2}$$

$$= \left[\frac{(2190)^2 \text{ ft}^2}{\text{sec}^2} + (2)\frac{0.240 \text{ Btu}}{\text{lbm-F}} \times (40 - 384)\text{ F} \times \frac{778 \text{ ft-lbf}}{\text{Btu}} \times \frac{32.2 \text{ lbm}}{\text{slug}} \times \frac{\text{slug-ft}}{\text{lbf-sec}^2}\right]^{1/2}$$

$V_2 = 812 \text{ ft/sec}$ ⟵ $\qquad\qquad V_2$

Continuity reduces to $G = \rho_1 V_1 = \rho_2 V_2$, so

$$\rho_2 = \rho_1 \frac{V_1}{V_2} = (0.054 \text{ lbm/ft}^3)\left(\frac{2190}{812}\right) = 0.146 \text{ lbm/ft}^3 \qquad\qquad \rho_2$$

The pressure can be obtained two ways:
(1) from the gas law

$$p_2 = \rho_2 R T_2 = \frac{0.146 \text{ lbm}}{\text{ft}^3} \times \frac{53.3 \text{ ft-lbf}}{\text{lbm-R}} \times 844 \text{ R} \times \frac{\text{ft}^2}{144 \text{ in.}^2} = 45.6 \text{ psia}$$

(2) from the momentum equation
Basic equation:

$$= 0(8) \qquad = 0(4)$$
$$F_{S_x} + \cancel{F_{B_x}} = \cancel{\frac{\partial}{\partial t} \int\int\int_{CV} V_x \rho \, d\forall} + \int\int_{CS} V_x \rho \vec{V} \cdot d\vec{A} \qquad (4.19a)$$

Assumption: (8) $F_{B_x} = 0$

EXAMPLE PROBLEM 10.11 (continued)

Then

$$p_1 A - p_2 A = V_1\{-|\rho_1 V_1 A|\} + V_2\{|\rho_2 V_2 A|\}$$
$$= \dot{m}(V_2 - V_1)$$

or

$$p_1 - p_2 = \rho_1 V_1 (V_2 - V_1)$$

Solving,

$$p_2 = p_1 - \rho_1 V_1 (V_2 - V_1)$$

$$= 10 \frac{\text{lbf}}{\text{in.}^2} - 0.054 \frac{\text{lbm}}{\text{ft}^3} \times \frac{2190 \text{ ft}}{\text{sec}} \times \frac{(812 - 2190) \text{ ft}}{\text{sec}} \times \frac{\text{slug}}{32.2 \text{ lbm}} \times \frac{\text{lbf-sec}^2}{\text{slug-ft}} \times \frac{\text{ft}^2}{144 \text{ in.}^2}$$

$$p_2 = 10 + 35.1 = 45.1 \text{ psia} \longleftarrow \hspace{4cm} p_2$$

which is fair agreement.
For adiabatic flow, $T_0 = $ constant. Thus

$$T_{0_2} = T_{0_1} = 900 \text{ R} \longleftarrow \hspace{5cm} T_{0_2}$$

$$\frac{p_{0_2}}{p_2} = \left(\frac{T_{0_2}}{T_2}\right)^{k/(k-1)} = \left(\frac{900}{844}\right)^{3.5} = 1.252;$$

$$p_{0_2} = 1.252 p_2 = 56.5 \text{ psia} \longleftarrow \hspace{4cm} p_{0_2}.$$

Comparing, we see that

$$T_{0_2} = T_{0_1}$$
$$p_{0_2} < p_{0_1}$$
$$T_2 > T_1$$
$$p_2 > p_1$$

and

$$V_2 < V_1$$

in accord with Table 10.3.
    The entropy change may be computed from the $T \, ds$ equation

$$T \, ds = dh - v \, dp$$

For an ideal gas

$$ds = c_p \frac{dT}{T} - R \frac{dp}{p}$$

Integrating, for constant specific heat

$$s_2 - s_1 = c_p \ln \frac{T_2}{T_1} - R \ln \frac{p_2}{p_1}$$

Since $s_{0_2} - s_{0_1} = s_2 - s_1$, this is easiest to evaluate at stagnation conditions, because $T_{0_2} = T_{0_1}$. At stagnation conditions, we have

$$s_{0_2} - s_{0_1} = c_p \ln \frac{\overset{0}{\cancel{T_{0_2}}}}{T_{0_1}} - R \ln \frac{p_{0_2}}{p_{0_1}}$$

$$= -\frac{53.3 \text{ ft-lbf}}{\text{lbm-R}} \times \frac{\text{Btu}}{778 \text{ ft-lbf}} \times \ln\left(\frac{56.5}{78.2}\right)$$

$$\Delta s = 0.0222 \frac{\text{Btu}}{\text{lbm-R}} \longleftarrow \qquad\qquad \Delta s$$

Finally, the Ts diagram may be sketched:

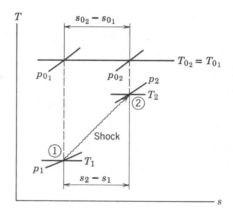

{Note once again that $s_{0_2} - s_{0_1} = s_2 - s_1$, since $s_{0_1} = s_1$ and $s_{0_2} = s_2$.}

**10–6.2 TABLES FOR COMPUTATION OF NORMAL SHOCKS

Consideration of the basic equations for the flow through a normal shock has indicated that conditions downstream of the shock are unique for given conditions ahead of the shock. It should then be possible to develop expressions for the property ratios across the shock in terms of the Mach number, M_1, ahead of the shock; these results can then be tabulated as a function of M_1.

To obtain the results we proceed in three steps. First we obtain property ratios (e.g. T_2/T_1 and p_2/p_1) in terms of M_1 and M_2. Then we develop a relation between

---

** This section may be omitted without loss of continuity in the text material.

M_1 and M_2. Finally we use this relation to obtain expressions for the property ratios in terms of the upstream Mach number, M_1.

The temperature ratio can be expressed as

$$\frac{T_2}{T_1} = \frac{T_2}{T_{0_2}} \frac{T_{0_2}}{T_{0_1}} \frac{T_{0_1}}{T_1}$$

Since the stagnation temperature is constant, we have

$$\frac{T_2}{T_1} = \frac{1 + \dfrac{k-1}{2} M_1^2}{1 + \dfrac{k-1}{2} M_2^2} \tag{10.22a}$$

A velocity ratio may be obtained by using

$$\frac{V_2}{V_1} = \frac{M_2 c_2}{M_1 c_1} = \frac{M_2 \sqrt{kRT_2}}{M_1 \sqrt{kRT_1}} = \frac{M_2}{M_1} \sqrt{\frac{T_2}{T_1}}$$

or

$$\frac{V_2}{V_1} = \frac{M_2}{M_1} \left[\frac{1 + \dfrac{k-1}{2} M_1^2}{1 + \dfrac{k-1}{2} M_2^2} \right]^{1/2} \tag{10.22b}$$

A ratio of densities may be obtained from the continuity equation

$$\rho_1 V_1 = \rho_2 V_2 \tag{10.21a}$$

On substituting from Eq. 10.22b

$$\frac{\rho_2}{\rho_1} = \frac{V_1}{V_2} = \frac{M_1}{M_2} \left[\frac{1 + \dfrac{k-1}{2} M_2^2}{1 + \dfrac{k'-1}{2} M_1^2} \right]^{1/2} \tag{10.22c}$$

Finally, we can obtain a pressure ratio from the momentum equation

$$p_1 A - p_2 A = \dot{m} V_2 - \dot{m} V_1 \tag{10.21b}$$

or

$$p_1 + \rho_1 V_1^2 = p_2 + \rho_2 V_2^2$$

Substituting $\rho = p/RT$ and factoring out pressures,

$$p_1 \left[1 + \frac{V_1^2}{RT_1} \right] = p_2 \left[1 + \frac{V_2^2}{RT_2} \right]$$

Noting that

$$\frac{V^2}{RT} = k\frac{V^2}{kRT} = kM^2$$

then

$$p_1[1 + kM_1^2] = p_2[1 + kM_2^2]$$

and finally

$$\frac{p_2}{p_1} = \frac{1 + kM_1^2}{1 + kM_2^2} \qquad (10.22d)$$

In order to solve for M_2 in terms of M_1, we need to obtain another expression for one of the property ratios given by Eqs. 10.22a–d.

From the ideal gas equation of state the temperature ratio may be written as

$$\frac{T_2}{T_1} = \frac{p_2/\rho_2 R}{p_1/\rho_1 R} = \frac{p_2}{p_1}\frac{\rho_1}{\rho_2}$$

Substituting from Eqs. 10.22c and 10.22d

$$\frac{T_2}{T_1} = \left[\frac{1 + kM_1^2}{1 + kM_2^2}\right]\frac{M_2}{M_1}\left[\frac{1 + \dfrac{k-1}{2}M_1^2}{1 + \dfrac{k-1}{2}M_2^2}\right]^{1/2} \qquad (10.23)$$

Equations 10.22a and 10.23 are two equations for T_2/T_1. We can combine them and solve for M_2 in terms of M_1. Combining and canceling gives

$$\left[\frac{1 + \dfrac{k-1}{2}M_1^2}{1 + \dfrac{k-1}{2}M_2^2}\right]^{1/2} = \frac{M_2}{M_1}\left[\frac{1 + kM_1^2}{1 + kM_2^2}\right]$$

Squaring, we obtain

$$\frac{1 + \dfrac{k-1}{2}M_1^2}{1 + \dfrac{k-1}{2}M_2^2} = \frac{M_2^2}{M_1^2}\left[\frac{1 + 2kM_1^2 + k^2M_1^4}{1 + 2kM_2^2 + k^2M_2^4}\right]$$

which may be solved explicitly for M_2^2. Two solutions are obtained:

$$M_2^2 = M_1^2 \qquad (10.24a)$$

and

$$M_2^2 = \frac{M_1^2 + \dfrac{2}{k-1}}{\dfrac{2k}{k-1}M_1^2 - 1} \qquad (10.24\text{b})$$

Obviously the first of these is trivial. The second expresses the unique dependence of M_2 on M_1. This relation is tabulated in Appendix B, Table B.4. Now, having a relationship between M_2 and M_1, we can solve for the property ratios across a shock knowing the upstream Mach number. Knowing M_1, M_2 is obtained from Eq. 10.24b. The property ratios can subsequently be determined from Eqs. 10.22a–d. These property ratios are tabulated as functions of M_1 in Table B.4 of Appendix B.

Since the stagnation temperature remains constant, the stagnation temperature ratio across the shock is unity. The ratio of stagnation pressures is evaluated as

$$\frac{p_{0_2}}{p_{0_1}} = \frac{p_{0_2}}{p_2}\frac{p_2}{p_1}\frac{p_1}{p_{0_1}} = \frac{p_2}{p_1}\left[\frac{1 + \dfrac{k-1}{2}M_2^2}{1 + \dfrac{k-1}{2}M_1^2}\right]^{k/(k-1)} \qquad (10.25)$$

Combining Eqs. 10.22d and 10.24b, we obtain (after considerable algebra)

$$\frac{p_2}{p_1} = \frac{1 + kM_1^2}{1 + kM_2^2} = \frac{2k}{k+1}M_1^2 - \frac{k-1}{k+1} \qquad (10.26)$$

Using Eqs. 10.24b and 10.26, Eq. 10.25 becomes

$$\frac{p_{0_2}}{p_{0_1}} = \frac{\left[\dfrac{k+1}{2}M_1^2\right]^{k/(k-1)}}{\left[\dfrac{2k}{k+1}M_1^2 - \dfrac{k-1}{k+1}\right]^{1/(k-1)}} \qquad (10.27)$$

The stagnation pressure ratio as a function of upstream Mach number is included in Table B.4 of Appendix B.

Use of the tables in the solution of problems involving a normal shock is illustrated in Example Problem 10.12.

Example 10.12

A normal shock stands in a duct. The fluid is air, which can be considered an ideal gas. Properties upstream from the shock are $p_1 = 10$ psia, $T_1 = 40$ F, and

$V_1 = 2190$ ft/sec. Determine the properties downstream and $s_2 - s_1$. Include a Ts plot.

(Note that these upstream conditions are the same as those given in Example Problem 10.11. However, with the tables available, we do not need to know any properties downstream to complete a solution.)

EXAMPLE PROBLEM 10.12

GIVEN:

Normal shock in a duct as shown:

$T_1 = 40$ F

$p_1 = 10$ psia

$V_1 = 2190$ ft/sec

Flow →   Flow is air

① ②

FIND:

(a) Properties at ②
(b) $s_2 - s_1$
(c) Ts plot

SOLUTION:

To use the shock tables, we need to know M_1. For an ideal gas,

$$c_1 = \sqrt{kRT_1} = 1095 \frac{\text{ft}}{\text{sec}}; \quad M_1 = \frac{V_1}{c_1} = \frac{2190}{1095} = 2.0$$

The stagnation properties at station ① may be evaluated using isentropic flow tables. From Table B.1, Appendix B, at $M = 2.0$,

$$\frac{T}{T_0} = \frac{T_1}{T_{0_1}} = 0.556; \quad T_{0_1} = \frac{T_1}{0.556} = 900 \text{ R}$$

$$\frac{p}{p_0} = \frac{p_1}{p_{0_1}} = 0.128; \quad p_{0_1} = \frac{p_1}{0.128} = 78.1 \text{ psia}$$

Normal shock data are given in Table B.4, Appendix B. At $M_1 = 2.0$,

M_1	M_2	p_2/p_1	T_2/T_1	p_{0_2}/p_{0_1}
2.0	0.577	4.500	1.687	0.721

From these data

$$T_2 = 1.687\, T_1 = 844\ R \qquad\qquad T_2$$

$$p_2 = 4.500 p_1 = 45.0\ \text{psia} \qquad\qquad p_2$$

$$p_{0_2} = 0.721 p_{0_1} = 56.3\ \text{psia} \qquad\qquad p_{0_2}$$

The stagnation temperature is constant in adiabatic flow. Thus

$$T_{0_2} = T_{0_1} = 900\ R \qquad\qquad T_{0_2}$$

For an ideal gas,

$$c_2 = \sqrt{kRT_2} = 1424\,\frac{\text{ft}}{\text{sec}}; \qquad M_2 = \frac{V_2}{c_2}$$

so

$$V_2 = M_2 c_2 = (0.577)(1424\ \text{ft/sec}) = 822\ \text{ft/sec} \qquad\qquad V_2$$

and

$$\rho_2 = \frac{p_2}{RT_2} = \frac{45\ \text{lbf}}{\text{in.}^2} \times \frac{144\ \text{in.}^2}{\text{ft}^2} \times \frac{\text{lbm-R}}{53.3\ \text{ft-lbf}} \times \frac{1}{844\ R} = 0.144\ \text{lbm/ft}^3 \qquad \rho_2$$

The entropy change across the shock may be found from the $T\,ds$ equation

$$T\,ds = dh - v\,dp$$

For an ideal gas,

$$ds = c_p\frac{dT}{T} - R\frac{dp}{p}$$

Integrating for constant specific heats,

$$s_2 - s_1 = c_p\ln\frac{T_2}{T_1} - R\ln\frac{p_2}{p_1}$$

But $s_{0_2} - s_{0_1} = s_2 - s_1$, so

$$s_{0_2} - s_{0_1} = c_p\ln\overset{0}{\cancel{\frac{T_{0_2}}{T_{0_1}}}} - R\ln\frac{p_{0_2}}{p_{0_1}} = -53.3\,\frac{\text{ft-lbf}}{\text{lbm-R}} \times \frac{\text{Btu}}{778\ \text{ft-lbf}} \times \ln(0.721)$$

and

$$s_{0_2} - s_{0_1} = s_2 - s_1 = 0.0224\ \text{Btu/lbm-R} \qquad\qquad s_2 - s_1$$

EXAMPLE PROBLEM 10.12 (continued)

The *Ts* diagram is

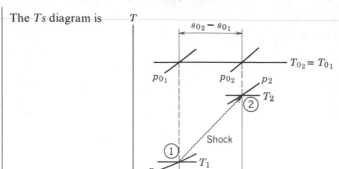

Comparing the solutions presented in Example Problems 10.11 and 10.12, we see that using the tables simplifies the computations appreciably. However, one must be careful to check his work as he proceeds to avoid making errors. Each calculated result should be examined to be sure both its trend and magnitude are reasonable. This process is simplified and made almost automatic when a *Ts* diagram is drawn for each problem.

10–6.3 FLOW IN A CONVERGING–DIVERGING NOZZLE

Since we have considered normal shocks, we are now in a position to complete our discussion of flow in a converging-diverging nozzle operating under varying back pressures, discussed previously in Section 10–3.5. The pressure distribution through the nozzle for different back pressures is shown in Fig. 10.20.

Four different regimes of flow are possible. In Regime I the flow is subsonic throughout. The flowrate increases with decreasing back pressure. At condition (*iii*), which forms the dividing line between regimes I and II, the flow at the throat is sonic, that is, $M_t = 1$.

As the back pressure is lowered below that of condition (*iii*), a normal shock appears downstream of the throat. There is a pressure rise across the shock. Since the flow is subsonic behind the shock ($M < 1$), the flow decelerates, with an accompanying increase in pressure, through the diverging portion of the channel. As the back pressure is lowered further the shock moves down the nozzle until it appears at the exit plane of the nozzle (condition *vii*). In Regime II, as in Regime I, the exit flow is subsonic and consequently $p_e = p_b$. Since the flow properties at the throat are constant for all conditions in Regime II, the flowrate in Regime II does not vary with back pressure.

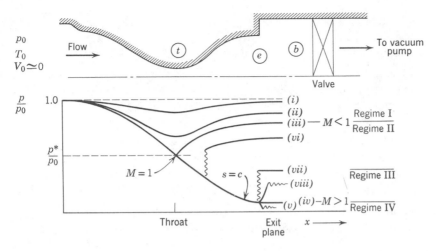

p_0
T_0
$V_0 \simeq 0$

Flow

t

e

b

Valve

To vacuum
pump

$\dfrac{p}{p_0}$

1.0

(i)
(ii)
(iii) — $M < 1$
(vi)

Regime I
Regime II

$\dfrac{p^*}{p_0}$

$M = 1$

$s = c$

(vii)
$(viii)$
(v) (iv)–$M > 1$

$\overline{\text{Regime III}}$

$\overline{\text{Regime IV}}$

Throat

Exit
plane

$x \longrightarrow$

Fig. 10.20 *Pressure distributions for flow in a converging-diverging nozzle as a function of back pressure.*

In Regime III, as exemplified by condition (*viii*), the back pressure is higher than the exit pressure, but not sufficiently high to sustain a normal shock in the exit plane. The flow adjusts to the back pressure through a series of oblique compression shocks that cannot be treated by one-dimensional theory.

As was previously noted in Section 10–3.5, condition (*iv*) represents the design condition[3]. In Regime IV the flow adjusts to the lower back pressure through a series of oblique expansion waves that cannot be treated by one-dimensional theory.

problems

10.1   Air flows steadily and isentropically through a passage. At section ① where the cross-section area is 2 ft², the air is at 6 psia, 600 R, and the Mach number is 2.0. At a section ② downstream, the velocity is 1000 ft/sec.

(a) Calculate the Mach number at section ②.
(b) Sketch the shape of the passage between sections ① and ②. Explain why the shape is as you show it.

10.2   Air flows steadily and isentropically through an insulated passage at a rate of 1.10 slug/sec. At the section where the area is 0.2 ft², $M = 3$, $T = 40$ F and $p = 10$ psia. Determine the velocity and cross-section area downstream where $T = 340$ F. Sketch the flow passage.

[3] Flow behavior in converging-diverging nozzles and applications to supersonic wind tunnels are shown in the film *Channel Flow of a Compressible Fluid*, D. Coles, principal.

10.3 Air at 10 psia and 80 F enters a passage at 1600 fps. The cross-section area at the entrance is 0.20 ft². At section ②, further downstream, the pressure is 18 psia. Assuming isentropic flow, calculate the Mach number at section ②.

10.4 A nozzle is designed to expand air isentropically to atmospheric pressure from a large tank in which properties are held constant at $p = 44.1$ psia, $T = 40$ F. The desired flow rate is 2 lbm/sec. Determine the exit area of the nozzle. Prepare a plot of Mach number and pressure as a function of distance along the nozzle.

10.5 At a section upstream of the throat in a converging-diverging nozzle the flow is at a pressure of 30 psia, a temperature of 90 F, and its velocity is 575 ft/sec. For the isentropic flow of air determine the Mach number at the point where the pressure is 12 psia.

10.6 A stream of air flowing in a duct (area = 1 in.²) is at a pressure of 20 psia, has a Mach number of 0.6, and flows at a rate of 0.5 lbm/sec.

(a) Compute the stagnation temperature.
(b) If the cross-section area of the passage were reduced downstream, what is the maximum percentage reduction of area allowable without reducing the flowrate (assume isentropic flow).
(c) Determine the velocity and pressure at the minimum area location.

10.7 Air flows isentropically through a converging nozzle into a receiver in which the pressure is 36 psia. The air enters the nozzle with negligible velocity at a pressure of 60 psia and a temperature of 200 F. Determine the flowrate through the nozzle for a nozzle throat area of 0.1 ft².

10.8 Air flows from a large tank ($p = 100$ psia, $T = 1000$ F) through a converging nozzle, with a throat area of 1 in.², and discharges to the atmosphere. Determine the mass rate of flow for isentropic flow through the nozzle.

10.9 A large tank supplies air to a convergent nozzle. The atmospheric pressure to which the nozzle discharges is 14.7 psia. Assume the flow to be reversible and adiabatic.

(a) For what range of tank pressures will the flow at the nozzle exit be sonic ($M = 1$)?
(b) If the tank pressure is 100 psia and the temperature is 500 R, what is the mass flowrate through the nozzle if the exit area is 2 in.²?

10.10 Air, with a stagnation temperature of 600 R, flows isentropically through a converging nozzle. At the point in the flow where the temperature is 571 R, the pressure is 100 psia. Determine the velocity and the stagnation pressure at the downstream position where $T = 532$ R.

10.11 Air in a very large tank is maintained at 100 psia, 140 F. Air leaves the tank steadily and isentropically through a converging nozzle that discharges to the atmosphere at a pressure of 14.7 psia. The nozzle throat area is 0.02 ft². Determine the air density at the nozzle throat.

10.12 Air at 300 psia and 240 F discharges isentropically from a large tank through a converging nozzle whose throat area is 0.25 ft². Find the temperature at the exit and the pressure difference between nozzle exit and atmosphere.

10.13 Air, with a stagnation pressure of 100 psia and a stagnation temperature of 600 R, is flowing isentropically through a converging nozzle. At the section in the nozzle where the area is 4 in.² the Mach number is 0.5. The nozzle discharges to a back pressure of 40 psia. Determine the exit area of the nozzle.

10.14 A converging nozzle is bolted to the side of a large tank. Air inside the tank is maintained at a constant pressure of 50 psia and a temperature of 100 F. The inlet area of the nozzle is 1 ft² and the exit area is 1 in.² The nozzle discharges to the atmosphere. For isentropic flow in the nozzle determine the total force on the bolts, and indicate whether the bolts are in tension or compression.

10.15 Air flowing isentropically through a converging nozzle discharges to the atmosphere. At the section where the pressure is 26 psia, the temperature is 100 F and the air velocity is 600 ft/sec. Determine the nozzle throat pressure.

10.16 A nozzle discharges into a receiver where the pressure is 18 psia. Conditions at station ① in the nozzle result in a pressure of 50 psia, a temperature of 190 F, and a velocity of 400 ft/sec. The weight rate of flow of air is 0.30 lbf/sec. Assume the flow is frictionless and adiabatic and determine for the converging nozzle:

(a) The Mach number at the nozzle exit.
(b) The exit area of the nozzle.

10.17 Air flows isentropically through a converging nozzle into a receiver where the pressure is 33 psia. If the pressure is 50 psia, and the velocity is 500 ft/sec at the nozzle location where the Mach number is 0.4, determine the pressure, velocity, and Mach number at the nozzle throat.

10.18 A supersonic diffuser decelerates a flow isentropically from a Mach number of 3 to a Mach number of 1. If the static pressure at the diffuser inlet is 2 psia, calculate:
(a) The static pressure rise in the diffuser.
(b) The ratio of inlet to outlet area of the diffuser.

10.19 Air enters a converging-diverging nozzle with negligible velocity at a pressure of 150 psia and a temperature of 140 F. If the flow is isentropic and the exit temperature is 10 F, what is the Mach number at exit?

10.20 Air enters a converging-diverging nozzle with negligible velocity at 160 psia and a temperature of 120 F. For a frictionless, adiabatic process, what is the Mach number at a section where the pressure is 50 psia?

10.21 Consider the isentropic flow of air through a converging-diverging channel. The stagnation pressure at the nozzle entrance is 100 psia and the stagnation temperature is 600 R. At a section downstream of the throat the pressure is 78.4 psia and the area is 1.888 in.² Determine the Mach number, the temperature, and the stagnation pressure at this section.

10.22 A converging-diverging nozzle is attached to a very large tank in which the pressure is 20 psia and the temperature is 40 F. The nozzle exhausts to the atmosphere where the pressure is 14.7 psia. The exit area of the nozzle is 2 in.² What is the flowrate through the nozzle? Assume the air flow to be isentropic.

10.23 A convergent-divergent nozzle with a throat area of 2 in.² is connected to a tank in which air is kept at a pressure of 80 psia at a temperature of 60 F. If the nozzle is to operate at design conditions (flow is isentropic) and the ambient pressure outside the nozzle is 12.9 psia, calculate the exit area of the nozzle, and the mass rate of flow.

10.24 Air is to be expanded through a converging-diverging nozzle by a frictionless adiabatic process from a pressure of 160 psia and a temperature of 240 F to a pressure of 20 psia. Determine the throat and exit areas for a well-designed shockless nozzle if the weight rate of flow is 4.5 lbf/sec.

10.25 Air, with a stagnation pressure of 100 psia and a stagnation temperature of 580 F flows isentropically through a converging-diverging nozzle having a throat area of 8 in.² Determine the velocity at the downstream section where the Mach number is 2 and also the mass rate of flow.

10.26 A large tank supplies air to a converging-diverging nozzle. The pressure in the tank remains constant at 115 psia. The tank temperature remains constant at 1000 R. The flow throughout the nozzle is isentropic. The nozzle is designed to discharge to atmospheric pressure with an exit Mach number, $M = 2$. The exit area of the nozzle is 6 in.² What is the mass flowrate through the nozzle?

10.27 Air flows isentropically through a converging-diverging nozzle attached to a large tank in which the pressure is 100 psia and the temperature is 500 R. The nozzle is operating at design conditions for which the nozzle exit pressure, p_e, is equal to the surrounding atmospheric pressure, p_a. The exit area of the nozzle, $A_e = 4.0$ in.²

(a) Calculate the flowrate through the nozzle in slug/sec.
(b) If the temperature of the air in the tank is increased to 2000 R (all pressures remaining the same), how will the flow rate in part (a) be affected?

10.28 Air at a pressure and temperature of 44 psia and 250 F enters a nozzle with negligible velocity. The exhaust jet is directed against a large flat plate that is perpendicular to the jet axis. The flow leaves the nozzle at atmospheric pressure ($p_{atm} = 14.7$ psia). The exit area is 0.03 ft². Find the force required to hold the plate.

10.29 At a point upstream of the throat in a converging-diverging nozzle, the air velocity is 600 ft/sec, $p = 20$ psia, and $T = 60$ F. The flow is isentropic and is supersonic at the nozzle exit. If the nozzle throat area is 0.1 ft^2, determine the flowrate.

10.30 Air is flowing isentropically in a converging-diverging nozzle. At the section in the converging portion where the area is 2 in.^{2}, the pressure is 100 psia, the temperature 500 R, and the Mach number 0.50. Determine the area in the diverging section where the Mach number is 2.

10.31 Air flows steadily and isentropically through a converging-diverging nozzle. At the throat the air is at 20 psia, 140 F. The throat cross-section area if 0.5 ft^2. At a certain section in the diverging part of the nozzle, the pressure is 10 psia. Calculate the velocity and the area at this section.

10.32 Consider compressible flow in a long straight duct. The inlet to the duct is from the atmosphere where $T = 80$ F and $p = 14.7$ psia. The flow is considered to be adiabatic and the pipe is sufficiently long that the flow is choked. Find the velocity and temperature at the pipe exit.

10.33 Air flows adiabatically with friction through a duct of 1 ft. square cross section. At one section, the pressure, temperature, and velocity are 10 psia, 40 F, and 500 ft/sec, respectively. At the exit from the duct, the pressure and temperature are respectively 4.9 psia and -10 F. Calculate the exit Mach number and the stagnation temperature midway between the first section and the exit.

10.34 Air flows steadily and adiabatically in a horizontal 2 in. diameter pipe. At section ① the pressure is 50 psia and the temperature is 140 F. At section ② (downstream) the temperature is 240 F and the velocity is 1950 fps. Determine the velocity and Mach number at section ①.

10.35 Air flows steadily and adiabatically through a constant area duct. The density at the inlet is 0.81 lbm/ft^3. If the exit Mach number is unity and the exit temperature and pressure are 1000 R and 10 psia respectively, determine the inlet Mach number.

10.36 Air is flowing through a well-insulated constant area channel. At a section where the velocity is 585 ft/sec, $T = 571$ R and $p = 84.3$ psia. Determine the temperature and stagnation pressure at the downstream section where the density is 0.233 lbm/ft^3.

10.37 A converging-diverging nozzle supplies air to a well-insulated constant area duct. At the inlet to the duct, $M = 2.0$, $p = 19.4$ psia, $T = 278$ R. At the duct exit the Mach number is unity and the stagnation pressure is 90 psia. Determine the pressure and temperature at the duct exit.

10.38 Consider adiabatic flow in a constant area pipe with friction. At one section of the pipe, $p_0 = 100$ psia, $T_0 = 500$ R, and $M = 0.70$. If the cross-section area of the pipe is 1 ft^2 and the Mach number at the exit is $M_2 = 1$, find the friction force exerted on the fluid by the pipe.

575   **problems**

10.39 Air flows steadily and adiabatically from a large tank through a convergent nozzle connected to a constant area duct. The nozzle itself may be considered frictionless. Air in the tank is at $p = 100$ psia, $T = 200$ F. The pressure at the nozzle exit (duct inlet) is 71.6 psia. Determine the pressure at the end of the duct length, L, if the temperature there is 100 F.

10.40 A converging-diverging nozzle discharges air into a constant area insulated pipe (area $= 1$ in.^{2}). At the pipe inlet, $p = 12.78$ psia, $T = 333$ R, $M = 2.00$.

(a) For a shockless flow to a Mach number of unity at the pipe exit, calculate the exit temperature.

(b) Calculate the net force of the fluid on the pipe.

10.41 Air flows through a well-insulated 4 in. diameter pipe at a rate of 600 lbm/min. At one section the air is at 100 psia, 80 F. Determine the minimum pressure and the maximum velocity that can occur in the pipe.

10.42 Air is flowing in an insulated duct with a velocity of 500 fps. The temperature and pressure are 540 F and 200 psia, respectively.

(a) Find the temperature in this duct where the pressure has dropped to 150 psia as a result of friction.

(b) If the duct has a diameter of 6 in. and if the friction factor is 0.02, find the distance between the two points.

10.43 Consider the frictionless flow of air in a constant area duct. At section ①, $M_1 = 0.50$, $p_1 = 158$ psia, $T_{0_1} = 600$ R. Through the effect of heat transfer, the Mach number at station ② is $M_2 = 0.90$ and the stagnation temperature, $T_{0_2} = 861$ R.

(a) Determine the amount of heat transfer per unit mass to or from the fluid between stations ① and ②.

(b) Determine the pressure difference, $p_1 - p_2$.

10.44 Air is flowing in a constant area duct without friction. At section ① in the duct, $p_1 = 97.25$ psia, $T_1 = 992$ R, and the velocity is $V_1 = 308$ ft/sec. Through the effect of heat transfer the velocity at section ② downstream is $V_2 = 651$ ft/sec. Determine:

(a) The pressure and temperature at section ②.

(b) The stagnation pressure and stagnation temperature at section ②.

(c) The heat transfer per unit mass between sections ① and ②.

10.45 At a position 10 ft from the exit of a constant area duct (area $= 2$ ft^2), air is at a pressure of 18 psia and a temperature of 500 F. The air flows steadily through the duct (without friction) at the rate of 40 lbm/sec. The air leaves the duct subsonically at a pressure of 14.7 psia. Determine:

(a) The Mach number, temperature, and stagnation temperature at the exit of the duct.

(b) The heat transfer over the 10 ft of duct length.

576   **10/one-dimensional compressible flow**

10.46 A constant area duct is fed with air from a converging-diverging nozzle. At the entrance to the duct the following properties are known: $p_{0_1} = 64.0$ psia, $T_{0_1} = 600$ R, $M_1 = 3.0$. A short distance down the duct (at section ②) the density is 0.024 lbm/ft^3. Assuming frictionless flow, determine:

(a) The velocity, pressure and Mach number at section ②.
(b) The heat transfer between the inlet and section ②.

10.47 Air flows through a short tube without friction. Heat is supplied to increase the initial Mach number of 0.3 at a temperature of 100 F to a final Mach number of 0.6. Determine the heat supplied per unit mass.

10.48 In the frictionless air flow through a 4 in. diameter duct, 4.0 lbm/sec enters at $T = 30$ F, $p = 10$ psia. How much heat, Btu/lbm, can be added without choking the flow?

10.49 Air flows through a constant area duct. At the duct inlet, $p_1 = 40$ psia, $M_1 = 0.2$, and $T_{0_1} = 500$ R. Over a length of the duct heat addition is 80 Btu/lbm when the flowrate is 5 lbm/sec. Neglecting the effect of friction determine the Mach number and the stagnation pressure following the heat addition.

10.50 Air flows steadily in a constant area, frictionless duct of 0.5 ft^2 cross-section area. If the inlet conditions are $M_1 = 0.3$, $T_{0_1} = 400$ R, $p_{0_1} = 10$ psia, and the exit stagnation temperature is 800 R, determine:

(a) The amount of heat transfer.
(b) The exit Mach number and pressure.

10.51 An air stream, with a velocity, $V_1 = 1575$ ft/sec, static pressure, $p_1 = 10$ psia, and static temperature, $T_1 = 0$ F, undergoes a normal shock. The following information is known

$$\frac{T_2}{T_1} = 1.32 \qquad \frac{p_2}{p_1} = 2.46$$

Calculate:

(a) The velocity, V_2, and Mach number, M_2, behind the shock.
(b) The stagnation pressure behind the shock.

10.52 Consider the flow through a normal shock. Upstream of the shock the following properties are known:

$$p_{0_1} = 100 \text{ psia} \qquad T_{0_1} = 600 \text{ R} \quad \text{and} \quad M_1 = 2$$

If the velocity behind the shock is $V_2 = 671$ ft/sec, determine the static pressure behind the shock.

10.53 Air approaches a normal shock with a velocity of 3300 ft/sec. The pressure and temperature are 20 psia and 45 F, while the stagnation temperature and Mach number of the flow are 1414 R and 3, respectively. The pressure behind the shock is 206.6 psia. Determine the velocity and temperature behind the shock.

10.54 A normal shock stands in a constant area duct in which air is flowing. Immediately upstream of the shock the following properties are known:

$$T_{0_1} = 1000 \text{ R} \qquad p_{0_1} = 100 \text{ psia} \quad \text{and} \quad M_1 = 3$$

If the Mach number immediately downstream of the shock is 0.475, determine the static pressure downstream of the shock.

10.55 Air approaching a normal compression shock has the following properties: ,

$$V_1 = 2300 \text{ ft/sec} \qquad p_1 = 25.56 \text{ psia} \quad \text{and} \quad T_1 = 555 \text{ R}$$

If the density behind the shock is 0.331 lbm/ft³, determine the temperature, T_2, and the stagnation pressure, p_{0_2}.

10.56 An air stream with $M = 3$, $\rho = 0.146$ lbm/ft³, and $p_0 = 1000$ psia undergoes a normal shock. If the ratio of static to stagnation pressure immediately behind the shock is 0.853, determine the static temperature behind the shock.

10.57 Air flows steadily through a long insulated constant area pipe. At section ①, the Mach number is 2, the temperature is 140 F and the density is 0.005 slug/ft³. At section ② downstream from a normal shock (the shock lies between points ① and ②), the velocity is 1000 fps.

(a) Calculate the Mach number and the density at section ②.
(b) Make a qualitative sketch of the pressure distribution along the pipe.

10.58 A normal shock occurs in air at a section where the flow velocity is 3000 ft/sec. The temperature at this point is 500 R and the pressure is 5 psia. Determine the velocity and Mach number behind the shock.

10.59 Air approaches a normal compression shock with a pressure of 14.7 psia, a temperature of 60 F and a velocity of 2500 fps.

(a) Determine the velocity behind the shock.
(b) Determine the pressure change across the shock.
(c) Considering a frictionless, shockless, deceleration between the same velocities, calculate the corresponding pressure change.

10.60 Air enters the inlet of an insulated converging-diverging nozzle with negligible velocity, stagnation pressure of 100 psia and a stagnation temperature of 600 R.

(a) If the velocity at the throat is sonic and the throat area is 1 in.², calculate the flowrate through the nozzle.
(b) A normal shock occurs downstream of the throat. At a point ②, downstream from the shock the temperature is 561 R. Calculate the Mach number at point ②.

578   **10/one-dimensional compressible flow**

10.61 Air flows adiabatically from a reservoir at 90 psia and 140 F through a converging-diverging nozzle. A normal shock occurs at a section where the Mach number is 2. Assuming isentropic flow to the shock, find:

(a) The pressure change across the shock.
(b) The temperature change across the shock.

10.62 A normal shock occurs in the diverging portion of a converging-diverging nozzle at a section where the area is 4 in.^{2} and the Mach number is 2.5. The stagnation pressure is 100 psia and the stagnation temperature is 1000 F. The nozzle exit area is 6 in.^{2} Determine:

(a) The mass flowrate through the nozzle.
(b) The nozzle exit pressure.

10.63 A converging-diverging nozzle is attached to a large tank in which the pressure is maintained at 100 psia and the temperature is constant, $T = 600$ R. The nozzle throat area is 1 in.^{2} and the exit area is 1.568 in.^{2} The nozzle is insulated and the flow is frictionless.

(a) If the nozzle is designed to operate at a back pressure of 14.7 psia, calculate the exit Mach number at design conditions. Sketch the pressure distribution for this case.
(b) Determine the minimum back pressure required to obtain isentropic flow throughout the nozzle with sonic velocity at the throat. What is the exit Mach number? Sketch the pressure distribution.
(c) Determine the back pressure for which a normal shock stands in the exit plane of the nozzle. Sketch the pressure distribution and calculate the velocity and Mach number after the shock.
(d) For what range of back pressures is $p_e = p_b$?
(e) For what range of back pressures is $p_e > p_b$?

10.64 Air is flowing through a constant area duct. At a certain point in the flow (point ①) the temperature, $T_1 = 992$ R, the pressure, $p_1 = 97.25$ psia, and the Mach number, $M_1 = 0.20$. At a second point in the duct the following information is given: $T_2 = 1962$ R, $p_2 = 91$ psia, $M_2 = 0.30$.

(a) Is the flow isentropic? Justify your answer.
(b) Calculate the stagnation enthalpy, stagnation temperature, and stagnation pressure at points ① and ②.

10.65 Air flows steadily through a constant area duct. At section ①, the air is at 60 psia, 600 R, with a velocity of 500 ft/sec. As a result of heat transfer and friction, the air at section ② downstream is at 40 psia, 800 R. Calculate:

(a) The heat transfer per pound of air between sections ① and ②.
(b) The stagnation pressure at section ②.

10.66 In long constant area pipelines, such as those used for natural gas, the temperature may be considered constant. Assume gas leaves a pumping station at 50 psia and 70 F with a Mach number of 0.10 and a density $\rho = 0.255$ lbm/ft^3. At the section along the pipe where the pressure has dropped to 20 psia:

(a) Calculate the Mach number of the flow at this section.
(b) Is heat added to or removed from the gas over the length between the pressure taps? Justify your answer.
(c) Sketch the process on a Ts diagram. Indicate (qualitatively) T_{0_1}, T_{0_2}, p_{0_2}.

10.67 Natural gas (molecular weight, 18, $k = 1.3$) is to be pumped through a 36 in. i.d. pipe connecting two compressor stations 40 miles apart. At the upstream station the pressure is not to exceed 90 psig and at the downstream station it is to be at least 10 psig. Calculate the maximum allowable rate of flow (ft^3/day at 70 F and 1 atm) assuming that there is sufficient heat transfer through the pipe to maintain the gas at 70 F.

10.68 Air enters a 6 in. diameter pipe at 200 psia and 60 F with a velocity of 190 fps. The friction coefficient, f, is 0.016. What is the Mach number and distance from entrance at the section where the pressure is 75 psia for isothermal flow?

appendix A

fluid property data

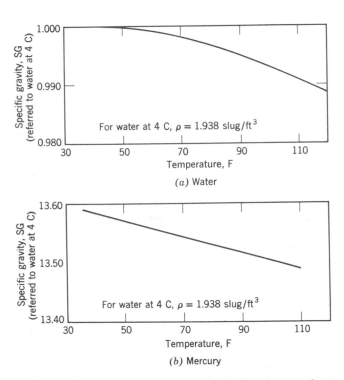

Fig. A.1 *Variation in specific gravity of water and mercury as a function of temperature (data from Ref. 1).*

581

table A.1 *Specific Gravity Data for Several Common Manometer Fluids (data from Refs. 1, 2)*

FLUID	SPECIFIC GRAVITY	TEMPERATURE
Meriam red oil	0.827	60 F
Benzene	0.879	0 F
Dibutyl phthalate	1.04	83 F
Monochloronaphthalene	1.20	77 F
Carbon tetrachloride	1.595	68 F
Bromoethylbenzene	1.75	55 F
Tetrabromoethane	2.95	78 F
Mercury	13.54	72 F

table A.2 *Physical Properties of Common Liquids (data from Refs. 2, 3)*

LIQUID	TEMPERATURE (F)	BULK MODULUS (lbf/in^2)	VAPOR PRESSURE (psia)	SPECIFIC GRAVITY (−)
Benzene	68	150,000	1.50	0.879
Carbon tetrachloride	73	162,000	1.93	1.595
Castor oil	59			0.969
Gasoline	68			0.66–0.75
Glycerine	59	669,000	2×10^{-5}	1.26
Kerosene	68			0.82
Mercury	59	3,910,000	2.3×10^{-5}	13.55
Oil, SAE 10		250,000		0.92
Oil, SAE 30		250,000		0.92
Water	68	320,000	0.34	1.00

table A.3 *Surface Tension Data for Common Liquids at 68 F*
(data from Refs. 2, 3)

LIQUID	SURFACE TENSION σ(lbf/ft)	CONTACT ANGLE, θ(degrees)
(a) In contact with air		Air ⟍ Liquid ∖θ
Benzene	0.00197	
Carbon tetrachloride	0.0184	
Glycerin	0.0043	
Kerosene	0.0017	
Lube oil	0.0025	
Mercury	0.0320	
Water	0.0050	~0
(b) In contact with water		Water ⟍ Liquid ∖θ
Carbon tetrachloride	0.0308	
Mercury	0.0257	130

Surface Tension

The values of surface tension, σ, for most organic compounds are remarkably similar at room temperature: the typical range is 25 to 40 dyne/cm. Water is higher, at about 73 dyne/cm at 20 C. Liquid metals have values in the range between 300 and 600 dyne/cm; liquid mercury has a value of about 476 dyne/cm at 20 C. Surface tension decreases with temperature; the decrease is nearly linear with absolute temperature. Surface tension at the critical temperature is zero.

Values of σ are usually reported for surfaces in contact with the pure vapor of the liquid being studied or with air. At low pressures both values are about the same.

table A.4 *Properties of the U.S. Standard Atmosphere (data from Ref. 4)*

GEOMETRIC ALTITUDE (feet)	TEMPERATURE (F)	$\dfrac{p}{p_0}$	$\dfrac{\rho}{\rho_0}$
− 500	60.8	1.022	1.015
0	59.0	1.000*	1.000†
500	57.2	0.982	0.985
1,000	55.4	0.964	0.971
2,000	51.9	0.929	0.943
3,000	48.3	0.896	0.915
4,000	44.7	0.864	0.888
5,000	41.2	0.832	0.862
6,000	37.6	0.801	0.836
7,000	34.0	0.772	0.811
8,000	30.5	0.743	0.786
9,000	26.9	0.715	0.762
10,000	23.4	0.688	0.739
12,000	16.2	0.636	0.693
14,000	9.1	0.588	0.650
16,000	2.0	0.542	0.609
18,000	− 5.1	0.499	0.570
20,000	− 12.3	0.460	0.533
25,000	− 30.0	0.371	0.449
30,000	− 47.8	0.297	0.375
35,000	− 65.6	0.236	0.311
40,000	− 69.7	0.186	0.247
45,000	− 69.7	0.149	0.195
50,000	− 69.7	0.115	0.153
60,000	− 69.7	0.071	0.095
70,000	− 67.4	0.044	0.059
80,000	− 62.0	0.028	0.036
90,000	− 56.5	0.017	0.022
100,000	− 51.1	0.011	0.014
150,000	19.4	0.001	0.001
200,000	− 2.7	—	—
250,000	− 107.8	—	—

* p_0 = 14.696 psia = 2116.2 psfa.
† ρ_0 = 0.002378 slug/ft^3 = 0.07651 lbm/ft^3.

A/fluid property data

The Physical Nature of Viscosity

Viscosity is a measure of internal fluid friction, that is resistance to deformation. The mechanism of gas viscosity is reasonably well understood, but the theory is poorly developed for liquids. We can gain some insight into the physical nature of viscous flow by exploring these mechanisms briefly.

The viscosity of a Newtonian fluid is fixed by the state of the material. Thus $\mu = \mu(T, p)$. Temperature is the more important variable, so let us consider it first. Excellent empirical equations for prediction of viscosity as a function of temperature are available.

a. EFFECT OF TEMPERATURE ON VISCOSITY

1. Gases

All gas molecules are in continuous random motion. When there is bulk motion due to flow, the bulk motion is superimposed on the random motions. It is then distributed throughout the fluid by molecular collisions. Theoretical analyses based on these simple considerations predict

$$\mu \propto \sqrt{T}$$

which is in excellent agreement with experimental trends, but the constant of proportionality and one or more correction factors must be determined. This limits practical application of this simple equation.

If two or more experimental datum points are available, data may be correlated using the empirical equation

$$\mu = \frac{bT^{1/2}}{1 + S/T} \tag{A.1}$$

Equation A.1 is known as the Sutherland correlation. The constants b and S may be determined most simply by writing

$$\mu = \frac{bT^{3/2}}{S + T}$$

or

$$\frac{T^{3/2}}{\mu} = \left(\frac{1}{b}\right)T + \frac{S}{b}$$

(Compare this with $y = mx + c$.) From a plot of $T^{3/2}/\mu$ versus T, one obtains the slope $1/b$ and the intercept S/b. For air,

$$b = 2.22 \times 10^{-8} \frac{\text{lbf-sec}}{\text{ft}^2 - R^{1/2}}$$

$$S = 180 \, R$$

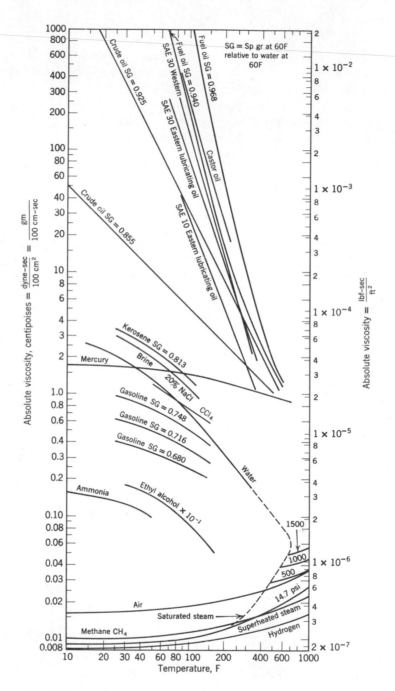

Fig. A.2 *Dynamic (absolute) viscosity of common fluids as a function of temperature (data from Refs. 3 and 5).*

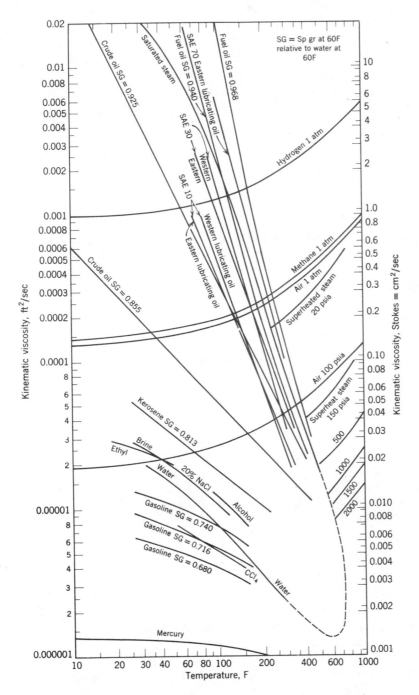

Fig. A.3 *Kinematic viscosity of common fluids (at atmospheric pressure) as a function of temperature (data from Refs. 3 and 5).*

2. Liquids

Viscosities for liquids cannot be estimated well theoretically. The phenomenon of momentum transfer by molecular collisions seems overshadowed in liquids by the effects of the interacting force fields between the closely packed liquid molecules.

Liquid viscosities are affected drastically by temperature. This temperature dependence is well represented by the empirical equation

$$\mu = Ae^{B/T} \tag{A.2}$$

Equation A.2 requires two datum points to fit A and B. This may be done easily by plotting log μ versus $1/T$ in the form

$$\log \mu = \log A + \left(\frac{B}{2.3}\right)\frac{1}{T}$$

Log A may be identified as the intercept and $(B/2.3)$ as the slope of the resulting plot.

b. EFFECT OF PRESSURE ON VISCOSITY

1. Gases

The viscosity of gases is essentially independent of pressure for values between a few hundredths of an atmosphere and a few atmospheres. However, viscosity at high pressures increases with pressure (or density).

2. Liquids

The viscosities of most liquids are not affected by moderate pressures, but large increases have been found at very high pressures. For example, the viscosity of water at 10,000 atm is twice the value at 1 atm. More complex compounds show a viscosity increase of several orders of magnitude over the same pressure range.

More information may be found in Reference 6.

Lubricating Oils

Engine and transmission oils are classified by viscosity according to standards established by the Society of Automotive Engineers (Ref. 7). The allowable ranges of viscosity are given in Table A.5.

Viscosity numbers with W (e.g. 20 W) are classified by viscosity at 0 F. Those without W are classified by viscosity at 210 F.

Multigrade oils (e.g. 10W − 40) are formulated to minimize viscosity variation with temperature. High polymer "viscosity index improvers" are used in blending these multigrade oils. Such additives are highly non-Newtonian; they may suffer permanent viscosity loss due to shearing.

table A.5 *Viscosity Ranges for SAE Lubricant Classifications (data from Ref. 7)*

| LUBRICANT TYPE | SAE VISCOSITY NUMBER | VISCOSITY RANGE (centistokes) | | | |
| | | At 0 F | | At 210 F | |
		Minimum	Maximum	Minimum	Maximum
Crankcase	5 *W*		1,200	3.9	
	10 *W*	1,200	2,400	3.9	
	20 *W*	2,400	9,600	3.9	
	20			5.7	9.6
	30			9.6	12.9
	40			12.9	16.8
	50			16.8	22.7
Transmission and axle	75		15,000		
	80	15,000	100,000		
	90			75	120
	140			120	200
	250			200	
Automatic transmission fluid	Type *A*	39*	43*	7	8.5

* At 100 F.

Molecular Weight, Specific Heat and Specific Heat Ratio for Common Gases at STP (data from Ref. 8)

GAS	CHEMICAL SYMBOL	MOLECULAR WEIGHT, M	R $\dfrac{\text{ft-lbf}}{\text{lbm} - \text{R}}$	c_p $\dfrac{\text{Btu}}{\text{lbm} - \text{R}}$	c_v $\dfrac{\text{Btu}}{\text{lbm} - \text{R}}$	$k = \dfrac{c_p}{c_v}$
Hydrogen	H_2	2	766.8	3.42	2.43	1.40
Helium	He	4	386.3	1.25	0.754	1.66
Methane	CH_4	16	96.4	0.601	0.477	1.26
Steam	H_2O	18	85.8*	0.439	0.335	1.28
Carbon monoxide	CO	28	55.2	0.243	0.172	1.41
Nitrogen	N_2	28	55.2	0.247	0.176	1.40
Air	—	29	53.3	0.240	0.171	1.40
Oxygen	O_2	32	48.3	0.217	0.155	1.40
Carbon dioxide	CO_2	44	35.1	0.195	0.151	1.28

* Water vapor behaves as an ideal gas when superheated by 100 F or more.

references

1. "Meriam Standard Indicating Fluids," Pamphlet No. 920GEN: 430–1. *The Meriam Instrument Co.,* 10920 Madison Ave., Cleveland, Ohio 44102.
2. *Handbook of Chemistry and Physics,* 35th ed., Cleveland, Ohio: Chemical Rubber Publishing Co., 1953, p. 1950.
3. *ASCE,* "Hydraulic Models," *Manual of Engineering Practice No. 25.* New York: American Society of Civil Engineers, 1942.
4. *The U.S. Standard Atmosphere (1962).* Washington, D.C.: U.S. Government Printing Office, 1962, Tables IV, pp. 113–137, and VI, pp. 197–221.
5. R. L. Daugherty, "Some Physical Properties of Water and Other Fluids," *Trans., ASME, Hydraulics Division,* **57**, 2, 1934, pp. 193–196.
6. R. C. Reid, and T. K. Sherwood, *The Properties of Gases and Liquids,* 2nd ed. New York: McGraw-Hill, 1966, p. 415.
7. "Crankcase Oil Viscosity Classification, Recommended Practice SAE J300a," *SAE Handbook,* 1971 ed. New York: Society of Automotive Engineers, 1971, p. 394.
8. NASA, *Compressed Gas Handbook.* Washington, D.C.: National Aeronautics and Space Administration, SP-3054, 1971.

tables for computations in compressible flow

table B.1   *Isentropic Flow Functions (Uniform Flow, Ideal Gas, k = 1.4)*

M	T/T_0	p/p_0	ρ/ρ_0	A/A^*
0.00	1.0000	1.0000	1.0000	∞
0.02	0.9999	0.9997	0.9998	28.94
0.04	0.9997	0.9989	0.9992	14.48
0.06	0.9993	0.9975	0.9982	9.666
0.08	0.9987	0.9955	0.9968	7.262
0.10	0.9980	0.9930	0.9950	5.822
0.12	0.9971	0.9900	0.9928	4.864
0.14	0.9961	0.9864	0.9903	4.182
0.16	0.9949	0.9823	0.9873	3.673
0.18	0.9936	0.9777	0.9840	3.278
0.20	0.9921	0.9725	0.9803	2.964
0.22	0.9904	0.9669	0.9762	2.708
0.24	0.9886	0.9607	0.9718	2.496
0.26	0.9867	0.9541	0.9670	2.317
0.28	0.9846	0.9470	0.9619	2.166
0.30	0.9823	0.9395	0.9564	2.035
0.32	0.9799	0.9315	0.9506	1.922
0.34	0.9774	0.9231	0.9445	1.823
0.36	0.9747	0.9143	0.9380	1.736
0.38	0.9719	0.9052	0.9313	1.659
0.40	0.9690	0.8956	0.9243	1.590

M	T/T_0	p/p_0	ρ/ρ_0	A/A^*
0.40	0.9690	0.8956	0.9243	1.590
0.42	0.9659	0.8857	0.9170	1.529
0.44	0.9627	0.8755	0.9094	1.474
0.46	0.9594	0.8650	0.9016	1.425
0.48	0.9560	0.8541	0.8935	1.380
0.50	0.9524	0.8430	0.8852	1.340
0.52	0.9487	0.8317	0.8766	1.303
0.54	0.9449	0.8201	0.8679	1.270
0.56	0.9410	0.8082	0.8589	1.240
0.58	0.9370	0.7962	0.8498	1.213
0.60	0.9328	0.7840	0.8405	1.188
0.62	0.9286	0.7716	0.8310	1.166
0.64	0.9243	0.7591	0.8213	1.145
0.66	0.9199	0.7465	0.8115	1.127
0.68	0.9154	0.7338	0.8016	1.110
0.70	0.9108	0.7209	0.7916	1.094
0.72	0.9061	0.7080	0.7814	1.081
0.74	0.9013	0.6951	0.7712	1.068
0.76	0.8964	0.6821	0.7609	1.057
0.78	0.8915	0.6691	0.7505	1.047
0.80	0.8865	0.6560	0.7400	1.038
0.82	0.8815	0.6430	0.7295	1.030
0.84	0.8763	0.6300	0.7189	1.024
0.86	0.8711	0.6170	0.7083	1.018
0.88	0.8659	0.6041	0.6977	1.013
0.90	0.8606	0.5913	0.6870	1.009
0.92	0.8552	0.5785	0.6764	1.006
0.94	0.8498	0.5658	0.6658	1.003
0.96	0.8444	0.5532	0.6551	1.001
0.98	0.8389	0.5407	0.6445	1.000
1.00	0.8333	0.5283	0.6339	1.000
1.02	0.8278	0.5160	0.6234	1.000
1.04	0.8222	0.5039	0.6129	1.001
1.06	0.8165	0.4919	0.6024	1.003
1.08	0.8108	0.4801	0.5920	1.005
1.10	0.8052	0.4684	0.5817	1.008
1.12	0.7994	0.4568	0.5714	1.011
1.14	0.7937	0.4455	0.5612	1.015
1.16	0.7880	0.4343	0.5511	1.020
1.18	0.7822	0.4232	0.5411	1.025
1.20	0.7764	0.4124	0.5311	1.030
1.22	0.7706	0.4017	0.5213	1.037
1.24	0.7648	0.3912	0.5115	1.043
1.26	0.7590	0.3809	0.5019	1.050
1.28	0.7532	0.3708	0.4923	1.058
1.30	0.7474	0.3609	0.4829	1.066

M	T/T_0	p/p_0	ρ/ρ_0	A/A^*
1.30	0.7474	0.3609	0.4829	1.066
1.32	0.7416	0.3512	0.4736	1.075
1.34	0.7358	0.3417	0.4644	1.084
1.36	0.7300	0.3323	0.4553	1.094
1.38	0.7242	0.3232	0.4463	1.104
1.40	0.7184	0.3142	0.4374	1.115
1.42	0.7126	0.3055	0.4287	1.126
1.44	0.7069	0.2969	0.4201	1.138
1.46	0.7011	0.2886	0.4116	1.150
1.48	0.6954	0.2804	0.4032	1.163
1.50	0.6897	0.2724	0.3950	1.176
1.52	0.6840	0.2646	0.3869	1.190
1.54	0.6783	0.2570	0.3789	1.204
1.56	0.6726	0.2496	0.3711	1.219
1.58	0.6670	0.2423	0.3633	1.234
1.60	0.6614	0.2353	0.3557	1.250
1.62	0.6558	0.2284	0.3483	1.267
1.64	0.6502	0.2217	0.3409	1.284
1.66	0.6447	0.2152	0.3337	1.301
1.68	0.6392	0.2088	0.3266	1.319
1.70	0.6337	0.2026	0.3197	1.338
1.72	0.6283	0.1966	0.3129	1.357
1.74	0.6229	0.1907	0.3062	1.376
1.76	0.6175	0.1850	0.2996	1.397
1.78	0.6121	0.1794	0.2931	1.418
1.80	0.6068	0.1740	0.2868	1.439
1.82	0.6015	0.1688	0.2806	1.461
1.84	0.5963	0.1637	0.2745	1.484
1.86	0.5911	0.1587	0.2686	1.507
1.88	0.5859	0.1539	0.2627	1.531
1.90	0.5807	0.1492	0.2570	1.555
1.92	0.5756	0.1447	0.2514	1.580
1.94	0.5705	0.1403	0.2459	1.606
1.96	0.5655	0.1360	0.2405	1.633
1.98	0.5605	0.1318	0.2352	1.660
2.00	0.5556	0.1278	0.2301	1.688
2.02	0.5506	0.1239	0.2250	1.716
2.04	0.5458	0.1201	0.2200	1.745
2.06	0.5409	0.1164	0.2152	1.775
2.08	0.5361	0.1128	0.2105	1.806
2.10	0.5314	0.1094	0.2058	1.837
2.12	0.5266	0.1060	0.2013	1.869
2.14	0.5219	0.1027	0.1968	1.902
2.16	0.5173	0.0996	0.1925	1.935
2.18	0.5127	0.0965	0.1882	1.970
2.20	0.5081	0.0935	0.1841	2.005

M	T/T_0	p/p_0	ρ/ρ_0	A/A^*
2.20	0.5081	0.0935	0.1841	2.005
2.22	0.5036	0.0906	0.1800	2.041
2.24	0.4991	0.0878	0.1760	2.078
2.26	0.4947	0.0851	0.1721	2.115
2.28	0.4903	0.0825	0.1683	2.154
2.30	0.4859	0.0800	0.1646	2.193
2.32	0.4816	0.0775	0.1610	2.233
2.34	0.4773	0.0751	0.1574	2.274
2.36	0.4731	0.0728	0.1539	2.316
2.38	0.4689	0.0706	0.1505	2.359
2.40	0.4647	0.0684	0.1472	2.403
2.42	0.4606	0.0663	0.1440	2.448
2.44	0.4565	0.0643	0.1408	2.494
2.46	0.4524	0.0623	0.1377	2.540
2.48	0.4484	0.0604	0.1347	2.588
2.50	0.4444	0.0585	0.1317	2.637
2.52	0.4405	0.0567	0.1288	2.687
2.54	0.4366	0.0550	0.1260	2.737
2.56	0.4328	0.0533	0.1232	2.789
2.58	0.4289	0.0517	0.1205	2.842
2.60	0.4252	0.0501	0.1179	2.896
2.62	0.4214	0.0486	0.1153	2.951
2.64	0.4177	0.0471	0.1128	3.007
2.66	0.4141	0.0457	0.1103	3.065
2.68	0.4104	0.0443	0.1079	3.123
2.70	0.4068	0.0430	0.1056	3.183
2.72	0.4033	0.0417	0.1033	3.244
2.74	0.3998	0.0404	0.1010	3.306
2.76	0.3963	0.0392	0.0989	3.370
2.78	0.3928	0.0380	0.0967	3.434
2.80	0.3894	0.0369	0.0946	3.500
2.82	0.3860	0.0357	0.0926	3.567
2.84	0.3827	0.0347	0.0906	3.636
2.86	0.3794	0.0336	0.0887	3.706
2.88	0.3761	0.0326	0.0867	3.777
2.90	0.3729	0.0317	0.0849	3.850
2.92	0.3697	0.0307	0.0831	3.924
2.94	0.3665	0.0298	0.0813	3.999
2.96	0.3633	0.0289	0.0796	4.076
2.98	0.3602	0.0281	0.0779	4.155
3.00	0.3571	0.0272	0.0762	4.235

M	T/T_0	p/p_0	ρ/ρ_0	A/A^*
3.00	0.3571	0.0272	0.0762	4.235
3.05	0.3496	0.0253	0.0723	4.441
3.10	0.3422	0.0234	0.0685	4.657
3.15	0.3351	0.0218	0.0650	4.884
3.20	0.3281	0.0202	0.0617	5.121
3.25	0.3213	0.0188	0.0585	5.369
3.30	0.3147	0.0175	0.0555	5.629
3.35	0.3082	0.0163	0.0527	5.900
3.40	0.3019	0.0151	0.0501	6.184
3.45	0.2958	0.0141	0.0476	6.480
3.50	0.2899	0.0131	0.0452	6.790
3.55	0.2841	0.0122	0.0430	7.113
3.60	0.2784	0.0114	0.0409	7.450
3.65	0.2729	0.0106	0.0389	7.802
3.70	0.2675	0.0099	0.0370	8.169
3.75	0.2623	0.0092	0.0352	8.552
3.80	0.2572	0.0086	0.0335	8.951
3.85	0.2522	0.0081	0.0320	9.366
3.90	0.2474	0.0075	0.0304	9.799
3.95	0.2427	0.0070	0.0290	10.25
4.00	0.2381	0.0066	0.0277	10.72
4.05	0.2336	0.0062	0.0264	11.21
4.10	0.2293	0.0058	0.0252	11.71
4.15	0.2250	0.0054	0.0240	12.24
4.20	0.2208	0.0051	0.0229	12.79
4.25	0.2168	0.0047	0.0219	13.36
4.30	0.2129	0.0044	0.0209	13.95
4.35	0.2090	0.0042	0.0200	14.57
4.40	0.2053	0.0039	0.0191	15.21
4.45	0.2016	0.0037	0.0182	15.87
4.50	0.1980	0.0035	0.0174	16.56
4.55	0.1945	0.0032	0.0167	17.28
4.60	0.1911	0.0031	0.0160	18.02
4.65	0.1878	0.0029	0.0153	18.79
4.70	0.1846	0.0027	0.0146	19.58
4.75	0.1814	0.0025	0.0140	20.41
4.80	0.1783	0.0024	0.0134	21.26
4.85	0.1753	0.0023	0.0129	22.15
4.90	0.1724	0.0021	0.0123	23.07
4.95	0.1695	0.0020	0.0118	24.02
5.00	0.1667	0.0019	0.0113	25.00

table B.2 *Fanno Line Functions (Uniform Flow, Ideal Gas, k = 1.4)*

M	T/T*	p/p*	p_0/p_0^*	V/V*	$\bar{f}L_{max}/D_h$
0.00	1.200	∞	∞	0.0000	∞
0.02	1.200	54.77	28.94	0.0219	448
0.04	1.200	27.38	14.48	0.0438	110
0.06	1.199	18.25	9.666	0.0657	48.3
0.08	1.199	13.68	7.262	0.0876	26.7
0.10	1.198	10.94	5.822	0.109	16.7
0.12	1.197	9.116	4.864	0.131	11.4
0.14	1.195	7.809	4.182	0.153	8.13
0.16	1.194	6.829	3.673	0.175	6.05
0.18	1.192	6.066	3.278	0.197	4.64
0.20	1.191	5.456	2.964	0.218	3.63
0.22	1.189	4.955	2.708	0.240	2.90
0.24	1.186	4.538	2.496	0.261	2.35
0.26	1.184	4.185	2.317	0.283	1.92
0.28	1.182	3.882	2.166	0.304	1.59
0.30	1.179	3.619	2.035	0.326	1.32
0.32	1.176	3.389	1.922	0.347	1.11
0.34	1.173	3.185	1.823	0.368	0.938
0.36	1.170	3.004	1.736	0.389	0.795
0.38	1.166	2.842	1.659	0.410	0.676
0.40	1.163	2.696	1.590	0.431	0.575
0.42	1.159	2.563	1.529	0.452	0.493
0.44	1.155	2.443	1.474	0.473	0.423
0.46	1.151	2.333	1.425	0.494	0.383
0.48	1.147	2.231	1.380	0.514	0.311
0.50	1.143	2.138	1.340	0.535	0.267
0.52	1.138	2.052	1.303	0.555	0.229
0.54	1.134	1.972	1.270	0.575	0.197
0.56	1.129	1.898	1.240	0.595	0.168
0.58	1.124	1.828	1.213	0.615	0.144
0.60	1.119	1.763	1.188	0.635	0.123
0.62	1.114	1.703	1.166	0.654	0.104
0.64	1.109	1.646	1.145	0.674	0.0883
0.66	1.104	1.592	1.127	0.693	0.0745
0.68	1.098	1.541	1.110	0.713	0.0624
0.70	1.093	1.493	1.094	0.732	0.0520
0.72	1.087	1.448	1.081	0.751	0.0430
0.74	1.082	1.405	1.068	0.770	0.0353
0.76	1.076	1.365	1.057	0.788	0.0286
0.78	1.070	1.326	1.047	0.807	0.0229
0.80	1.064	1.289	1.038	0.825	0.0181
0.82	1.058	1.254	1.030	0.843	0.0140
0.84	1.052	1.221	1.024	0.861	0.0106
0.86	1.045	1.189	1.018	0.879	0.0077
0.88	1.039	1.158	1.013	0.897	0.0055
0.90	1.033	1.129	1.0089	0.915	0.0036

M	T/T^*	p/p^*	p_0/p_0^*	V/V^*	fL_{max}/D_h
0.90	1.033	1.129	1.0089	0.915	0.0036
0.92	1.026	1.101	1.0056	0.932	0.0022
0.94	1.020	1.074	1.0031	0.949	0.0012
0.96	1.013	1.049	1.0014	0.966	0.0005
0.98	1.007	1.024	1.0003	0.983	0.0001
1.00	1.000	1.000	1.0000	1.000	0.0000
1.02	0.993	0.977	1.0003	1.017	0.0001
1.04	0.987	0.955	1.0013	1.033	0.0004
1.06	0.980	0.934	1.0029	1.049	0.0010
1.08	0.973	0.913	1.0051	1.065	0.0016
1.10	0.966	0.894	1.0079	1.081	0.0025
1.12	0.959	0.875	1.011	1.097	0.0034
1.14	0.952	0.856	1.015	1.113	0.0046
1.16	0.946	0.838	1.020	1.128	0.0057
1.18	0.939	0.821	1.025	1.143	0.0070
1.20	0.932	0.804	1.030	1.158	0.0084
1.22	0.925	0.788	1.037	1.173	0.0099
1.24	0.918	0.773	1.043	1.188	0.0114
1.26	0.911	0.757	1.050	1.203	0.0129
1.28	0.904	0.743	1.058	1.217	0.0146
1.30	0.897	0.728	1.066	1.231	0.0162
1.32	0.890	0.715	1.075	1.245	0.0179
1.34	0.883	0.701	1.084	1.259	0.0196
1.36	0.876	0.688	1.094	1.273	0.0214
1.38	0.869	0.676	1.104	1.286	0.0231
1.40	0.862	0.663	1.115	1.300	0.0249
1.42	0.855	0.651	1.126	1.313	0.0267
1.44	0.848	0.640	1.138	1.326	0.0285
1.46	0.841	0.628	1.150	1.339	0.0304
1.48	0.834	0.617	1.163	1.352	0.0322
1.50	0.828	0.606	1.176	1.365	0.0340
1.52	0.821	0.596	1.190	1.377	0.0358
1.54	0.814	0.586	1.204	1.389	0.0377
1.56	0.807	0.576	1.219	1.402	0.0395
1.58	0.800	0.566	1.234	1.414	0.0413
1.60	0.794	0.557	1.250	1.425	0.0431
1.62	0.787	0.548	1.267	1.437	0.0450
1.64	0.780	0.539	1.284	1.449	0.0467
1.66	0.774	0.530	1.301	1.460	0.0484
1.68	0.767	0.521	1.319	1.471	0.0502
1.70	0.760	0.513	1.338	1.482	0.0520
1.72	0.754	0.505	1.357	1.494	0.0537
1.74	0.747	0.497	1.376	1.504	0.0554
1.76	0.741	0.489	1.397	1.515	0.0571
1.78	0.735	0.481	1.418	1.526	0.0588
1.80	0.728	0.474	1.439	1.536	0.0605

M	T/T^*	p/p^*	p_0/p_0^*	V/V^*	fL_{max}/D_h
1.80	0.728	0.474	1.439	1.536	0.0605
1.82	0.722	0.467	1.461	1.546	0.0621
1.84	0.716	0.460	1.484	1.556	0.0638
1.86	0.709	0.453	1.507	1.566	0.0654
1.88	0.703	0.446	1.531	1.576	0.0670
1.90	0.697	0.439	1.555	1.586	0.0686
1.92	0.691	0.433	1.580	1.596	0.0702
1.94	0.685	0.427	1.606	1.605	0.0717
1.96	0.679	0.420	1.633	1.615	0.0732
1.98	0.673	0.414	1.660	1.624	0.0748
2.00	0.667	0.408	1.688	1.633	0.0763
2.02	0.661	0.402	1.716	1.642	0.0777
2.04	0.655	0.397	1.745	1.651	0.0792
2.06	0.649	0.391	1.775	1.660	0.0806
2.08	0.643	0.386	1.806	1.668	0.0821
2.10	0.638	0.380	1.837	1.677	0.0835
2.12	0.632	0.375	1.869	1.685	0.0849
2.14	0.626	0.370	1.902	1.694	0.0862
2.16	0.621	0.365	1.935	1.702	0.0876
2.18	0.615	0.360	1.970	1.710	0.0889
2.20	0.610	0.355	2.005	1.718	0.0902
2.22	0.604	0.350	2.041	1.726	0.0915
2.24	0.599	0.346	2.078	1.734	0.0928
2.26	0.594	0.341	2.115	1.741	0.0941
2.28	0.588	0.336	2.154	1.749	0.0953
2.30	0.583	0.332	2.193	1.756	0.0966
2.32	0.578	0.328	2.233	1.764	0.0978
2.34	0.573	0.323	2.274	1.771	0.0990
2.36	0.568	0.319	2.316	1.778	0.100
2.38	0.563	0.315	2.359	1.785	0.101
2.40	0.558	0.311	2.403	1.792	0.102
2.42	0.553	0.307	2.448	1.799	0.104
2.44	0.548	0.303	2.494	1.806	0.105
2.46	0.543	0.300	2.540	1.813	0.106
2.48	0.538	0.296	2.588	1.819	0.107
2.50	0.533	0.292	2.637	1.826	0.108

M	T/T^*	p/p^*	p_0/p_0^*	V/V^*	fL_{max}/D_h
2.50	0.533	0.292	2.637	1.826	0.108
2.52	0.529	0.289	2.687	1.832	0.109
2.54	0.524	0.285	2.737	1.839	0.110
2.56	0.519	0.282	2.789	1.845	0.111
2.58	0.515	0.278	2.842	1.851	0.112
2.60	0.510	0.275	2.896	1.857	0.113
2.62	0.506	0.271	2.951	1.863	0.114
2.64	0.501	0.268	3.007	1.869	0.115
2.66	0.497	0.265	3.065	1.875	0.116
2.68	0.493	0.262	3.123	1.881	0.117
2.70	0.488	0.259	3.183	1.887	0.118
2.72	0.484	0.256	3.244	1.892	0.119
2.74	0.480	0.253	3.306	1.898	0.120
2.76	0.476	0.250	3.370	1.903	0.121
2.78	0.471	0.247	3.434	1.909	0.122
2.80	0.467	0.244	3.500	1.914	0.122
2.82	0.463	0.241	3.567	1.919	0.123
2.84	0.459	0.239	3.636	1.925	0.124
2.86	0.455	0.236	3.706	1.930	0.125
2.88	0.451	0.233	3.777	1.935	0.126
2.90	0.447	0.231	3.850	1.940	0.127
2.92	0.444	0.228	3.924	1.945	0.127
2.94	0.440	0.226	3.999	1.950	0.128
2.96	0.436	0.223	4.076	1.954	0.129
2.98	0.432	0.221	4.155	1.959	0.130
3.00	0.429	0.218	4.235	1.964	0.131
3.50	0.348	0.169	6.79	2.064	0.147
4.00	0.286	0.134	10.72	2.138	0.158
4.50	0.238	0.108	16.56	2.194	0.167
5.00	0.200	0.0894	25.00	2.236	0.173

table B.3   *Rayleigh Line Functions (Uniform Flow, Ideal Gas, k = 1.4)*

M	T_0/T_0^*	T/T^*	p/p^*	p_0/p_0^*	V/V^*
0.00	0.0000	0.0000	2.400	1.268	0.0000
0.02	0.0019	0.0023	2.399	1.268	0.0010
0.04	0.0077	0.0092	2.395	1.267	0.0038
0.06	0.0171	0.0205	2.388	1.265	0.0086
0.08	0.0302	0.0362	2.379	1.262	0.0152
0.10	0.0468	0.0560	2.367	1.259	0.0237
0.12	0.0666	0.0797	2.353	1.255	0.0339
0.14	0.0895	0.107	2.336	1.251	0.0458
0.16	0.115	0.137	2.317	1.246	0.0593
0.18	0.143	0.171	2.296	1.241	0.0744
0.20	0.174	0.207	2.273	1.235	0.0909
0.22	0.206	0.245	2.248	1.228	0.1088
0.24	0.239	0.284	2.221	1.221	0.1279
0.26	0.274	0.325	2.193	1.214	0.1482
0.28	0.310	0.367	2.163	1.206	0.1696
0.30	0.347	0.409	2.131	1.199	0.1918
0.32	0.384	0.451	2.099	1.190	0.2149
0.34	0.421	0.493	2.066	1.182	0.2388
0.36	0.457	0.535	2.031	1.174	0.2633
0.38	0.493	0.576	1.996	1.165	0.2883
0.40	0.529	0.615	1.961	1.157	0.3137
0.42	0.564	0.653	1.925	1.148	0.3395
0.44	0.597	0.690	1.888	1.139	0.3656
0.46	0.630	0.725	1.852	1.131	0.3918
0.48	0.661	0.759	1.815	1.122	0.4181
0.50	0.691	0.790	1.778	1.114	0.4445
0.52	0.720	0.820	1.741	1.106	0.4708
0.54	0.747	0.847	1.704	1.098	0.4970
0.56	0.772	0.872	1.668	1.090	0.5230
0.58	0.796	0.896	1.632	1.083	0.5489
0.60	0.819	0.917	1.596	1.075	0.5745
0.62	0.840	0.936	1.560	1.068	0.5998
0.64	0.859	0.953	1.525	1.061	0.6248
0.66	0.877	0.968	1.491	1.055	0.6494
0.68	0.894	0.981	1.457	1.049	0.6737
0.70	0.909	0.993	1.424	1.043	0.6975
0.72	0.922	1.003	1.391	1.038	0.7209
0.74	0.934	1.011	1.359	1.033	0.7439
0.76	0.945	1.017	1.327	1.028	0.7665
0.78	0.955	1.022	1.296	1.023	0.7885
0.80	0.964	1.025	1.266	1.019	0.8101
0.82	0.972	1.028	1.236	1.016	0.8313
0.84	0.978	1.029	1.207	1.012	0.8519
0.86	0.984	1.028	1.179	1.010	0.8721
0.88	0.988	1.027	1.152	1.007	0.8918
0.90	0.992	1.025	1.125	1.005	0.9110

M	T_0/T_0^*	T/T^*	p/p^*	p_0/p_0^*	V/V^*
0.90	0.992	1.025	1.125	1.005	0.9110
0.92	0.995	1.021	1.098	1.003	0.9297
0.94	0.997	1.017	1.073	1.002	0.9480
0.96	0.999	1.012	1.048	1.001	0.9658
0.98	1.000	1.006	1.024	1.000	0.9831
1.00	1.000	1.000	1.000	1.000	1.0000
1.02	1.000	0.993	0.977	1.000	1.016
1.04	0.999	0.986	0.955	1.001	1.032
1.06	0.998	0.978	0.933	1.002	1.048
1.08	0.996	0.969	0.912	1.003	1.063
1.10	0.994	0.960	0.891	1.005	1.078
1.12	0.991	0.951	0.871	1.007	1.092
1.14	0.989	0.942	0.851	1.010	1.106
1.16	0.986	0.932	0.832	1.012	1.120
1.18	0.982	0.922	0.814	1.016	1.133
1.20	0.979	0.912	0.796	1.019	1.146
1.22	0.975	0.902	0.778	1.023	1.158
1.24	0.971	0.891	0.761	1.028	1.171
1.26	0.967	0.881	0.745	1.033	1.182
1.28	0.962	0.870	0.729	1.038	1.194
1.30	0.958	0.859	0.713	1.044	1.205
1.32	0.953	0.848	0.698	1.050	1.216
1.34	0.949	0.838	0.683	1.056	1.226
1.36	0.944	0.827	0.669	1.063	1.237
1.38	0.939	0.816	0.655	1.070	1.247
1.40	0.934	0.805	0.641	1.078	1.256
1.42	0.929	0.795	0.628	1.086	1.266
1.44	0.924	0.784	0.615	1.094	1.275
1.46	0.919	0.773	0.602	1.103	1.284
1.48	0.914	0.763	0.590	1.112	1.293
1.50	0.909	0.753	0.578	1.122	1.301
1.52	0.904	0.742	0.567	1.132	1.310
1.54	0.899	0.732	0.556	1.142	1.318
1.56	0.894	0.722	0.545	1.153	1.325
1.58	0.889	0.712	0.534	1.164	1.333
1.60	0.884	0.702	0.524	1.176	1.340
1.62	0.879	0.692	0.513	1.188	1.348
1.64	0.874	0.682	0.504	1.200	1.355
1.66	0.869	0.673	0.494	1.213	1.361
1.68	0.865	0.663	0.485	1.226	1.368
1.70	0.860	0.654	0.476	1.240	1.375
1.72	0.855	0.645	0.467	1.255	1.381
1.74	0.850	0.635	0.458	1.269	1.387
1.76	0.846	0.626	0.450	1.284	1.393
1.78	0.841	0.618	0.442	1.300	1.399
1.80	0.836	0.609	0.434	1.316	1.405

M	T_0/T_0^*	T/T^*	p/p^*	p_0/p_0^*	V/V^*
1.80	0.836	0.609	0.434	1.316	1.405
1.82	0.832	0.600	0.426	1.332	1.410
1.84	0.827	0.592	0.418	1.349	1.416
1.86	0.823	0.584	0.411	1.367	1.421
1.88	0.818	0.575	0.403	1.385	1.426
1.90	0.814	0.567	0.396	1.403	1.431
1.92	0.810	0.559	0.390	1.422	1.436
1.94	0.806	0.552	0.383	1.442	1.441
1.96	0.802	0.544	0.376	1.462	1.446
1.98	0.797	0.536	0.370	1.482	1.450
2.00	0.793	0.529	0.364	1.503	1.455
2.02	0.789	0.522	0.358	1.525	1.459
2.04	0.785	0.514	0.352	1.547	1.463
2.06	0.782	0.507	0.346	1.569	1.467
2.08	0.778	0.500	0.340	1.592	1.471
2.10	0.774	0.494	0.335	1.616	1.475
2.12	0.770	0.487	0.329	1.640	1.479
2.14	0.767	0.480	0.324	1.665	1.483
2.16	0.763	0.474	0.319	1.691	1.487
2.18	0.760	0.467	0.314	1.717	1.490
2.20	0.756	0.461	0.309	1.743	1.494
2.22	0.753	0.455	0.304	1.771	1.497
2.24	0.749	0.449	0.299	1.799	1.501
2.26	0.746	0.443	0.294	1.827	1.504
2.28	0.743	0.437	0.290	1.856	1.507
2.30	0.740	0.431	0.286	1.886	1.510
2.32	0.736	0.426	0.281	1.917	1.513
2.34	0.733	0.420	0.277	1.948	1.517
2.36	0.730	0.415	0.273	1.979	1.520
2.38	0.727	0.409	0.269	2.012	1.522
2.40	0.724	0.404	0.265	2.045	1.525
2.42	0.721	0.399	0.261	2.079	1.528
2.44	0.718	0.394	0.257	2.114	1.531
2.46	0.716	0.388	0.253	2.149	1.533
2.48	0.713	0.384	0.250	2.185	1.536
2.50	0.710	0.379	0.246	2.222	1.539

M	T_0/T_0^*	T/T^*	p/p^*	p_0/p_0^*	V/V^*
2.50	0.710	0.379	0.246	2.222	1.539
2.52	0.707	0.374	0.243	2.259	1.541
2.54	0.705	0.369	0.239	2.298	1.543
2.56	0.702	0.365	0.236	2.337	1.546
2.58	0.700	0.360	0.233	2.377	1.548
2.60	0.697	0.356	0.229	2.418	1.551
2.62	0.695	0.351	0.226	2.459	1.553
2.64	0.692	0.347	0.223	2.502	1.555
2.66	0.690	0.343	0.220	2.545	1.557
2.68	0.687	0.338	0.217	2.589	1.559
2.70	0.685	0.334	0.214	2.634	1.561
2.72	0.683	0.330	0.211	2.680	1.563
2.74	0.680	0.326	0.209	2.727	1'565
2.76	0.678	0.322	0.206	2.775	1.567
2.78	0.676	0.319	0.203	2.824	1.569
2.80	0.674	0.315	0.200	2.873	1.571
2.82	0.672	0.311	0.198	2.924	1.573
2.84	0.670	0.307	0.195	2.975	1.575
2.86	0.668	0.304	0.193	3.028	1.577
2.88	0.665	0.300	0.190	3.081	1.578
2.90	0.664	0.297	0.188	3.136	1.580
2.92	0.662	0.293	0.186	3.191	1.582
2.94	0.660	0.290	0.183	3.248	1.583
2.96	0.658	0.287	0.181	3.306	1.585
2.98	0.656	0.283	0.179	3.365	1.587
3.00	0.654	0.280	0.176	3.424	1.588
3.50	0.616	0.214	0.132	5.328	1.620
4.00	0.589	0.168	0.103	8.227	1.641
4.50	0.570	0.135	0.082	12.50	1.656
5.00	0.556	0.111	0.067	18.63	1.667

Normal Shock Functions (Uniform Flow, Ideal Gas, k = 1.4)

M_1	M_2	p_2/p_1	T_2/T_1	ρ_2/ρ_1	p_{0_2}/p_{0_1}
1.00	1.000	1.000	1.000	1.000	1.000
1.02	0.981	1.047	1.013	1.033	1.000
1.04	0.962	1.095	1.026	1.067	1.000
1.06	0.944	1.144	1.039	1.101	1.000
1.08	0.928	1.194	1.052	1.135	0.999
1.10	0.912	1.245	1.065	1.169	0.999
1.12	0.897	1.297	1.078	1.203	0.998
1.14	0.882	1.350	1.090	1.238	0.997
1.16	0.868	1.403	1.103	1.272	0.996
1.18	0.855	1.458	1.115	1.307	0.995
1.20	0.842	1.513	1.128	1.342	0.993
1.22	0.830	1.570	1.141	1.376	0.991
1.24	0.818	1.627	1.153	1.411	0.988
1.26	0.807	1.686	1.166	1.446	0.986
1.28	0.796	1.745	1.178	1.481	0.983
1.30	0.786	1.805	1.191	1.516	0.979
1.32	0.776	1.866	1.204	1.551	0.976
1.34	0.766	1.928	1.216	1.585	0.972
1.36	0.757	1.991	1.229	1.620	0.968
1.38	0.748	2.055	1.242	1.655	0.963
1.40	0.740	2.120	1.255	1.690	0.958
1.42	0.731	2.186	1.268	1.724	0.953
1.44	0.723	2.253	1.281	1.759	0.948
1.46	0.716	2.320	1.294	1.793	0.942
1.48	0.708	2.389	1.307	1.828	0.936
1.50	0.701	2.458	1.320	1.862	0.930
1.52	0.694	2.529	1.334	1.896	0.923
1.54	0.687	2.600	1.347	1.930	0.917
1.56	0.681	2.673	1.361	1.964	0.910
1.58	0.675	2.746	1.374	1.998	0.903
1.60	0.668	2.820	1.388	2.032	0.895
1.62	0.663	2.895	1.402	2.065	0.888
1.64	0.657	2.971	1.416	2.099	0.880
1.66	0.651	3.048	1.430	2.132	0.872
1.68	0.646	3.126	1.444	2.165	0.864
1.70	0.641	3.205	1.458	2.198	0.856
1.72	0.635	3.285	1.473	2.230	0.847
1.74	0.631	3.366	1.487	2.263	0.839
1.76	0.626	3.447	1.502	2.295	0.830
1.78	0.621	3.530	1.517	2.327	0.822
1.80	0.617	3.613	1.532	2.359	0.813
1.82	0.612	3.698	1.547	2.391	0.804
1.84	0.608	3.783	1.562	2.422	0.795
1.86	0.604	3.870	1.577	2.454	0.786
1.88	0.600	3.957	1.592	2.485	0.777
1.90	0.596	4.045	1.608	2.516	0.767

M_1	M_2	p_2/p_1	T_2/T_1	ρ_2/ρ_1	p_{0_2}/p_{0_1}
1.90	0.596	4.045	1.608	2.516	0.767
1.92	0.592	4.134	1.624	2.546	0.758
1.94	0.588	4.224	1.639	2.577	0.749
1.96	0.584	4.315	1.655	2.607	0.740
1.98	0.581	4.407	1.671	2.637	0.730
2.00	0.577	4.500	1.687	2.667	0.721
2.02	0.574	4.594	1.704	2.696	0.712
2.04	0.571	4.689	1.720	2.725	0.702
2.06	0.567	4.784	1.737	2.755	0.693
2.08	0.564	4.881	1.754	2.783	0.684
2.10	0.561	4.978	1.770	2.812	0.674
2.12	0.558	5.077	1.787	2.840	0.665
2.14	0.555	5.176	1.805	2.868	0.656
2.16	0.553	5.277	1.822	2.896	0.646
2.18	C.550	5.378	1.839	2.924	0.637
2.20	0.547	5.480	1.857	2.951	0.628
2.22	0.544	5.583	1.875	2.978	0.619
2.24	0.542	5.687	1.892	3.005	0.610
2.26	0.539	5.792	1.910	3.032	0.601
2.28	0.537	5.898	1.929	3.058	0.592
2.30	0.534	6.005	1.947	3.085	0.583
2.32	0.532	6.113	1.965	3.110	0.575
2.34	0.530	6.222	1.984	3.136	0.566
2.36	0.527	6.331	2.002	3.162	0.557
2.38	0.525	6.442	2.021	3.187	0.549
2.40	0.523	6.553	2.040	3.212	0.540
2.42	0.521	6.666	2.059	3.237	0.532
2.44	0.519	6.779	2.079	3.261	0.523
2.46	0.517	6.894	2.098	3.285·	0.515
2.48	0.515	7.009	2.118	3.310	0.507
2.50	0.513	7.125	2.137	3.333	0.499
2.52	0.511	7.242	2.157	3.357	0.491
2.54	0.509	7.360	2.177	3.380	0.483
2.56	0.507	7.479	2.198	3.403	0.475
2.58	0.506	7.599	2.218	3.426	0.468
2.60	0.504	7.720	2.238	3.449	0.460
2.62	0.502	7.842	2.259	3.471	0.453
2.64	0.500	7.965	2.280	3.494	0.445
2.66	0.499	8.088	2.301	3.516	0.438
2.68	0.497	8.213	2.322	3.537	0.431
2.70	0.496	8.338	2.343	3.559	0.424
2.72	0.494	8.465	2.364	3.580	0.417
2.74	0.493	8.592	2.386	3.601	0.410
2.76	0.491	8.721	2.407	3.622	0.403
2.78	0.490	8.850	2.429	3.643	0.396
2.80	0.488	8.980	2.451	3.664	0.389

M_1	M_2	p_2/p_1	T_2/T_1	ρ_2/ρ_1	p_{0_2}/p_{0_1}
2.80	0.488	8.980	2.451	3.664	0.389
2.82	0.487	9.111	2.473	3.684	0.383
2.84	0.485	9.243	2.496	3.704	0.376
2.86	0.484	9.376	2.518	3.724	0.370
2.88	0.483	9.510	2.540	3.743	0.364
2.90	0.481	9.645	2.563	3.763	0.358
2.92	0.480	9.781	2.586	3.782	0.352
2.94	0.479	9.918	2.609	3.801	0.346
2.96	0.478	10.06	2.632	3.820	0.340
2.98	0.476	10.19	2.656	3.839	0.334
3.00	0.475	10.33	2.679	3.857	0.328
3.05	0.472	10.69	2.738	3.902	0.315
3.10	0.470	11.05	2.799	3.947	0.301
3.15	0.467	11.41	2.860	3.990	0.288
3.20	0.464	11.78	2.922	4.031	0.276
3.25	0.462	12.16	2.985	4.072	0.265
3.30	0.460	12.54	3.049	4.112	0.253
3.35	0.457	12.93	3.114	4.151	0.243
3.40	0.455	13.32	3.180	4.188	0.232
3.45	0.453	13.72	3.247	4.225	0.222
3.50	0.451	14.13	3.315	4.261	0.213
3.55	0.449	14.54	3.384	4.296	0.204
3.60	0.447	14.95	3.454	4.330	0.195
3.65	0.446	15.38	3.525	4.363	0.187
3.70	0.444	15.81	3.596	4.395	0.179
3.75	0.442	16.24	3.669	4.426	0.172
3.80	0.441	16.68	3.743	4.457	0.164
3.85	0.439	17.13	3.817	4.487	0.158
3.90	0.438	17.58	3.893	4.516	0.151
3.95	0.436	18.04	3.969	4.544	0.145
4.00	0.435	18.50	4.047	4.571	0.139
4.05	0.434	18.97	4.125	4.598	0.133
4.10	0.432	19.45	4.205	4.624	0.128
4.15	0.431	19.93	4.285	4.650	0.122
3.20	0.430	20.41	4.367	4.675	0.117
4.25	0.429	20.91	4.449	4.699	0.113
4.30	0.428	21.41	4.532	4.723	0.108
4.35	0.427	21.91	4.616	4.746	0.104
4.40	0.426	22.42	4.702	4.768	0.099
4.45	0.425	22.94	4.788	4.790	0.096
4.50	0.424	23.46	4.875	4.812	0.092
4.55	0.423	23.99	4.963	4.833	0.088
4.60	0.422	24.52	5.052	4.853	0.085
4.65	0.421	25.06	5.142	4.873	0.081
4.70	0.420	25.61	5.233	4.893	0.078
4.75	0.419	26.16	5.325	4.912	0.075
4.80	0.418	26.71	5.418	4.930	0.072
4.85	0.417	27.28	5.512	4.948	0.069
4.90	0.417	27.85	5.607	4.966	0.067
4.95	0.416	28.42	5.703	4.983	0.064
5.00	0.415	29.00	5.800	5.000	0.062

films and film loops for fluid mechanics

Listed below by supplier are titles of 16 mm sound films and Super 8 silent film "loops" (silent films enclosed in plastic cartridges designed for use in the Technicolor Instant Movie Projector).

1. Encyclopaedia Britannica Educational Corporation
   425 North Michigan Avenue
   Chicago, Illinois 60611
   (Films available for rental or purchase)

a. 16 mm sound films (length as noted)
   Aerodynamic Generation of Sound
       (44 min, principals: J. L. Lighthill, J. E. Ffowcs Williams)
   Boundary Layer Control (25 min, principal: D. C. Hazen)
   Cavitation (31 min, principal: P. Eisenberg)
   Channel Flow of a Compressible Fluid (29 min, principal: D. E. Coles)
   Deformation of Continuous Media (38 min, principal: J. L. Lumley)
   Eulerian and Lagrangian Descriptions in Fluid Mechanics
       (27 min, principal: J. L. Lumley)
   Flow Instabilities (27 min, principal: E. L. Mollo-Christensen)
   Flow Visualization (31 min, principal: S. J. Kline)

The Fluid Mechanics of Drag (4 parts, 118 min, principal: A. H. Shapiro)
Fundamentals of Boundary Layers (24 min, principal: F. H. Abernathy)
Low-Reynolds-Number Flows (33 min, principal: Sir G. I. Taylor)
Magnetohydrodynamics (27 min, principal: J. A. Shercliff)
Pressure Fields and Fluid Accelerations (30 min, principal: A. H. Shapiro)
Rarified Gas Dynamics (33 min, principals: F. C. Hurlbut, F. S. Sherman)
Rheological Behavior of Fluids (22 min, principal: H. Markovitz)
Rotating Flows (29 min, principal: D. Fultz)
Secondary Flow (30 min, principal: E. S. Taylor)
Stratified Flow (26 min, principal: R. R. Long)
Surface Tension in Fluid Mechanics (29 min, principal: L. M. Trefethen)
Turbulence (29 min, principal: R. W. Stewart)
Vorticity (2 parts, 44 min, principal: A. H. Shapiro)
Waves in Fluids (33 min, principal: A. E. Bryson)

b. Super 8 film loops (all are 3–5 min in length)

S-FM001	*Some Regimes of Boundary Layer Transition*
S-FM002	*Structure of the Turbulent Boundary Layer*
S-FM003	*Shear Deformation of Viscous Fluids—Dye Marking in Glycerine*
S-FM004	*Separated Flows, Part I—$TiCl_4$ Visualization in Air*
S-FM005	*Separated Flows, Part II*
S-FM006	*Boundary Layer Formation*
S-FM008	*The Occurrence of Turbulence*
S-FM010	*Generation of Circulation and Lift for an Airfoil*
S-FM011	*The Magnus Effect*
S-FM012	*Flow Separation and Vortex Shedding*
S-FM013	*The Bathtub Vortex*
S-FM014A	*Visualization of Vorticity with Vorticity Meter, Part I*
S-FM014B	*Visualization of Vorticity with Vorticity Meter, Part II*
S-FM015	*Incompressible Flow Through Area Contractions and Expansions*
S-FM016	*Flow from a Reservoir to a Duct*
S-FM017	*Flow Patterns in Venturis, Nozzles and Orifices*
S-FM018	*Secondary Flow in a Teacup*
S-FM019	*Secondary Flow in a Bend*
S-FM020	*The Horseshoe Vortex*
S-FM021	*Techniques of Visualization for Low Speed Flows, Part I*
S-FM022	*Techniques of Visualization for Low Speed Flows, Part II*
S-FM024	*Wing Tip Vortex*
S-FM026	*Tornadoes in Nature and the Laboratory*
S-FM028	*Transonic Flow Past a Symmetrical Airfoil*
S-FM029	*Occurrence of Supersonic Zones on Airfoils in Subsonic Flow*
S-FM031	*Instabilities in Circular Couette Flow*

609

S-FM116	*Swimming Propulsion*
S-FM117	*Subsonic Flow Patterns and Pressure Distributions for an Airfoil*
S-FM118	*Laminar-Flow Versus Conventional Airfoils*
S-FM119	*Reduction of Airfoil Friction Drag by Suction*
S-FM120	*Some Methods for Increasing Lift Coefficient*
S-FM122	*Non-Linear Shear Stress Behavior in Steady Flows*
S-FM123	*Normal Stress Effects in Viscoelastic Fluids*
S-FM124	*Memory Effects in Viscoelastic Fluids*
S-FM125	*Examples of Cavitation*
S-FM126	*Cavitation of Hydrofoils*
S-FM127	*Cavitation Bubble Dynamics*
S-FM128	*Cavity Flows*
S-FM129	*Compressible Flow Through Convergent-Divergent Nozzle*
S-FM130	*Starting of Supersonic Wind Tunnel*
S-FM134	*Laminar and Turbulent Pipe Flow*
S-FM135	*Averages and Transport in Turbulence*
S-FM139	*Small-Amplitude Gravity Waves in an Open Channel*
S-FM140	*Flattening and Steepening of Large Amplitude Gravity Waves*
S-FM141	*The Hydraulic Surge Wave*
S-FM142	*The Hydraulic Jump*
S-FM143	*Free-Surface Flow Over a Towed Obstacle*
S-FM144	*Flow of a Two-Liquid Stratified Fluid Past an Obstacle*
S-FM146	*Examples of Flow Instability, Part I*
S-FM147	*Examples of Flow Instability, Part II*
S-FM148	*Experimental Study of a Flow Instability*

2. University of Iowa
   Audiovisual Center Media Library
   Iowa City, Iowa 52240
   (Films available for rental or purchase)
   Characteristics of Laminar and Turbulent Flow (26 min, principal: H. Rouse)
   Effects of Fluid Compressibility (17 min, principal: H. Rouse)
   Fluid Motion in a Gravitational Field (23 min, principal: H. Rouse)
   Form Drag, Lift and Propulsion (24 min, principal: H. Rouse)
   Fundamental Principles of Flow (23 min, principal: H. Rouse)
   Introduction to the Study of Fluid Motion (24 min, principal: H. Rouse)

3. Ohio State University
   Film Distribution Supervisor
   Motion Picture Division
   1885 Neil Avenue
   Columbus, Ohio 44223

(The following film contains spectacular original footage from this disaster, which occurred in 1940.)
Tacoma Narrows Bridge Collapse

4. Shell Film Library   or   430 Peninsular Avenue
 450 N. Meridian Street   San Mateo, California 94401
 Indianapolis, Indiana 46205   149-07 Northern Blvd.
                                          Flushing, New York 11354

(Films available for free loan; borrower pays only return postage.)

Approaching the Speed of Sound. This film includes a good discussion of the critical Mach number for high-subsonic speed flight, including a description of the effects of profile thickness and wing sweepback. Shock formation at the critical Mach number is shown with color Schlieren visualization. The Mach cone is introduced with an animated sequence which illustrates it very clearly.

Beyond the Speed of Sound. The film includes a good discussion of drag in both transonic and supersonic flow.

Schlieren. The principles of optical visualization techniques for compressible flows are discussed. The color Schlieren is used to analyze and discuss flow fields about a variety of shapes. This is an excellent introductory film.

Transonic Flight. This film deals specifically with transonic flight, where subsonic, critical, and supersonic flow are present simultaneously. Early problems with transonic flight are reviewed, and some popular misconceptions, for example, the "sound barrier" are dispelled. The "area rule" is explained.

review of vector concepts and operations

This appendix is included to serve as a quick review of vector concepts and manipulations. Applications are illustrated by giving derivations or proofs of some results used in the text.

D–1 Symbols

Scalars are represented by a, b, and so on. Vector quantities are denoted by $\vec{a}$, $\vec{b}$, and so on. The unit vectors in the x, y, and z coordinate directions are given by $\hat{\imath}, \hat{\jmath}$, and $\hat{k}$, respectively. A vector is written in terms of its components as

$$\vec{a} = a_x\hat{\imath} + a_y\hat{\jmath} + a_z\hat{k} \tag{D.1a}$$

or

$$\vec{b} = b_x\hat{\imath} + b_y\hat{\jmath} + b_z\hat{k} \tag{D.1b}$$

In cylindrical coordinates, the unit vectors in the r, θ, and z coordinate directions are denoted by $\hat{\imath}_r, \hat{\imath}_\theta$, and $\hat{\imath}_z$, respectively.

A unit vector in the direction of $\vec{a}$ may be obtained by writing

$$\hat{\imath}_a = \frac{\vec{a}}{|\vec{a}|}$$

where $|\vec{a}|$ is the magnitude of $\vec{a}$. From Fig. D.1, we can obtain an expression for $|\vec{a}|$ in terms of components, a_x, a_y, and a_z. From trigonometry,

$$c^2 = (a_x)^2 + (a_z)^2$$

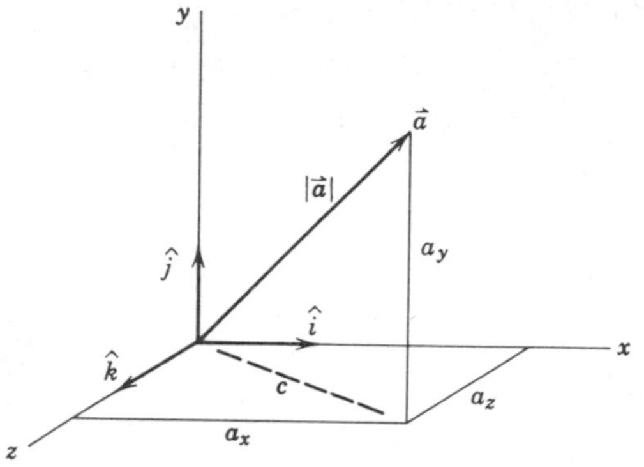

Fig. D.1 *Magnitude of Vector $\vec{a}$.*

and

$$[|\vec{a}|]^2 = c^2 + (a_y)^2$$

or

$$|\vec{a}| = \sqrt{(a_x)^2 + (a_y)^2 + (a_z)^2} \qquad \text{(D.2)}$$

Vectors are added by accounting for both magnitude and direction, using the parallelogram rule, Fig. D.2. Thus

$$\vec{c} = \vec{a} + \vec{b}$$

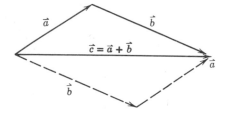

Fig. D.2 *Addition of vectors by parallelogram rule.*

and

$$\vec{c} = c_x\hat{\imath} + c_y\hat{\jmath} + c_z\hat{k}$$
$$= a_x\hat{\imath} + a_y\hat{\jmath} + a_z\hat{k} + b_x\hat{\imath} + b_y\hat{\jmath} + b_z\hat{k}$$
$$\vec{c} = (a_x + b_x)\hat{\imath} + (a_y + b_y)\hat{\jmath} + (a_z + b_z)\hat{k}$$

Hence

$$c_x = a_x + b_x \qquad c_y = a_y + b_y \quad \text{and} \quad c_z = a_z + b_z \qquad \text{(D.3)}$$

If a vector, $\vec{d}$, equals zero, then $d_x = d_y = d_z = 0$.

D–2 Products of Vectors

D–2.1 THE DOT PRODUCT (OR SCALAR PRODUCT)

The dot product of two vectors, $\vec{a}$ and $\vec{b}$, is written $\vec{a} \cdot \vec{b}$, and defined as

$$\vec{a} \cdot \vec{b} = |\vec{a}|\,|\vec{b}| \cos \theta_{ab} \qquad \text{(D.4)}$$

where θ_{ab} is the angle between the vectors. Physically, it represents the product of $|\vec{a}|$ with the component of $|\vec{b}|$ in the direction of $\vec{a}$. The dot product is a scalar; if $\theta_{ab} < \pi/2$, its magnitude is positive, whereas for $\theta_{ab} > \pi/2$, it is negative. If $\theta_{ab} = \pi/2$, $\vec{a} \cdot \vec{b} = 0$.

The dot products of unit vectors in the xyz coordinate system are:

$$\hat{\imath} \cdot \hat{\imath} = 1 \qquad \hat{\jmath} \cdot \hat{\imath} = 0 \qquad \hat{k} \cdot \hat{\imath} = 0$$

$$\hat{\imath} \cdot \hat{\jmath} = 0 \qquad \hat{\jmath} \cdot \hat{\jmath} = 1 \qquad \hat{k} \cdot \hat{\jmath} = 0$$

$$\hat{\imath} \cdot \hat{k} = 0 \qquad \hat{\jmath} \cdot \hat{k} = 0 \qquad \hat{k} \cdot \hat{k} = 1$$

Then

$$\vec{a} \cdot \vec{b} = (a_x\hat{\imath} + a_y\hat{\jmath} + a_z\hat{k}) \cdot (b_x\hat{\imath} + b_y\hat{\jmath} + b_z\hat{k})$$

$$\vec{a} \cdot \vec{b} = a_x b_x + a_y b_y + a_z b_z \qquad \text{(D.5)}$$

Using Eq. D.5, we can show:

(a) The dot product is *commutative*, that is, $\vec{a} \cdot \vec{b} = \vec{b} \cdot \vec{a}$. This follows from Eq. D.5 because $a_x b_x = b_x a_x$, and so on.

(b) The dot product is *distributive*, that is, $\vec{a} \cdot (\vec{b} + \vec{c}) = \vec{a} \cdot \vec{b} + \vec{a} \cdot \vec{c}$. This follows from Eqs. D.3 and D.5, since

$$\vec{a} \cdot (\vec{b} + \vec{c}) = (a_x\hat{\imath} + \cdots) \cdot [(b_x + c_x)\hat{\imath} + \cdots]$$

$$= a_x(b_x + c_x) + \cdots$$

$$\vec{a} \cdot (\vec{b} + \vec{c}) = a_x b_x + a_x c_x + \cdots = \vec{a} \cdot \vec{b} + \vec{a} \cdot \vec{c}$$

(c) The dot product is *not associative*, that is, $\vec{a}(\vec{b} \cdot \vec{c}) \neq (\vec{a} \cdot \vec{b})\vec{c}$. This follows, because $\vec{a}(\vec{b} \cdot \vec{c}) = \alpha\vec{a}$, since $\vec{b} \cdot \vec{c}$ is a scalar. Likewise $(\vec{a} \cdot \vec{b})\vec{c} = \beta\vec{c}$. Since vectors have both magnitude and direction, $\alpha\vec{a} \neq \beta\vec{c}$.

Applications of the dot product may be illustrated using work or flowrate as examples. The work done by a force, $\vec{F}$, moving through a distance, $d\vec{s}$, is a scalar, equal to the component of the force in the direction of $d\vec{s}$ times the distance moved. Thus

$$\delta W = |\vec{F}| \cos \theta_{Fs}| \, d\vec{s}| = \vec{F} \cdot d\vec{s}$$

The volumetric flowrate is given by the component of velocity, $\vec{V}$, normal to an element of area, dA, times the area of the element. Taking $d\vec{A}$ as the vector normal to dA,

$$dQ = |\vec{V}| \cos \theta_{V \, dA}| d\vec{A}| = \vec{V} \cdot d\vec{A}$$

D–2.2 CROSS PRODUCT (OR VECTOR PRODUCT)

The cross product of two vectors, $\vec{a}$ and $\vec{b}$, is written $\vec{a} \times \vec{b}$. It is a vector which has the properties:

1. $|\vec{a} \times \vec{b}| = |\vec{a}| |\vec{b}| \sin \theta_{ab}$
2. $\vec{a} \times \vec{b}$ is a vector perpendicular to *both* $\vec{a}$ and $\vec{b}$.
3. The *sense* of $\vec{a} \times \vec{b}$ is given by the right-hand rule, that is, as $\vec{a}$ is rotated into $\vec{b}$, then $\vec{a} \times \vec{b}$ points in the direction of the right thumb. This is shown in Fig. D.3.

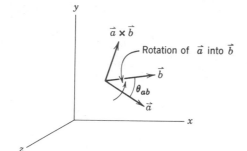

Fig. D.3 *Cross product of vectors.*

If $\vec{a}$ and $\vec{b}$ are parallel, then $\sin \theta_{ab} = 0$ and $\vec{a} \times \vec{b} = 0$.

Applying these rules, we obtain the cross products among unit vectors:

$$\hat{\imath} \times \hat{\imath} = 0 \qquad \hat{\jmath} \times \hat{\imath} = -\hat{k} \qquad \hat{k} \times \hat{\imath} = \hat{\jmath}$$
$$\hat{\imath} \times \hat{\jmath} = \hat{k} \qquad \hat{\jmath} \times \hat{\jmath} = 0 \qquad \hat{k} \times \hat{\jmath} = -\hat{\imath}$$
$$\hat{\imath} \times \hat{k} = -\hat{\jmath} \qquad \hat{\jmath} \times \hat{k} = \hat{\imath} \qquad \hat{k} \times \hat{k} = 0$$

The cross product, $\vec{a} \times \vec{b}$, is conveniently evaluated from the determinant

$$\vec{a} \times \vec{b} = \begin{vmatrix} \hat{\imath} & \hat{\jmath} & \hat{k} \\ a_x & a_y & a_z \\ b_x & b_y & b_z \end{vmatrix} \qquad\qquad \text{(D.6a)}$$

Expanding in minors of the first row,

$$\vec{a} \times \vec{b} = \hat{\imath}(a_y b_z - a_z b_y) + (-\hat{\jmath})(a_x b_z - a_z b_x) + \hat{k}(a_x b_y - a_y b_x)$$

or

$$\vec{a} \times \vec{b} = \hat{\imath}(a_y b_z - a_z b_y) + \hat{\jmath}(a_z b_x - a_x b_z) + \hat{k}(a_x b_y - a_y b_x) \qquad \text{(D.6b)}$$

Using Eqs. D.6, we can show:

(a) The cross product is *not commutative*, that is, $\vec{a} \times \vec{b} \neq \vec{b} \times \vec{a}$. Exchanging the order of two rows or columns changes the sign of a determinant. Thus $\vec{a} \times \vec{b} = -\vec{b} \times \vec{a}$.

(b) The cross product is *distributive*, that is, $\vec{a} \times (\vec{b} + \vec{c}) = \vec{a} \times \vec{b} + \vec{a} \times \vec{c}$. This follows by direct substitution of the components of $(\vec{b} + \vec{c})$ into Eq. D.6a.

(c) The cross product is *not associative*, that is, $\vec{a} \times (\vec{b} \times \vec{c}) \neq (\vec{a} \times \vec{b}) \times \vec{c}$. This may be shown by choosing $\vec{b} = \vec{c}$. In the left side, then $\vec{a} \times (\vec{b} \times \vec{c}) = \vec{a} \times \vec{0} = \vec{0}$. But $\vec{a} \times \vec{b} \neq \vec{0}$, and the right side, $(\vec{a} \times \vec{b}) \times \vec{c} \neq \vec{0}$. Thus $\vec{a} \times (\vec{b} \times \vec{c}) \neq (\vec{a} \times \vec{b}) \times \vec{c}$.

One application of the cross product is in formulation of the torque or moment applied to a body. Physically, the torque is $|\vec{r}|$ times the component of $\vec{F}$ perpendicular to $\vec{r}$. Thus for the body shown in Fig. D.4.

$$|\vec{T}| = |\vec{r}||\vec{F}| \sin \theta_{rF}$$

Fig. D.4 *Formulation of torque.*

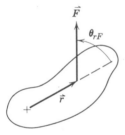

or

$$\vec{T} = \vec{r} \times \vec{F}$$

according to the right-hand rule.

D–3 *Differentiation of Vectors*

The conventional definition of a derivative may be used in differentiation of vectors. Consider a vector $\vec{a} = \vec{a}(t)$. Then $a_x = a_x(t)$, etc., and

$$\frac{d\vec{a}}{dt} = \lim_{\Delta t \to 0} \frac{\vec{a}(t + \Delta t) - \vec{a}(t)}{\Delta t}$$

$$\frac{d\vec{a}}{dt} = \lim_{\Delta t \to 0} \frac{[a_x(t + \Delta t) - a_x(t)]\hat{\imath} + \cdots}{\Delta t}$$

The limiting process applies to each term, so

$$\frac{d\vec{a}}{dt} = \lim_{\Delta t \to 0} \frac{a_x(t + \Delta t) - a_x(t)}{\Delta t}\hat{\imath} + \lim_{\Delta t \to 0} \frac{a_y(t + \Delta t) - a_y(t)}{\Delta t}\hat{\jmath} + \cdots$$

or

$$\frac{d\vec{a}}{dt} = \frac{da_x}{dt}\hat{\imath} + \frac{da_y}{dt}\hat{\jmath} + \frac{da_z}{dt}\hat{k} \cdot \tag{D.7}$$

Similarly, if $\vec{a} = \vec{a}(x, y, z)$, that is, $a_x = a_x(x, y, z)$, etc., then

$$\frac{\partial \vec{a}}{\partial x} = \lim_{\Delta x \to 0} \frac{\vec{a}(x + \Delta x, y, z) - \vec{a}(x, y, z)}{\Delta x}$$

$$\frac{\partial \vec{a}}{\partial x} = \frac{\partial a_x}{\partial x}\hat{\imath} + \frac{\partial a_y}{\partial x}\hat{\jmath} + \frac{\partial a_z}{\partial x}\hat{k} \tag{D.8}$$

D–4 The Vector Operator, ∇

D–4.1 DEFINITION OF ∇

Certain operations occur so frequently in mathematical formulations that a short-hand notation is desirable. The vector operator del, ∇, has been defined as

$$\left.\begin{array}{l} \nabla \equiv \hat{\imath}\dfrac{\partial}{\partial x} + \hat{\jmath}\dfrac{\partial}{\partial y} + \hat{k}\dfrac{\partial}{\partial z} \qquad \text{(Cartesian coordinates)} \\[3mm] \nabla \equiv \hat{\imath}_r\dfrac{\partial}{\partial r} + \hat{\imath}_\theta\dfrac{1}{r}\dfrac{\partial}{\partial \theta} + \hat{\imath}_z\dfrac{\partial}{\partial z} \quad \text{(Cylindrical coordinates)} \end{array}\right\} \qquad \text{D.9}$$

The terms in the definition of ∇ are written $\hat{\imath}(\partial/\partial x)$, and so on, to emphasize that ∇ is an operator. Thus the partial derivatives, such as $\partial/\partial x$, *operate* on any function that follows them.

∇ is a vector operator, so three possible "products" can be formed with other functions. These are discussed in the next three sections.

D–4.2 THE GRADIENT

When ∇ operates on a differentiable scalar function (or field), the result is called the gradient. Thus, if $\phi = \phi(x, y, z)$ is a scalar field, then

$$\nabla\phi = \text{gradient } \phi = \text{grad } \phi = \hat{\imath}\frac{\partial\phi}{\partial x} + \hat{\jmath}\frac{\partial\phi}{\partial y} + \hat{k}\frac{\partial\phi}{\partial z} \tag{D.10}$$

Note that ϕ is a scalar field, but grad ϕ is a vector field. The gradient, $\nabla\phi$, is a vector perpendicular to a surface of constant ϕ. Its magnitude and direction are those of the greatest space rate of increase of ϕ.

The pressure gradient, ∇p, is important in fluid mechanics. ∇p for a pressure field, $p = p(x, y, z)$, is evaluated and discussed in Section 3–1.

D–4.3 THE DIVERGENCE

When the dot product is formed between ∇ and a vector function (or field), the result is called the *divergence*. In Cartesian coordinates, if $\vec{V} = \vec{V}(x, y, z)$, then

$$\nabla \cdot \vec{V} = \text{divergence } \vec{V} = \text{div } \vec{V} = \left(\hat{i}\frac{\partial}{\partial x} + \hat{j}\frac{\partial}{\partial y} + \hat{k}\frac{\partial}{\partial z} \right) \cdot (V_x\hat{i} + V_y\hat{j} + V_z\hat{k})$$

$$= \hat{i} \cdot \frac{\partial V_x}{\partial x}\hat{i} + \hat{i} \cdot V_x\overset{=0}{\cancel{\frac{\partial \hat{i}}{\partial x}}} + \hat{j} \cdot \frac{\partial V_y}{\partial y}\hat{j} + \hat{j} \cdot V_y\overset{=0}{\cancel{\frac{\partial \hat{j}}{\partial y}}} + \hat{k} \cdot \frac{\partial V_z}{\partial x}\hat{k} + \hat{k} \cdot V_z\overset{=0}{\cancel{\frac{\partial \hat{k}}{\partial z}}}$$

$$\nabla \cdot \vec{V} = \frac{\partial V_x}{\partial x} + \frac{\partial V_y}{\partial y} + \frac{\partial V_z}{\partial z} \quad \text{(Cartesian coordinates)} \tag{D.11}$$

In cylindrical coordinates, if $\vec{V} = \vec{V}(r, \theta, z)$, then

$$\nabla \cdot \vec{V} = \left(\hat{i}_r\frac{\partial}{\partial r} + \hat{i}_\theta\frac{1}{r}\frac{\partial}{\partial \theta} + \hat{i}_z\frac{\partial}{\partial z} \right) \cdot (\hat{i}_r V_r + \hat{i}_\theta V_\theta + \hat{i}_z V_z)$$

$$= \hat{i}_r : \overset{=0}{\cancel{\frac{\partial \hat{i}_r}{\partial r}}} V_r + \hat{i}_r \cdot \hat{i}_r\frac{\partial V_r}{\partial r} + \hat{i}_\theta \cdot \frac{1}{r}\frac{\partial \hat{i}_r}{\partial \theta} V_r + \hat{i}_\theta \cdot \hat{i}_\theta\frac{1}{r}\frac{\partial V_\theta}{\partial \theta} + \hat{i}_z \cdot \overset{=0}{\cancel{\frac{\partial \hat{i}_z}{\partial z}}} V_z + \hat{i}_z \cdot \hat{i}_z\frac{\partial V_z}{\partial z}$$

$$\nabla \cdot \vec{V} = \frac{\partial V_r}{\partial r} + \hat{i}_\theta \cdot \frac{1}{r}\frac{\partial \hat{i}_r}{\partial \theta} V_r + \frac{1}{r}\frac{\partial V_\theta}{\partial \theta} + \frac{\partial V_z}{\partial z}$$

But $\partial \hat{i}_r/\partial \theta = \hat{i}_\theta$, as shown in Example Problem 4.12. Thus

$$\nabla \cdot \vec{V} = \frac{\partial V_r}{\partial r} + \frac{V_r}{r} + \frac{1}{r}\frac{\partial V_\theta}{\partial \theta} + \frac{\partial V_z}{\partial z}$$

But

$$\frac{\partial V_r}{\partial r} + \frac{V_r}{r} = \frac{1}{r}\frac{\partial}{\partial r}(rV_r)$$

so finally

$$\nabla \cdot \vec{V} = \frac{1}{r}\frac{\partial}{\partial r}(rV_r) + \frac{1}{r}\frac{\partial V_\theta}{\partial \theta} + \frac{\partial V_z}{\partial z} \quad \text{(Cylindrical coordinates)} \quad \text{(D.12)}$$

$\nabla \cdot \vec{V}$, where $\vec{V}$ is the velocity field (Section 2–2), represents the net volume rate of fluid flow from an infinitesimal control volume, per unit volume. For an incompressible flow, $\nabla \cdot \vec{V} = 0$, that is, the net rate of flux of fluid from any infinitesimal control volume must vanish.

D–4.4 THE CURL

When the cross product is formed between ∇ and a vector field, the result is called the *curl*. If $\vec{V} = \vec{V}(x, y, z)$, then in Cartesian coordinates

$$\nabla \times \vec{V} = \text{curl } \vec{V} = \begin{vmatrix} \hat{\imath} & \hat{\jmath} & \hat{k} \\ \dfrac{\partial}{\partial x} & \dfrac{\partial}{\partial y} & \dfrac{\partial}{\partial z} \\ V_x & V_y & V_z \end{vmatrix}$$

$$\nabla \times \vec{V} = \hat{\imath}\left(\frac{\partial V_z}{\partial y} - \frac{\partial V_y}{\partial z}\right) + \hat{\jmath}\left(\frac{\partial V_x}{\partial z} - \frac{\partial V_z}{\partial x}\right) + \hat{k}\left(\frac{\partial V_y}{\partial x} - \frac{\partial V_x}{\partial y}\right) \quad \text{(D.13)}$$

For cylindrical coordinates, $\vec{V} = \vec{V}(r, \theta, z) = \hat{\imath}_r V_r + \hat{\imath}_\theta V_\theta + \hat{\imath}_z V_z$. Since the unit vectors change with angle, θ, we must proceed with caution. Thus

$$\nabla \times \vec{V} = \left(\hat{\imath}_r \frac{\partial}{\partial r} + \hat{\imath}_\theta \frac{1}{r}\frac{\partial}{\partial \theta} + \hat{\imath}_z \frac{\partial}{\partial z}\right) \times (\hat{\imath}_r V_r + \hat{\imath}_\theta V_\theta + \hat{\imath}_z V_z)$$

$$= \hat{\imath}_r \frac{\partial}{\partial r} \times \hat{\imath}_r V_r + \hat{\imath}_r \frac{\partial}{\partial r} \times \hat{\imath}_\theta V_\theta + \hat{\imath}_r \frac{\partial}{\partial r} \times \hat{\imath}_z V_z$$

$$+ \hat{\imath}_\theta \frac{1}{r}\frac{\partial}{\partial \theta} \times \hat{\imath}_r V_r + \hat{\imath}_\theta \frac{1}{r}\frac{\partial}{\partial \theta} \times \hat{\imath}_\theta V_\theta + \hat{\imath}_\theta \frac{1}{r}\frac{\partial}{\partial \theta} \times \hat{\imath}_z V_z$$

$$+ \hat{\imath}_z \frac{\partial}{\partial z} \times \hat{\imath}_r V_r + \hat{\imath}_z \frac{\partial}{\partial z} \times \hat{\imath}_\theta V_\theta + \hat{\imath}_z \frac{\partial}{\partial z} \times \hat{\imath}_z V_z$$

$$\nabla \times \vec{V} = \hat{i}_r \times \hat{i}_r \frac{\partial V_r}{\partial r} + \hat{i}_r \times V_r \frac{\partial \hat{i}_r}{\partial r}^{=0} + \hat{i}_r \times \hat{i}_\theta \frac{\partial V_\theta}{\partial r} + \hat{i}_r \times V_\theta \frac{\partial \hat{i}_\theta}{\partial r}^{=0}$$

$$+ \hat{i}_r \times \hat{i}_z \frac{\partial V_z}{\partial r} + \hat{i}_r \times V_z \frac{\partial \hat{i}_z}{\partial r}^{=0}$$

$$+ \hat{i}_\theta \times \hat{i}_r \frac{1}{r}\frac{\partial V_r}{\partial \theta} + \hat{i}_\theta \times V_r \frac{1}{r}\frac{\partial \hat{i}_r}{\partial \theta}^{=\hat{i}_\theta} + \hat{i}_\theta \times \hat{i}_\theta \frac{1}{r}\frac{\partial V_\theta}{\partial \theta}^{=0}$$

$$+ \hat{i}_\theta \times V_\theta \frac{1}{r}\frac{\partial \hat{i}_\theta}{\partial \theta}^{=-\hat{i}_r} + \hat{i}_\theta \times \hat{i}_z \frac{1}{r}\frac{\partial V_z}{\partial \theta} + \hat{i}_\theta \times V_z \frac{1}{r}\frac{\partial \hat{i}_z}{\partial \theta}^{=0}$$

$$+ \hat{i}_z \times \hat{i}_r \frac{\partial V_r}{\partial z} + \hat{i}_z \times V_r \frac{\partial \hat{i}_r}{\partial z}^{=0} + \hat{i}_z \times \hat{i}_\theta \frac{\partial V_\theta}{\partial z} + \hat{i}_z \times V_\theta \frac{\partial \hat{i}_\theta}{\partial z}^{=0}$$

$$+ \hat{i}_z \times \hat{i}_z \frac{\partial V_z}{\partial z}^{=0} + \hat{i}_z \times V_z \frac{\partial \hat{i}_z}{\partial z}^{=0}$$

or

$$\nabla \times \vec{V} = \hat{i}_z \frac{\partial V_\theta}{\partial r} - \hat{i}_\theta \frac{\partial V_z}{\partial r} - \hat{i}_z \frac{1}{r}\frac{\partial V_r}{\partial \theta} + \hat{i}_z \frac{V_\theta}{r} + \hat{i}_r \frac{1}{r}\frac{\partial V_z}{\partial \theta} + \hat{i}_\theta \frac{\partial V_r}{\partial z} - \hat{i}_r \frac{\partial V_\theta}{\partial z}$$

and finally

$$\nabla \times \vec{V} = \hat{i}_r \left(\frac{1}{r}\frac{\partial V_z}{\partial \theta} - \frac{\partial V_\theta}{\partial z} \right) + \hat{i}_\theta \left(\frac{\partial V_r}{\partial z} - \frac{\partial V_z}{\partial r} \right) + \hat{i}_z \frac{1}{r}\left(\frac{\partial}{\partial r}(rV_\theta) - \frac{\partial V_r}{\partial \theta} \right)$$

(Cylindrical coordinates) (D.14)

Equation D.13 was derived in Section 5–5 by evaluating the rates of rotation of three pairs of mutually perpendicular line segments in a fluid particle. In Section 5–5, we showed that

$$\nabla \times \vec{V} = 2\vec{\omega}$$

where $\vec{V}$ is the velocity field, and $\vec{\omega}$ is the rotation vector for a fluid particle. Thus the curl of a velocity field is related to the rotation of the field.

D–4.5 THE LAPLACIAN

Another operator that occurs frequently is obtained by evaluating the dot product $\nabla \cdot \nabla$. This new operator is called the Laplacian, and is given the symbol ∇^2. Thus in Cartesian coordinates.

$$\nabla^2 = \nabla \cdot \nabla = \frac{\partial^2}{\partial x^2} + \frac{\partial^2}{\partial y^2} + \frac{\partial^2}{\partial z^2} \tag{D.15}$$

In cylindrical coordinates,

$$\nabla^2 = \nabla \cdot \nabla = \left(\hat{\imath}_r \frac{\partial}{\partial r} + \hat{\imath}_\theta \frac{1}{r} \frac{\partial}{\partial \theta} + \hat{\imath}_z \frac{\partial}{\partial z} \right) \cdot \left(\hat{\imath}_r \frac{\partial}{\partial r} + \hat{\imath}_\theta \frac{1}{r} \frac{\partial}{\partial \theta} + \hat{\imath}_z \frac{\partial}{\partial z} \right)$$

$$= \frac{\partial^2}{\partial r^2} + \hat{\imath}_\theta \cdot \frac{1}{r} \left(\overset{=\hat{\imath}_\theta}{\frac{\partial \hat{\imath}_r}{\partial \theta}} \frac{\partial}{\partial r} + \hat{\imath}_r \frac{\partial^2}{\partial \theta \partial r} + \overset{=-\hat{\imath}_r}{\frac{\partial \hat{\imath}_\theta}{\partial \theta}} \frac{1}{r} \frac{\partial}{\partial \theta} + \hat{\imath}_\theta \frac{1}{r} \frac{\partial^2}{\partial \theta^2} \right) + \frac{\partial^2}{\partial z^2}$$

$$\nabla^2 = \frac{\partial^2}{\partial r^2} + \frac{1}{r} \frac{\partial}{\partial r} + \frac{1}{r^2} \frac{\partial^2}{\partial \theta^2} + \frac{\partial^2}{\partial z^2}$$

But

$$\frac{\partial^2}{\partial r^2} + \frac{1}{r} \frac{\partial}{\partial r} = \frac{1}{r} \frac{\partial}{\partial r} \left(r \frac{\partial}{\partial r} \right)$$

so

$$\nabla^2 = \frac{1}{r} \frac{\partial}{\partial r} \left(r \frac{\partial}{\partial r} \right) + \frac{1}{r^2} \frac{\partial^2}{\partial \theta^2} + \frac{\partial^2}{\partial z^2} \tag{D.16}$$

in cylindrical coordinates.

As pointed out in Sections 5–3 and 5–6.1, both the stream function, ψ, and velocity potential, ϕ, satisfy Laplace's equation,

$$\nabla^2 \psi = \nabla^2 \phi = 0$$

for two-dimensional, irrotational, incompressible flow. Laplace's equation also describes many other physical phenomena, including steady state heat conduction, for which $\nabla^2 T = 0$.

D–5 Vector Identities

D–5.1 $\nabla \times \nabla\phi = 0$

In Section 5–5.1, we noted that $\nabla \times \nabla\phi \equiv 0$, and used that relation to derive the velocity potential, ϕ. This relation may be verified by expanding it into components.

Thus, in Cartesian coordinates,

$$\nabla \times \nabla\phi = \nabla \times \left(\hat{\imath}\frac{\partial\phi}{\partial x} + \hat{\jmath}\frac{\partial\phi}{\partial y} + \hat{k}\frac{\partial\phi}{\partial z} \right)$$

$$= \begin{vmatrix} \hat{\imath} & \hat{\jmath} & \hat{k} \\[4pt] \dfrac{\partial}{\partial x} & \dfrac{\partial}{\partial y} & \dfrac{\partial}{\partial z} \\[8pt] \dfrac{\partial\phi}{\partial x} & \dfrac{\partial\phi}{\partial y} & \dfrac{\partial\phi}{\partial z} \end{vmatrix}$$

$$\nabla \times \nabla\phi = \hat{\imath}\left(\frac{\partial^2\phi}{\partial y\partial z} - \frac{\partial^2\phi}{\partial z\partial y} \right) + \hat{\jmath}\left(\frac{\partial^2\phi}{\partial z\partial x} - \frac{\partial^2\phi}{\partial x\partial z} \right) + \hat{k}\left(\frac{\partial^2\phi}{\partial x\partial y} - \frac{\partial^2\phi}{\partial y\partial x} \right)$$

If $\phi = \phi(x, y, z)$ is a continuous, differentiable function, then the order of repeated differentiation is not important, that is,

$$\frac{\partial^2\phi}{\partial x\partial y} = \frac{\partial^2\phi}{\partial y\partial x}, \quad \frac{\partial^2\phi}{\partial x\partial z} = \frac{\partial^2\phi}{\partial z\partial x} \quad \text{and} \quad \frac{\partial^2\phi}{\partial y\partial z} = \frac{\partial^2\phi}{\partial z\partial y}$$

Consequently, $\nabla \times \nabla\phi \equiv 0$.

Although a more lengthy process, this identity also may be proved in cylindrical coordinates.

D–5.2 $(\vec{V} \cdot \nabla)\vec{V} = \frac{1}{2}\nabla(\vec{V} \cdot \vec{V}) - \vec{V} \times (\nabla \times \vec{V})$

This relation was used in Section 6–5.2 in developing the Bernoulli integral in rectangular coordinates. It may also be proved by expanding both sides into components. Thus, in Cartesian coordinates,

$$\vec{V} \cdot \nabla = V_x\frac{\partial}{\partial x} + V_y\frac{\partial}{\partial y} + V_z\frac{\partial}{\partial z}$$

and

$$(\vec{V} \cdot \nabla)\vec{V} = \hat{\imath}\left(V_x\frac{\partial V_x}{\partial x} + V_y\frac{\partial V_x}{\partial y} + V_z\frac{\partial V_x}{\partial z} \right)$$
$$+ \hat{\jmath}\left(V_x\frac{\partial V_y}{\partial x} + V_y\frac{\partial V_y}{\partial y} + V_z\frac{\partial V_y}{\partial z} \right)$$
$$+ \hat{k}\left(V_x\frac{\partial V_z}{\partial x} + V_y\frac{\partial V_z}{\partial y} + V_z\frac{\partial V_z}{\partial z} \right) \qquad \text{(D.17)}$$

Also

$$\vec{V} \cdot \vec{V} = V_x^2 + V_y^2 + V_z^2$$

so

$$\tfrac{1}{2}\nabla(\vec{V}\cdot\vec{V}) = \hat{\imath}\left(V_x\frac{\partial V_x}{\partial x} + V_y\frac{\partial V_y}{\partial x} + V_z\frac{\partial V_z}{\partial x}\right)$$

$$+ \hat{\jmath}\left(V_x\frac{\partial V_x}{\partial y} + V_y\frac{\partial V_y}{\partial y} + V_z\frac{\partial V_z}{\partial y}\right)$$

$$+ \hat{k}\left(V_x\frac{\partial V_x}{\partial z} + V_y\frac{\partial V_y}{\partial z} + V_z\frac{\partial V_z}{\partial z}\right) \qquad \text{(D.18a)}$$

and

$$-\vec{V}\times(\nabla\times\vec{V}) = -\begin{vmatrix} \hat{\imath} & \hat{\jmath} & \hat{k} \\ V_x & V_y & V_z \\ \left(\dfrac{\partial V_z}{\partial y}-\dfrac{\partial V_y}{\partial z}\right) & \left(\dfrac{\partial V_x}{\partial z}-\dfrac{\partial V_z}{\partial x}\right) & \left(\dfrac{\partial V_y}{\partial x}-\dfrac{\partial V_x}{\partial y}\right) \end{vmatrix}$$

$$= \hat{\imath}\left(V_z\frac{\partial V_x}{\partial z} - V_z\frac{\partial V_z}{\partial x} - V_y\frac{\partial V_y}{\partial x} + V_y\frac{\partial V_x}{\partial y}\right)$$

$$+ \hat{\jmath}\left(-V_z\frac{\partial V_z}{\partial y} + V_z\frac{\partial V_y}{\partial z} + V_x\frac{\partial V_y}{\partial x} - V_x\frac{\partial V_x}{\partial y}\right)$$

$$+ \hat{k}\left(-V_x\frac{\partial V_x}{\partial z} + V_x\frac{\partial V_z}{\partial x} + V_y\frac{\partial V_z}{\partial y} - V_y\frac{\partial V_y}{\partial z}\right) \qquad \text{(D.18b)}$$

Collecting terms in Eqs. D.18

$$\tfrac{1}{2}\nabla(\vec{V}\cdot\vec{V}) - \vec{V}\times(\nabla\times\vec{V}) = \hat{\imath}\left(V_x\frac{\partial V_x}{\partial x} + V_y\frac{\partial V_x}{\partial y} + V_z\frac{\partial V_x}{\partial z}\right)$$

$$+ \hat{\jmath}\left(V_x\frac{\partial V_y}{\partial x} + V_y\frac{\partial V_y}{\partial y} + V_z\frac{\partial V_y}{\partial z}\right)$$

$$+ \hat{k}\left(V_x\frac{\partial V_z}{\partial x} + V_y\frac{\partial V_z}{\partial y} + V_z\frac{\partial V_z}{\partial z}\right) \qquad \text{(D.19)}$$

Comparing Eqs. D.17 and D.19 shows that

$$(\vec{V}\cdot\nabla)\vec{V} = \tfrac{1}{2}\nabla(\vec{V}\cdot\vec{V}) - \vec{V}\times(\nabla\times\vec{V})$$

This identity could also be proved in cylindrical coordinates; the latter proof would be extremely lengthy!

index